BIBLIOTHECA MYCOLOGICA

HERAUSGEGEBEN VON

J. CRAMER

BAND 31

NORTH AMERICAN SPECIES OF MYCENA

BY

ALEXANDER H. SMITH

AUTHORIZED REPRINT 1971

3301 LEHRE
VERLAG VON J. CRAMER

WHELDON & WESLEY, LTD · STECHERT-HAFNER SERVICE AGENCY, INC.
CODICOTE, HERTS. NEW YORK, N.Y.

PRINTED IN GERMANY

University of Michigan Studies

SCIENTIFIC SERIES

VOLUME XVII

NORTH AMERICAN SPECIES OF MYCENA

NORTH AMERICAN SPECIES OF MYCENA

ALEXANDER H. SMITH

ANN ARBOR: UNIVERSITY OF MICHIGAN PRESS

LONDON: GEOFFREY CUMBERLEGE, OXFORD UNIVERSITY PRESS

1947

PREFACE

AT THE present time generic concepts in the gill fungi may be said to be in a state of transition. The genera of the Friesian classification have been critically evaluated in the light of information obtained on microscopic characters and as a result of the discovery of many interesting species from other parts of the world, and it has become evident that considerable regrouping throughout the agarics is desirable. As yet the systems proposed since that of Fries have not been subjected to the rigid test of extensive use, and consequently their faults have not been brought to light in as detailed a manner. Few of the new genera or extensive regroupings have been treated monographically, and until such treatments are available any new system of classification or regrouping of species among now accepted genera must be regarded as tentative.

At present the trend is toward natural as opposed to artificial genera, and a so-called natural arrangement of species within genera. Closeness of relationship, which is what we are dealing with, is based on degree of similarity in characters accepted as fundamental. Since the taxonomist is working almost entirely from circumstantial evidence, there is not much to be gained by insisting that a classification be based on relationship when the point has been reached where affinities are difficult to ascertain and must be determined on secondary characters. Consequently, in my estimation, the most satisfactory arrangement of species will always involve some compromise between artificiality and naturalness.

At the generic level, however, more emphasis can rightly be placed on relationship, and this is the modern trend. This does not necessarily involve splitting up large genera into smaller ones, as some investigators seem to think. At least in the gill fungi, the results have been the creation of such large genera as Rhodophyllus and Psathyrella as well as small ones such as Laccaria, Cystoderma, and others. In Mycena it appeared logical to accept a broader rather than a narrower concept than that of Fries. It is evident that in this genus speciation has proceeded largely on microscopic characters, such as those of spores and cystidia, and because of this I have not seen fit to use these at the generic level. By adopting a broad concept the relationships of the species, particularly those which formerly were regarded as borderline species between Mycena and Collybia or Omphalia, could be more satisfactorily expressed. At the same time Mycena, as a concept, is maintained for a group of fungi so similar in general appearance that anyone with a small amount of experience can nearly always recognize it at sight in the field. To me this is one of the criteria of a truly natural genus.

Many investigators are discouraged from starting extensive revisions of this type because they know that, after the work has been completed, the cost of publishing it is likely to prove an insurmountable obstacle. I have been exceedingly fortunate in this respect, and wish to express my deep appreciation to the University of Michigan Press for accepting this manuscript.

ALEXANDER H. SMITH

CONTENTS

LIST OF PLATES

(The plates face page 510)

PLATES

FIGURES IN THE TEXT

(Drawings showing microscopic characters of the various species)

INTRODUCTION

THE need for a comprehensive treatment of the North American species of Mycena has been recognized by every investigator who has attempted to study them. In recent times it has been accentuated by the emphasis given to microscopic characters in the arrangement as well as the delimitation of species. Both Atkinson and Kauffman not only were aware of this need but also had started work on monographs. At the time of his death Atkinson had the largest collection of species of Mycena in the United States and, probably, in the world, since in times past European investigators apparently did not attempt to keep many dried agarics as specimens. From the outline which Atkinson left behind it was evident that he planned to utilize fully the microscopic characters now in general use. However, he had not studied his material critically and had not organized his notes and other information.

Kauffman, using the same approach as Atkinson, had a tentative manuscript drawn up in 1928–29, but did not live to complete the work. He had not studied the types of American species, and for this reason felt that his manuscript was still in the preliminary stage. Because the studies of Atkinson and Kauffman were never finished, students of the North American species of Mycena have been forced to rely on the publications of Peck, scattered through periodicals and the *Annual Reports of the New York State Museum*, on Kauffman's *Agaricaceae of Michigan*, Beardslee and Coker's article on the species of North Carolina, and the *North American Flora*. Each of these served a very useful purpose but, considered from the standpoint of North America as a whole, they are all incomplete.

It was my good fortune to be in the field with Kauffman in 1929, when he was concentrating on Mycena, and to profit by his many years of experience and his accumulated knowledge of the fruiting habits and peculiarities of agarics. Although I was studying Cortinarius primarily at the time, the frequent stimulating discussions we had on the value of the numerous characters used to delimit species in Mycena aroused my interest in the genus to such an extent that I subsequently undertook its study.

The work of classifying any group of agarics consists of two phases. The first is concerned with the recognition of the various species as they occur in nature and the variations they exhibit from year to year or in different habitats. The second has to do with the nomenclature, and it is here that one encounters the greatest difficulties. Neither Atkinson nor Kauffman made a thorough study of the nomenclature for many of the species. Because American investigators saved the type specimens of the new species they described, it has been possible to ascertain the microscopic details now considered essential in delimiting a species, and thus avoid the confusion of concepts that has characterized European agaricology since the time of Fries. However, since many American students did not pay close attention to microscopic details, numerous misidentifications have appeared in our literature, and a multiplication of names for some species has resulted.

Besides the need for more detailed information on the characters of American Mycenae, the question of the limits of the genus was also very pertinent because of some disagreement among investigators over the position of many "borderline" species. Thanks to the work of Kühner (1938), the limits of the genus have now been made much more precise. Although the arrangement presented here differs in some particulars from that of Kühner, it is in close agreement with his.

DEFINITION OF MYCENA

In his most widely used work Fries (1874) considered Mycena a subgenus of Agaricus and described it as follows (p. 129): ". . . stipes fistulosus, *cartilagineus*. Pileus submembranaceus, plus minus striatus, primitus conico- l. parabolico-cylindricus *ob marginem primitus rectum, stipitem sursum attenuatum amplectentem l. parallelo-adpressum*. Lamellae haud decurrentes (tantum denticulo uncinatae). *Epiphyti l. radicati, graciles, subcampanulati, vix umbilicati.*"

Kühner (1938) has given the following detailed characters of the genus: Carpophores gymnocarpic (always ?), small, slender, delicate, putrescent; hyphae with clamp connections, with nonaeriferous intercellular passages, the pigment in general vacuolar. Pileus thin, submembranous, more or less translucent striate or sometimes plicate, very often conic, campanulate or furnished with a pointed mammillate umbo, the margin at first straight or more rarely somewhat incurved. Flesh of pileus with the upper hyphae generally short and inflated (the hypoderm). Covering of the pileus with superficial hyphae almost always furnished with short protuberances or with the ends straightened to form numerous hairs. Stipe central, cartilaginous, more or less fistulose, with the base frequently bristling with rhizoids, the fundamental hyphae often containing numerous nuclei. Lamellae confluent with the flesh of the pileus, thin, often ascending and more or less adnate, more rarely arcuate-decurrent; medial strand regular or interwoven; subhymenium ramose, generally thin, lacking calloses at the transverse walls or these not strongly basophile; basidioles cylindric, the nuclei superposed; mature basidia more or less clavate, the fusion nucleus characteristically undergoing three successive divisions in the normal form; cystidia almost always present, smooth or roughened. Spores white in mass or rarely tinged yellowish (latter observation by R. Maire), smooth or nodulose, almost always with a thin wall, generally uninucleate in four-spored forms. Saprophytic species habitually epiphytic or lignicolous.

The definition adopted in the present work is as follows: Stipe cartilaginous, tubular or hollow, glabrous or appressed-fibrillose; pileus variously shaped but usually conic or convex, margin usually straight and appressed against the stipe when young but sometimes incurved at first, fragile, fleshy, or somewhat cartilaginous in consistency but hardly reviving when remoistened (*M. corticola* excepted), usually membranous to submembranous; lamellae usually ascending-adnate or hooked, often arcuate and occasionally decurrent (if decurrent, the margin of the pileus *not* incurved); spores white in deposit, thin-walled, smooth, falsely echinulate (see Fayodia), or nodulose, amyloid reaction either positive or negative; cystidia usually present, at least on the gill edges though often embedded and difficult to locate; gill trama usually interwoven; pileus trama characterized by some differentiation either of the hyphae of the cuticle or of those beneath it.

As one can readily see, this definition is somewhat broader than that of Fries and, in addition, places emphasis on microscopic characters. Certain species formerly included in Collybia and Omphalia are here transferred to Mycena. Those with an incurved margin and white or bright-colored pigments (in contrast to the gray, brownish-gray, bluish-gray, or black forms) are admitted to Mycena if cheilocystidia are present and the gills are adnate. If the gills are decurrent and the cap margin is incurved at first, the species are retained in Omphalia, provided the stipe is not viscid. The latter look like small species of Clitocybe and have their relationships with that genus. Grayish or dark-colored species with an incurved margin and adnate gills are retained in Mycena if their pilei possess a differentiated hypoderm (this excludes such fungi as *Collybia myriadophylla* and *C. familia*). All the white species of the section Mycenarii of Omphalia have been transferred to Mycena as well as the colored forms having cheilocystidia or amyloid spores (with the exception of *Omphalia campanella*, *O. picta*, etc., which are referred to the genus Xeromphalina [Omphalopsis] by some authors and which in any case form a series more closely related to Marasmius than to Mycena). The series of Collybiae found on old cones—*C. albipilata*, *C. myosura*, and the aberrant American *C. conigenoides* on Magnolia fruits—are excluded from Mycena. The greatest difficulties in establishing the limits of Mycena are encountered in the sections Hydropus and Corticatae. In the latter the hypoderm is not differentiated, and the margin of the pileus may be somewhat incurved. Because of their Mycena-like appearance, they are included in Mycena, but they can hardly be separated from the series of Collybiae with corticated pilei. Singer (1942) recognizes Hydropus as a distinct genus. In keeping with the broad concept of Mycena employed here, it has been included as a section. Although rather distinctive as a group, its affinities appear to be in part in the section Lactipedes and in part in the Omphaliariae. Critical studies need to be made on fresh material of several species in order to determine their relationships more clearly.

These limits differ from those given by Kühner (1938) in excluding such species as *Omphalia maura*, *Collybia myosura*, and *C. myriadophylla*, and in including those with falsely echinulate spores.

HISTORY OF THE GENUS IN NORTH AMERICA

The first significant list of the North American species now placed in Mycena was published by Schweinitz in 1822, the year following the publication of Volume I of *Systema Mycologicum*, by Elias Fries. This is of some interest because the work of Fries has since been selected as the starting point in agaric nomenclature. Schweinitz listed under the generic name Agaricus such well-known species as *M. galericulata*, *M. polygramma*, *M. haematopus*, *M. epipterygia*, *M. elegans*, *M. stylobates*, *M. vulgaris*, and *M. corticola*. Some which are less common and some now referred to Marasmius, as well as to other genera, were also included. In 1834 Schweinitz published a more extensive list of the North American species.

In his catalogue of North Carolina plants Curtis (1867) included a list of agarics. It contained twenty-one names assigned to the subgenus Mycena of Agaricus. Some, such as *A. iocephalus* and *A. cohaerens*, are now placed in Marasmius. Because of the contributions of Schweinitz and Curtis the Mycena flora of the Carolinas became particularly well known at an early date. Peck in 1872 published a synopsis of the subgenus Mycena of Agaricus for New York. In it he described five new species, three of which are still recognized as valid in Mycena. One, *A. paluster*, has been transferred to Collybia (Lyophyllum or Tephrophana of some authors). It has been necessary to exclude one because the type has been lost and the meager description fails to characterize the fungus sufficiently in the difficult group to which it belongs. For the next forty years Peck continued to contribute to our knowledge of the genus, as is evidenced by the number of species ascribed to him in the following text. At the time Peck worked little attention had been given to microscopic characters of agarics by the more prominent European investigators and, like most of them, Peck followed the Friesian system.

In 1916 W. A. Murrill published an account of the group in the *North American Flora* under the generic names Prunulus and Galactopus. (He used the name Mycena for Bolbitius of the Friesian system.) He described many new species, particularly from the Pacific Coast and from tropical North America. He did not place sufficient emphasis on microscopic characters and, unfortunately, many of those given have been found to be incorrect. In addition, his descriptions are frequently sketchy and do not convey a sufficiently clear picture of the plant to the reader. A study of his type specimens, however, indicates clearly that Murrill has collected and described many unusual fungi.

Kauffman in 1918 published his *Agaricaceae of Michigan*, in which thirty-seven species in Mycena were recognized and described. He gave information on the cystidia as well as the spores of many species. From conversations with Kauffman, I know that he was well aware that his treatment of the Michigan species was very incomplete, but in the light of what was known at the time it was an outstanding contribution.

Beardslee and Coker in 1924 brought up to date the information on the species of Mycena occurring in North Carolina. Their article contains accounts of forty-two and gives many valuable data on the microscopic characters of the group. Numerous photographs of the carpophores and line drawings of the microscopic characters made it our most completely illustrated contribution on Mycena up to that time.

Many lists, of course, have been published for various parts of North America. That of Harkness and Moore (1881) was the first to record a substantial number from the Pacific Coast. Most lists contain *M. galericulata*, *M. corticola*, *M. Leaiana*, and *M. haematopus*. It has been impossible to verify the published reports of all species by an examination of the specimens. Without any question many are erroneous. Most reports of *M. galericulata* probably are based on *M. inclinata*. *M. corticola*, *M. haematopus*, and *M. Leaiana* are the three species most likely to be identified correctly. However, the report of *M. corticola* by Harkness and Moore needs further confirmation. I have searched in vain for this species along the Pacific coast, but found only *M. madronicola*, a fungus very similar in appearance. It is likely that most early American records of *M. amicta* or *M. Iris* for the eastern United States are based on *M. subcaerulea*.

SUMMARY

In the following account 218 species are recognized from the United States and Canada and 19 from the American tropics. Those from the United States and Canada are distributed in the subgenera as follows: Pseudomycena, 6; Eumycena, 185; Glutinipes, 21; Mycenella, 6. In Eumycena the distribution in sections is: Cyanescentes, 2; Corticolae, 5; Deminutivae, 37; Lactipedes, 11; Adonidae, 27; Calodontes, 18; Typicae, 66; Omphaliariae, 14; Floccipedes, 3; Corticatae, 3; Hydropus, 5. Naturally the Typicae of Eumycena form a center about which to group the other sections and subgenera.

Kühner in his monograph recognized 143 species in Mycena for Europe. This figure should be compared with the 218 known to occur in the United States and Canada. Even after one allows for the usual differences of opinion among investigators as to what constitutes a species, it is still evident that North America contains a much more diverse Mycena flora than Europe. This is in line with observations on other genera such as Leucopaxillus (see Singer and Smith, 1943), and also with mine on Cortinarius. Studies of the Mycena flora of other parts of the world have not been extensive enough to justify comparison. Cleland (1934) recognized 17 species from South Australia, but that flora must still be regarded as relatively unexplored as compared with those of Europe and North America. The small number recorded from South America, Asia, and Africa can be similarly explained. The distribution by states and provinces in the United States and Canada has not been summarized because of insufficient knowledge about it in both countries. Large numbers of species have been reported for the Pacific Coast and for North Carolina, Tennessee, New York, and Michigan in the East because intensive collecting has been carried on in these regions. During the course of this study over three thousand collections of Mycenae have been studied.

ACKNOWLEDGMENTS

No extensive work such as this could have been undertaken without the help of many people and institutions. The field work in regions distant from Ann Arbor was carried out by means of grants from the Horace H. Rackham School of Graduate Studies of the University of Michigan. The notes and photographs taken on these expeditions make up the bulk of the descriptive material and illustrations. The accurate identification of the specimens, however, would have been impossible without the coöperation of those in charge of herbaria in which the types of American species have been deposited. It is a pleasure to acknowledge the wholehearted support received from Professor H. M. Fitzpatrick, of Cornell University; Dr. Homer D. House, State Botanist, New York State Museum; and Dr. Fred J. Seaver, of the New York Botanical Garden. Through their aid the types of the species described by Atkinson, Peck, and Murrill were studied and many errors have been corrected. I also wish to express my thanks for the help given me by Dr. W. A. Murrill, now of Gainesville, Florida, who has very kindly allowed me to examine type material of the species he has described from Florida.

I am under obligation to still other individuals and institutions for aid given during the time this investigation was in progress. Dr. David H. Linder,* Curator, Farlow Herbarium, Harvard University, made available to me the large number of Mycenae deposited in that herbarium. Professor L. O. Overholts,* of Pennsylvania State College, placed at my disposal all the species in his private herbarium. Dean L. R. Hesler, of the University of Tennessee, has not only allowed me to examine the plants in the herbarium of that institution, but has given a great deal of time to collecting material in eastern Tennessee and the Great Smoky Mountains National Park. Through his efforts a number of interesting species have been found. To

* Died November 10, 1946.

Assistant Dean Clyde Leavitt and Dr. J. L. Lowe, of the New York State College of Forestry at Syracuse, New York, I am indebted for assistance in the field and facilities furnished for my studies both at the Archer and Anna Huntington Wild Life Forest Station near Newcomb, New York, and at the Pack Forest near Warrensburg. Professor H. S. Jackson, of the University of Toronto, Toronto, Ontario, placed the equipment of the Bear Island Laboratory at Lake Timagami, Ontario, at my disposal during the late summer of 1936. Mr. Preston P. Macy, Superintendent of the Olympic National Park, encouraged my field work in that area. Mr. Arthur Stupka, Park Naturalist in the Great Smoky Mountains National Park in Tennessee and North Carolina, did much to make my collecting there successful. Dr. R. P. Burke, of Montgomery, Alabama, sent to me many collections from Alabama during the season of 1942. Mr. Maurice B. Walters, Cleveland, Ohio, contributed specimens, and Mr. H. C. Beardslee, Perry, Lake County, Ohio, who has had a lifelong interest in Mycena, aided the project with both suggestions and material. Dr. James Walton Groves, of the Division of Botany and Plant Pathology, Central Experimental Farm, Ottawa, Ontario, supplied fungi from the region around Ottawa. Dr. G. R. Bisby, formerly of the University of Manitoba, Canada, and now of the Imperial Mycological Institute, Kew, Surrey, England, gave me Mycenae from Manitoba and western Ontario. Dr. S. M. Zeller, of Oregon State College, and Mr. Harold Parks, of Trinidad, California, did much to make my expeditions in Oregon and California both profitable and pleasant. Dr. D. E. Stuntz, of the Department of Botany, University of Washington, Seattle, Washington, Mr. J. B. Flett, Zenith, Washington, Mr. Wm. B. Gruber, Portland, Oregon, Miss Elizabeth E. Morse, of Berkeley, California, and Mr. Paul Marshall Rea, of Santa Barbara, California, all donated collections and field notes. Dr. A. W. Slipp, of the School of Forestry, University of Idaho, gathered material from northern Idaho during the course of his ecological investigations in that area. Professor E. B. Mains, Director of the University of Michigan Herbarium, has not only collected plants for this study in various parts of the United States, but has encouraged and aided the project in many other ways. Professor D. V. Baxter, of the School of Forestry and Conservation of the University of Michigan, made many valuable suggestions during the preparation of the manuscript.

Dr. Robert Kühner, Lyon, France, Mr. A. A. Pearson, Hindhead,

Surrey, England, and Dr. Rolf Singer, now at the Farlow Herbarium, Harvard University, aided, both by exchanges of specimens and by correspondence, in the study and identification of European species which occur in North America.

The war prevented me from making a study of the materials preserved in European herbaria. As a result many of the species described by Montagne and Berkeley from material collected in North America still await interpretation on the basis of modern microscopic studies. Undoubtedly some of the species now in the excluded and doubtful list can be adequately characterized by such studies if the original materials can still be found.

DIAGNOSTIC CHARACTERS OF THE SPECIES

SEASONAL OCCURRENCE

SPECIES of Mycena fruit abundantly at any time when a combination of moisture and temperature conditions is favorable. In most parts of temperate North America there is a spring and early summer as well as a fall fruiting period. Relatively few species fruit in hot wet summer weather in temperate regions. I have not studied their fruiting habits in the tropics. In the mountains they are abundant in the wake of the melting snow, and they are also among the first agarics to appear at the end of a dry season. Although to date a few species have been found only during certain seasons, the time of fruiting is apparently not taxonomically significant. In a local spruce plantation *M. vulgaris* is often the last to fruit in the fall and the first to appear in the spring if the weather is cool and wet. Each species has its own requirements of temperature, moisture, and release from dormancy. When the right sequence of events has occurred, the fungus fruits.

HABIT AND HABITAT

Habit and habitat are useful taxonomically, but must be cautiously interpreted. There is a striking difference between the species which grow scattered though in profusion over the fallen needles of conifers and those which grow in clusters on decaying logs. However, typically lignicolous species may come (and often do, to the confusion of the taxonomist) from buried wood and present an aspect quite different from that of typical forms. Buller (1934) noted this for *M. galericulata*, and I have verified his observation many times in such species as *M. alcalina*, *M. rugulosiceps*, and *M. maculata* in addition. *M. alnicola* is one Mycena in which the lignicolous habitat definitely aids in distinguishing a species from its close relatives. Carpophores of many terrestrial species are often attached to small twigs or bits of wood. These should not be interpreted as lignicolous. Certain species, such as *M. rubromarginata*, fruit on the smaller twigs of fallen treetops or the dead lower branches of conifers and frequently can be collected in great quantities in slashings. Some, such as *M. purpureofusca*, that usually fruit on the wood of conifers, can be found singly, springing from buried cones. Others habitually occur on the remains of particular groups of higher plants. *M. pterigena*, for example, is found on dead fern fronds (those of the previous year), and *M. juncicola* inhabits old sedge culms. Species living in the needle beds under conifers often fruit in great profusion, whereas those growing in wet mossy areas such as stream banks usually produce only a small number of carpophores. In general, the lignicolous species appear to be somewhat restricted to wood of either conifer or broad-leaved trees, but this separation cannot be used as a specific distinction with much assurance. Peck has even reported finding *M. Leaiana* on hemlock. For several species related to *M. galericulata* I have noted a marked ability to grow on either type of wood.

It can readily be seen that the species of Mycena exhibit all degrees of affinity for higher plants, from the highly specialized type represented by *M. pterigena* to the most generalized terrestrial type very well represented by *M. pura*. Within these two extremes all stages of specialization can be found. Many of the new species I have described have been named for the substratum upon which they usually occur, for example, *M. madronicola*, *M. piceicola*, and *M. ulmicola*.

Many of the terrestrial species show some affinity for habitats produced or dominated by certain genera of our common forest trees. *M. clavicularis*, *M. monticola*, *M. rosella*, *M. alcaliniiformis*, *M. metata*, and *M. pseudoclavicularis* nearly always grow on the needle beds of various pines. *M. pseudoclavicularis* appears to be a characteristic species of the ponderosa-pine forests. *M. parabolica* and *M. borealis* generally are found on the wood and debris of ponderosa and white pine respectively, and occasionally give the impression of being terrestrial. *M. tenax*, *M. piceicola*, and *M. vulgaris* frequently occur on the needle beds under spruce. Differences caused by affinities for particular sites are rather difficult to establish in the various mixtures of balsam-fir and spruce types, particularly in northern regions. However, *M. delectabilis* is abundant in these habitats and also under Douglas fir (*Pseudotsuga taxifolia* Britt.) in the West. *M. atroalboides* is common in bogs under black spruce in the northeastern United States and also under silver fir (*Abies grandis* Lindbl.) and Douglas fir in the West. *M. galopus* is one of the most characteristic species of the redwood agaric flora of California, but in other parts of the continent is found under either hardwoods or conifers. The same thing is true to a lesser extent of *M. scabripes*. *M. quinaultensis* and *M. elegans* show a pronounced affinity for stands of western red cedar (*Thuja plicata* Donn.). The forests of *Thuja occidentalis* L. of northern Michigan appear, however, to lack a characteristic species of Mycena. *M. alcalina* frequently occurs on cedar, but, to judge from my own experience, is more commonly found on larch. *M. laevigata* has a strong affinity for the wood of larch in the western United States, but can be expected to grow on almost any decaying conifer log.

The various oak forests support a large fungous flora of both lignicolous and terrestrial species, a feature which may be considered characteristic of them. *M. algeriensis* and *M. Kauffmanii* are nearly always on or associated with either elm or ash. *M. radicatella* is usually on the wood of either aspen or maple. *M. floccipes*, according to my experience, is commonly found on the bark of hickory or elm which has sloughed off fallen trees. Other instances of the association of species of Mycena with certain higher plants are indicated in the text.

Most species do not give a collector any clue to their presence until the fruiting bodies are produced, but *M. insignis* is an exception. It produces telltale whitish patches over the areas where the fruiting bodies will appear. When collecting in different regions the mycologist frequently must adjust his ideas of habitats to new conditions. In the wet, dense cedar forests of the Oregon coast, where some accumulation of dead needles often occurs on the lower branches of the cedars, one can find *M. rorida* growing luxuriantly anywhere from six to twenty feet above the ground. Ordinarily it occurs on the ground under conifers.

The position which terrestrial species of Mycena occupy in the life of a forest is difficult to ascertain, but without question they fruit more abundantly at some stages of forest development than at others. The lignicolous forms, of course, can be expected to occur wherever the substratum and the environmental conditions are favorable. The species characteristic of the needle beds under conifers appear to attain their optimum development in stands where ground cover is lacking and a fairly deep layer of needles has accumulated. In forests with a heavy ground cover the number of fruiting bodies produced by a given Mycena as well as the number of species present is likely to be small. This has been found to be true of hardwood forests. The most luxuriant fruiting I have observed for *M. Abramsii*, *M. stannea*, and *M. scabripes* was in the relatively young and clean, rather than mature and brushy, hardwood forests. On the other hand, the only luxuriant fruiting of *M. fagetorum* and *M. flavescens* encountered was in an overmature stand of red oak. Here, however, a great deal of young growth was also present, so that the observation cannot be given much emphasis. When one considers the number of variable factors that must be taken into account in evaluating ecological observations on agarics, data collected over a ten-year period actually represent a very small amount of information and must be regarded as preliminary in nature. In addition to the usual hazards of temperature and moisture conditions, a factor involving not only the invasion of a site by a fungus but the exhaustion of the site and the consequent disappearance of the fungus must be considered. When the collector is visiting widely separated localities, observations on the ecology of agarics cannot be made most advantageously. The studies upon which my remarks are based are those carried out in the Olympic Mountains of Washington between 1935 and 1941 and in the vicinity of Ann Arbor, from 1930 to 1943.

In general, species of Mycena may occur singly, gregariously, or cespitosely (clustered) on or over the substratum. All intergradations between these habits can be found for many, depending on the conditions at the time of fruiting or the amount of food available. The lignicolous species are more inclined to be cespitose and the terrestrial ones gregarious (that is, to have the fruiting bodies scattered over the substratum, but limited to a rather definite area). The term "scattered" has been used to indicate a condition in which the carpophores are rather widely dispersed throughout a more or less indefinite area. The occurrence of single fruiting bodies of most species of Mycena appears to be the result of conditions unfavorable for normal fruiting. Some such fruiting bodies, however, are undoubtedly stragglers which come up after a period of luxuriant fruiting.

It may be said in summary that the characteristic responses to the environment which are exhibited by many species have taxonomic value only in special cases, and that no simple rule can be given for ascertaining their importance. This is true of the manner of growth and of fruiting. In the Typicae of Eumycena the cespitose lignicolous habit serves to distinguish a group of apparently closely related species.

SIZE AND STATURE OF CARPOPHORES

Size and stature are so likely to be influenced by local conditions that one must allow for considerable variation in both. In the following taxonomic treatment both have been used to some extent to characterize species, but only after careful study. Although Mycenae are small fungi, there is a surprising range of variation in both size and stature. Despite the use here of size as a means of separating the diminutive species from the others of the subgenus Eumycena, its limitations are clearly realized. It does not appear advisable to erect new taxonomic units other than forms with size alone as the major distinction, but as a character of general value in distinguishing a troublesome group of species comprised of several phylogenetic lines, size serves a very useful purpose.

Certain species of Mycena are variable in stature, the variations apparently being caused by local factors which frequently dominate a habitat. *M. erubescens*, for instance, may be short and squatty if it grows in the open on bare soil, or it may develop a greatly elongated stipe when growing alongside logs where there is an accumulation of old leaves with high moisture content. Under certain conditions *M. pura* may have a cap barely 1 cm. broad, and under others a cap 5–6 cm. broad, with all other parts in proportion. In the descriptions the usual range in size has been given, and the variations from it have been indicated in parentheses. The smallest species of the Deminutivae have pilei approximately 1 mm. broad, whereas the largest species of the Typicae have caps up to 7 cm. If one seeks for the most accurate gauge for measuring stature, he will find it in the diameter of the stipe. Even the smallest and slenderest carpophores of typically robust species will have thicker stipes than those which belong to the minute Mycenae, although the pilei may be exactly the same in both size and shape. Atkinson and Kauffman recognized thickness of the stipe as the least variable of all the parts concerned, and my field studies have verified their observations. One word of caution should be added: Be careful to take the measurements from fully matured specimens. The stems, particularly among the lignicolous species, are often rather thick in the immature carpophores, but become thinner as they elongate and the pileus expands.

COLOR

One cannot help being impressed by the diversity of color exhibited by the various members of the genus, and as a character it is very important. Its value, however, depends to a large extent on the degree of difference shown by the specimens in question and their nearest relatives. Color is of little value in the long series of grayish to grayish-brown members of the section Typicae of Eumycena. It goes without saying that color varies greatly and that the taxonomist must learn to recognize the degree of variation exhibited by each species. Some, such as *M. pura,* contain a mixture of pigments, and can be characterized by their great variability of color. Such species must be recognized by other more stable characters. In some, such as *M. monticola,* a particular tint is constant. Colors usually fade as the specimens age and in some may change completely, as in *M. strobilinoides.* The distribution of the various pigments through the carpophore is also important. The pigments of some, most of which are grouped in the subsection Calodontes of Eumycena, may be concentrated in the cell sap of the cheilocystidia, thus imparting a darker or a brighter color to the gill edge than to the faces. The color is usually about the same as that of the pileus and stipe, and its variability may be caused by variation in the number of cheilocystidia present on different fruiting bodies or on different gills of the same fruiting body, or by a greater or lesser concentration of pigment in the cystidia. The bright- and the dark-colored pigments may both be present in the same fruiting body, as in *M. aurantiomarginata* and *M. elegans,* and in such cases the dominant color tone of the cap is determined by the concentration of each. Many of the white species have a tendency to develop yellowish pigments as the carpophores mature, and some, like *M. amabilissima,* which are very brightly colored, fade to nearly pure white. The exposure of the carpophore to intense light during development will often influence color, as in *M. citrinomarginata.* In intense light the dark pigments become most prominent. In both *M. amicta* and *M. subcaerulea* the bright-blue pigments fade rapidly, and frequently the only evidences remaining are in the hairs at the base of the stipe. The faces of the gills of many species are white or pallid or merely possess the same pigments found in the rest of the carpophore, though in smaller amounts. In these the color of the gills fades more quickly than that of the pileus or the stipe. Although color is an important character, it is difficult to determine when it should be considered significant enough to distinguish a species. The emphasis placed upon it depends on the differences exhibited by the members in the particular group to which the species concerned belongs. For example, in the Adonidae its use is generally recognized, but is of doubtful value and has been employed here with some hesitation to distinguish *M. aurantiidisca* Murrill from *M. Adonis* (Fr.) Quélet. The distinction in this instance is questionable because there are apparently no correlated differences. The colors belong to the same series of reds and appear to be nothing more than extreme variations in intensity.

MACROSCOPIC CHARACTERS OF THE SURFACE OF BOTH PILEUS AND STIPE

In thin-fleshed agarics like Mycena the gill outlines often show through the moist, somewhat translucent flesh of the pileus as conspicuous radially arranged lines or striations, and the cap is then described as "translucent-striate." Although the character is usually mentioned in descriptions, it is of distinctly secondary importance from a taxonomic standpoint. The amount of moisture present has much to do with determining whether the striations are obscure or conspicuous, and they usually disappear as the pileus fades. Pilei are frequently characterized by rather distinct radial wrinkles or folds and are then described as "sulcate-striate." In some larger species, such as *M. parabolica*, these are quite distinctive. They are much more frequently found in the very small species, however, and here they usually represent a somewhat faded condition, their development being correlated with the loss of moisture from the pileus. To be of any value, the character should be observed on young fresh specimens at the time of collection. In a few Mycenae, small as well as large, the pileus is irregularly wrinkled, particularly around the disc. This condition is described as "rugulose," and is exceedingly variable in most species, including *M. marginella* var. *rugosodisca*. It is illustrated for *M. niveipes*, Plate 68, and should be regarded as a secondary character. In *M. niveipes* the wrinkles frequently disappear as the pileus expands.

The character of the surface of the pileus is readily classified in three categories: dry, moist, and viscid. Representatives of the first group are rare. In general, those species with numerous pilocystidia or fibrils over the cap have a rather dry appearance. This is typical for *M. delicatella* and is an important diagnostic feature. However, by far the majority of the species fall into the second group, and hence, because of the frequency of its occurrence, the character of moistness is of secondary importance to the taxonomist. Most of the members of this group show a distinct change in their appearance with loss of moisture and are said to be "hygrophanous." As a rule, the fragile species fade quickly, thus accentuating the change in appearance. The cartilaginous forms fade more slowly and, because the change is not so abrupt, they are usually described as "subhygrophanous."

The viscidity of the pileus is a character of major importance, and, fortunately, one's observations can be checked by microscopic examination. In some species the hyphae of the cuticle of the pileus or stipe, or both, gelatinize and cause these surfaces to be very slimy or sticky. The technical terms for this condition, which are "viscid" and "glutinous," merely express different degrees of the gelatinizing process. If the pileus is viscid, the gelatinous pellicle can usually be removed, at least in part, and appears as a thin, transparent skin. In some, such as *M. quinaultensis*, it is too thin and too highly gelatinized to hold together and can be removed only in shreds. In others, for instance, *M. laevigata*, the pellicle may be adnate (not separable from the cap) and only subgelatinous, a condition causing the pileus to feel lubricous or slippery instead of truly viscid. As one might expect, weather conditions will have considerable influence on a character like this, but one soon learns to allow for them. In dry weather typically viscid layers may seem to be merely lubricous. In continued rainy weather they may be completely washed away, and the surface will then feel moist instead of sticky. This happens not infrequently in regions such as the rainy belt along our Pacific coast. When one is collecting under these conditions, it is necessary to obtain young fruiting bodies which have not had the gluten washed off the stipe or cap by continuous rain. This may seem obvious to many readers, but it should be pointed out that species like *M. clavicularis* often produce hundreds of fruiting bodies, all in more or less the same stage of development. When a collector finds such quantities of material, he is not likely to spend much time searching for immature carpophores, and thus may, because of the sheer abundance of carpophores, fail to collect those giving the clue to the identity of the species. Prolonged rainy weather will cause the pellicle of some agarics not ordinarily classed as viscid to be somewhat sticky to the touch. Here again, however, the young rapidly developing carpophores will not be misleading.

Many species have a hoary appearance when young. This character is found on the dry, moist, and viscid species alike, and is caused by delicate microscopic cells, which soon collapse and disappear. Lubricous, viscid, or glutinous species, when old, usually have a polished or varnished appearance because of the dried gluten. In others the pileus may appear pubescent because of projecting hairs of almost microscopic size. One, *M. stylobates*, has coarse projections made up of fascicles of hyphae.

The surface of the stipe exhibits the same variations in its characters as the pileus. It may be scurfy or pruinose owing to minute hairs or cystidia, glabrous and polished, or viscid. In some, such as *M. polygramma*, it may be appressed-fibrillose and develop distinct longitudinal striations or lines. The base may be naked or covered with strigose mycelial hairs. In one group the stipe is characterized by an abrupt basal disc or bulb, a character of major importance. In many small species it is perfectly glabrous at the base and is inserted abruptly onto the substratum. This type of attachment is significant. In most species the base of the stipe is somewhat strigose. The degree to which this development takes place depends almost entirely on local conditions. If a cluster of carpophores arises from a hard exposed surface, the stipes may be inserted directly on the substratum, with few hairs present. If the cluster arises from deep in the rotten wood, the buried portion will more than likely be conspicuously strigose.

The base of the stipe may be characterized by a long rootlike projection, termed "pseudorhiza" by Fayod (1889). This is typical of a number of species, such as *M. galericulata*, *M. megaspora*, *M. nodulosa*, and *M. Kauffmanii*. Buller (1934) has discussed this structure in detail, so that the following comments will be limited to its use as a taxonomic character. It is important, but certain species in which it is characteristically present may not show it if the fruiting bodies originate from a hard surface instead of coming from deep in the ground or from other substrata. In addition, the unobserving collector may fail to obtain it. The pseudorhiza is broadest at the point where it reaches the surface, and it tapers from there downward to a slender rootlike strand which, if one follows it, is found to be attached to an old root or some other piece of buried wood. Occasionally more than one fruiting body may develop from a single pseudorhiza, as is shown in *M. megaspora* (Plate 61). In Mycena this variation appears to be rare. Many species related to *M. galericulata* and *M. megaspora* have a tendency to develop a long, somewhat rooting, strigose base. *M. semivestipes* and *M. laevigata* are typical examples. This type of base cannot be classified as a pseudorhiza.

LAMELLAE

The manner in which the gills are attached to the stipe is a rather important character, but the spacing, color, width, and shape are of only secondary importance for most species. Marginate gills have been discussed under "Color" (p. 15). They are taxonomically important, but have been used as a varietal distinction when no other correlated differences were involved.

In the great majority of species the gills are ascending-adnate and slightly toothed or "hooked." In *M. pura* and *M. rutilantiformis* they are frequently uncinate, and in Omphaliariae the attachment varies from arcuate to truly decurrent. In the Deminutivae possessing gills the manner of attachment varies from free or nearly so to distinctly decurrent. For the very minute species the manner of attachment, whether adnate, arcuate, or decurrent, is an unreliable character, and I do not believe it will serve, unaided, to distinguish taxonomic units. For those in which the presence of lamellae is variable certainly no importance can be placed on the manner of their attachment. Because of the exceptionally large number of species in both the Deminutivae and the Omphaliariae which exhibit various degrees of attachment from adnate to decurrent, the old distinction between Mycena and Omphalia must be considered untenable.

The spacing of the lamellae is usually designated as "subdistant." Variations from it occasionally aid in recognizing some species, but the character cannot be considered of major importance. Depauperate specimens are likely to have truly distant gills. This is particularly true of *M. erubescens*, so far as is indicated by my collections. It is also true that occasional large specimens of a species may have more distant lamellae than is usual. One should note the gill spacing on freshly matured pilei which are fully expanded. Lamellae nearly always appear close in unexpanded pilei and are more likely to be distant in overmature caps. The best way to approach the problem of gill spacing is to count the number of lamellae which reach the stipe and the number of rows of lamellulae present. The gills which do not reach the stipe are referred to as "lamellulae," and they occur in a more or less distinct pattern. In the text this pattern is expressed in terms of rows or tiers. Each tier is identified by the distance which its members project toward the stipe. Thus in Plates 23 and 24, illustrating *Mycena pura*, it can be seen that between pairs of complete gills there occurs one short gill which extends one half to three fourths the distance toward the stipe. Since these are roughly equal in length and occur fairly regularly between pairs of complete gills, they are considered the first or primary row. A second row of still shorter lamellulae can also be seen in the plates referred to above. It consists of a very short gill inserted between each primary lamellula and the adjacent lamella. These are frequently no more than half the length of the primary lamellula. Thus it can be said that these lamellulae are in two rows or tiers. The degree to which they develop is likely to determine whether the gills appear distant or close. In Plate 24 they would be considered close; in Plate 23, subdistant. In this instance, however, part of the difference is caused by the more expanded condition of the pileus shown in Plate 23. Sometimes when the second row remains rudimentary and the primary row develops poorly, the lamellae may appear very distant. This often happens in *M. erubescens*. In some species considerable irregularity in the development of lamellulae is encountered, and at times it is even difficult to decide whether or not a single gill reaches the stipe. In Mycena lamellulae may be entirely lacking or up to three rows may be present. A knowledge of the variations in their development aids greatly when one is trying to decide whether the specimen in hand is characterized by close, subdistant, or distant gills.

The width of the lamellae is usually given as narrow or broad. This character has some importance, but is best expressed by giving measurements along with the description. It should always be remembered that a gill 2 mm. wide may be very narrow in a large cap and quite broad in a small one. It is essentially a relationship of length to width that one seeks to express. Ventricose gills (those abruptly widened in the middle) are rare in Mycena.

In some North Temperate species there is a strong tendency to develop veins in the spaces between the gills. It is most noticeable in those related to *M. galericulata* and also in *M. pura*. So far as these species are concerned, it does not appear to be an important character, but in the tropics the situation appears to be different. Van Overeem (1923–26) has described and illustrated some tropical forms with poroid hymenia but otherwise apparently related to *M. pura*.

FLESH

Certain characteristics of the flesh of the carpophore are important taxonomically. The odor and taste are among the foremost of these. *M. cinerella* has a very characteristic rancid taste. In most, however, the taste is merely mild or slightly farinaceous (mealy). Certain others, such as *M. pura*, are characterized by a distinct radishlike taste and odor, but the characters do not always appear to be constant. The odors of many species are peculiar. In addition to those with a raphanoid odor there are some in which the odor is alkaline or nitrous; some in which it is strongly farinaceous or reminiscent of raw cucumbers; some in which it resembles that of iodoform; and a few in which it is decidedly fragrant or fruity (resembling that of some esters). In the dried specimens of *M. odorifera* the odor persists for a long time. In some the odor is strongest at the time the carpophores are collected but fades quickly. Some have no odor when collected but one soon develops which is very distinctive, as in *M. iodiolens*. In some that are normally distinguished by the presence of an odor, odorless forms which apparently are genetically constant are known. Since the majority of the species do not have a distinctive odor or taste, the presence of one or both is an important character which should be noted.

Some Mycenae tend to stain sordid reddish when cut or bruised, or become spotted with sordid-reddish spots in age. The character has been given some prominence in the literature. I have studied it at every opportunity over a period of ten years and on the basis of field observations have found that in some it is very constant under all kinds of weather conditions, as it is in *M. maculata*, *M. rubrotincta*, and *M. viscosa*. In many, such as *M. rugulosiceps* and *M. galericulata*, the gills become spotted only in extreme age or in wet weather. In addition, some species have been described which appear to differ only in this one character. *M. hemisphaerica* is characterized by the absence of such spots and *M. maculata* by their presence. The former is found chiefly on the wood of hardwoods; the latter, on wood of conifers, an additional difference that should not be given much emphasis. It would be interesting to make a series of single-spore cultures from both and to try mating experiments. It is highly probable that such culture work would go far toward elucidating the relationships of these two as well as those of similar species. Kühner has cited some interesting observations on this subject in his discussion of *M. strobilicola*. Although I am inclined to question the value of this color change as a character of taxonomic importance, it has been used as an aid in recognizing species in several instances other than those mentioned above. In such instances, however, the color change has been correlated with some other secondary character.

The texture of the flesh is very difficult to describe accurately, but does deserve some consideration as a secondary character. Many of the large species, such as *M. rugulosiceps*, have a pliant cartilaginous texture, whereas others, such as *M. niveipes* and *M. parabolica*, are distinctly fragile. Because of the presence of thick gelatinous layers over the pilei and stipes of some, they seem to be pliant and tough. *M. tenax* was so named because of its toughness. The minute species may also be either pliant or fragile, but here there is much more intergradation.

LATEX

The presence of milklike or colored latex distinguishes the Lactipedes from other sections of Eumycena, and a watery hyaline fluid which exudes from cut surfaces of the carpophore distinguishes the section Hydropus, also of Eumycena. *M. Leaiana*, in Glutinipes, has a juice which stains the fingers and is watery-orange if pressed out. Because of its viscid stipe, however, it is not included in Lactipedes. The watery latex of Hydropus is best demonstrated by cutting the pileus slightly with a sharp instrument, such as a razor. The fluid oozes out in numerous droplets from fresh specimens. The milklike or colored latex is best demonstrated by cutting off the stipe with a sharp razor. A drop or two of the latex will ooze out. In *M. galopus* the latex is often scanty, and the average collector fails to notice it. It is most readily seen by observing the base of the stipe at the time the specimens are collected. The color, but not the quantity, is of diagnostic value within the section. Species with white, orange, and reddish latex are known from North America.

MACROSCOPIC CHARACTERS OF DRIED SPECIMENS

Many agarics, including some in Mycena, assume a distinctive appearance when dried. In some the colors of the fresh specimens are well preserved in the dried material, and in others the carpophores assume a distinctive color in drying. Among the members of the section Typicae of Eumycena, however, nearly all the dried material has much the same appearance. *M. monticola*, *M. strobilinoides*, and *M. oregonensis* are distinct when dried because the colors of the fresh specimens are preserved to a remarkable extent. In *M. fuliginaria*, however, all the specimens I have seen dried black, the result, no doubt, of the characteristic color change exhibited by the fresh material. The white or pale species usually vary when dry from dead white to pale yellowish. Very faded (hence whitish) specimens of *M. niveipes* usually dry whitish, but young, unfaded carpophores which are dark gray will retain this color in drying, and hence in some cases one must learn to recognize the variation in the appearance of dried specimens just as one learns to recognize color variations of fresh carpophores. Dried material, of course, is of greatest value in studies of the microscopic characters.

MICROSCOPIC CHARACTERS

Readers interested in the anatomical details of the various types of fruiting bodies grouped in Mycena should consult Kühner's (1938) excellent account. The present discussion is limited to the characters I have emphasized in the arrangement and description of species in order to avoid needless repetition.

Spores

Size, shape and markings, and reaction to iodine (Melzer's reagent) are of major importance in recognizing species. Spore size should be determined from spore deposits taken from perfectly fresh fruiting bodies. My favorite method is to place a cap, gills down, on a glass slide for five, ten, or twenty minutes and then to measure the spores that have been discharged. In following this procedure one should always remember that, as the cap becomes wilted, the discharged spores are likely to be atypical. The length of the spores of the various species ranges from as low as 3–4 μ to 15–18 μ. The majority have spores 6–11 μ long. Slight differences in spore size and shape, such as those existing between *M. amicta* and *M. subcaerulea* or between *M. galericulata* and *M. hemisphaerica*, have been used to distinguish species only after many collections have been studied and the differences have been found to be constant. Where they are not distinctive, as in *M. capillaripes* and its forms, they have not been emphasized. This creates an apparent inconsistency in the text, but I believe the difficulty is one that needs to be elucidated by culture work and mating experiments rather than by continued field observations. Arnold (1935) showed that in *Marasmius elongatipes* Peck long- and short-spored forms were interfertile. Such a study as this should be made for *M. subcaerulea* and *M. amicta*.

Shape is also an important character. If one examines the line drawings in the text figures, he will note that the more or less ellipsoid spore is characteristic of most of the section Typicae. In the series of small white species one frequently finds subfusoid spores or the long, narrow type (see *M. gypsea* and *M. delicatella*). There does not appear to be much correlation between shape and iodine reaction. Those of *M. paucilamellata* are more or less aciculate and amyloid.

It is now generally recognized that the spores from two-spored basidia are larger than those from four-spored individuals of the same species and that this difference and the two-spored condition together are not of sufficient taxonomic importance to justify the erection of new species. The increase in spore size in two-spored forms appears to be the result of the distribution of the contents of the basidium to two instead of four spores. If only one spore is produced, it is much larger than either of the pair if two are produced; if three spores are formed, the size is usually intermediate between that of typically two- and four-spored forms. In fruiting bodies in which the various basidia of a pileus produce one, two, three, and four spores, the collector is certain to obtain a very confusing spore deposit. Consequently, in measuring the spores of a Mycena care must be taken to examine the basidia of the *cap from which the deposit was obtained*. The character of producing two instead of the normal four sterigmata on a basidium seems to be constant in some forms but variable in others. Since it is found in forms in which the nuclear behavior appears to be normal as well as in the parthenogenetic forms, it cannot be regarded as correlated with a definite type of nuclear history.

In general, the spores from four-spored and two-spored forms are of the same shape. In *M. supina*, however, this is not true. In the four-spored form they are globose or nearly so; in the other they are broadly ellipsoid. The spores from monosporous basidia are frequently quite misshapen.

The spores are smooth in all but a few species. In the section Nodulosae of Mycenella they are nodulose; in the section Pseudoechinulatae they appear echinulate under ordinary magnifications, but in reality are smooth. The thickened inner wall is filled with fine pores which create an echinulate appearance. Some investigators recognize these two sections as distinct genera (Mycenella and Fayodia). No species with truly echinulate spores are included in Mycena.

The iodine reaction of the spore wall is another character of major importance, but I have minimized its use in the keys as much as possible because of the difficulty of determining it on fresh material. It is a character of considerable value to the specialist, but one likely to be difficult for the beginner to interpret. The chief reason for this situation lies in the fact that the various species show all degrees of reaction from very positive to completely negative. The details of the technique employed to determine the character are discussed later (p. 35). In a strong positive reaction the spore becomes dark violet, and the change is clear either in daylight or under a daylight bulb. However, as one examines specimens in which the reaction is weaker, it is necessary to use daylight. In a weak reaction the spores become very pale bluish gray, and not all of them react. As Kühner has pointed out, groups of immature spores along the gills frequently give a clear reaction when others fail to do so. An excellent way to get a quantitative test is to make a spore print, preferably heavy, on a slide, and when it is thoroughly dry, to test it with a drop of iodine solution. If amyloid, the spore deposit will become dark or pale gray or violet; if nonamyloid, it will remain practically hyaline or become slightly yellowish when viewed against white paper.

Basidia

The size and shape of the basidia are of little taxonomic significance except in some of the small species, where they are relatively obese (15–18 × 7–8 μ) or relatively narrow (15–24 × 5–6 μ). The number of spores borne on a basidium is also of little help to the taxonomist except as it applies to the spore size of one-, two-, and three-spored forms (see the discussion on p. 24).

Cystidia

The markings, shape, and distribution of the cystidia are among the most important characters used to distinguish species of Mycena. A cystidium in Mycena may be defined as any sterile cell in the hymenium or along the edges of the gills. In the dark-spored agarics, where the paraphyses or pseudoparaphyses are often differentiated from the basidia, this definition would not apply. The cystidia that occur on the faces of the gills are termed "pleurocystidia"; those on the edges, "cheilocystidia." For convenience the sterile cells projecting from the surface of the pileus are termed "pilocystidia"; those projecting from the surface of the stipe, "caulocystidia." When studying the pilocystidia one should be careful to distinguish between differentiated hyphal cells and the short, often somewhat contorted processes which arise from the walls of the hyphae forming the pellicle or even from the cells of the hypoderm if the pellicle is not well developed. These processes are comparable to the rodlike projections found on many of the pleurocystidia and cheilocystidia. In this classification setae are regarded as cystidia. They are usually long, pointed, thick-walled structures, and in Mycena occur chiefly as pilocystidia or caulocystidia. In *Marasmius cohaerens*, an agaric which in the past was often placed in Mycena, they occur as pleurocystidia and are characterized by dark-brown thickened walls. A glance through the text figures will illustrate the great diversity in form and markings that the cystidia of Mycena exhibit.

The simplest cystidia are the smooth clavate to cylindric type present in *M. cineraria* and *M. arenaria*. These can ordinarily be distinguished from immature basidia only by their greater size. Another simple type is found in *M. amicta* and *M. subcaerulea*, in which the cheilocystidia are filamentous, giving the impression that they are merely outgrowths of the hyphae of the gill trama and not otherwise differentiated. In *M. speirea* they are at first hardly distinguishable from young basidia, but as they mature they elongate greatly and become more or less filamentous. In a large number of species the cystidia are fusoid-ventricose and smooth, but the apex may be pointed, obtuse, or forked. Sometimes two or more long fingerlike protuberances develop from the upper part of the enlarged portion. The term "fusoid-ventricose" is here applied to *any cystidium which has the midportion distinctly enlarged and which tapers to the apex as well as to the pedicel.* There are all degrees of variation between this and the clavate type, and frequently one finds many of the variations on different parts of the same gill progressively from the stipe to the margin of the cap. The same variation can be found if one compares the lamellulae with the complete gills. Some species have cystidia in which the widest portion is near the apex and the tip is greatly reduced in size, forming only a small protuberance. Such cystidia are described as "mucronate." In *M. pectinata* the majority of the cheilocystidia are of this shape. The mucronate cystidia may, like the fusoid-ventricose individuals, give rise to fingerlike projections over the apex, as is shown in the text figures. In some species the cystidia remain more or less clavate and develop numerous fingerlike prolongations on or near the apex, and in others these projections may become variously branched or contorted. As the number of projections on the clavate cystidium increases, they tend to become smaller, and finally one arrives at the clavate-echinulate or verrucose type, in which the enlarged portion is studded with many very small blunt or pointed projections, as it is in *M. strobilinoides.* In a few species the cystidia may be cylindric or subventricose and be covered with small rodlike projections. In *M. borealis* the tips of the greatly elongated fusoid cystidia are covered with small echinulations. In *M. latifolia* the ventricose portion is roughened, and the elongated neck is smooth.

As is pointed out under *M. citrinomarginata* (p. 218), the type of cystidium is extremely variable in some species and constant in others. In this study variation has been determined first by an examination of many sections from a single pileus and then of a large number of pilei from a single collection. The range thus established was used to interpret subsequent collections. *M. citrinomarginata* and *M. stylobates* appear to have races characterized by a preponderance of smooth, rough, or intermediate cystidia. In *M. rosella*, *M. latifolia*, and *M. tenax* the clavate-roughened type occurs on the gill edges and the fusoid-ventricose type over the faces. This appears to be true whenever the pleurocystidia and the cheilocystidia are different. Mycenae with clavate-roughened pleurocystidia and smooth fusoid cheilocystidia are not known. Because the types of cystidia in Mycena are diverse and because they are not correlated in their distribution with the other major characters of the genus, it is not desirable to regard the type of cystidium as a major diagnostic character. Instead, it has been used as a means of separating stirpes (small groups of species) or other subdivisions of the various sections.

The distribution of cystidia is frequently taxonomically important, but there are numerous exceptions. If cheilocystidia are abundant, one can nearly always find at least a few near the gill edge as pleurocystidia. If the pleurocystidia are widely scattered or rare and similar to the cheilocystidia, it is likely that their position is not significant. The clavate-roughened types as well as the fusoid types are found on both the faces and the edges of the gills. There appears to be no correlation between the shape of the cystidia on the gills and those on the pileus and stipe. In general, on the surfaces of the pileus and stipe these organs are of a simplified type. In the setose species and in *M. apiculata*, however, there are highly specialized pilocystidia or caulocystidia different from any on the gills.

Size appears to be of some importance in such species as *M. flavescens* and *M. macrocystidia.* Generally speaking, however, cystidia vary greatly in size. This is particularly true if one compares young and old carpophores. In young material they may not have elongated. In old specimens they are frequently greatly elongated, somewhat irregular in outline, or unduly contorted or branched, and sometimes partly collapsed.

Most cystidia have a hyaline homogeneous content, but in some species they are yellow, pink, red, wine color, or dark sordid brown. The pigment is dissolved in the cell sap. The brighter colors fade rapidly, but the browns often persist even when the carpophores are dried and later revived in KOH. The pigment may be present in only the cheilocystidia or throughout all the cystidia of the carpophore. The caulocystidia and pilocystidia are often colored in species in which the pleurocystidia and cheilocystidia are hyaline. *M. floccipes* is an example. In *M. erubescens* and *M. Sabali* the content is frequently not homogeneous but, instead, consists of rather highly refractive droplets.

Structure of Gills and Pileus

The gill trama usually is composed of subparallel to interwoven more or less elongated hyphae. It furnishes little aid in classifying the species. The subhymenium in certain members of Glutinipes is gelatinous to subgelatinous, and furnishes a character of some importance. In some members of Glutinipes the gill edge also gelatinizes and becomes greatly enlarged. This is a valuable diagnostic character, but at the same time it usually obscures the details of the cheilocystidia and makes a study of them difficult. In such species one should use immature fruiting bodies to ascertain the type of cheilocystidium.

The pileus of a typical Mycena is characterized by a pellicle, a hypoderm, and a mass of floccose tissue making up the body of the trama. The pellicle is usually formed by narrow hyphae more or less radially arranged. From the walls of these hyphae there usually arise numerous short rodlike projections. Because of the compactness of the layer and the small size of the elements involved, detailed observations of them are a bit tedious to make and are best carried out with the aid of an oil-immersion lens. The degree to which the pellicle is developed is, however, an important character. In some species it is absent or almost absent, and in others it is thick, and the hyphae have gelatinous walls. The viscid character of the surface discussed previously is caused by the gelatinization of the pellicle. This can be easily demonstrated from sections of dried material revived in KOH. In such mounts the pellicle swells appreciably and has a characteristic translucent appearance. Since a gelatinous pellicle is usually readily separable from the pileus, one must be careful not to tear it off when cutting sections.

The hypoderm, when present, usually lies next to the pellicle, but it is not characteristic of all Mycenae. In a vertical section of a pileus it can be seen as a layer one or several cells thick in which the hyphae are much larger than those forming the remainder of the flesh. There are all degrees of differentiation of the hypoderm. In some it can be recognized only with difficulty, whereas in others it occupies most of the thickness of the pileus, the filamentous trama then being reduced to a narrow band. Gelatinous layers of narrow hyphae, even though they may occur just beneath a nongelatinous pellicle, are not classified as hypodermal tissue. In such forms, if a hypoderm is present, it will be found beneath the gelatinous layer. Gelatinous layers which occur in the tissue of the pileus instead of on the surface are very important taxonomically. These can be observed only with the aid of a microscope.

Iodine Reactions of the Flesh

Iodine reactions, which have been used by recent investigators as a rather important character, have been included in this study. The examinations must be made from dried material, and for this reason the reactions are most useful to the specialist. They are often weak, however, and are interpreted with some difficulty. A positive amyloid reaction for the tissue of the pileus and gills is not the same as one for the spores. These tissues turn a dark vinaceous red or almost purple in a good positive reaction and remain hyaline or become yellowish to dull brown in one that is negative. Gelatinous layers remain hyaline or become faintly yellowish. According to my experience, a positive reaction may be distinctive only in limited areas, or be weak and inconclusive throughout. In many species the tissue of the stipe will give an exceedingly strong positive reaction.

Viscidity of the Stipe

Viscidity of the stipe has been used to distinguish a subgenus of Mycena. It can be ascertained by means of a microscopic study of dried specimens almost as easily as from fresh material. Cross sections of the stipe of a viscid species show the usual fundamental and connective hyphae in the interior, but the peripheral layer is a sheath of narrow hyphae with gelatinous walls. These have the characteristic translucent or "glassy" appearance and appear to be widely separated. The chief source of error in ascertaining this character from dried material is that the gelatinous layer may be torn away in the process of sectioning. The sections should be floated on a razor. In *M. insignis* the viscidity is caused, not by a layer, but by numerous individual hyphae projecting from the stipe and possessing gelatinous tips. They are sticky in wet weather but soon dry out and leave the stipe perfectly dry to the touch. *M. laevigata* will create some difficulty because the outer layer of hyphae on its stipe gelatinizes only slightly, causing it to be slippery or lubricous to the touch. In wet weather the hyphae may gelatinize enough to cause the stipe to be covered with a thin layer of gluten. In 1929 this led me to describe the fungus as a variety of *M. vulgaris*.

TECHNIQUE AND MATERIAL EXAMINED

FIELD WORK

THE field work which forms the basis of this monograph was carried out largely in two areas, one extending across the continent from Nova Scotia to the Olympic Peninsula of Washington, and the other reaching in a north–south direction from Lake Timagami in Ontario to the Great Smoky Mountains of Tennessee and North Carolina. Obviously such coverage is by no means complete for the entire area of the United States and Canada, and many forest types still remain almost completely unexplored. However, the regions in which field work was conducted were those in which, to judge from the studies of previous investigators, it was evident that an abundant Mycena flora occurred, and the places neglected are largely the ones in which species of Mycena form only a very minor part of the known agaric flora. Because of the difficulties encountered in collecting agarics it should not be assumed that the neglected areas will not, under favorable conditions, produce an interesting Mycena flora of considerable diversity. This applies particularly to regions adjacent to the Gulf of Mexico. Since species of Mycena appear to be more numerous in cool climates, and are very abundant under conifers, a great deal of the collecting was carried on in the mountains, particularly in the northern United States, and in sphagnum bogs in Michigan and the Pacific Northwest.

As many different forest types as possible were investigated in the regions surveyed. In the vicinity of Ann Arbor, for instance, local conifer plantations were studied intensively, and a systematic effort was made to investigate the various forest types of the Central Hardwood Region. The same procedure was applied to the forest types in northern Michigan although the studies were not so extensive as in the vicinity of Ann Arbor. The other regions which were intensively studied are northwestern Washington, the Cascade Mountains in central Oregon, and northern California. One period of good collecting was experienced in the mountains of central Idaho and another in Nova Scotia. Collecting in New York State and in the Lake Timagami district of Ontario was somewhat unsuccessful because of dry weather. There were periods of both good and poor collecting in the Great Smoky Mountains of Tennessee and North Carolina. As can readily be seen, the emphasis was placed on the North Temperate species. There appears to be a rather sharp difference between these and the tropical Mycenae. A study of the latter would bring up many questions which can be answered only by a rather complete survey of the tropical species of Marasmius as well as of Mycena. This in itself is a task more than equal to the one I have undertaken.

The collecting, photographing, and studying of the fresh material was carried out according to an established routine. Because of the delicate nature of most species I found that best results could be obtained by collecting the material in the morning and studying it in the afternoon of the same day. When the specimens were collected they were carefully wrapped in waxed paper to prevent loss of moisture or damage by rain. Those to be photographed were usually wrapped separately and placed upside down in coffee cans. Placing the specimens cap downwards in the cans prevented further elongation of the stipes, and the carpophores remained just as they were when collected. If stood more or less upright, the stipes usually curved in such a manner as to orient the pilei for spore discharge, a condition in which they were not so desirable for photographs.

Whenever the material was adequate, a complete description was made from the collection. A few pilei were usually placed on slides, gills down, and by the time most of the description was written good spore deposits had been made. The chart employed for color notes was that of Ridgway (1912), and all color terms in quotation marks are from that source. The photographs of the very delicate forms were taken as soon as the notes were recorded. The more durable species were usually left in their packages until evening, when photographs were taken with the aid of artificial light. A 9 × 12 cm. film-pack camera with a Schneider f. 4.5 lens was used. A Weston exposure meter served to determine the time of exposure. The specimens were arranged against either a white cardboard or a black velvet background, depending on the color of the specimens to be photographed. Film packs made it possible to take a large number of photographs at one time and to do the developing at one's convenience.

As soon as the notes and photographs were completed the collection was placed on a drying apparatus. The most satisfactory apparatus for drying large numbers of specimens consisted of the racks (illustrated in Plate 99) with electricity as the source of heat. The frames of the racks have interchangeable parts held together by one-quarter-inch bolts with winged nuts. The racks can be set up or taken down in a few minutes, and when demounted they can be conveniently packed in an automobile. The shelves for the specimens were built by soldering wire screening between two metal frames. The metal frames make the shelves compact and easy to stack in the back of a car. The screens are rather fine-meshed because specimens shrink considerably in drying and are easily lost. The secret of the drying process is to have a steady current of warm air (not too hot) circulating through the piles of specimens on the shelves. The lowest shelf is kept far enough above the source of heat to prevent it from becoming too hot to touch. I usually put the tough woody fungi or the pliant types like Marasmius on the lowest shelf and the watery ones on the uppermost. Mycenae can easily be dried on any shelf. When dry, a collection with its label is wrapped in waxed paper in the same manner as a fresh collection and stored in a carton. When removed from the drier the specimens are very brittle, but, although wrapped in waxed paper, they soon absorb enough moisture from the atmosphere to make them pliant. In this condition they can be packed tightly in cartons without danger of breakage. They are shipped to the laboratory at frequent intervals, where they are fumigated with paradichlorbenzene and sealed up until such time as the expedition has been completed and they can be restudied.

LABORATORY TECHNIQUE

Methods of studying herbarium specimens vary considerably with the individual, and so I shall briefly describe those used in this work. In the first place, microscopic studies of fresh specimens are checked by a reëxamination of the collections in the laboratory. The procedure for studying dried specimens is as follows: The part to be sectioned is moistened with 70 per cent alcohol to wet the surface and drive out the air. Then it is immersed in water until it softens. The softened material is next placed between two pieces of pith, and freehand sections are made with a sharp razor. These are cut in such a way as to give cross sections of the gills as well as sections through the pileus showing the structure of the flesh. They are mounted on a slide in a drop or two of 2 per cent KOH. If they are to be tested with iodine, the sections are placed in a drop or two of the reagent and allowed to stand for a short time (5–20 minutes) before the reaction is recorded. The sections placed in KOH will revive rapidly, and the details of the spores, cystidia, and basidia can be obtained as easily as from fresh material. The trama of the pileus may revive more slowly, but by careful manipulation the details can be seen very nicely on sections of properly dried specimens. Sections of the stipe can be made as readily as those of the gills and pileus. In KOH the gelatinous layers have a characteristic translucent or glassy appearance, and the individual hyphae often appear to be widely separated. When one is making critical studies it is helpful to add a drop of a 1 per cent solution of aqueous basic phloxine to the mount. By the use of this stain small spores become more readily visible, and the details of the wall can be more clearly discerned. Occasionally the spores of some smooth-spored species appear rough under ordinary magnifications of the compound microscope. This is caused by the accumulation of many small highly refractive oil droplets in a layer next to the spore wall. Because of the relative invisibility of the adjacent smooth spore wall, it then appears as if the oil droplets were the wall instead, and because of their uneven appearance the impression of roughness is created. The use of phloxine will prevent one from being misled.

In order to bring the cystidia, particularly the cheilocystidia, completely into view (in sections the pedicels and lower portions are generally obscured by the other cystidia or by the hymenium), it is often necessary to crush the sections by pressure on the cover glass. The location of the cystidia should be carefully studied first. The cystidia shown in the text figures were separated from the hymenium and from the gill edge by this method. In order to study the details of the cuticle of the pileus, one should use radial sections (parallel to instead of across the gills). When these are stained and crushed the individual hyphae of the cuticle can easily be seen and their details observed.

The drawings for the text figures were made with the aid of a camera lucida. The material for the drawings was revived in 2.5 per cent KOH. The cystidia and cells from the gills, pileus, and stipe are reproduced about 750 times natural size unless otherwise stated and the spores 1650 times.

CHEMICAL TESTS

The only chemical test applied in this work was the iodine reaction. A solution of five parts chloral hydrate, two parts water, and an excess of iodine was employed. It is highly desirable that this test be conducted on thoroughly dried material, and it is essential that dried specimens be used for critical work when the reaction is weak. For ordinary purposes one can mount the parts to be studied in a drop or two of the solution, place a cover glass on the mount, and make the needed observations. Where my results have not checked with Kühner's, however, I have employed the procedure he outlined. The following solution is used: KI, 1.5 gr.; iodine, 0.5 gr.; water, 20 cc. Mix one part of this solution with one part chloral hydrate. Place the specimens to be examined in a drop on a slide and allow them to stand several minutes. Wash out the iodine solution with an aqueous solution of chloral hydrate and make the examination by means of daylight.

MATERIAL EXAMINED

The manner of citing the collections requires some explanation. The descriptions, unless otherwise qualified, have been drawn from fresh material that I collected. My collections have been cited first in order to allow the reader to ascertain quickly my acquaintance with the species as it occurs in nature. The number of collections cited, however, is not necessarily a true criterion of the completeness of the data on a species. One large collection may be infinitely more valuable than a dozen in which each contains only a few fruiting bodies. *M. subincarnata* and *M. monticola* furnish an example of this. The number of collections listed for both is the same, but to me *M. monticola* is much more accurately known because of the large number of carpophores studied. An additional point should also be made: Even though the number of collections is large, it does not mean that all the specimens I have seen and studied are recorded. One does not preserve all the material of the well-known and common species that he has collected.

After the citation of the material studied in the fresh condition, other material that has been identified or confirmed is listed. The names of collectors are arranged in alphabetical order. Much material, particularly Atkinson's and Peck's collections, still remains unidentified and is not cited here. Most of these collections fall in the confusing series of species in the Typicae of Eumycena, in which it is very difficult to identify even fresh collections.

As a rule, it is not advisable to describe new species of Mycena from dried specimens and notes sent in by collectors. As a result of studying the types of various species of agarics described by previous investigators, I found that in some instances certain authors had described the same species as new more than once. This happened most frequently when they relied on notes accompanying the specimens. It is the possibility of getting misinformation from the notes that creates the need for caution. In general, in the following text all new species have been based on studies of fresh specimens. However, in such clear-cut cases as that of *M. cayugaensis*, described from Atkinson's collections, I have made an exception to the rule because the character by which the plant is distinguished, namely, the lack of pleurocystidia, can be determined as easily from herbarium specimens as from fresh material.

All my own collections and those of Brown, Burke, Kauffman, Mains, and Povah are deposited in the University of Michigan Herbarium. The collections of Hesler are for the most part deposited in the Herbarium of the University of Michigan and the Herbarium of the University of Tennessee. Atkinson's are in the Atkinson Herbarium of Cornell University, Ithaca, New York, and Peck's are at the New York State Museum, Albany, New York. The type specimens of the new species described by Dr. W. A. Murrill while he was at the New York Botanical Garden are deposited in the herbarium of that institution in New York City. Those he has described recently from Florida are in the Herbarium of the Florida Agricultural Station, Gainesville, Florida.

SUBDIVISIONS OF MYCENA

FOUR subgenera in Mycena—Mycenella, Pseudomycena, Glutinipes, and Eumycena—have been recognized. The traditional concept of Mycenella has been enlarged here to include one section with species having falsely echinulate spores, and the other with those having nodulose spores. If these sections were to be recognized as genera, they would then be Mycenella, characterized by species with nodulose spores, and Fayodia, distinguished by the falsely echinulate type of spore. The species of the latter have their closest relationships in the Omphaliariae of Eumycena, and those of the former in the Typicae. Pseudomycena has been used to include the species of the section Basipedes of Fries (1872) characterized by the presence of a basal disc. Glutinipes is defined to include all species of both Mycena and Omphalia characterized by viscid stipes. Earle (1909) gave the generic name Collopus to the viscid-stiped species of Mycena. Kühner (1938) has put these in various groups under his section Mycenopsis of Eumycena, an arrangement which, in my estimation, does not place sufficient emphasis on the viscidity of the stipe. The series of more or less cespitose lignicolous forms in Glutinipes is related to members of the Typicae through *M. laevigata*. The gray terrestrial species are related to the members of either the Omphaliariae or the Typicae—*M. quinaultensis* and *M. tenax* to the latter and *M. clavicularis*, *M. odorifera*, and *M. vulgaris* to the former. The relationships of those grouped in the Viscosae are not so evident as in those mentioned above, but they appear to be with the members of Eumycena, possibly in the Calodontes. If one were to divide Mycena into several genera, a procedure I do not now believe to be desirable, Glutinipes would be one of the logical divisions.

Kühner (1938) distinguished two subgenera, Eumycena for species with amyloid spores and Paramycena for those with nonamyloid spores. Singer, in 1938, also proposed an arrangement in which several genera were recognized. Mycena was limited to species with amyloid spores; the rough-spored species (they have nonamyloid spores) were placed in Mycenella; and those remaining were divided between Hemimycena and Pseudohiatula. My arrangement differs sharply from the two mentioned above in putting secondary emphasis on the amyloid reaction of the spores. Because of the difference in the degree of the reaction of many species and because of rather numerous exceptions which occur throughout the genus, the character does not appear to be of greater practical or theoretical value than the other major characters of the fruiting body. *M. pseudotenax* is an example of a typical gray Mycena with nonamyloid spores. *M. monticola* is a species with echinulate cheilocystidia and nonamyloid spores; *M. hymenocephala* has amyloid spores but belongs in the Corticatae, the other species of which have nonamyloid spores, and *M. lilacifolia* of Glutinipes has nonamyloid spores. (Singer has transferred the latter to Clitocybe.) I do not believe that the small white species of Deminutivae with amyloid spores (*M. cylindrospora*, *M. paucilamellata*, *M. littoralis*, *M. kalalochensis*, and *M. pusillissima*) can justifiably be placed in a different genus from the other small white species of Mycena which they resemble in most respects. The genus Delicatula would be the most logical place to put them if they were excluded from Mycena, but their development is gymnocarpic, as in Mycena, rather than hemiangiocarpic, as in Delicatula. In addition, they are more closely related to Mycena because of their cystidia. In the arrangement proposed in the text I have regarded the amyloid character as secondary when it conflicted sharply with all the other characters. It is evident in all groups of agarics that each has developed along particular lines which have emphasized the value of individual characters in the delimitation of species and genera. In both Leucopaxillus and Melanoleuca (the latter in the restricted sense of Patouillard) the amyloid reaction of the spores appears to be a character which, when correlated with certain others, delimits a group of related species of generic rank. In Mycena the amyloid character of the spores and flesh of the carpophore does not appear to be sufficiently correlated with any other characters to justify separating groups of generic or subgeneric rank.

In the following outline an attempt has been made to place related species in small groups or stirpes, each of which is named for its central species. This arrangement is within that of the subgenera, sections, and subsections as given in the text.

CLASSIFICATION INTO STIRPES

I. SUBGENUS PSEUDOMYCENA

A. Section Tenerrimae

1. Stirps tenerrima: *M. tenerrima, M. osmundicola*

B. Section Basipedes
  1. Stirps Mucor: *M. Mucor, M. Gaultheri*
  2. Stirps longiseta: *M. longiseta*
  3. Stirps stylobates: *M. stylobates, M. bulbosa, M. clavularis*

II. Subgenus Eumycena

A. Section Cyanescentes
  1. Stirps amicta: *M. amicta, M. subcaerulea*

B. Section Corticolae
  1. Stirps corticola: *M. corticola, M. pseudocorticola, M. corticalis, M. madronicola*
  2. Stirps subcana: *M. subcana*

C. Section Deminutivae
  1. Stirps minutula: *M. minutula, M. cylindrospora*
  2. Stirps mauretanica: *M. mauretanica*
  3. Stirps albicolor: *M. albicolor, M. filiformis, M. albissima*
  4. Stirps paucilamellata: *M. paucilamellata, M. litoralis, M. kalalochensis, M. pusillissima*
  5. Stirps setulosa: *M. setulosa, M. apiculata*
  6. Stirps gracilis: *M. gracilis, M. subimmaculata, M. McMurphyi, M. albidula, M. ignobilis*
  7. Stirps papillata: *M. subvestita, M. papillata*
  8. Stirps pterigena: *M. juncicola, M. pterigena, M. tenella, M. Smithiana, M. Lohwagii*
  9. Stirps capillaris: *M. minutissima, M. capillaris, M. mirata*
  10. Stirps supina: *M. supina, M. Quercus-Ilicis*
  11. Stirps subsupina: *M. subsupina, M. subcucullata*
  12. Stirps acicula: *M. acicula, M. oregonensis, M. fibula, M. Swartzii, M. pseudogrisella*

D. Section Lactipedes
  1. Stirps galopus: *M. galopus, M. cayugaensis, M. erubescens, M. copiosa* (*M. Sabali* ?)
  2. Stirps anomala: *M. anomala, M. crocata*
  3. Stirps haematopus: *M. haematopus, M. fagicola, M. subsanguinolenta, M. sanguinolenta*

E. Section Adonidae
  1. Subsection Albidae
    a) Stirps lineata: *M. lineata*
    b) Stirps gypsea: *M. gypsea, M. olida* var. *americana* (*M. hiemalis* ?)
    c) Stirps translucentipes: *M. translucentipes*

   d) Stirps delicatella: *M. delicatella, M. pseudolactea*
   e) Stirps delectabilis: *M. Rickenii, M. delectabilis*
2. Subsection Typicae
   a) Stirps flavifolia: *M. flavifolia*
   b) Stirps carolinensis: *M. carolinensis*
   c) Stirps luteopallens: *M. luteopallens*
   d) Stirps flavoalba: *M. flavoalba, M. leptophylla, M. roseipallens*
   e) Stirps Adonis: *M. fusipes, M. aurantiidisca, M. Adonis*
   f) Stirps amabilissima: *M. roseocandida, M. amabilissima*
   g) Stirps monticola: *M. monticola, M. subincarnata*
3. Section Ianthinae
   a) Stirps pura: *M. pura, M. Kuehneriana, M. subaquosa*
   b) Stirps pelianthina: *M. pelianthina, M. rutilantiformis*

F. Section Calodontes
1. Subsection Granulatae
   a) Stirps rosella: *M. rosella* (*M. subincarnata* ?)
   b) Stirps strobilinoides: *M. strobilinoides*
   c) Stirps elegans: *M. elegans, M. flavescens, M. aurantiomarginata*
2. Subsection Ciliatae
   a) Stirps capillaripes: *M. debilis, M. capillaripes, M. olivaceobrunnea, M. citrinomarginata, M. avenacea, M. cheboyganensis*
   b) Stirps elegantula: *M. purpureofusca, M. elegantula, M. rubromarginata, M. viridimarginata, M. atromarginata*

G. Section Typicae
1. Stirps scabripes: *M. trichoderma, M. scabripes*
2. Stirps leptocephala: *M. pseudotenax, M. Abramsii, M. griseiconica, M. leptocephala, M. subvitrea, M. stannea, M. pectinata, M. atrocyanea*
3. Stirps praelonga: *M. plumbea, M. praelonga, M. fusco-ocula, M. fragillima, M. Kauffmaniana, M. subfusca, M. subfumosa* (*M. syringescens* ?)
4. Stirps latifolia: *M. latifolia*
5. Stirps pusilla: *M. constans, M. pusilla, M. alcaliniformis, M. urania, M. psammicola, M. atroalboides*
6. Stirps metata: *M. metata, M. hudsoniana, M. plicosa, M. piceicola, M. subplicosa*
7. Stirps filipes: *M. iodiolens, M. filopes* (*M. atroalboides*)
8. Stirps vitilis: *M. vitilis, M. pullata, M. peltata*
9. Stirps polygramma: *M. polygramma*
10. Stirps fagetorum: *M. fagetorum*

11. Stirps megaspora: *M. megaspora, M. magna, M. longipes*
12. Stirps borealis: *M. borealis*
13. Stirps algeriensis: *M. excisa, M. tenuiceps, M. algeriensis, M. niveipes, M. macrocystidiata, M. alcalina, M. incarnatifolia*
14. Stirps laevigata: *M. collybiiformis, M. laevigata, M. semivestipes, M. ochraceicinerea*
15. Stirps parabolica: *M. parabolica, M. alnicola, M. rubrotincta, M. intertexta*
16. Stirps inclinata: *M. inclinata, M. pseudoinclinata*
17. Stirps galericulata: *M. maculata, M. occidentalis, M. hemisphaerica, M. subinclinata, M. galericulata, M. rugulosiceps, M. radicatella*

H. Section Omphaliariae
1. Stirps speirea: *M. speirea*
2. Stirps misera: *M. Brownii, M. misera, M. concolor*
3. Stirps cinerella: *M. cinerella, M. serotina*
4. Stirps subconcolor: *M. subconcolor*
5. Stirps pseudogrisea: *M. praedecurrens, M. fuliginella, M. pseudoclavicularis, M. arenaria, M. turbinata, M. pseudogrisea*
6. Stirps thujina: *M. thujina*

I. Section Floccipes
1. Stirps floccipes: *M. floccipes, M. Kauffmanii, M. ulmicola*

J. Section Corticatae
1. Stirps hymenocephala: *M. hymenocephala*
2. Stirps ludoviciana: *M. wyomingensis, M. ludoviciana*

K. Section Hydropus
1. Stirps fuliginaria: *M. fuliginaria, M. taxodii*
2. Stirps marginella: *M. marginella, M. umbrina, M. seriflua*

III. Subgenus Glutinipes

A. Section Diversiformes
1. Stirps rorida: *M. rorida*
2. Stirps clavicularis: *M. clavicularis*
3. Stirps insignis: *M. insignis*

B. Section Caespitosae
1. Stirps Leaiana: *M. texensis, M. Leaiana, M. hondurensis, M. euspeirea, M. glutinosa, M. Austinii*
2. Stirps subepipterygia: *M. subepipterygia*
3. Stirps lilacifolia: *M. lilacifolia*

C. Section Viscosae
1. Stirps epipterygia: *M. epipterygia, M. viscosa, M. epipterygioides*
2. Stirps griseoviridis: *M. griseoviridis*

D. Section Fuliginellae

1. Stirps tenax: *M. tenax, M. quinaultensis*
2. Stirps pelliculosa: *M. pelliculosa, M. odorifera*
3. Stirps vulgaris: *M. vulgaris, M. militaris*

IV. SUBGENUS MYCENELLA

1. Stirps bryophila: *M. margaritispora, M. trachyspora, M. bryophila, M. nodulosa*
2. Stirps cineraria: *M. bisphaerigera, M. cineraria*

MYCENA S. F. GRAY

Fries, Elias, Syst. Myc., 1: 140 (as a tribe of Agaricus). 1821.
Gray, S. F. Nat. Arr. Brit. Plants, 1: 619. 1821.
(Not Mycena in Murrill, North Am. Flora, 10: 190. 1917[1])

SYNONYMY

? Gloiocephala Masse, Grevillea, 21: 33. 1892.
Mycenula Karsten, Medd. Soc. Faun. Fl. Fenn., 16: 89. 1889.
Eomycenella Atkinson, Bot. Gaz., 34: 37. 1902.
Basidopus Earle, Bull. New York Bot. Gard., 5: 426. 1909.
Collopus Earle, *ibid.*,
Galactopus Earle, *ibid.*
Insiticia Earle, *ibid.*, p. 425.
Linopodium Earle, *ibid.*, p. 427.
Stereopodium Earle, *ibid.*, p. 426.
Pseudomycena Cejp, Publ. Fac. Sci. Univ. Charles, p. 138. 1930.
Hemimycena Singer, Rev. Mycol., 3: 194. 1938.
Mycenella (Lange) Singer, *ibid.*, p. 195.
Fayodia Kühner em. Singer, Lloydia, 5: 128. 1942.
Hydropus (Kühner) Singer, *ibid.*

Type of the genus: *Agaricus* (*Mycena*) *galericulatus* Fries

KEY TO THE SUBGENERA OF MYCENA

1. Spores rough, nodulose, or echinulate or appearing so when stained (*M. luteopallens* with very slightly wrinkled spores is keyed out in Eumycema) .... Mycenella, p. 442
1. Spores smooth 2
2. Stipe with basal disc or bulb, or, if without a bulb, pileus and stipe granulose under a lens because of numerous inflated cells with verrucose walls; stipe separable from pileus Pseudomycena, p. 44
2. Stipe without a distinct basal bulb or disc, not readily separable from pileus (small species have slightly clavate stipes inserted on substratum) 3
3. Stipe in cross section showing outer gelatinous layer; hence usually viscid to the touch when fresh ... Glutinipes, p. 401
3. Stipe not as above .... Eumycena, p. 60

[1] Murrill used the name "Mycena" for the fungi belonging to Bolbitius of the Friesian system, and "Prunulus S. F. Gray" for the species here placed in Mycena.

SUBGENUS PSEUDOMYCENA

IN THIS subgenus are grouped all the minute, delicate species characterized by the presence of a small but distinct bulb or disc at the base of the stipe and having the stipe readily separable from the pileus. In one section the fruiting bodies are very coarsely granular (as seen under a lens) owing to the covering of verrucose globular cells. The two sections together represent a remarkably homogeneous group of species. The limits of the subgenus, as given here, exclude such species as *M. pterigena* and *M. corticola* and, in fact, are very closely in line with the limits of the section Basipedes of Fries. *M. Brownii*, with a very distinct bulb but with the stipe not separable from the pileus, is placed in the Omphaliariae of Eumycena.

KEY TO SECTIONS

Verrucose inflated cells covering pileus, fruiting bodies generally with chalky appearance .... Section Tenerrimae

Pileus setulose, glabrous, or slightly pruinose at first, or soon becoming glabrous and moist .... Section Basipedes

TENERRIMAE

1. Cheilocystidia with apical needlelike projections ...... 1. *M. tenerrima*
1. Cheilocystidia merely verrucose .... 2. *M. osmundicola*

BASIPEDES

1. Pileus and stipe at first covered with long, thick-walled hyaline setae 8. *M. longiseta*
1. Pileus and stipe not having thick-walled setae 2
2. Pileus yellowish, bulb bright yellow .... 4. *M. Gaultheri*
2. Not as above 3
3. Cheilocystidia 18–25 × 7–12 μ, ovoid to clavate, apices echinulate 3. *M. Mucor*
3. Cheilocystidia, if rough, with fingerlike projections, their outlines often irregular; sometimes smooth and fusoid-ventricose 4
4. Spores globose to subglobose .... 7. *M. clavularis*
4. Spores ellipsoid 5

5. Gill edges gelatinous; cheilocystidia narrow (up to 7.5 μ), with one to several apical projections .... 6. *M. bulbosa*
5. Gill edges not gelatinous; cheilocystidia 8–13 μ broad, fusoid ventricose or with fingerlike projections .... 5. *M. stylobates*

SECTION TENERRIMAE

1. Mycena tenerrima (Berk.) Quélet

Champ. Jura et Vosges, p. 109. 1872

Agaricus (*Mycena*) *tenerrimus* Berkeley, in Smith, English Flora, 5: 61. 1836.
Prunulus tenerrimus Murrill, North Am. Flora, 9: 322. 1916.
Pseudomycena tenerrima Cejp, Publ. Fac. Sci. Univ. Charles, 104: 151. 1930.

Illustrations:
Plate 1 D; Text fig. 1, nos. 1–3 (p. 46).
Kühner, Encyc. Myc., 10: 207, fig. 66.
Lange, Flora Agar. Dan., 2, pl. 57 C.

Pileus 2.2–3.5 mm. broad, obtuse or cucullate, soon convex, the margin straight, becoming broadly convex, appearing as if sprinkled with sugar when viewed under a lens, pallid gray with a whitish margin when young, soon white over all, sulcate-striate to the disc, margin entire; flesh membranous and fragile; lamellae free or attached to the stipe by a line, broad, distant, sometimes adhering to each other and thus forming a slight collar around the stipe, white at all stages; stipe 5–15 (20) mm. long, usually curved and filiform, base enlarged into a slight bulb, which is globular at first but becomes flattened in age, setose under a lens at first but in age merely pruinose over all, including the bulb, hyaline white.

Spores 8–10 × 5–6.5 (7) μ, broadly ellipsoid, amyloid; basidia two-spored; pleurocystidia rare or absent, if present similar to the cheilocystidia; cheilocystidia abundant, 28–44 × 8–12 μ, variable in shape, often fusoid-ventricose or with 2–3 needlelike projections arising from the apex, the projections forked at times, the ventricose portion covered with short rodlike protuberances or merely verrucose; gill trama vinaceous brown in iodine; pileus trama made up of greatly enlarged cells, the surface covered with clavate to almost globular cells 25–40 × 20 ± μ, their walls finely verrucose, all but the verrucose cells vinaceous brown in iodine.

Habit, habitat, and distribution.—Scattered to gregarious on twigs, bark, and debris of conifers and hardwoods, spring and fall; Washing-

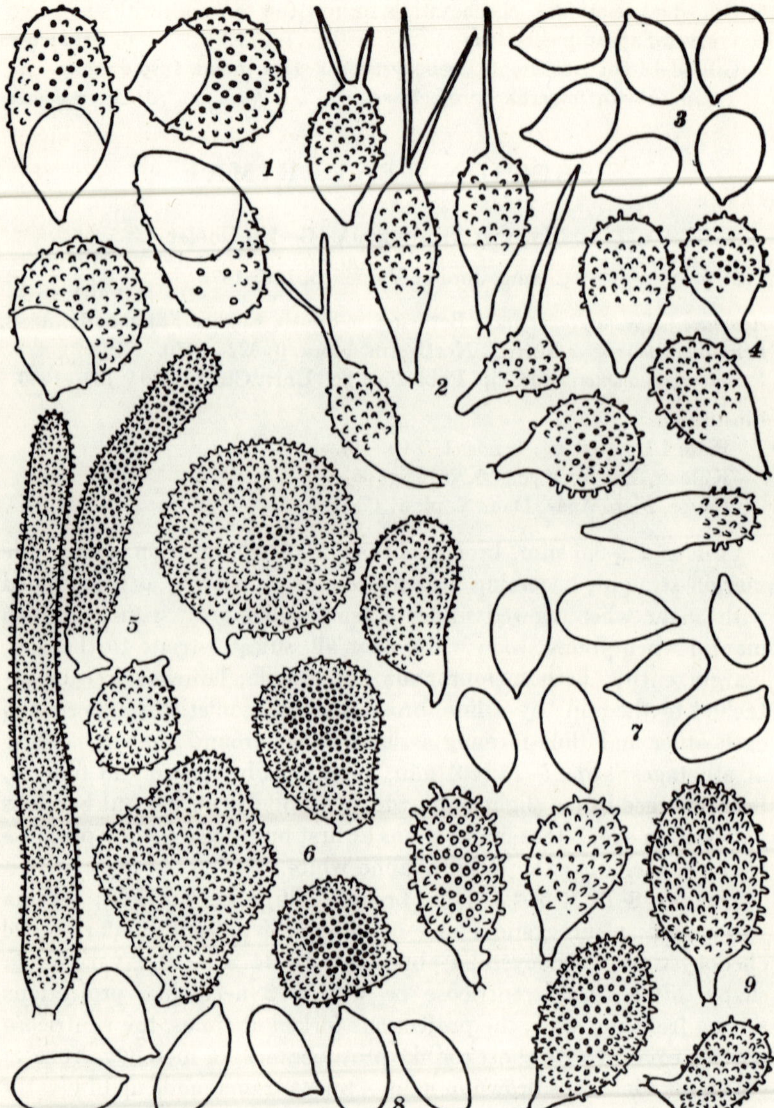

FIG. 1. *M. tenerrima:* 1, cells from surface of pileus; 2, cheilocystidia; 3, spores. *M. osmundicola:* 4, cheilocystidia; 5, cells from surface of stipe; 6, cells from surface of pileus; 7, spores. *M. Gaultheri:* 8, spores; 9, cheilocystidia

ton to California. It is not uncommon along the Pacific coast, but I have not found it elsewhere in North America.

Material studied.—Smith, 2505, 3511, 3645, 3668, 3674, 3705, 3771, 8401, 9368, 13561, 13963, 13999. Gruber, 20, Mt. Scott.

Observations.—The collections I reported (1935a) from Michigan under this name are now referred to *M. osmundicola. M. tenerrima* is characterized by its short stipe, its cheilocystidia, which nearly always develop long needlelike projections from their apices, and the verrucose cells scattered over the surface of the pileus. The cells, as Kühner has pointed out, often have the upper part filled solidly with what appears to be mucilaginous wall material. Lange reports both a two- and four-spored form from Denmark. I have collected only the two-spored form. For additional comments see *M. osmundicola.*

2. Mycena osmundicola Lange

Dansk Bot. Arkiv, 1 (5): 35. 1914

Illustrations:

Plate 1 B; Text fig. 1, nos. 4–7.

Beardslee & Coker, Journ. Elisha Mitchell Sci. Soc., 40, pl. 15, upper left; pl. 16, lower left.

Lange, Flora Agar. Dan., 2, pl. 57 A, 1936 (also B. as *tenerrima* var. *carpophila*).

Smith, Am. Journ. Bot., 22, pl. 4, fig. 4 (as *M. tenerrima*).

Pileus (2) 3–6 mm. broad, convex to obtuse in button stages, the margin appressed, becoming either conic or convex, the margin flaring or recurved in age on some, surface appearing dry and powdery owing to the dense sugarlike covering, moist and striate when pruinosity is removed, color pale grayish at first beneath the white powdery coating, fading to chalk white in age, sometimes becoming sulcate; flesh very thin and flaccid, tender, odor and taste not distinctive; lamellae distant to subdistant, narrow, sometimes slightly ventricose at maturity, free or attached by a slight tooth, white, edges even; stipe 2–3 cm. long, up to 1 mm. thick, slightly thickened toward the base (which was abruptly inserted on fallen larch needles), tubular, densely white villous from the base upward (the hairs shorter toward the apex), in age merely finely pruinose or chalky in appearance, white to whitish.

Spores 7–9 (10) × 4–5 μ, ellipsoid, weakly amyloid; basidia usually four-spored, rarely two-spored; no pleurocystidia seen; cheilocystidia abundant, 18–28 × 9–15 μ, clavate, the enlarged portion covered with small blunt projections or densely verrucose, hyaline; gill trama vinaceous brown in iodine; pileus trama covered with a layer or coating of inflated or globular readily detachable verrucose cells 18–24 × 10–30 μ, tissue beneath colored vinaceous brown in iodine; stipe tissue yellow to hyaline in iodine, surface covered with long or short verrucose cells or filaments 60–300 × 9–20 μ, their apices obtuse.

Habit, habitat, and distribution.—Single to scattered on conifer needles, fern debris, or old leaves during spring, summer, and fall. It is known from Alabama, North Carolina, Tennessee, New York, Ohio, Michigan, and Washington in the United States and from Ontario and British Columbia in Canada. It has also been found in British Honduras. Because of its size and delicate texture it is easily overlooked or neglected. Mr. Walters found it on Osmunda fibers in a greenhouse near Cleveland, Ohio.

Material studied.—Smith, 32–19, 32–20, 32–230, 33–576, 33–597, 33–933, 715, 4713, 7026, 7471, 10741, 10906. Atkinson, New York (as *M. tenerrima*). Burke, Alabama. Mains, 3740, 3858, 4050. Sharp, 5170. Walters, November 15, 1940, Ohio.

Observations.—I have never observed the cheilocystidia of this fungus to have the long needlelike projections which characterize western collections of typical *M. tenerrima*. In regard to the identity of *M. osmundicola* and *M. tenerrima* var. *carpophila* Lange there seems to be no doubt that the variety should be referred to the former as a synonym. At least I have found material on beech mast which checks perfectly with *M. osmundicola* and which does not have the characteristic cheilocystidia of *M. tenerrima*. In his first publication (1914, p. 35) Lange stated, of the variety, "for the rest microscopically like 50a" (*M. tenerrima*). But for the latter (which is no. 50a, referred to) he described the cheilocystidia as "*apex either obtuse, obtuse with hairlike appendix or gradually tapering off into a hairlike appendix.*" Since he makes no definite statement whether this variation was noted on a single cap or on different collections, it is impossible to be certain in regard to the shape of the cheilocystidia in the variety.

Lange believed that *M. osmundicola* was an introduced species, and thus was obviously not expecting to find it native to Denmark. In North America it is apparent that it has adapted itself to a wide range of substrata and is very widely distributed. It is possible, of course, that the fungus on beech mast in Denmark is different from that on the same substratum in North America, but this does not seem probable in view of the descriptions and illustrations.

There are several other forms in this group that deserve consideration. I have one collection on old oak leaves from near Ann Arbor. It had all the characters of *M. osmundicola* except for the spores, which are 8–10 × 3–4 μ on four-spored basidia. There was a beautiful almost globular bulb at the base of the stipe. Its cheilocystidia lacked the awl-shaped apical projections, and all other characters were as given in the foregoing description.

At Pinckney, Michigan, E. B. Mains made one collection of a two-spored form with spores 10–12 × 5–6 μ. He found the material on debris of ferns in a bog. As in the previous collection, all other characters were as given in the foregoing description.

At Ravenel Lake, Highlands, North Carolina (Hesler, 5170), and in the Great Smoky Mountains National Park in a locality near Gatlinburg, Tennessee; a form was found (Hesler and Smith, 7471) in which the stipe measured (1) 2–6 cm. long and had a rounded bulb which disappeared as the stipe elongated. Its spores measure 7–9 (10) × 4–5 μ and were borne on four-spored basidia. The pileus was a delicate yellowish-green color beneath the sugary coating and faded to pale yellowish and finally white. Again, the cystidia are similar to those of *M. osmundicola*. The hairs on the stem are longer than usual for the latter, but when they are examined and compared under the microscope, that difference does not appear taxonomically significant. This form was quite abundant in one locality along a stream under rhododendrons near Gatlinburg.

The difficulties of working with these small fungi are such that at present I am inclined to regard the collections mentioned above as local variations about which we do not have enough information to justify giving them separate taxonomic rank. I suspect that *M. osmundicola* nearly always has a small bulb at first, and that its absence is mostly a matter of the elongation of the stipe. The variations in spore size of 7–8, 7–9.5, or 8–10 μ do not appear significant, and the number of spores borne on a basidium has been shown to be of no taxonomic value as a distinguishing character. The greenish yellow colors and long stipe of the Tennessee and North Carolina collections may be more significant, and may serve to distinguish a southern variety of the species.

The specimens Kauffman (1918) referred to *M. setosa* were found to be the typical form of *M. osmundicola.* I suspect that the same is true of the material reported as *M. setosa* by Beardslee and Coker (1924). Kühner does not regard *M. setosa* as sufficiently well known to be recognized. Fries placed it in the section Insititiae. Rea lists *M. tenerrima* Berk. as a synonym of *M. setosa.* There seems little hope of stabilizing the application of names in such a group of small fungi as this unless one arbitrarily accepts those names of species which have been clearly described, and abandons those over which there has been much disagreement, even though the latter fall within the time limits established by the *International Rules of Nomenclature.*

SECTION BASIPEDES

3. Mycena Mucor (Fr.) Quélet

Champ. Jura et Vosges, p. 436 (8 in reprint). 1875

Agaricus (Mycena) Mucor Fries, Syst. Myc., 1: 155. 1821.
Pseudomycena Mucor Cejp, Publ. Fac. Sci. Univ. Charles, 104: 144. 1930.

Illustrations:
Plate 2 A.
Lange, Flora Agar. Dan., 2, pl. 56 A.

Pileus 2–5 mm. broad, ellipsoid at first, becoming broadly conic and later expanding to convex or plane, when moist conspicuously striate, glabrous but appearing pruinose at first, pallid watery grayish over the disc and striations, the remainder watery white, fading to whitish, sometimes hyaline white over all when moist; flesh very thin and soft, pileus readily separable from the stipe, odor and taste not recorded; lamellae rather distant, 6–9 reach the stipe, one or two tiers of lamellulae, ascending-adnate but becoming horizontally adnate, seceding but adhering to each other to form a collar around the stipe, white, edges even; stipe 1–3 (4) cm. long, filiform, straight or flexuous, equal above a distinct rounded bulb, bulb disappearing somewhat in age and visible only as an inconspicuous flattened disc, bulb and lower portion of the stipe slightly pruinose-pubescent, sometimes glabrous, sometimes with scattered flexuous hairs, translucent white over all.

Spores 7–10 × 3–4 μ, subcylindric to somewhat pear-shaped, smooth, amyloid; basidia four-spored; pleurocystidia not differentiated; cheilocystidia 18–22 × 7–12 μ, ovoid to clavate with rounded apices which are covered with scattered slender cylindric prolongations.

Habit, habitat, and distribution.—Gregarious to scattered on oak leaves, usually late in the fall in company with *M. capillaris*. I have records of it only from Michigan, but it apparently is widely distributed.

Material studied.—Smith, 31–668, 32–660, 1169, 1176, 1268. Kauffman, October 30, 1929, Michigan.

Observations.—This is a very small delicate species easily distinguished from *M. capillaris* by the slight basal bulb of the stipe. I have not studied the trama of the gills and pileus in detail. For a more complete account of these characters see Kühner (1938). It is difficult to make specimens of such a small agaric as this since, by the time the fruiting bodies have been dried, they are practically invisible. The gills may appear to be free in small specimens, and in fact sometimes are, but the character has little taxonomic significance. Many characters which are very constant in larger species are found to be extremely variable in the minute forms, and one must be very careful to avoid describing each collection as a new species or variety. Spore size is a good example. In a single deposit a great range of variation can be found. Deposits from fresh, unwilted pilei are hard to get, and the last spores deposited are apt to be rather irregular in size and shape. Similar differences can be found between collections, and it is here that one is most likely to be misled. In mounts of revived material the spore size is likely to be smaller than that obtained from spore deposits of perfect specimens. This is apparently caused by the larger percentage of immature spores that occur in the mounts of the revived material.

4. **Mycena Gaultheri,** sp. nov.

Illustrations: Text fig. 1, nos. 8–9 (p. 46).

Pileus 1–3 (4) mm. latus, convexus vel conicus, melleus, striatus; lamellae adnatae, distantes, latae, pallidae; stipes 2–4 cm. longus, flexuosus vel strictus, bulbosus, deorsum luteus, sursum pallidus; sporae 8–9 (10) × 4 5 μ; cheilocystidia clavata vel capitata, echinulata. Specimen typicum in Herb. Univ. Mich. conservatum. Legit

A. H. Smith, n. 17390, prope Heart O'Hills, Mt. Angeles, Olympic Mts., Wash., Sept. 29, 1941.

Pileus 1–3 (4) mm. broad, obtusely conic to convex, margin straight at first, in age plane or the margin somewhat uplifted, glabrous, moist, unpolished, disc "Isabella color" (sordid yellowish brown), paler and more buff-colored toward the margin ("cream buff"), when moist conspicuously translucent-striate to the disc with broad, dark striations, somewhat plicate when faded; flesh membranous, not fragile but not reviving when remoistened, odor and taste not distinctive; lamellae adnate but soon seceding and adhering to each other to form a collar around the stipe, distant, 7–9 reach the stipe, lamellulae lacking or one tier present, moderately broad, pallid (whitish in contrast to cap), edges even and not differently colored; stipe 2–4 cm. long, more or less filiform, straight or flexuous, seated on a "pale orange-yellow" flat strigose-hairy basal disc, glabrous and polished toward the apex, minutely pruinose toward the basal disc, lower portion pale sordid yellow, pallid to whitish above.

Spores 8–9 (10) $\times$ 4–5 μ, narrowly ellipsoid, smooth, amyloid, basidia four- or occasionally two-spored; pleurocystidia not differentiated; cheilocystida abundant, 18–36 $\times$ 9–16 μ, clavate to subcapitate, the enlarged portion verrucose, hyaline in water mounts of fresh material; pileus trama very distinctive, the surface corticated with a layer of cells similar to the cheilocystida or larger, the layer beneath these (including their pedicels) gelatinous, central body of the trama vesiculose, the subhymenium of short hyphae with rather broad cells.

Habit, habitat, and distribution.—Singly or in groups of two or more carpophores on dead leaves of *Gaultheria shallon* Pursh. The species at present is known only from the type locality, but will more than likely be found in Oregon and northern California as well.

Material studied.—Smith, 17390.

Observations.—The sordid yellowish-brown pileus, yellowish stipe, and pale orange-yellow basal disc distinguish the fungus macroscopically. Microscopically the plant is very similar to *M. Mucor* as described by Kühner (1938), particularly in the organization of the trama of the pileus. However, the degree of its organization in *M. Gaultheri* appears to be higher than that of *M. Mucor.* The former also differs from the latter in consistency. *M. Gaultheri* is rather pliant and somewhat like *M. corticola* in this respect, whereas the European species, at least as I know it, is very delicate.

5. Mycena stylobates (Fr.) Quélet

Champ. Jura et Vosges, p. 109. 1872

Agaricus stylobates Fries, Syst. Myc., 1: 153. 1821.
Pseudomycena stylobates Cejp, Publ. Fac. Sci. Univ. Charles, 104: 150. 1930.

Illustrations:

Plate 1 C, E–F; Text fig. 2, nos. 3–4 (p. 58).
Beardslee, Mycologia, 9, pl. 4, fig. 1.
Beardslee and Coker, Journ. Elisha Mitchell Sci. Soc., 40, pl. 15, lower left; pl. 16, upper figures.
Lange, Flora Agar. Dan., 2, pl. 54 C and also B (as *M. clavularis*).
Ricken, Die Blätterpilze, 2, pl. 109, fig. 10.

Pileus 3–15 mm. broad, obtusely conic to convex, the margin straight or curved in slightly, becoming campanulate, umbonate or plane, at times the margin flaring or recurved, surface smooth or spinulose under a lens (especially around the disc), soon glabrous, moist and glistening somewhat, translucent-striate, color evenly pale watery gray except for a whitish margin, somewhat hygrophanous and fading to pallid or nearly white and then more or less uneven or sulcate; flesh very thin, pallid, no odor or taste; lamellae appearing close in unexpanded caps and distant at times in old individuals or sometimes close in large pilei when mature, 8–16 reach the stipe, one or two tiers of lamellulae, narrow but becoming ventricose and sometimes very broad in age, attached by a line or very narrowly adnate, sometimes seceding and adhering to each other and thus forming a collar around the stipe, very pale gray but soon becoming whitish, edges even; stipe separable from the pileus, (5) 10–60 mm. long, 0.5–1 mm. thick, equal above a flat circular disc, not markedly fragile for such a small fungus, covered with fine white scattered fibrils or delicately pruinose, glabrescent, bluish gray when very fresh but soon fading through watery gray to whitish, sometimes whitish from the first and occasionally somewhat grayish when faded, basal disc striate (from gill impressions) and pruinose or finely pubescent, soon glabrescent.

Spores 6–8 or 8–10 × 3.5–4.5 μ, narrowly ellipsoid, faintly amyloid; basidia four-spored, rarely two-spored; pleurocystidia not differentiated; cheilocystidia abundant and variable, usually clavate with thick obtuse projections (2–5) arising from near the apex, sometimes more or less covered with numerous protuberances over the enlarged portion and the neck more or less contorted, 26–38 × 8–13 μ, hyaline;

gill trama of greatly enlarged cells, pale vinaceous in iodine; pileus trama with a pellicle which usually gelatinizes in KOH or water mounts, the surface hyphae covered with short rodlike projections, occasionally some of the hyphae become aggregated into peglike structures (which project from the surface and cause the appearance of scattered coarse spines on the cap when viewed under a 10× lens), tissue beneath the pellicle entirely of greatly enlarged cells, which are pale vinaceous in iodine.

Habit, habitat, and distribution.—Scattered to gregarious on oak leaves or coniferous needles, spring and summer or early fall. It is common during warm, wet seasons. I have collected it in Tennessee, Michigan, Idaho, and Washington in the United States and in Nova Scotia and Ontario in Canada. Dried material from Pennsylvania and from Manitoba, Canada, has also been studied.

Material studied.—Smith, 32–468, 32–472, 32–493, 33–437, 33–579, 33–908, 1615, 1701, 4824, 7025, 9668, 13589, 14111, 14680, 14819, 15751, 15858. Kauffman, Pennsylvania. Wehmeyer, 691.

Observations.—Although I have complete descriptions of over half a dozen collections, and have examined many more, the limits of the species do not appear to be sharply defined, and the best that can be done at present is to include under this name a series of very closely related forms. The description given here has been drawn from material which appeared in great quantity in a local pine plantation (Smith, 1615), but which was collected with equal abundance both on oak leaves that had become lodged among the needles and on the needles themselves. The variation in the cystidia appears to be as great as that of *M. citrinomarginata*—a conclusion drawn from studying the collection mentioned above. In individual collections from other localities the cystidia of some were fusoid-ventricose or with the apices furnished with two or more fingerlike projections. In others the cystidia were clavate with short rodlike projections or with obtuse, blunt contorted processes covering their apices. In regard to gill attachment and pubescence of the bulb no clear-cut distinctions were found in my material. The gills may or may not form a collar around the stipe—depending upon the degree of attachment at first. If broadly attached they tend to secede and form a collar; if the attachment is slight they may not secede at all or, if they do, they may not adhere to each other. In some small caps they are usually more or less distant at maturity, whereas in robust carpophores they may be quite crowded.

The spore size is also variable. In deposits from a number of collections they measure 6–8 × 3–4 μ, 7–9 × 3.5–4 μ, 6–8 × 3.5–4 μ, 8–10 × 4–4.5 μ, 8–10 × 3 μ, and 6–7.5 × 3.5 μ. In one form with two-spored basidia the spores measured 11–14 × 2.5–3 μ. Scattered four-spored basidia on the same sections had spores 8–10 × 3–5.5 μ. Between some species the difference of 6–8 and 8–10 μ in spore length is significant, but I cannot so regard it here. Kühner gives the length of the spores as 7–10 μ long, and illustrates the cheilocystidia as decidedly variable in shape, as in American collections.

A form with a strongly plicate pileus was found on mossy alder logs in the Hoh River Valley of Washington (A. H. Smith, 13589, 14111). The flesh was soft and, in general, the fruiting bodies had the aspect of a Pseudocoprinus. Two collections were made from the same log within a month. In view of the spores (8–10 × 4–5 μ), the cheilocystidia (intermediate between the fusoid-ventricose and the clavate-roughened types), and the gelatinous pellicle, I am placing the collections here. More observations are needed to determine just how constant the striations and the soft consistency are before giving them further consideration. They could very easily be chance variations. Typical *M. stylobates* was also found in the same vicinity on debris under willows in the river's flood plain. *M. coprinoides* Karsten (1883) appears to be somewhat similar to collections 13589 and 14111, but differs in the color of the pileus. Karsten recognized the relationship of *M. coprinoides* to *M. stylobates*.

6. Mycena bulbosa (Cejp) Kühner

Encyc. Myc., 10: 176. 1938

Pseudomycena bulbosa Cejp, Publ. Fac. Sci. Univ. Charles, 104: 149. 1930.
(The following description is adapted from Kühner.)

Pileus 3.5–6 mm. broad, hemispheric or convex, 2.5–5 mm. across the base, obtuse or submammillate, radially striate and somewhat plicate (the creases not over the backs of the large lamellae), grayish brown, paler with age but the striae always strikingly colored, hygrophanous, fading from the center, glabrous and not spinose as in *M. stylobates*, with an entirely gelatinous separable pellicle; flesh thin, odor and taste mild; lamellae 9–12 reach the stipe, one to three tiers of lamellulae, whitish or with a hyalinc-grayish tint, ventricose, nearly free, separating from the stipe in a star-shaped arrangement, slightly intervenose; stipe 4–10 mm. long, 0.2–0.5 mm. thick, subequal, flexuous, often ascending and curved, hyaline, more or less soiled grayish brown, glistening puberulent under a lens, sometimes glabrescent at least toward the base, fistulose, with a basal disc 1–1.5 mm. broad, convex, pubescent, white, striate to plicate on the upper side, sometimes arising from a sclerotium with brown walls, which is hidden in the mesophyll of the leaves that support it.

Spores elliptic, 8–10 × 3.2–4.5 μ, nonamyloid, germinating on the gills; basidia four-spored, 18–24 × 7.5–8.5 μ; cheilocystidia elongated and hardly inflated (4–7.5 μ in diameter), or ventricose to clavate, not covered with bristles but often furnished at the summit with a few (2–3) short projections; gill trama with subglobular hyphae (35–40 μ thick); subhymenium very thin, loosely branched, and not gelatinous; the gill edge more or less gelatinous and swollen; flesh of the pileus distinctly amyloid, the hyphae radially arranged, somewhat interwoven, and with a violaceous-brown pigment; pellicle of pileus gelatinous, 35–58 μ thick, the surface hyphae with short granulose projections.

Habit, habitat, and distribution.—On Juncus and rarely on Carex, August and September; Europe. I have not been able to recognize this fungus with certainty in North America. Occasional single specimens on Carex from Michigan may possibly belong here. The pellicle did not gelatinize in water mounts, however, and no sclerotium was found in the host tissue. No specimens have been preserved.

7. Mycena clavularis (Fr.) non Lange

Mycena globispora Kühner, Bull. Soc. Linn. Lyon, 10: 125. 1931.
(Translated and adapted from Kühner, Encyc. Myc., 10: 180. 1938)

Pileus 4–7 mm., hemispheric to convex (2.5–4 mm. across the base), radially striate and more or less plicate, hyaline and pale, with striae and center grayish or of a clear grayish brown, densely and distinctly but very finely villous-pubescent under a lens (10×), the pellicle gelatinous; flesh thin; lamellae 8–11 reach the stipe, one or two tiers of lamellulae, distant, whitish or a whitish gray with the edge white, ventricose, ascending, free because they separate from the stipe forming a star-shaped collar around it. Stipe 4–10 mm. long, 0.2–0.5 mm. thick, filiform, equal or subattenuated from the base upward, hyaline, translucent and shining except toward the base, where it is white floccose-villous, with a narrow or often broad basal disc, which is white and villous-tomentose.

Spores somewhat spheric or slightly elliptic-ovoid, rarely notched, and pear-shaped at the apex, 8–10.5 μ, perhaps a little compressed dorsiventrally; basidia remarkably short and inflated, 21–31 × 12.2–13.7 μ, with four sterigmata; cheilocystidia clavate or ovoid, crowned with very slender flexuous hairs; gill trama with rather large hyphae (20–34 μ thick), distinctly amyloid; subhymenium indistinct; pellicle of the pileus gelatinous, about 50 μ thick, its hyphae filiform (1–2.5 μ thick), the surface aeriferous, not granular-roughened but completely covered with filiform, elongated, flexuous, obtuse, very slender hairs.

Habit, habitat, and distribution.—In small groups on mossy trunks of hardwood trees, October to January; Europe. So far as I know, this fungus has not been reported in North America, but it very likely occurs here. Most of these minute forms are very widely distributed though seldom identified because the collector rarely has enough literature at his disposal at the time the material is fresh.

8. Mycena longiseta von Höhnel

Stizb. K. Acad. der Wiss. Wien, 118: 282. 1909 (Frag. zur Myk., 186)
Mycena codoniceps var. *aciculata* Smith, Mycologia, 29: 344. 1937.

Illustrations: Text fig. 2, nos. 1–2, 9 (p. 58).

Pileus 1–5 mm. broad, obtusely conic, margin flaring in age, sometimes becoming convex, surface at first pubescent from a rather dense covering of setae, becoming sparsely pubescent and finally glabrous as the setae become appressed, translucent-striate but soon sulcate, color dark bluish gray or brownish gray, becoming sordid pallid gray to whitish, margin entire; flesh very delicate and fragile, no odor or distinctive taste; lamellae narrow, narrowly attached or nearly free, subdistant, faces and edges grayish or pallid; stipe 1–3 cm. long, filiform, very soft and delicate, with a small rounded bulb at the base, gray but whitish in age, covered with setae or nearly glabrous, bulb also setose.

Spores 6–8 × 3–4 μ, ellipsoid, pale yellowish in iodine; basidia two- or four-spored; pleurocystidia not differentiated; cheilocystidia abundant, 25–40 × 7–15 μ, clavate to fusoid-ventricose, apex drawn out to a needlelike projection or studded with several needlelike projections, some with the enlarged portion more or less echinulate in addition; gill trama of vesiculose cells, vinaceous brown in iodine; pileus trama composed of a thin gelatinous pellicle (often quite thick when revived in KOH) and a tramal body of vesiculose cells, the

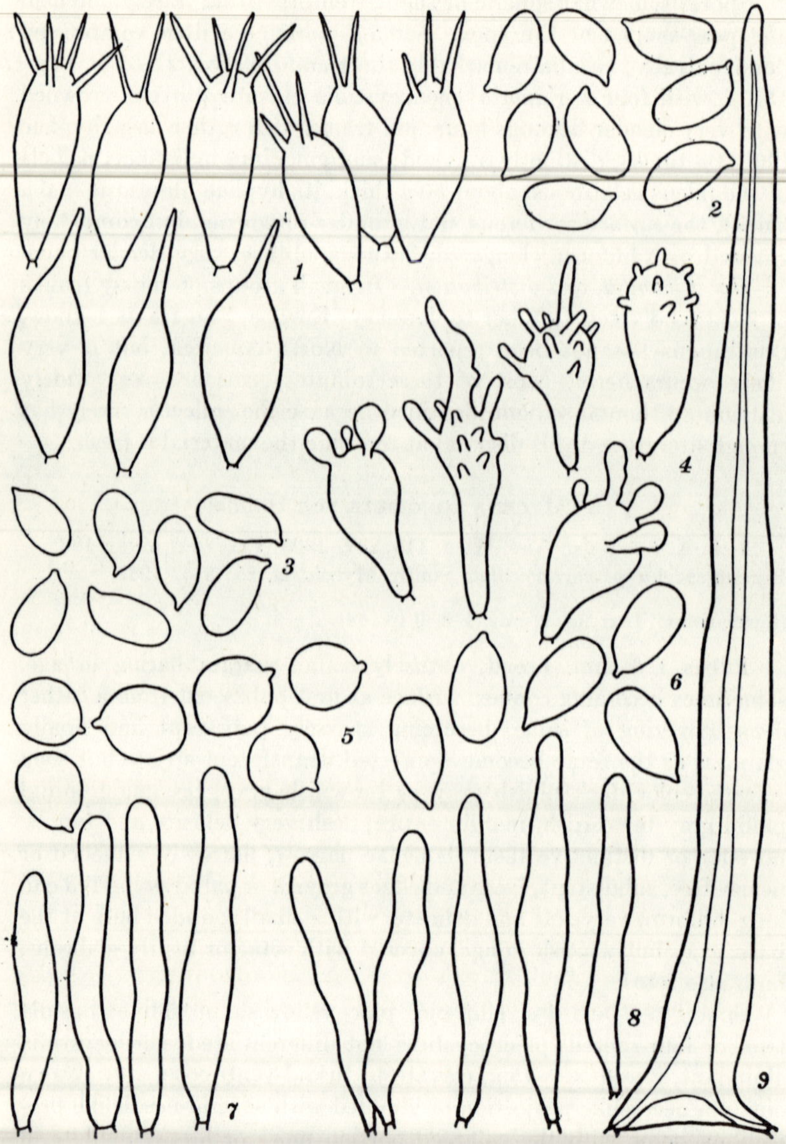

FIG. 2. *M. longiseta:* 1, cheilocystidia; 2, spores; 9, seta from pileus. *M. stylobates:* 3, spores; 4, cheilocystidia. *M. subcaerulea:* 5, spores; 7, cheilocystidia. *M. amicta:* 6, spores; 8, cheilocystidia

latter vinaceous brown in iodine, numerous long, pointed, hyaline, thick-walled setae 150–200 × 8–14 μ arising from the pellicle; stipe and bulb covered with flexuous thick-walled setae up to 300 μ long; stipe readily separable from the pileus.

Habit, habitat, and distribution.—Singly to gregarious on fallen leaves, needles, or cones; North Carolina, Tennessee, New York, Michigan, Washington and California in the United States and in Ontario in Canada. It fruits during the spring and fall, but one seldom finds it in sufficient quantity to make a herbarium specimen. However, on one occasion it was found fruiting prolifically on the northeast slope of Mt. Angeles in the Olympic Mountains of Washington. A carpet of fir needles at an elevation of about 3500 feet was literally covered with hundreds of the delicate little carpophores.

Material studied.—Smith, 33–537, 33–912, 352, 702, 714, 3782, 4544, 8231, 10770, 10917, 13954, 14650.

Observations.—A hand lens is often necessary to detect the setae on the pileus. Sometimes they become appressed to the cap and are visible only when sections are mounted for microscopic study. Such specimens, which are apparently glabrous, are usually difficult to place. However, a few setae can nearly always be found around the bulb. The fungus is so delicate that it is very difficult to get it to the laboratory in good enough condition to photograph. The cheilocystidia are rather variable, and when they possess a single apical projection they often appear to be fusoid-ventricose. Here, as in *M. citrinomarginata*, a great deal of variation can be found on a single pileus, and hence the number of projections borne on a cystidium is not of as much taxonomic significance as was at first supposed. The bluish-gray colors are not always correlated with the roughened cystidia and should be regarded as simple variation. In water mounts of fresh material the pellicle does not always appear gelatinous.

As a result of finding the fungus in quantity on Mt. Angeles it became apparent that *M. codoniceps* var. *aciculata* is merely a variation of the species and not deserving of separate designation. In my first account of the fungus I used the name in the sense of Kühner (1926). Kühner has since decided that the name given by von Höhnel is a more logical choice because his description is much more applicable to the fungus. Cooke did not mention a basal bulb on the stipe of *M. codoniceps*. Since I am accepting the concept of the fungus as Kühner established it, I have also accepted his change in names.

SUBGENUS EUMYCENA

IN EUMYCENA, the largest of the four subgenera, are grouped all the species with smooth spores and nonviscid more or less equal stipes. The diversity of forms included is indicated by the number of sections into which the subgenus is divided. As is to be expected, many evolutionary trends are evident; some are represented by large numbers of species, and others are quite fragmentary. As has been previously stated, the limits between this subgenus and Omphalia are not sharp so far as the white and gray species are concerned. The section Typicae represents the center of the genus around which the other eleven sections and three subgenera are arranged.

KEY TO SECTIONS

1. Stipe with milklike or brightly colored latex .... Lactipedes
1. Stipe with copious watery juice, in some staining the fruiting body blackish .... Hydropus
1. Stipe with merely scant watery juice, not staining black or blackish when bruised 2

2. Densely gregarious on naked bark of standing trees or fallen trunks, occasionally on bark of branches; mostly small species with base of stipe naked or nearly so .... Corticolae
2. Not as above 3

3. Carpophore usually with bright to sordid-blue or greenish-blue colors, at least at base of stipe; stipe densely pubescent; pileus with thick gelatinous pellicle .... Cyanescentes
3. Not as above 4

4. Pileus minute, 1–6 (10) mm. broad; stipe usually about 0.5 mm. thick, filiform .... Deminutivae
4. Pileus larger (1–5 cm.); stipe 1 mm. ± thick or occasionally down to 0.5 mm. in depauperate forms 5

5. Gills not marginate; carpophore white or brightly colored (red, orange, yellow, etc.) .... Adonidae
5. Gills marginate; carpophore brightly colored or colors mixed with fuscous and, consequently, dull (see *M. Kauffmanii* and *M. marginella* also) .... Calodontes

5. Not as above; usually dark gray to blackish, some bluish gray, some cinnamon brown, some tinged or spotted red or pink at least in gills (certain rather large whitish lignicolous species also key out here) 6

6. Pileus corticated (see fig. 48, no. 3, p. 389) by inflated cells; stipe smooth Corticatae
6. Pileus not corticated; if appearing so (because of exposed hypoderm), the stipe floccose 7

7. Gills horizontal to arcuate or decurrent .... Omphaliariae
7. Gills ascending-adnate to hooked, occasionally bluntly adnate in expanded caps 8

8. Spores nonamyloid; stipe floccose, at least when young ..... Floccipedes
8. Spores amyloid or if nonamyloid, stipe smooth at first (*M. scabripes* has scurfy stipe and amyloid spores; *M. pseudotenax* has nonamyloid spores and polished stipe) .... Typicae

SECTION CYANESCENTES

Mycena amicta and *M. subcaerulea* are the American representatives of this section. Old faded specimens of both of these in which all traces of blue or greenish blue have vanished can be placed in the proper section by the following combination of characters: a thick gelatinous pellicle over the pileus, smooth slender cheilocystidia, and finely pubescent stipes. A number of European species are keyed out with the American forms. They are included to show the relationships of the section to the small species of the sections Corticolae and Deminutivae.

KEY TO SPECIES

1. Cheilocystidia with obtuse to rounded smooth heads; contents yellow; $30 \pm \times 20 \pm \mu$ .... *M. cyanescens*
1. Cheilocystidia roughened with short rodlike projections 2
1. Cheilocystidia smooth, narrow (almost filamentous) 3

2. Spores subglobular, $7.5–9.5 \times 6–8.5\ \mu$ .... *M. pachyderma*
2. Spores ellipsoid, $5.7–9.5 \times 4–5\ (6)\ \mu$ .... *M. cyanorhiza*

3. Spores ellipsoid, $7–9\ (10) \times 4–5\ (6)\ \mu$ .... 9. *M. amicta*
3. Spores subglobose, $7–8 \times 6–8\ \mu$ .... 10. *M. subcaerulea*

9. Mycena amicta (Fr.) Quélet

Champ. Jura et Vosges, p. 243. 1872

Agaricus amictus Fries, Syst. Myc., 1: 141. 1821.
Agaricus (*Mycena*) *Iris* Berkeley, in Smith, English Flora, 5[2]: 56. 1836.

Mycena Iris Quélet, Champ. Jura et Vosges, p. 243. 1872.
Mycena calorhiza Bresadola, Fungi Trid., 1: 9. 1881.
Mycena caerulescens Schroeter, Pilze Schles., p. 634. 1889.
Mycena vestita Velenovský, České Houby, p. 307. 1920.

Illustrations:
Plate 3 A–B; Text fig. 2, nos. 6, 8 (p. 58).
Bresadola, Icon. Mycol., 6, pl. 252, fig. 2.
Fries, Icon. Sel. Hymen., 1, pl. 82, fig. 3.
Lange, Flora Agar. Dan., 2, pl. 50 C (as *M. Iris*).
Ricken, Die Blätterpilze, 2, pl. 111, fig. 8 (too green).
Smith, Mycologia, 28, 416, fig. 2, no. 3.

Pileus 5–25 mm. broad, obtusely conic, becoming conic-campanulate or convex, usually with an obtuse umbo, margin appressed against the stipe by a narrow sterile band, surface faintly pruinose at first, in age appearing somewhat granulose or minutely scaly because of the breaking up of the cuticle, intermediate stages polished to lubricous or subviscid, pellicle separable and tenacious, color when young nearly "olivaceous black (1)" on the disc and "light mineral gray" on the margin, often with strong aeruginous to bluish tints pervading through the olive gray, disc in large specimens "dark orient blue" and the margin "light glaucous blue," soon fading to livid gray, in age all forms becoming "wood brown" to "avellaneous" (pale grayish brown) on the disc and "tilleul buff" (pallid) on the margin, in some the disc becoming tinged vinaceous gray; flesh thin, cartilaginous, pallid, odor and taste mild; lamellae close, 20–30 reach the stipe, narrow (1.5–2.5 mm.), narrowly adnate, sometimes free or nearly free, whitish, becoming pale avellaneous; stipe 3–8 cm. long, 1–2.5 mm. thick, pallid to sordid brownish gray (sometimes bright blue when young), hoary from a dense pruinose-pubescent covering, color beneath the pubescence dark greenish to bluish gray and soon fading to brownish gray, base somewhat strigose with blue or white mycelioid strands, equal, hollow, fairly fragile.

Spores 7–9 (10) × 4–5 μ, narrowly ellipsoid, smooth, amyloid; basidia four-spored; pleurocystidia not differentiated, cheilocystidia abundant, hyaline, narrowly fusoid with only slightly tapered necks and obtuse to somewhat acute apices, becoming subfilamentous in age, 28–44 × 5–7 μ; gill trama homogeneous, pale vinaceous brown in iodine; pileus trama with a thick gelatinous pellicle, tissue beneath homogeneous and pale vinaceous brown in iodine.

Habit, habitat, and distribution.—Scattered to densely gregarious on conifer needles and debris or on rotting conifer wood; New York,

Washington, Oregon, and California in the United States and in British Columbia and Ontario in Canada. This apparently is not an uncommon species along the Pacific coast but is very rare in northeastern North America.

Material studied.—Smith, 2515, 2525, 2581, 3277, 3330, 3584, 3675, 3703, 3738, 3917, 4762, 7783, 7938, 8178, 8330, 9372, 13062, 13248, 13419, 13516, 13594, 13656, 13716, 13882, 13923, 13997, 14476, 16160, 16168, 16298, 16479. Arnold, July 16, 1932; Atkinson, New York. Three collections as *M. Iris*, spores 7–11 × 3.5–5 μ; Kauffman, two collections, 1925; Wehmeyer, 1922.

Observations.—The synonymy given for this species is taken from Kühner (1938). As frequently happens in the case of a variable species, *M. amicta* has received many names. A second study of Murrill's type of *M. caesiialba* proved the type collection to be a mixture of two species. The only well-preserved specimen is identical in every respect with the large lignicolous form of *M. amicta.* For additional comments on *M. caesiialba* see page 474.

The fine pubescence of the stipe, the thick tenacious separable pellicle of the pileus, the close gills, ellipsoid spores, and narrow cheilocystidia form a group of characters which enable one to identify specimens that have lost all traces of the bluish or greenish colors. When the plants are growing on conifer logs and stumps the fruiting bodies are often quite large and highly colored, but there appear to be no other characters correlated with these that would serve to distinguish a taxonomic unit.

10. Mycena subcaerulea (Pk.) Saccardo

Syll. Fung., 5: 263. 1887

Agaricus (Mycena) subcaeruleus Peck, Bull. Buffalo Soc. Nat. Sci., 1: 47. 1873.
Mycena cyaneobasis Peck, Ann. Rep. New York State Mus., 51: 284. 1898.
Mycena cyanothrix Atkinson, Stud. Am. Fungi, p. 98. 1900.
Prunulus cyaneobasis Murrill, North Am. Flora, 9: 323. 1916.

Illustrations:

Plate 4; Text fig. 2, nos. 5, 7 (p. 58).
Atkinson, Stud. Am. Fungi, fig. 99.
Beardslee and Coker, Journ. Elisha Mitchell Sci. Soc., 40, pls. 21–22, upper figs.
Peck, Ann. Rep. New York State Mus., 51, pl. B, figs. 1–7.
White, Conn. State Geol. and Nat. Hist. Surv. Bull., 3, pl. 7 (as *M. cyanothrix*).

Pileus (3) 5–15 (25) mm. broad, more or less ovoid with an appressed or slightly incurved margin, becoming obtusely conic to campanulate, sometimes broadly convex to nearly plane, small individuals sometimes papillate, surface lubricous to subviscid, glabrous or appearing somewhat granulose near the margin owing to the slight checking of the cuticle, translucent-striate on the margin when moist, pellicle tenacious and completely separable, colors extremely variable, when perfectly young and fresh a beautiful pale blue or greenish blue, soon tinged with brown and assuming various degrees of bluish, greenish, or grayish brown with a pallid margin, often sordid yellowish in age, bluish tints often lingering on the margin; flesh thin, pallid, pliant, odor and taste mild, lamellae close to crowded, 18–25 reach the stipe, two or three tiers of lamellulae, ascending-adnate, sometimes narrowly adnate or practically free, narrow to moderately broad, white or tinged grayish, edges slightly fimbriate; stipe 3–8 cm. long, 1–2 (2.5) mm. thick, equal, terete, flexuous or strict, tubular, cartilaginous, elastic, at first densely pruinose or minutely pubescent over all from a dense coating of caulocystidia, somewhat glabrescent, base mycelioid, the mycelium blue at first but soon fading to white, bluish to greenish blue above at first, soon fading to grayish or finally sordid brownish.

Spores 6–8 × 6–7 (8) μ, globose or subglobose, amyloid; basidia four-spored; pleurocystidia not differentiated; cheilocystidia abundant, 32–60 × 5–8 μ, subfusoid with obtuse apices but becoming more or less cylindric, sometimes flexuous, smooth, hyaline; gill trama homogeneous, very faintly brownish in iodine; pileus trama characterized by a thick gelatinous pellicle (blue color located along the surface of the pellicle in incompletely gelatinized hyphae), homogeneous beneath the pellicle and faintly vinaceous in iodine; stipe covered with numerous cystidia similar to those on the gill edges or more elongated and flexuous, the tissue of the stipe becoming deep vinaceous red in iodine.

Habit, habitat, and distribution.—Single, scattered or gregarious on debris, decaying wood, or on the bark around the bases of live trees of oak in particular, but also occurring quite frequently on decaying wood of basswood, elm, beech, and other hardwoods. I have seldom collected it in great quantity, but have found it regularly every season near Ann Arbor. It fruits during both the spring and fall, though it is more abundant locally in the spring. It is widely distributed in eastern North America; I have examined material from Alabama, North Carolina, Tennessee, Pennsylvania, New York, and Michigan in the United States and from Nova Scotia, Ontario, and Manitoba in Canada. Although I have watched for it along the Pacific coast, I have never been able to find it there. Kauffman failed to collect it either on the West Coast or in the Rocky Mountains.

Material studied.—Smith, 32–114, 32–363, 32–624, 33–162, 33–218, 33–378, 33–379, 33–517, 33–530, 33–700, 415, 457, 689, 768, 840, 905, 979, 1314, 1349, 1364, 1384, 4700, 6272, 6322, 6350, 6373, 9563, 10054, 15068. Atkinson, 2382, 2389, 9686, 14679, 15138, 15335, 18443, 21234, 22677, 23037, 23304, 24813. Bisby, July 10, 1932, Manitoba. Burke, 1942, Alabama. Hesler, 10054, 14183. Kelly, 1185. Kauffman, New York and Michigan. Mains, 32–72, 33–517, 33–700. Overholts, 1528, 3919, 7610.

Observations.—Except for its globose spores and habitat near or on hardwood debris this species has all the diagnostic characters of *M. amicta*. The remarks about color which are given for *M. amicta* apply equally well to *M. subcaerulea*. I have had the latter in culture. On malt agar it produces a fine white mycelium, the aerial hyphae of which develop a beautiful pale-blue tint if they are grown in diffused daylight.

Mycena pachyderma, *M. cyanorhiza*, and *M. cyanescens* are not known from North America. Kühner placed *M. pachyderma* in this section because of its pubescent stipe and the gelatinous pellicle of the pileus. It apparently lacks the bluish colors exhibited by the others. The species might also be put in the Corticolae. Its habitat, stature, late occurrence, and globose spores relate it to *M. corticola*. Through *M. pachyderma* the sections Corticolae and Cyanescentes appear to be closely related.

Mycena cyanescens is said to turn blue when wounded. Its pileus is 8–10 mm. broad, the stipe 5–8 cm. long, 1 mm. ± thick, and covered entirely with a pruinose mealy coating. The gills are crowded, and the spores measure 8–9 × 5–6 μ. If the cheilocystidia actually have a yellow content, it is very probable that the gill edges are bordered with yellow. The species is apparently very rare in Europe.

Mycena cyanorhiza has been studied critically by both Kühner and Josserand. Its pileus is 3–9 mm. broad across the base; the stipe is 1–3 cm. long and 0.2–1 mm. thick. The pileus has the typical gelatinous pellicle. The stipe is entirely pubescent and looks as if it had been powdered with blue near the base. The gills are described as being fairly close and as having gelatinous edges, an unusual and distinctive character for a species of this group.

SECTION CORTICOLAE

The members of this section form a series of closely related species with *M. corticola* at one end and *M. subcana* at the other. In *M. corticola* and *M. pseudocorticola* the spores are globose and the cheilocystidia are of the roughened type. In *M. subcana* the spores are ellipsoid and the cheilocystidia are fusoid-ventricose, with an occasional one having fingerlike projections. Intermediate between them is *M. madronicola*, with the exact habit of *M. corticola* but with ellipsoid spores and cheilocystidia which vary from the typically clavate-roughened type to almost fusoid-ventricose. *M. corticalis* has cheilocystidia with numerous contorted fingerlike projections and almost globose spores, but is larger in stature than *M. corticola* and, along with *M. subcana*, approaches the fragile gray species of the section Typicae. *M. supina* is not included in this series. As I have found it, it appears to be more closely related to those species here grouped in the section Deminutivae.

KEY TO SPECIES

1. Cheilocystidia fusoid-ventricose or only a few clavate individuals present, sometimes the neck greatly elongated, occasionally with fingerlike prolongations; carpophores often growing singly on low-hanging dead branches or in tops of fallen trees usually well off the ground, but also over bark of standing trees .... 15. *M. subcana*
1. Cheilocystidia clavate, apices echinulate or covered with short or long, often contorted, projections 2

2. Spores globose 3
2. Spores ellipsoid or broadly ellipsoid 5

3. Pellicle gelatinous; stipe minutely pubescent .... *M. pachyderma*, p. 65
3. Pellicle not gelatinous 4

4. Color of carpophore gray to blackish ...... *M. supina* (Deminutivae)
4. Color of fruiting body decidedly purplish to reddish brown when young .... 11. *M. corticola*
4. Color bluish gray to gray when young; pigment in vacuoles 12. *M. pseudocorticola*

5. Spores 4.5–6 μ wide .... 13. *M. madronicola*
5. Spores 7–9 μ wide .... 14. *M. corticalis*

11. Mycena corticola (Fr.) S. F. Gray

Nat. Arr. Brit. Plants, 1: 621. 1821

Agaricus corticola Fries, Syst. Myc., 1: 159. 1821.
Prunulus corticalis Murrill, North Am. Flora, 9: 328. 1916.
Agaricus Meliigena Berkeley and Cooke, Grevillea, 6: 129. 1878.
Mycena Meliigena Saccardo, Syll. Fung., 5: 302. 1887.
Prunulus Meliigena Murrill, North Am. Flora, 9: 324. 1916.
Pseudomycena corticola Cejp, Publ. Fac. Sci. Univ. Charles, 104: 142. 1930.

Illustrations:

Plate 5; Text fig. 3, nos. 1–4 (p. 68).
Beardslee and Coker, Journ. Elisha Mitchell Sci. Soc., 40, pl. 10, upper figs.
Bresadola, Icon. Mycol., 5, pl. 248, fig. 2.
Fries, Icon. Sel. Hymen., I, pl. 85, fig. 2.
Lange, Flora Agar. Dan., 2, pl. 57 E.
Hard, The Mushrooms, Edible and Otherwise, fig. 93.

Pileus 3–10 mm. broad, convex to nearly globose when young, becoming broadly convex in age, the disc often becoming slightly depressed, margin straight when young, surface densely and coarsely pruinose (under a lens) when young, somewhat glabrescent in age except for the disc and grooves, margin faintly sulcate at first, soon sulcate or plicate to the center, canescent but soon polished, dark purplish, soon fading to some shade of vinaceous brown ("army brown" or "vinaceous brown" and fading to near "wood brown") and finally pale grayish brown; flesh thin, fragile, concolorous with the pileus, odor and taste none; lamellae bluntly adnate or slightly decurrent in age, distant, 6–10 reach the stipe, one or two tiers of lamellulae, broad, concolorous with the pileus or a paler and brighter vinaceous brown, edges even; stipe up to 1 cm. ± long, 0.5 mm. ± thick, equal, fragile, densely pruinose when young and pallid because of the pruinosity, more or less naked in age and then concolorous with the gills or pileus, with only a few white hairs at the base, inserted on the substratum.

Spores globose, 9–11 μ, amyloid; basidia four-spored, 30–34 × 8–10 μ, the sterigmata stout; cheilocystidia 26–33 × 8–12 μ, clavate, the apices echinulate or with contorted fingerlike projections; pleurocystidia none; gill trama purplish red in iodine; pileus trama with a well-differentiated hypoderm, in tangential section the "cells" 10–20 μ thick, their walls reddish brown, occupying half the thickness

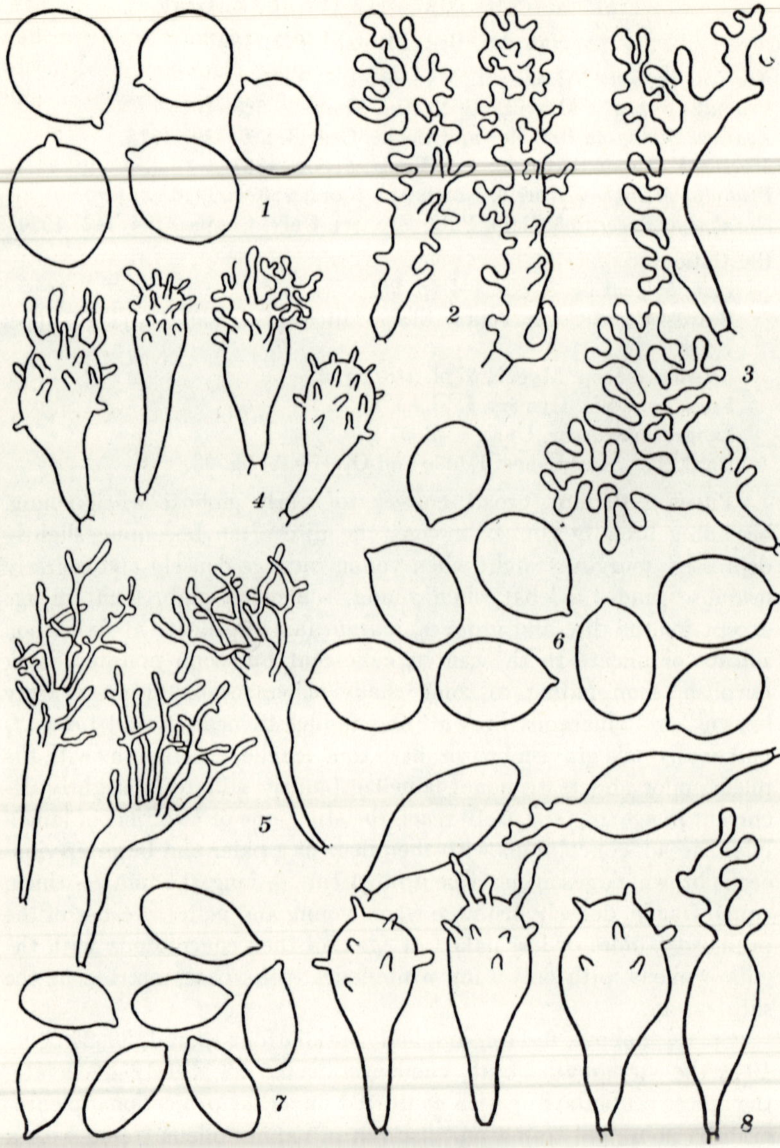

Fig. 3. *M. corticola:* 1, spores; 2, caulocystidia; 3, pilocystidia; 4, cheilocystidia. *M. corticalis:* 5, cheilocystidia; 6, spores. *M. madronicola:* 7, spores; 8, cheilocystidia

of the trama, the tissue below it of loosely interwoven floccose hyphae, the pellicle poorly differentiated, its hyphae 4–6 μ thick and furnished with short branches or variously and irregularly roughened with obtuse simple or branched projections; caulocystidia abundant and similar to the cells covering the surface of the pileus.

Habit, habitat, and distribution.—Densely gregarious on dead or live trees, often covering large areas of the bark. Common in late fall throughout eastern North America on all kinds of trees but particularly on oak, pine, hop hornbeam, and basswood. It also appears in the early spring but usually is not so abundant at that time. I have not yet collected it along the Pacific coast, and I know of no authentic collections from that area. Overholts (3145) has collected it on the umbrella pine (*Sciadopitys verticillata* Sieb. & Zucc.) at State College, Pennsylvania, during August. Material from New York, Pennsylvania, Ohio, Michigan, Tennessee, Missouri, and Texas has been examined.

Material studied.—Smith, 32–279, 32–605, 1745, 4752, 6961, 8981, 14974, November 18, 1931, Michigan. Atkinson, 3967, 16192. Ellis and Everhardt, North Am. Fungi, 2007. Hesler, 10204. One collection, by L. N. Johnston, Michigan. Four collections by Kauffman, New York, Michigan. Overholts, 1297, 1527, 2912, 2959, 3043, 3109, 3113, 3131, 3145, 5769, 13139, 13147, 14998. Patterson, 1941. Routien, 655. E. A. Smith, 1935, Texas.

Observations.—The dark-purplish color of the whole fruiting body is at first obscured by the canescent covering of cystidia. The purple soon fades, and before the pileus becomes naked the colors have changed to vinaceous brown.

I have not examined authentic material of *M. Meliigena,* but from the description it appears that there are no characters by which it could be separated from *M. corticola.* The "papillate" spores seem to be the only remote possibility, and I strongly suspect that the authors were describing the apiculus on the spore rather than the character of its surface.

12. Mycena pseudocorticola Kühner

Le Botaniste, 26: 368. 1934

Illustrations: Lange, Flora Agar. Dan., 2, pl. 57 E[1], E[2]. 1936.

Pileus 3–14 mm. (2–8 mm. across the base), globose when young, then conic, campanulate, or hemispheric, expanding less than in *M.*

corticola but at times with a tendency to be recurved on the margin, which becomes split in a star-shaped manner, smooth at first, then long-striate or sulcate, deep lead gray or bluish gray when young, fading especially on the margin to a bluish or ashy gray or subfuscous, never purplish rose, white-pruinose under a lens (10×); flesh thin; lamellae distant, 8–11 reach the stipe, one to three tiers of lamellulae, lamellae white or whitish gray at the base, sometimes tinted with the color of the fruiting body, the edges whitish crenulate under a lens, broadly adnate and often uncinate but usually more or less sinuate-ascending or ventricose-sinuate; stipe 3–20 mm. long, 0.3–1 mm. thick, equal, ordinarily curved and ascending, concolorous with the pileus (grayish blue, then grayish fuliginous), often paler above, entirely covered with a fine pruina, the base distinctly bristly with white hairs.

Spores globose, 9–14 μ, content granular, amyloid; basidia two- or three-spored or, rarely, four-spored; cheilocystidia 6–8 μ broad, with rodlike projections over their apices; pellicle of the pileus well developed (21–26 μ thick), the hyphae 2–7.5 μ thick, radially arranged and somewhat separated from each other by the gelatinization of their walls, the walls without pigment, the surface cells furnished with short projections; flesh of the pileus amyloid.

Habit, habitat, and distribution.—Gregarious on the bark of standing trees, October to February; Europe. I have no herbarium specimens of this species, but have found it rarely in Michigan late in the fall (November) during prolonged rainy weather. The description above is translated and adapted from Kühner because my own notes are incomplete.

13. Mycena madronicola A. H. Smith

Mycologia, 31: 269. 1939

Illustrations:

Plate 6 A, B; Text fig. 3, nos. 7–8 (p. 68).
Smith, Mycologia, 31, fig. 1, A (spores). 1939.

Pileus 5–12 mm. broad, obtusely campanulate to convex when young, becoming broadly convex and the disc somewhat flattened or even slightly depressed, surface hoary at first, soon naked and in age more or less polished, moist and translucent-striate to the disc, becoming sulcate in age, margin at first appressed against the stipe, color "hair brown" to "cinnamon brown" and fading to "avellaneous," in age sometimes "vinaceous buff" or grayish; flesh concolorous, thin, pliant, reviving somewhat when moistened as in *M. corticola*, odor farinaceous but soon fading, taste mild; lamellae broadly adnate but becoming toothed or somewhat decurrent in age, subdistant to distant, 14–20 reach the stipe, narrow to moderately broad (2–2.5 mm.), color "tilleul buff" to pallid at all stages, edges even; stipe short, 1–2 (3.6) cm. long, about 1 mm. thick, equal above a basal subovoid bulb, which soon disappears, tubular, delicately frosted over all at first, soon polished and translucent, base pruinose and inserted on the bark (as in *M. corticola*), concolorous with the pileus or paler, the apex pallid.

Spores narrowly ellipsoid and pointed at one end, 9–11 × 5–6 μ from deposits, 7.5–9.5 × 4–5 μ on dried material, amyloid; basidia four-spored, 28–32 × 6–7 (8) μ; cheilocystidia forming a sterile band along the gill edge or embedded and inconspicuous, 20–36 × 5–9 μ, clavate to capitate, upper portion varying from smooth to echinulate or the apex prolonged into a much-branched, contorted neck; no pleurocystidia; gill trama faintly vinaceous brown in iodine; pileus trama made up of a thin pellicle of very narrow hyphae, beneath this a region of enlarged hyphal cells with dull-brown contents (the hypoderm), the remainder of narrower hyphae forming a floccose filamentous tissue, all but the pellicle sordid vinaceous brown in iodine.

Habit, habitat, and distribution.—Densely gregarious by the hundreds on the scaly bark of old madroña trees in open places after prolonged wet weather; Oregon. It was found late in November and early in December.

Material studied.—Smith, 9224, 9286.

Observations.—Because of the large number of immature spores in mounts of revived material the average size is smaller than that obtained from material taken from a deposit, and for this reason measurements from both deposits and dried material are included. In most other species the size can be quite accurately determined from herbarium specimens. The variability of the markings of the cheilocystidia in this species is interesting. All of those shown in text figure 3, no. 8, are from the type collection taken from a single tree. The clavate-echinulate cystidia are most numerous and should be considered typical. Smooth cystidia are usually found on young gills or lamellulae. In age the echinulations on the clavate type may become elongated into slender, crooked fingerlike projections. Sometimes the smooth cystidia elongate at the apex instead of becoming echinulate.

14. Mycena corticalis A. H. Smith

Mycologia, 31: 268. 1939

Illustrations:

Plate 7 D; Text fig. 3, nos. 5–6 (p. 68).
Smith, Mycologia, 31, fig. 1–B (spores).

Pileus 6–12 (15) mm. broad, cylindric, conic or convex, remaining unexpanded, the disc often becoming slightly flattened, margin appressed against the stipe when young, glabrous, hygrophanous, translucent-striate when moist, pale to dark watery gray on the disc and the margin whitish, fading to cinereous and becoming sulcate; flesh thin, fragile, grayish, odor and taste mild; lamellae arcuate or broadly adnate with a distinct decurrent tooth, rather distant, 11–13 reach the stipe, broad (3 mm. ±), whitish to pale gray; stipe 1–2.5 cm. long, about 1 mm. thick, equal, straight or curved, tough and cartilaginous, glabrous, moist, concolorous with the pileus or paler toward the apex, slightly strigose at the base.

Spores broadly ovoid and pointed at one end, 9–11 × 7–9 μ, nonamyloid; basidia four-spored, 34–46 × 8–10 μ; cheilocystidia embedded and very inconspicuous, 20–25 × 6–8 μ, with simple or branched projections (5–15 × 1.5–2 μ) scattered over the surface of the upper (enlarged) portion; pleurocystidia not present; gill trama vinaceous brown to wine red in iodine; pileus trama covered with a very thin adnate pellicle made up of very slender hyphae, below it a region of somewhat inflated irregularly shaped cells rather compactly arranged, the remainder of loosely arranged hyphae with rather wide cells, all of the tissue below the pellicle becoming wine red to vinaceous brown in iodine.

Habit, habitat, and distribution.—Scattered on cedar bark (*Thuja plicata*) during October and November; Oregon. It has been found only on logs and trees that lacked a covering of Bryophytes. It is apparently quite rare.

Material studied.—Smith, 8046, 8159, 8918.

Observations.—The spores were originally described as hyaline to pale bluish in iodine, hence amyloid. These observations were made on freshly dried material. The type has been rechecked, and the spores have been found to be very pale yellow or hyaline, and consequently are only very weakly or not at all amyloid. The pileus and gill trama gave the typical reddish reaction previously observed. Both mature and immature fruiting bodies were studied. The most reliable characters for this species are the broad arcuate gills, which develop a distinct decurrent tooth in age, the spores, the habitat, and the cheilocystidia. The colors are not at all distinctive.

Some might consider this fungus to be more closely related to the species in the section Omphaliariae than to the others of this section. It should be remembered, however, that all the Corticolae except *M. subcana* are characterized by broadly adnate gills, so that any relationships determined by that character would also apply to the others. The group represents an interesting phylogenetic problem. It can be interpreted either as a line of evolution or as a level of evolution influenced in some measure by the nature of the habitat.

15. **Mycena subcana,** sp. nov.

Illustrations:

Plate 7 A, C, E; Text fig. 4, nos. 1–2 (p. 74).

Smith, Mycologia, 29, 353, fig. 3, b (as *M. brevipes*).

Pileus (0.5) 1–2.5 (3) cm. latus, demum campanulatus vel convexus, pruinosus, persistens subcanus, fuligineus demum cinereus, striatus; lamellae adnatae, angustae, vel sublatae, subdistantes vel distantes, pallidae; stipes 1.5–3.5 cm. longus, 1.5–3 mm. crassus, aequalis vel subbulbosus, pallide fuligineus, pruinosus, basi strigosus; sporae (7) 8–10 × 5–6 μ, late ellipsoideae; cheilocystidia (30) 40–64 × 10–18 μ, fusoide ventricosa vel subclavata. Specimen typicum in Herb. Univ. Mich. conservatum. Legit A. H. Smith, n. 8752, Siskiyou National Forest, Calif., Nov. 15, 1937.

Pileus (0.5) 1–2.5 (3) cm. broad, ovoid when young, the margin straight, expanding to obtusely conic, campanulate or convex, hoary-pruinose when young, becoming naked but retaining a somewhat hoary appearance at maturity, glabrous, moist, hygrophanous, translucent-striate almost to the disc, "deep neutral gray" and fading to "pale neutral gray" on the disc, sometimes "pale drab gray" over all, the margin usually "pallid neutral gray" and radially rugulose (dark ashy gray on the disc with a paler ashy-gray margin and fading to pale ash-gray over all), flesh thin, fragile, pallid gray, not changing when bruised, odor and taste not distinctive; lamellae adnate, narrow to moderately broad, subdistant to distant at maturity, intervenose at times, whitish to pale cinereus, edge concolorous with the sides and even; stipe 1.5–3.5 cm. long, 1.5–3 mm. thick, equal or the base subbulbous, concolorous with the pileus, pale cinereous or nearly white,

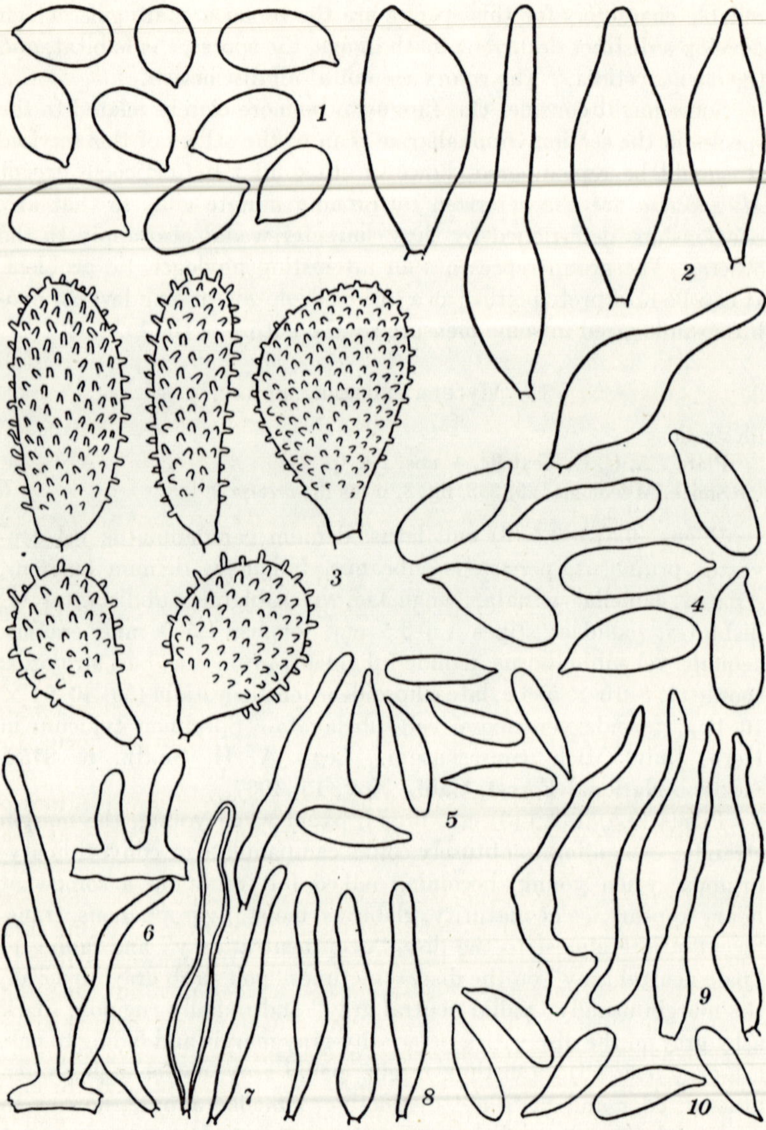

FIG. 4. *M. subcana:* 1, spores; 2, cheilocystidia. *M. cylindrospora:* 3, cells from margin of pileus; 4, spores. *M. albicolor:* 5, spores; 6, caulocystidia; 7, pilocystidia; 8, cheilocystidia. *M. mauretanica:* 9, caulocystidia; 10, spores

at first covered with a powdery bloom but soon polished, the base surrounded by a mass of white-strigose filaments or echinulate with white hairs radiating from the point of attachment, not rooting, occasionally the base stained sordid purplish brown in age.

Spores (7) 8–10 × 5–6 μ, broadly ellipsoid, smooth, amyloid; basidia four-spored; pleurocystidia not differentiated; cheilocystidia (30) 40–64 × 10–18 μ, fusoid-ventricose with obtuse apices or clavate, some abruptly ventricose with a long-drawn-out neck, smooth; gill trama homogeneous, pale vinaceous brown in iodine; pileus trama with a thin adnate pellicle, a well-defined hypoderm and the remainder filamentous, vinaceous brown in iodine.

Habit, habitat, and distribution.—Single to gregarious on dead sticks, branches, or trunks of coniferous trees, particularly cedar; occasionally on alder and tanbark oak; California, Oregon, and Washington. Rather common during the wet fall season, rare during the spring in 1939 in Washington. The type was collected on twigs of Port Orford cedar, but the species is frequently found on western red cedar also.

Material studied.—Smith, 2512, 2550, 3351, 3514, 3532, 3612, 3643, 3742, 3755, 7774, 8174, 8179, 8473, 9096, 16496, 18177; November 19, 1935, Oregon. Kauffman, 1925, Washington.

Observations.—I originally identified this fungus as *M. brevipes* Murr. (for an account of which see page 396). The characters of *M. subcana* are the typically short stipe and rather broad pileus, the habit on sticks, branches, and trunks that are usually above the forest floor, its broadly ellipsoid spores, lack of pleurocystidia, and lack of any distinctive odor or taste. It has been difficult to arrive at conclusions in regard to this species. It appears to be just another fragile gray Mycena of the *M. leptocephala* type, but it does not seem to be referable to either *M. leptocephala* or *M. stannea.* During three seasons of collecting along the Pacific coast I recognized it readily. In many of its characters it is similar to *M. fragillima*, but the structure of the pileus in the two is different, and there is a distinct difference in habit and habitat as well as in consistency. I have compared collections of fresh specimens of both. *M. fragillima* tends to have a delicately pubescent stipe at first and the stature of *M. filopes.* The stipe of *M. subcana* is merely frosted, and the stature is decidedly squatty, a character, which, along with the habitat, has led me to place the fungus in the Corticolae. When growing from branches that are heavily covered with Bryophytes, *M. subcana* does develop a moderately long stipe, but this, according to my experience, is not a common occurrence.

In northern California I have collected a form (8174) on redwood twigs which is watery white, or merely tinged very pale watery gray around the disc, and which fades to a dead white. Its cheilocystidia are 40–54 × 10–16 μ, clavate with the neck elongated, or with several long fingerlike projections arising from the apex. Its spores are 7–9 (10) × 4–5 μ. I have but one collection and, as yet, insufficient data to dispose of it satisfactorily. It appears to be very close to *M. subcana* and may be only a chance pale variation. I have observed similar variations for such well-known species as *M. megaspora.*

SECTION DEMINUTIVAE

As is to be expected, a section based on size or stature of the fruiting body is almost certain to be artificial. The present grouping, however, is much less artificial than one might suppose. In the series of small white species we find three groups which appear quite distinct but, nevertheless, are closely related. The first is distinguished by the absence of cheilocystidia. The other two are distinguished by the presence of cheilocystidia, which are smooth in one and clavate-roughened in the other. These groups intergrade with the white species of the section Adonidae. Here, as in the Adonidae, no sharp distinction exists between Mycena and Omphalia. For practical purposes I have included in the key most of the American species of both groups. The only forms omitted are those that resemble small species of Clitocybe, that is, have an inrolled margin and truly decurrent gills.

The small white species grouped here are among the most difficult to work with that I have found anywhere in the Agaricaceae. The difficulty rests largely in obtaining enough specimens in a single locality to characterize the species in question. One cannot, or certainly should not, pick up scattered fruiting bodies during the course of a days' collecting and consider them all to belong to one species simply because they look alike and have the same microscopic characters. The iodine reaction of the spores must also be obtained, and that can be done only on dried specimens. It is very seldom that one finds fruiting bodies with such distinctive microscopic characters that he can describe a species from a small number of carpophores. *M. setulosa,* however, is such an example.

The vinaceous-gray to bright-colored species grouped in this section impress me as being an isolated group which for the most part was derived from small forms of the typical species rather than from the Adonidae The gray forms intergrade imperceptibly with the smaller species of the section Typicae. In fact, one must reconcile himself to trying both groups when keying out intermediate forms. To include all these in both keys would make both unwieldy and would give a false impression of the species, particularly to investigators not experienced in the study of small agarics.

KEY TO SPECIES

1. Pileus white or whitish 2
1. Pileus colored—grayish, violaceous, vinaceous, bright red, yellowish, orange, or some other color; some fading to white in age 24

2. Cheilocystidia with roughened apices or with numerous projections over apices (but see *M. papillata* also) 3
2. Cheilocystidia, if differentiated, smooth (or only a few forked) 5

3. Spores 7–8 × 5.5–7.5 μ (2 spored; 4-spored form not known) 34. *M. minutula*
3. Spores longer (up to 10 μ or more in 4-spored forms) 4

4. Spores 10–13 μ long (4-spored); gills lacking (faded forms of *M. capillaris* key out here, but its gills are well formed) 33. *M. cylindrospora*
4. Spores 7–10 × 3–4.5 μ; stipe 3–24 mm. long (see Kühner, 1938, for description) .... *M. polyadelpha*

5. Cheilocystidia not differentiated or not known 6
5. Cheilocystidia well differentiated 15

6. Spores ellipsoid 8
6. Spores globose, cheilocystidia not known 7

7. Stipe glabrous .... 28. *M. papillata*
7. Stipe pubescent .... 26. *M. subvestita*

8. Spores 10–12 × 2.5–3 μ .... 21. *M. mauretanica*
8. Spores 7–9 × 2.5–3 μ (4-spored) .... 22. *M. gracilis*
8. Spores broader than in choices above 9

9. Gills narrow, almost foldlike at times 10
9. Gills broader and more or less triangular, at least in age 12

10. Spores 8–11 × 3.5–4 (12–14 × 4–5 μ; 2-spored). 23. *M. subimmaculata*
10. Spores up to 9 μ long 11

11. Spores ellipsoid .... 32. *M. pusillissima*
11. Spores ventricose in middle, subfusoid........... *Delicatula integrella*

12. Spores 12–16 × 7–8 μ......................... 24. *M. McMurphyi*
12. Spores smaller 13

13. Caulocystidia abundant, the walls slightly thickened; pilocystidia similar, scattered to abundant .... 20. *M. crispula*
13. Caulocystidia absent or a few thin-walled flexuous hairs present 14

14. Hyphae of pellicle with short projections .... 25. *M. albidula*
14. Pellicle not differentiated, surface cells smooth ...... 27. *M. ignobilis*

15. Spores three times as long as wide or longer 16
15. Spores broader in relation to their length 18

16. Cheilocystidia 9–14 μ broad, ventricose with obtuse apices 29. *M. paucilamellata*
16. Cheilocystidia narrower 17

17. Spores 7–9 × 2.5–3 μ, 4-spored (see *M. delicatella* also) 17. *M. albicolor*
17. Spores 9–12 × 2.5–3 μ, 4-spored .... 18. *M. filiformis*

18. Spores 3–4.5 × 2.5 μ .... *Omphalia Rhododendri*[1]
18. Spores larger 19

19. Pileus with long pilocystidia (80–150 × 8–12 μ), seta-like but thin-walled and hyaline .... 16. *M. setulosa*
19. Pilocystidia, when present, much shorter and, if seta-like, the walls slightly thickened 20

20. Spores 5.5–7 × 3.5–4 μ; pilocystidia with slightly thickened walls 20. *M. crispula*
20. Not as above 21

21. Spores 10–12 × 4–5 μ, slightly ventricose .... 19. *M. albissima*
21. Spores smaller 22

22. Spores amyloid 23
22. Spores nonamyloid .... 173. *M. hiemalis*

23. Spores 8–11 × 5–6 μ .... 31. *M. kalalochensis*
23. Spores 6–8 × 3–4 μ .... 30. *M. litoralis*

24. Cheilocystidia clavate-roughened 25
24. Cheilocystidia fusoid-ventricose 35

[1] These spore measurements were obtained from the type. Murrill gave them as 7–5 × 2 μ. I suspect that *Marasmius resinosus* has been confused with *Omphalia Rhododendri*, and it may be that the two are not distinct. Small spores are frequent in mounts of Marasmii, and they may account for my observation. At all events the two species should be in the same genus, *Marasmius*.

25. Pileus brown, reddish brown, pink, or bright red 26
25. Pileus essentially blackish to grayish or occasionally developing sordid-brown colors 30

26. On dead remains of Typha or Carex 27
26. On dead parts of ferns 28
26. On dead parts of other plants 29

27. Spores 11.5–14 × 3.5–5 μ .... *M. tubarioides*, p. 106
27. Spores 9–11 × 5–6 μ .... 36. *M. juncicola*

28. Pileus white, with brownish disc .... *M. Lohwagii*, p. 108
28. Pileus grayish to grayish vinaceous (not bright red) ... 38. *M. tenella*
28. Pileus coral red at first, gradually fading .... 35. *M. pterigena*

29. Spores narrowly ellipsoid, with tapered and curved apiculus 37. *M. Smithiana*
29. Spores broadly ovoid, apiculus not prominent .... 38. *M. tenella*

30. Spores globose to subglobose or very broadly ellipsoid 31
30. Spores ellipsoid (gray species with slender stipes, but arcuate or more or less decurrent gills should be sought for in the Omphaliariae) .. 32

31. Pileus pale, margin whitish .... *M. Quercus-Ilicis*, p. 116
31. Pileus dark to sordid fuscous .... 42. *M. supina*

32. Cheilocystidia hairlike but contorted .... 176. *M. Brownii*
32. Cheilocystidia clavate-echinulate 33

33. Spores 5–6 × 3–3.5 μ (small forms of *M. constans*, *M. atroalboides*, or *M. alcaliniformis* may key out here) .... 39. *M. minutissima*
33. Spores larger (8–10 μ or more long) 34

34. Spores 9–12 × 5–7 μ (2-spored); base of stipe strongly strigose-echinulate .... 41. *M. mirata*
34. Spores 8–10 × 4–5 μ (4-spored); 11–13 × 5–6 μ (2-spored) 40. *M. capillaris*

35. Pileus brightly colored 36
35. Pileus with dull colors (violet brown, gray or olive-yellowish brown) 37

36. Spores 9–11 × 3.5–4 μ (4-spored); pileus scarlet at first, on hardwood debris .... 44. *M. acicula*
36. Usually on moss; spores 4.5–6 × 2.5–3 μ; gills long-decurrent 45. *M. fibula*
36. Under conifers; spores 7–8 × 2.5–3 μ.............. 43. *M. oregonensis*

37. Pleurocystidia abundant and conspicuous 38
37. Pleurocystidia, if present, not conspicuous 40

38. Taste very bitter; at least some cystidia with highly refractive content in fresh condition .... **54. *M. erubescens***
38. Cystidia sharply pointed and with homogeneous content; taste probably mild (forms of *M. subsupina* may key out here) **50. *M. corticaticeps***
38. Not as above (cystidia obtuse) **39**

39. Spores 7–8 × 3–3.5 μ........................ **47. *M. pseudogrisella***
39. Spores 4–5 × 2–2.5 μ .... **46. *M. Swartzii***

40. Gills arcuate to more or less decurrent .... **174. *M. speirea***
40. Gills ascending-adnate to hooked **41**

41. Spores 6–7 (8) × 5–6 μ (4-spored) (*M. olivaceobrunnea* with pale-yellow gill edges may key out here also) .... **48. *M. subcucullata***
41. Spores (8) 9–14 × 5–7 μ (2-, 3-, 4-spored) **42**

42. Rather cartilaginous species usually on decaying wood (the 2-spored form of *M. subcucullata* keys out here also) ...... **49. *M. subsupina***
42. Fragile species on humus or debris **43**

43. Spores 7–9 × 5–6 μ (4-spored); stature of *M. filopes* but small (*M. subfumosa* may key out here also) .... **122. *M. subfusca***
43. Spores 9–14 × 5–7 μ (2-, 3-, 4-spored); stature of *M. sanguinolenta* **51. *M. debilis***

Subsection Albidae

16. **Mycena setulosa,** sp. nov.

Illustrations: Text fig. 5, nos. 8–10.

Pileus 1–3 mm. latus, conicus vel convexus, setulosus, griseo-albus demum albus; lamellae adnatae, distantes, angustae, albidae; stipes 5–8 mm. longus, filiformis, insiticius, albidus; sporae 7–9 (10) × 3.5–5.5 μ, subellipsoideae; basidia tetraspora; pleurocystidia et cheilocystidia 36–48 × 10–14 μ, fusoide ventricosa, apice subacuta. Specimen typicum in Herb. Univ. Mich. conservatum. Legit A. H. Smith, n. 9739, prope Husky Gap Trail, Great Smoky Mountains, National Park, Tennessee, Aug. 4, 1938.

Pileus 1–3 mm. broad, conic to convex, the margin appressed against the stipe, remaining unexpanded, surface pubescent under a lens from numerous projecting setae, surface even at first, pale watery grayish white, hygrophanous, fading to dead white, membranous; lamellae adnate, distant, narrow to lacking (in very small caps), white, edges pruinose from projecting cystidia; stipe up to 5–8 mm. long, filiform, inserted on the substratum, base not enlarged or very

FIG. 5. *M. kalalochensis:* 1, spores; 2, cheilocystidia; 3, pleurocystidia; 4, caulocystidia. *M. crispula:* 5, spores; 7, caulocystidia. *M. setulosa:* 8, pilocystidia; 9, pleurocystidia; 10, spores. *M. pusillissima:* 6, spores

slightly clavate (no bulb), white, densely to sparsely pubescent over all from projecting caulocystidia.

Spores 7–9 (10) × 3.5–4 μ, subellipsoid in one view, with an oblique apiculus and slightly inequilateral as seen attached to the sterigmata, smooth, nonamyloid; basidia four-spored; pleurocystidia and cheilocystidia abundant and similar, 36–48 × 10–14 μ, hyaline, fusoid-ventricose with tapered necks, apices subacute; gill trama yellowish in iodine; pileus trama yellowish in iodine; surface hyphae bearing numerous thin-walled, delicate, elongated pilocystidia 80–150 × 8–12 μ that collapse readily and are difficult to revive, basal portion ventricose, evenly tapered to an acute apex, smooth; caulocystidia numerous to scattered, similar to pilocystidia; stipe tissue yellowish in iodine.

Habit, habitat, and distribution.—Singly on bark of tulip tree (*Liriodendron Tulipifera* L.); Husky Gap Trail, Great Smoky Mountains National Park, Tennessee. Known only from the type locality.

Material studied.—Smith, 9739.

Observations.—The stipe did not separate readily from the pileus. This and the lack of a basal bulb indicate that any similarity to *M. longiseta* is purely superficial. The fungus resembles *M. mauretanica* in some respects, but is readily distinguished by its spores and abundant pleurocystidia and cheilocystidia.

17. **Mycena albicolor,** sp. nov.

Illustrations: Text fig. 4, nos. 5–8 (p. 74).

Pileus 2–5 mm. latus, conicus demum convexus, glaber, candidus; lamellae distantes, sublatae, late adnatae vel subdecurrentes, candidae; stipes 10–25 mm. longus, 0.5 ± crassus, candidus, pruinosus; sporae 7–9 (11.5) × 2.5–3 μ; cheilocystidia 26–38 × 5–9 μ. Specimen typicum in Herb. Univ. Mich. conservatum. Legit A. H. Smith, no. 7810, prope Blue River, Ore., Oct. 15, 1937.

Pileus 2–5 mm. broad, conic at first, becoming obtusely conic to convex, the margin appressed against the stipe at first, glabrous to the naked eye, shining white and pruinose when young, dead white in age, never conspicuously striate even when moist, becoming slightly sulcate in age, flesh thin and fragile, white, no odor or taste; lamellae distant, moderately broad, arcuate-decurrent, white, edges even; stipe 10–25 mm. long, about 0.5 mm. thick, shining white, equal, watery, solid, pruinose at first, base inserted on the substratum or with a few short white hairs.

Spores 7–9 (11.5) × 2.5–3 μ, subcylindric, smooth, nonamyloid; basidia four-spored (and two-spored ?); no pleurocystidia seen; cheilocystidia abundant, 26–38 × 5–9 μ, usually subfiliform and with irregular outlines, the base usually slightly enlarged; gill trama homogeneous, yellowish in iodine; pileus trama with a thin adnate pellicle made up of narrow (3–5 μ thick) hyphae the walls of which are covered with numerous short projections, from this layer arise numerous pilocystidia similar in size and shape to the cheilocystidia, homogeneous beneath the pellicle; caulocystidia abundant, similar to the cheilocystidia or longer (up to 50 μ).

Habit, habitat, and distribution.—Gregarious on cedar twigs (*Thuja plicata*); Blue River, Oregon. Known only from the type locality.

Material studied.—Smith, 7810.

Observations.—The spore size is apparently based on a mixture of two ranges, one very likely from four-spored basidia and the other from two-spored individuals. I was not able, however, to demonstrate conclusively the presence of the latter. Although the pileus was not papillate, the presence or the lack of a papilla is not of great significance in such small fungi as this and should not be emphasized. The important diagnostic character which separates *M. albicolor* from *M. gracilis* is the presence of numerous cystidia over the pileus and stipe and along the edges of the lamellae. The arcuate gills furnish an additional character to distinguish the two. *M. albicolor* was referred at first to *M. angustispora* (Joss.) Kühner. The latter, however, is reported as not having cheilocystidia and as having the base of the stipe somewhat fuscescent. In view of the fact that Josserand made an exhaustive anatomical study of his fungus, the difference noted in the cheilocystidia is important. The abundant cheilocystidia also distinguish *M. albicolor* from *M. pseudocrispula.*

18. **Mycena filiformis,** sp. nov.

Illustrations: Text fig. 6, nos. 5–6 (p. 85).

Pileus 1–5 mm. latus, convexus, membranaceus, candidus, glaber vel pruinosus; lamellae distantes, angustae, late adnatae vel subdecurrentes, candidae; stipes 10–15 mm. longus, filiformis, albidus, pruinosus; sporae 9–12 × 2.5–3 μ; basidia tetraspora; pleurocystidia et cheilocystidia 26–32 × 5–8 μ, levia. Specimen typicum in Herb. Univ. Mich. conservatum. Legit A. H. Smith, n. 15531, prope Milford, Mich., Nov. 5, 1940.

Pileus 1–5 mm. broad, convex with an inrolled margin, membranaceous, snow white and under a lens appearing as if frosted, surface moist but hardly translucent, usually distantly sulcate-striate and the surface somewhat rugose, no odor or taste; lamellae distant, 6–11 reach the stipe, narrow, one tier of lamellulae or none present, arcuate-adnate to subdecurrent, snow white, pruinose under a lens—especially along the edges; stipe 10–15 mm. long, filiform, hyaline watery white, under a lens faintly pruinose, inserted on old leaves by an equal, perfectly naked base.

Spores 9–12 × 2.5–3 μ, almost aciculate, smooth, hyaline, nonamyloid (tested a few days after drying); basidia four-spored; pleurocystidia similar to cheilocystidia, rare to scattered; cheilocystidia abundant, 26–32 × 5–8 μ, smooth, hyaline, subfusoid to subclavate, the upper portion only very slightly constricted and the apices obtuse; gill trama not distinctive; pileus trama homogeneous beneath a well-differentiated pellicle from which numerous pilocystidia project, the latter similar in outline to the cheilocystidia or a bit more irregular; caulocystidia present, similar to the cheilocystidia or more elongated and irregular.

Habit, habitat, and distribution.—Scattered to gregarious on fallen beech leaves; November 5, 1940, Milford, Michigan. The species was very widespread in one woods on November 5, but was not present in great quantity.

Material studied.—Smith, 15531.

Observations.—This species differs from *M. paucilamellata* in its narrow cystidia on the gills, cap, and stipe and in its well-formed lamellae. I hesitate to emphasize the nonamyloid character of the spores of *M. filiformis* because the tests made soon after the specimens were dried and again at the time this manuscript was completed were both inconclusive. *M. albicolor* is distinguished by its smaller spores, slightly different cystidia, and its habitat on cedar twigs instead of beech leaves. Many species seem to go from oak to pine without much difficulty, but the fungi that inhabit the wood and debris of cedar seem to be more restricted to their habitat than those inhabiting the wood and debris of other conifers. Thus the difference in habitat in this instance is considered significant.

19. **Mycena albissima**, sp. nov.

Illustrations: Text fig. 6, nos. 1–4.

Pileus 4–9 mm. latus, convexus demum subplanus, pruinosus, glaber, candidus, lamellae latae, distantes, late adnatae vel sub-

FIG. 6. *M. albissima:* 1, spores; 2, pilocystidia; 3, cheilocystidia; 4, caulocystidia. *M. filiformis:* 5, spores; 6, cheilocystidia. *M. paucilamellata:* 7, spores; 8, pleurocystidia. *M. kalalochensis:* 10, pilocystidia. *M. litoralis:* 9, spores; 11, cheilocystidia

decurrentes, albidae; stipes 1–3 cm. longus, 0.5 mm. crassus, aequalis, fragilis, candidus, pubescens, glabrescens; sporae 10–12 × 4–5 μ; basidia tetraspora; cheilocystidia 26–37 × 5–8 μ, levia. Specimen typicum in Herb. Univ. Mich. conservatum. Legit A. H. Smith, n. 3780, prope Orick, Calif., Dec. 5, 1935.

Pileus 4–9 mm. broad, convex with a connivent margin when young, nearly plane in age, surface delicately pruinose, glabrous, moist but scarcely translucent-striate, snow white over all, flesh thin but firm, no odor or taste; lamellae broad, distant, 6–8 reach the stipe, one or two tiers of lamellulae, bluntly adnate but soon arcuate and in age decurrent, white, edges white and even; stipe 1–3 cm. long, filiform or up to 0.5 mm. thick, equal, fragile, snow white, base slightly strigose, pubescent upward or merely pruinose near the apex, glabrescent.

Spores 10–12 × 4–5 μ, slightly inequilateral and tapered to the point of attachment, hyaline, smooth, nonamyloid; basidia four-spored; pleurocystidia rare (embedded and difficult to demonstrate), similar to the cheilocystidia; cheilocystidia abundant, forming a sterile band on the gill edge, 26–37 × 5–8 μ, narrowly fusoid with acute to subacute apices, when revived in KOH their apices frequently covered with a resinous secretion; gill trama homogeneous, yellowish in iodine; pileus trama with a thin pellicle the cells of which give off numerous rodlike projections, pilocystidia originating in the pellicle, measuring 18–27 × 4–7 μ and filiform to subfusiform in shape, hyaline, a well-differentiated hypoderm present beneath the pellicle, the remainder of the flesh filamentous, all parts yellowish in iodine; caulocystidia similar to the cheilocystidia or longer, abundant; stipe tissue yellowish in iodine.

Habit, habitat, and distribution.—Gregarious on a mossy redwood log; Orick, California, December 5, 1935. Known only from the type locality.

Material studied.—Smith, 3780.

Observations.—This species is very closely related to *M. delicatella*, but is readily distinguished by its large spores on four-spored basidia. It is also more fragile, but this character is difficult to use because of the variation exhibited by *M. delicatella*. *M. albissima* differs from *M. candida* (Bres.) Kühner in its broader cheilocystidia and less ventricose spores. From *M. Mairei* (Gilbert) Kühner it differs in its longer spores and generally smaller size as well as being snow-white over all. The abundant cheilocystidia and larger spores also distinguish *M. albissima* from *M. pseudogracilis* Kühner & Maire.

20. Mycena crispula (Quél.) Kühner

Encyc. Myc., 10: 642. 1938

Omphalia crispula Quélet, Champ. Jura et Vosges, Suppl. 10: 663. 1880.
Omphalia apiculata Kauffman, Pap. Mich. Acad. Sci., Arts, and Letters, 11: 201. 1930.

Illustrations:
Text fig. 5, nos. 5, 7 (p. 81).
Quélet, Champ. Jura et Vosges, Suppl. 10, pl. viii, fig. A. 1880.

Pileus 6–15 mm. broad, convex to obtusely conic, the margin connivent to the stipe at first, becoming broadly convex or with a sharp papillate umbo, sometimes the disc depressed, at times somewhat rugulose around the papilla, glabrous, when moist conspicuously translucent-striate to the disc, soon sulcate and the margin becoming plicate, pure translucent-white when moist, opaque when faded, flesh thin, white, fragile, taste subnauseous, no odor; lamellae distant, broad, decurrent, white, edges even; stipe 3–4 cm. long, 1 mm. or less thick, rather pliant, equal, inserted on fallen rhododendron leaves by a flat, thin inconspicuous disc, pure translucent white over all at first, opaque in age, minutely and evenly pubescent under a lens.

Spores 5.5–7 × 3.5–4 μ, ellipsoid, nonamyloid, smooth; basidia four-spored; pleurocystidia not differentiated; cheilocystidia resembling sterile basidia but slightly larger in age, scarcely differentiated; gill trama yellowish in iodine, homogeneous; pileus trama homogeneous, the walls of the surface cells bearing a few short projections, an occasional pilocystidium similar to the caulocystidia present, all parts yellowish in iodine; caulocystidia very abundant (28) 32–56 × 7–9 μ, with slightly thickened hyaline walls, broadest at the base and tapered evenly to a subacute apex, often with wavy outlines, stipe tissue yellowish in iodine.

Habit, habitat, and distribution.—Gregarious on conifer needles and rhododendron leaves. Known only from Oregon. Kauffman collected it at Takilma, and my material came from Florence.

Material studied.—Smith, 3476. Kauffman, Takilma, Oregon.

Observations.—In many respects large carpophores of this species closely resemble *M. delectabilis*. The fragile cap, somewhat pliant stipe, spores, stature, and color are similar in both. *M. crispula* differs in lacking pleurocystidia and well-differentiated cheilocystidia but in having abundant caulocystidia, in the absence of a distinctive odor, and in the presence of a subnauseous taste. The microscopic characters of the type of *O. apiculata* are as given in the description, with the exception that the cystidia on the stipe were not abundant. My description was taken entirely from collection 3476, Florence, Oregon.

Kauffman described the gills as narrow. As can readily be seen by comparing the descriptions of Kühner, Kauffman, and my no. 3476, the gills are broader and better developed in large carpophores and practically absent in the smallest. I do not regard the longer stipe, broader gills, and larger pilei of the American specimens as more than the usual variations among collections. The characters of diagnostic value are the pilocystidia and caulocystidia, nonamyloid spores, and lack of pleurocystidia. Kauffman (1930) described the spores as 4–5 × 3–3.5 μ, but my measurements from the type were 6–7.5 × 3–4 μ. Because of the similarity of *O. apiculata* to *M. crispula* in the characters just mentioned, Kauffman's species is regarded as synonymous with Quélet's.

21. MYCENA MAURETANICA (Maire) Kühner

Encyc. Myc., 10: 639. 1938

Omphalia mauretanica Maire, Bull. Soc. Myc. Fr., 44: 43. 1928.

Omphalia cuspidata Quélet var. *stenospora* Lange, Dansk Bot. Arkiv, 6 (5): 18. 1930.

Illustrations:

Text fig. 4, nos. 9–10 (p. 74).
Lange, Flora Agar. Dan., 2, pl. 62 B.
Kühner, Encyc. Myc., 10: 641.

Pileus 1–2 mm. broad, convex, sulcate-striate, pure white, membranous, fragile; lamellae remote, narrow to almost foldlike, adnate or slightly decurrent; stipe 1 cm. long, filiform, white, pruinose, inserted on dead leaves.

Spores 10–12 × 2.5–3 μ, smooth, subcylindric, nonamyloid; basidia two-spored; no pleurocystidia or cheilocystidia observed; the gill trama yellowish in iodine; pileus trama homogeneous, yellowish in iodine, with numerous hyaline pilocystidia 26–38 (50) × 6–9 μ, usually wavy in outline and broadest at the base, apex acute or subacute; caulocystidia similar to pilocystidia or longer and more flexuous; stipe tissue yellowish in iodine.

Habit, habitat, and distribution.—Scattered on dead leaves; Michigan and Washington.

Material studied.—Smith, 33–1065, 16607.

Observations.—The four-spored form was found in Washington by Mains in 1941.

The following are my notes from his collection: Pileus 1–3 mm. broad, convex or with a slight papilla, membranous and very fragile, white over all, somewhat sulcate-striate, margin incurved slightly at first, surface moist, unpolished when viewed under a lens; odor and taste not recorded; lamellae distant to subdistant, 6–11, narrow but not foldlike, spaces between the gills slightly wrinkled, arcuate-adnate to decurrent, edges even, lamellulae lacking or only one tier present; stipe filiform, up to 1 cm. long, equal, base inserted on substratum and not inflated, very finely pubescent over all from caulocystiida. Spores 7–9 × 2.2–2.7 μ, slightly allantoid in one view, nonamyloid; basidia four-spored; no pleurocystidia or cheilocystidia; pileus trama homogeneous, surface covered with contorted filamentous hyphal protuberances 30–44 × 3–6 μ; stipe covered with short hairs similar to those on the cap. Gregarious on alder bark; E. B. Mains, September 4, 1941, Baker Lake, Washington. A. H. Smith, 16607.

The material of both my own and Dr. Mains' collections was very scanty, and as a result my notes are not complete. The pubescence of the pileus was hardly noticeable, but the presence of the numerous pilocystidia made the cap appear unpolished under a lens. Because of the incompleteness of my own data the following description has been translated from Kühner (1938).

Pileus 0.3–4.5 mm. broad, campanulate, hemispheric, or often convex, obtuse or slightly depressed on the disc, sometimes mammillate, occasionally sulcate, densely pubescent (under a lens); lamellae 6–9 (11) reach the stipe, with only one tier of lamellulae or none at all in very small caps, narrow but not foldlike, sometimes forked, reaching the margin of the cap or disappearing 0.2–0.3 mm. from the edge, arcuate and somewhat decurrent, but not long-decurrent, interlamellar spaces sometimes netted with veins; stipe 1.5–5 mm. long, 0.04–0.24 mm. thick above, below 0.32 mm. thick, equal or very slightly attenuated from the base upward, with the base (sometimes a bit yellowish) slightly clavate but deprived of a disc and rhizoids (most often with cobwebby filaments extending over the substratum), strongly pubescent but less so from the base toward the apex, where it is hardly so at times.

Spores nonamyloid, narrow, 6.5–8 (9) × 1.7–2.5 (3) μ, rarely claviform, in general cylindric and slightly depressed on one side, obtuse at the apex, attenuated to the point of attachment, frequently adhering in groups of four; basidia four-spored (occasionally three- or two-spored), clavate, 16–18 × 4.2–5.5 μ; pleurocystidia and cheilocystidia lacking; gill trama interwoven in radial sections; pileus covering of appressed short hyphae (not filiform), 5–10 μ in diameter, irregularly echinulate with short rodlike projections but very dense and with numerous upright hairs 37–57 μ long, which are somewhat inflated at the base (4–7μ), gradually attenuated (or contracted) to a slender filament 1–2 μ thick, obtuse, with thin walls, and at the base often curved or with irregular spurs; flesh of pileus continuous with that of the stipe; covering of the stipe of hyphae bearing rodlike projections but not densely echinulate, and having numerous thin-walled projecting hairs similar to those of the pileus (28–45 × 4–6.5 μ).

22. Mycena gracilis (Quél.) Kühner

Encyc. Myc., 10: 650. 1938

Omphalia gracilis Quélet, Champ. Jura et Vosges, Suppl. 10: 662. 1880.
Agaricus (*Mycena*) *immaculata* Peck, Ann. Rep. New York State Mus., 38: 84. 1885.
Mycena immaculata Saccardo, Syll. Fung., 5: 264. 1887.
Omphalopsis immaculata Murrill, North Am. Flora, 9: 312. 1916.

Illustrations:
Plate 8 B, D; Text fig. 7, no. 1.
Bresadola, Icon. Mycol., 6, pl. 270, fig. 1 (as Omphalia).
Konrad et Maublanc, Icon. Sel. Fung., pl. 236, II.

Pileus 3–14 mm. broad, conic with an appressed margin at first, becoming campanulate to convex, seldom plane, often papillate, surface glabrous, moist, translucent-striate before losing moisture, snow white over all or occasionally developing a sordid-yellowish cast over the disc; flesh very thin and fragile, odor and taste not distinctive; lamellae adnate to hooked, soon developing a distinct decurrent tooth, subdistant (10–15 reach the stipe, two or three tiers of lamellulae), moderately broad (often nearly triangular in outline), sometimes seceding, pure white, edges even and white; stipe (1) 3–6 (7) cm. long, 0.5–1 mm. thick, equal, fragile, strict or flexuous, tubular, fragile, pure white, base attached to needles by distinct white hairs (sometimes densely strigose, at other times nearly glabrous), translucent and glabrous above, finely pruinose—especially toward the apex when viewed under a lens—pure white, usually watery white when perfectly fresh.

Spores 7–9 × 2.5–3 μ, nearly cylindric, smooth, nonamyloid; basidia either two- or four-spored, pleurocystidia not differentiated;

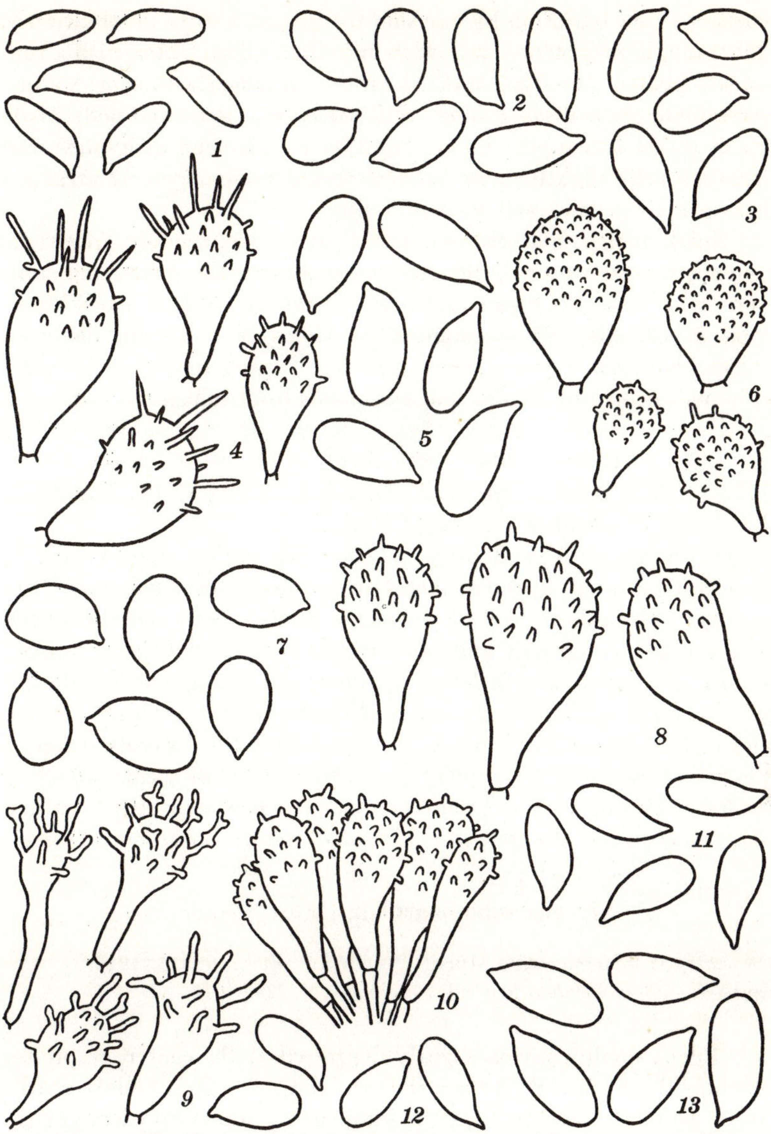

FIG. 7. *M. gracilis:* 1, spores. *M. albidula:* 2, spores. *M. pterigena:* 4, cheilocystidia; 5, spores. *M. juncicola:* 3, spores; 6, cheilocystidia. *M. tenella:* 7, spores; 8, cheilocystidia. *M. minutissima:* 9, cheilocystidia; 12, spores. *M. Smithiana:* 10, cheilocystidia; 11, spores. *M. capillaris:* 13, spores

cheilocystidia basidium-like or slightly longer, hardly differentiated; gill trama homogeneous, yellowish in iodine; pileus trama with a thin adnate pellicle, the hyphae of which are furnished with short rodlike projections; hypoderm scarcely differentiated, the tramal body made up of rather broad cells bound together by filamentous hyphae, the enlarged cells slightly more numerous and more regularly arranged beneath the pellicle, yellowish in iodine.

Habit, habitat, and distribution.—Scattered to densely gregarious under conifers, especially fir and spruce, during the spring and summer; New York, Michigan, and Idaho in the United States and Manitoba in Canada. It is common in Michigan and very likely so throughout northeastern North America. Bisby reports it from deciduous woods, but he also sent specimens from sphagnum bogs.

Material studied.—Smith, 32–11, 32–27, 32–382, 33–99, 33–331, 23, 1332, 1368, 1627, 4087, 6234, 6329, 8278, 13756, 13866, 14204, 15752, 15877, June 8, 18, 1932, August 21, 1929, Bisby, Manitoba. Kauffman, September 11, 1920, Colorado.

Observations.—The glabrous pileus, long narrow spores, lack of well-differentiated cystidia, and the white color distinguish the species. The size as well as the shape of the pileus and the length of the stipe vary greatly and can hardly be considered characteristic. Locally *M. gracilis* can be found in almost any sphagnum bog during late May or early June. It nearly always occurs under black spruce, and it most frequently fruits in the zone where the lower branches of the young trees touch the sphagnum or the surrounding chamadaphne bushes. During very wet periods it has also been collected in spruce plantations where the trees are old enough to form a fairly open stand.

23. **Mycena subimmaculata** (Murr.), comb. nov.

Omphalopsis subimmaculata Murrill, North Am. Flora, 9: 315. 1916.
Omphalia subimmaculata Murrill, Mycologia, 8: 220. 1916.

"Pileus small, convex, slightly depressed at the center, cespitose, reaching 8 mm. broad; surface smooth, glabrous, finely striate nearly to the center, pure-snow-white, margin entire, appressed when young: lamellae long-decurrent, distant, rather narrow, inserted, white: stipe very slender, cylindric, hollow, white, minutely whitish-pruinose to subglabrous, 2 cm. long, 0.5 mm. thick.

"Type collected on dead wood in woods near Seattle, Washington,

October 20–November 1, 1911, ***W. A. Murrill*** *688* (herb. N. Y. Bot. Gard.).

"Distribution: Known only from the type locality."

The spores of the type are variable in size; most of them measured 8–11 × 3.5–4 μ. A considerable number measuring 12–14 (16) × 4–5 (6) μ were found in groups of two. They were subfusiform, smooth, and hyaline. Unfortunately the iodine reaction was not obtained. The only basidia seen with sterigmata were two-spored. No differentiated pleurocystidia or cheilocystidia were observed. The trama of both the pileus and the gills appeared to be homogeneous and otherwise not distinctive.

Mycena albissima seems to be close to *M. subimmaculata*, but differs sharply in having abundant cheilocystidia. The spores of *M. subimmaculata* separate it from *M. gracilis* and other species with long, narrow spores. It seems quite probable that *M. pseudogracilis* Kühner & Maire is the same as Murrill's species. However, I have never seen either in the fresh condition.

24. **Mycena McMurphyi** (Murr.), comb. nov.

Omphalopsis McMurphyi Murrill, North Am. Flora, 9: 315. 1916.
Omphalia McMurphyi Murrill, Mycologia, 8: 220. 1916.

"Pileus rather small and delicate, convex, cespitose, 5–10 mm. broad; surface smooth, glabrous, white, margin undulate, furrowed, appressed when young: context thin, white, without characteristic taste or odor; lamellae decurrent, subdistant, about 20 in number, rather broad, white: stipe cylindric, slender, hollow, white, glabrous, mycelioid at the base, 1.5–3 cm. long, 0.5–1 mm. thick.

"Type collected in soil, probably attached to buried wood, on the bank of Madera Creek, near Stanford University, California, December 21, 1902, *James McMurphy 51* (herb. N. Y. Bot. Gard.).

"Distribution: Known only from the type locality."

As previously stated (Smith, 1938), the spores of the type measure 12–16 × 7–8 μ and are hyaline in iodine. No cystidia were found. This species appears to be distinct from *M. subimmaculata* because of its much broader spores borne on four-spored basidia.

25. **Mycena albidula** (Pk.), comb. nov.

Omphalia albidula Peck, Ann. Rep. New York State Mus., 49: 17. 1896.
Omphalopsis albidula Murrill, North Am. Flora, 9: 312. 1916.

Omphalopsis californiensis Murrill, *ibid.*, p. 315.
Omphalia californiensis Murrill, Mycologia, 8: 220. 1916.

Illustrations:
Text fig. 7, no. 2 (p. 91).
Beardslee, Mycologia, 9, pl. 4, fig. 2 (as *Omphalia gracillima*).
Kauffman, Agar. Mich., 2, pl. 171, lower figs. (as *Omphalia gracillima*).

Pileus (2.5) 3–10 mm. broad, convex or with a small sharp papilla, sometimes turbinate, the margin appressed against the stipe or connivent at first, surface moist, glabrous, translucent-striate, pure white, opaque when faded and then becoming sulcate, appearing minutely fibrillose under a lens when faded, flesh membranous and fragile, no odor or taste; lamellae decurrent, often unequally decurrent in age, some developing a long tooth, broad (2 mm.), distant, 8–11 reach the stipe, one or occasionally two tiers of lamellulae, pure white, edges even; stipe 2.5–4 cm. long, less than 1 mm. thick, equal, fragile, solid, hyaline-white, faintly frosted at first, soon glabrous, attached to fallen leaves by a flat mycelial disc so tightly that some of the leaf adheres to the base of the stipe when specimens are collected.

Spores 7–9 × (3) 4–5 μ, ellipsoid to narrowly ellipsoid, nonamyloid; basidia four-spored; pleurocystidia and cheilocystidia not differentiated; gill trama homogeneous, composed of enlarged cells (50–120 × 15–25 μ), yellowish in iodine; pileus with a thin surface pellicle from which arise numerous thin-walled filamentous simple or branched hyphal projections, the trama of enlarged sausage-shaped cells 50–200 × 15–25 μ, yellowish in iodine.

Habit, habitat, and distribution.—Scattered to gregarious on bark and fallen leaves of hardwoods during the late summer and autumn; Tennessee, New York, Michigan, Washington, and California.

Material studied.—Smith, 1346, 1375, 1651, 3274, 4414, 6969, 9837, 15197, 18309, June 21, 1935, Michigan. Hesler, 14477. Kauffman, September 3, 1914.

Observations.—The spores of the type measure 7–9 (10) × 3.5–5 μ, and no differentiated cystidia were found. The fungus Kauffman placed in this species had spores 8–10 × 2–4 μ and must remain doubtfully determined, since his specimens were lost and his measurements and comments on the spores do not check with the type. He probably had a form of *M. gracilis* in which the gills were more decurrent than usual. Kauffman's description of *Omphalia gracillima* more than likely applies to *M. albidula* also. He separated *O. albidula*

Peck from *O. gracillima* Fries on the basis of the straight margin of the pileus in the one and the incurved margin of the other. I have found specimens of *M. gracilis* which had a somewhat incurved margin, and hence I believe that that character is not against my identification of Kauffman's *O. albidula* as *M. gracilis*. Beardslee's report (1917) of *O. gracillima* belongs here also.

It is evident from a study of the type of *Omphalia californiensis* that it is merely a small specimen of *M. albidula*. The spores of the type are nonamyloid and measure 7–9 × 4–4.5 μ. Pleurocystidia and cheilocystidia are not differentiated. In addition, a sketch accompanying the type, aside from its small size, illustrates *M. albidula* perfectly.

26. **Mycena subvestita,** nom. nov.

(Not *M. vestita* Velenovský, České Houby, p. 307, 1920)

Omphalia vestita Peck, Bull. Torr. Bot. Club, 34: 345. 1907.

"Pileus thin, membranaceous, convex nearly plane or slightly depressed in the center, minutely pruinose or tomentose, white; lamellae few, very distant, adnate or decurrent, white or whitish; stem slender, short, solid or stuffed, pruinose-pubescent either wholly or on the basal half only, whitish or pallid, often becoming brownish with age; spores subglobose, 4–5 μ in diameter.

"Pileus 2–3 mm. broad; stem 6–10 mm. long, 0.5–1 mm. thick.

"Decaying vegetable matter in damp places. Horseshoe Island, Ontario, Canada. August. C. Guillet.

"This is a very small white species closely related to *O. integrella* Pers. and O. *pusillissima* Peck, from both of which it is separated by the minute tomentose covering of the pileus."

I have examined the type and can verify Peck's comments on the relationship of the species. It is also close to *M. delicatella*. I cannot understand why Murrill placed *Omphalia vestita* in synonymy with *O. umbellifera*. There is no resemblance whatever in either the macroscopic or the microscopic characters. I did not get iodine reactions on the spores of Peck's type because they were rare and difficult to find. Those seen confirmed Peck's description. Pleurocystidia and cheilocystidia were not observed but, because of the scanty material available, their presence or absence was not determined with any finality. The stipe is covered with slender filamentous projecting hyphae, and these are also present over the pileus.

It is the presence of these hyphae which indicates a relationship with *M. delicatella*. The spores, of course, readily distinguish the two.

27. Mycena ignobilis (Joss.) Kühner

Encyc. Myc., 10: 667. 1938

Omphalia ignobilis Josserand, Ann. Soc. Linn. Lyon, 80: 94. 1937.

Pileus 2–5 (7) mm. broad, flat or very broadly convex when young, soon turbinate-obconic or in age the disc broadly depressed (owing to the elevated margin), shining watery white and striate to the apex when moist, soon fading and then appearing minutely flocculose (under a lens), the margin at times crenate or lobed and usually slightly incurved at first; flesh membranous, odor and taste not distinctive; lamellae very distant, long-decurrent, triangular and broadest in the middle in age, pure white, one or two tiers of rudimentary lamellulae sometimes present; stipe 1–2 cm. long, about 0.5 mm. thick, equal, fragile, pure white, base naked.

Spores 7–9 × 3.5–4 μ, ellipsoid, smooth, nonamyloid (yellow to yellowish brown in iodine); basidia four-spored; cheilocystidia and pleurocystidia not differentiated; gill trama and pileus trama homogeneous, yellowish in iodine and not otherwise distinctive.

Habit, habitat, and distribution.—Gregarious on mud; California and Michigan. Rare, and apparently fruiting only during periods of very wet weather. I have not found it along the edges of swamps and on low ground in the woods, but, instead, it occurs in open pastures that under ordinary weather conditions are fairly dry.

Material studied.—Smith, 1392. C. O. Smith, June 25, 1935, Riverside, California.

Observations.—The spore size appeared to be quite variable in the collection (Smith, 1392) from which the description above was taken. In one cap they were 6–7 × 3.5–4 μ, but in the others 7–9 × 4 μ was the typical range. Kühner gives the size as 9–10 μ long, but Josserand has indicated a variability of 7–11 μ.

28. **Mycena papillata** (Peck), comb nov.

Omphalia papillata Peck, Ann. Rep. New York State Mus., 51: 285. 1898.
Omphalopsis papillata Murrill, North Am. Flora, 9: 311. 1916.

"Pileus membranaceous, conical or campanulate, nearly even, papillate at the apex, pure white; lamellae few, distant, arcuate and strongly decurrent, white; stem filiform, glabrous, white, attached to the matrix by a few radiating white filaments; spores broadly elliptic or subglobose, .00016 to .0002 in. long.

"Pileus 1 to 3 lines broad; stem about 1 in. long, scarcely thicker than a thread.

"Sticks and fallen leaves in woods. Gansevoort. July.

"The species is related to *Omphalia Fibula*. It should also be cautiously separated from *Mycena immaculata*."

Material studied.—Kauffman, August 20, 1919, Cabin John, Maryland.

Observations.—I have been unable to locate the type. The almost globose spores, however, should separate it readily from other small white agarics of similar appearance. It is closest, apparently, to *M. albidula*. The decurrent gills distinguish it from *M. Rickenii*.

29. **Mycena paucilamellata,** sp. nov.

Illustrations: Text fig. 6, nos. 7–8 (p. 85).

Pileus 1–1.5 mm. latus, conico-campanulatus, candidus, glaber; lamellae distantes, plicatae, subdecurrentes, candidae; stipes 1 cm. longus, filiformis, candidus, pruinosus; sporae (8) 9–11 × 3–3.5 μ; pleurocystidia et cheilocystidia fusoide ventricosa, 28–37 × 9–14 μ. Specimen typicum in Herb. Univ. Mich. conservatum. Legit A. H. Smith, n. 3720, prope Orick, Calif., December 2, 1935.

Pileus about 1–1.5 mm. broad, conic-campanulate, the margin connivent at first, pure shining white and opaque at all stages; flesh membranous, white, no odor or taste evident; lamellae none or evident only as 3–5 low ridges which are somewhat decurrent on the stipe; stipe 1 cm. long, filiform, white, pruinose above, base slightly pubescent, the apex usually enlarged slightly, base inserted on the redwood twigs and needles.

Spores (8) 9–11 × 3–3.5 μ, cylindric to subaciculate, tapered to a long point at the base, smooth, hyaline, amyloid; basidia four-spored; cystidia scattered through the hymenium and abundant over the apex of the stipe, 28–37 × 9–14 μ, fusoid-ventricose with obtuse apices; gill trama and pileus trama similar, yellowish in iodine; pileus surface covered with enlarged cells (15–30 × 10–16 μ), the upper surfaces of which give off numerous short rodlike projections, no true pellicle present.

Habit, habitat, and distribution.—On fallen twigs of *Sequoia sempervirens*; December 2, 1935, Orick, California. Known from the type locality, Prairie Creek State Park.

Material studied.—Smith, 3720.

Observations.—Although the spores are amyloid and the gills foldlike, the species does not appear to be congeneric with *Delicatula integrella*. The development is altogether different. The young fruiting bodies of *D. integrella* are characterized by the presence of a veil, and the development of the hymenium is at least hemiangiocarpic. In *M. paucilamellata* the development appears to be gymnocarpic. There are no traces of a veil. My study of the latter was made from freehand sections of fresh material and an examination of the whole button stages under the low power of a microscope as well as under a hand lens.

Mycena paucilamellata is very well characterized by its long, narrow amyloid spores and smooth obtuse ventricose cystidia as well as by its minute size and foldlike gills. It was found in great quantity, but, as one can readily imagine, the fruiting bodies make very poor herbarium specimens when dried.

30. **Mycena litoralis,** sp. nov.

Illustrations: Text fig. 6, nos. 9, 11 (p. 85).

Pileus 2–4 mm. latus, campanulatus vel cuspidatus, glaber, subrugulosus, striatus, candidus; lamellae distantes, plicatae vel angustae, decurrentes, albidae; stipes 1 cm. longus, 0.5 mm. crassus, fragilis, albidus, demum candidus, pruinosus; sporae 6–8 (9) × 3–4 μ, ellipsoideae; cheilocystidia 24–32 × 8–11 μ, fusoide ventricosa. Specimen typicum in Herb. Univ. Mich. conservatum. Legit A. H. Smith, n. 3470, prope Siltcoos Outlet Forest Camp, Florence, Ore., Nov. 14, 1935.

Pileus 2–4 mm. broad, campanulate or cuspidate, the umbo very prominent, surface moist and watery, shining white, somewhat rugulose around the disc, otherwise even, glabrous, distantly translucent-striate when moist, opaque and becoming sulcate when faded, margin appressed against the stipe at first, not discoloring, flesh thin, white, fragile, taste mild, no odor; lamellae distant, veinlike or very narrow, decurrent, hymenial tissue forming a collar on the stipe for some distance down, edges even, whitish; stipe 1 cm. long, 0.5 mm. thick, equal, fragile, translucent-white, soon opaque and shining white, pruinose to minutely pubescent under a lens, inserted on fern stalks by a naked base.

Spores 6–8 (9) × 3–4 μ, ellipsoid, smooth, amyloid; basidia four-spored; pleurocystidia scattered and difficult to find, similar to the cheilocystidia; cheilocystidia 24–32 × 8–11 μ, fusoid-ventricose, hyaline, with obtuse to subacute apices, the neck often almost entirely lacking; gill trama homogeneous, yellow in iodine; pileus trama homogeneous, yellowish in iodine, with a thin pellicle of narrow (3–5 μ thick) hyphae which give off numerous short rodlike projections.

Habit, habitat, and distribution.—Gregarious on fern debris, in the sand; November 14, 1935, Siltcoos Outlet Forest Camp, Florence, Oregon.

Material studied.—Smith, 3488, 3740.

Observations.—In its cystidia, spores, and foldlike gills this species resembles *M. paucilamellata*, but is at once distinguished by the size and shape of the spores. The gills vary between the type found in *M. paucilamellata* and the normal type. This discovery, of course, does not rule out the character "foldlike" in distinguishing species of agarics, but does indicate the need of caution in its use. Like most characters, it is variable, more so in some species than in others, and the range of variation must be determined for each.

31. **Mycena kalalochensis,** sp. nov.

Illustrations: Text figs. 5, nos. 1–4 (p. 81); 6, no. 10 (p. 85).

Pileus 3–8 mm. latus, convexus demum planus, albidus, pruinosus, non hygrophanus; lamellae adnatae, latae, distantes, albidae; stipes 3–7 mm. longus, 0.5 mm. crassus, albidus, pruinosus, aequalis; sporae 8–11 × 5–6 μ, amyloideae; pleurocystidia 22–30 × 5–9 μ, fusoide ventricosa; cheilocystidia 36–62 × 6–10 μ. Specimen typicum in Herb. Univ. Mich. conservatum. Legit A. H. Smith, n. 13035, prope Kalaloch, Wash., April 30, 1939.

Pileus 3–8 mm. broad, convex, remaining broadly convex, margin incurved at first, spreading in age, chalk white and appearing pruinose under a lens at first, glabrous and uneven in age but remaining chalky, slightly sulcate at maturity, not hygrophanous; flesh membranous and pliant (but not reviving), odor not distinctive, taste not recorded; lamellae adnate, broad, distant, 10–12 reach the stipe, two tiers of lamellulae, white over all, edges pruinose; stipe 3–7 mm. long, less than 0.5 mm. thick, equal or the base flanged slightly, strigose, the remainder pruinose like the pileus, chalky white over all.

Spores 8–11 × 5–6 μ, ellipsoid, hyaline, smooth, distinctly amyloid; basidia four-spored; pleurocystidia present but buried in the hymenium and difficult to locate, 22–30 × 5–9 μ, narrowly fusoid-ventricose or some with irregular walls; cheilocystidia very conspicuous, forming a sterile band, 36–62 × 6–12 μ, somewhat fusoid to subcylindric or nearly filamentous, hyaline, the walls often irregular; gill trama homogeneous, yellowish in iodine, hyphae slender; pileus trama homogeneous beneath a turflike covering of upright pilocystidia and filamentous projections, the pilocystidia 32–46 × 8–12 μ, clavate to ventricose, the apex often subcapitate, filamentous cells 3–4 μ thick, branched or variously contorted, no incrustations seen; caulocystidia similar to pilocystidia, stipe tissue vinaceous brown in iodine.

Habit, habitat, and distribution.—Scattered on debris of red alder and thimbleberry; Kalaloch, Washington. Known only from the type locality.

Material studied.—Smith, 13035.

Observations.—This fungus resembles small forms of *M. delicatella* in stature and in the flexibility of the pileus and stipe, but the spores separate it at once by both their shape and amyloid reaction. The stipe is not separable from the pileus, and its base is not truly bulbous; hence the species cannot be referred to Pseudomycena. The broad, well-formed gills distinguish it from the preceding species having amyloid spores. There are other differences as well, such as spore size and the nature of the hairs and cystidia covering the pileus. From the species in Kühner's section Lacteae it differs in its amyloid spores and stipe tissue.

32. **Mycena pusillissima** (Pk.), comb. nov.

Omphalia pusillissima Peck, Bull. New York State Mus., 116: 27. 1907.
Omphalopsis pusillissima Murrill, North Am. Flora, 9: 311. 1916.

Illustration: Text fig. 5, no. 6 (p. 81).

Pileus 1–2 mm. broad, conic to convex or somewhat cucullate, the disc flattened or slightly depressed, glabrous, pure white and opaque, not striate, membranous, odor and taste not distinctive; lamellae very distant, 3–5 reach the stipe, decurrent, white, becoming foldlike and often disappearing before reaching the margin of the pileus; stipe 1 cm. long, filiform, glabrous, hyaline white, inserted on the substratum.

Spores 6–7 (9) × 3–3.5 (4.5) μ, narrowly ellipsoid, smooth, bright blue in iodine (amyloid); basidia four-spored; pleurocystidia and cheilocystidia not differentiated; pileus trama homogeneous, surface hyphae with scattered rodlike projections over their walls, no pilocystidia or caulocystidia; tissue of pileus and stipe yellowish in iodine.

Habit, habitat, and distribution.—Gregarious on sticks of conifers (small twigs and branches which have fallen to the ground); November 24, 1935, Booth, Oregon. Known from New York and Oregon.

Material studied.—Smith, 3618, and the type.

Observations—The description was drawn from the Oregon collection cited. The type has been examined; the spores measured 7–8 × 4–5 μ. No differentiated cystidia were found. The iodine test was not applied because of the small amount of material available and because spores were rare in the one mount which was made. In view of the risk involved it seems more sensible to interpret Peck's species as having amyloid spores than to describe a new species just like Peck's but supposedly differing in the reaction of its spores to iodine. Peck's original description is given below:

"Pileus membranaceous, broadly convex or nearly plane, glabrous, umbilicate, slightly striate on the margin when dry, white; lamellae few, distant, decurrent, white; stem slender, filiform, flexuous, glabrous, white; spores subglobose or broadly elliptic, .0002–.00024 of an inch long, .00016–.0002 broad.

"Pileus 1–2 lines broad; stem 3–5 lines long. On humus and decaying twigs under pine trees. Delmar, Albany County, August.

"This is one of the smallest species of Omphalia known to me. The lamellae are very narrow, sometimes branched and sometimes absent. It is a smaller mushroom than *Omphalia integrella*, and differs from it in its umbilicate pileus. The stem is hollow but the cavity is minute."

33. **Mycena cylindrospora,** sp. nov.

Illustrations: Text fig. 4, nos. 3–4 (p. 74).

Pileus 1–2 mm. latus, convexus vel planus, candidus, nonlamellatus; stipes 10–15 mm. longus, filiformis, candidus; sporae 10–13 (14) × 4–5 (6) μ, cylindricae. Specimen typicum in Herb. Univ. Mich. conservatum. Legit A. H. Smith, n. 9848, prope Grassy Patch, Great Smoky Mountains National Park, Tenn., Aug. 6, 1938.

Pileus 1–2 mm. broad, convex to plane, snow white, glabrous or appearing frosted, translucent in age, opaque when faded, not striate but soon becoming wrinkled; flesh membranous, odor and taste not recorded; lamellae lacking; stipe 10–15 mm. long, filiform, hyaline white, pruinose, scarcely enlarged at the base and with only a few mycelial hairs at the point of insertion on the substratum.

Spores 10–13 (14) × 4–5 (6) μ, cylindric to subcylindric, smooth, hyaline, amyloid; basidia four-spored; margin of the pileus and apex of stipe studded with saccate to clavate sterile cells, 19–40 × 9–27 μ, with the enlarged portion echinulate; the tissue of the stipe and pileus remaining hyaline in iodine; trama of pileus composed of enlarged cells, the surface hyphae more or less covered with rodlike projections, pellicle, if it exists, very rudimentary.

Habit, habitat, and distribution.—Gregarious on old rhododendron leaves; Grassy Patch, Great Smoky Mountains National Park, Tennessee. The old leaves are frequently found more or less curled up on the forest floor, and the agaric usually fruits inside these curled-up leaves.

Material studied.—Smith, 9848.

Observations.—The spores were very abundant both in mounts of fresh material and also when revived in chloral hydrate to obtain the iodine reaction. It would not be in the least surprising to come upon specimens with gills and to find the gill edges furnished with cheilocystidia similar to the echinulate sterile cells described above. The characters that should be emphasized are the type of sterile cells, the very large cylindric spores on four-spored basidia, and the pure-white color of all parts.

34. Mycena minutula (Pk.) Saccardo

Syll. Fung., 5: 263. 1887

Agaricus (*Mycena*) *minutulus* Peck, Bull. Buffalo Soc. Nat. Sci., 1: 47. 1873.
Prunulus parvulus Murrill, North Am. Flora, 9: 323. 1916.
Mycena parvula Murrill, Mycologia, 8: 221. 1916.

"Gregarious, white throughout, pileus campanulate, becoming convex, smooth, striatulate, with a minute shiny umbo or apiculus at the apex; lamellae broad, subdistant, attached with a slight decurrent tooth; stem slender, hollow, smooth, about 1′ high, 2″–3″ broad, stem ¼″ thick. On bark of a prostrate maple or basswood trunk in hardwoods, Portville. Sept.

"A very small white species distinguished by the apiculus, the interspaces are slightly reticulate. Viewed carefully in a certain light the stem appears to be sprinkled with minute mealy particles when fresh."

This is a copy of Peck's notes taken from Book 4, page 52, of his field notes at the New York State Museum, Albany, New York. I have examined the type and made the following observations: The spores measure 7–8 × 5.5–7.5 μ, are smooth, broadly ellipsoid, subglobose, or subglobose with a prominent apiculus. The basidia measure 29–32 × 6–7 μ, and all were two-spored. The sections revived well and many individuals with sterigmata were seen. Pleurocystidia were lacking in my sections, but the cheilocystidia, though not conspicuous, were characteristic. They measured 31–39 × 8–11 μ, were cylindric to clavate or subventricose, and the surfaces were covered with short, obtuse rodlike projections. Details of the pileus trama were not obtained, and no tests were made with iodine.

I have not collected this species and know of no authentic material other than Peck's type. In all of Kauffman's collections I found the cheilocystidia to be smooth and fusoid-ventricose. Consequently, his account must be regarded as a misdetermination. To judge from his specimens, he probably had small forms of *M. olida* var. *americana*. *M. minutula*, by virtue of its cheilocystidia, should be very readily recognized among the small white species. Apparently it must be either very rare or very sporadic in its appearance.

Subsection Subincarnatae

35. Mycena pterigena (Fr.) Quélet

Champ. Jura et Vosges, p. 109. 1872

Agaricus pterigenus Fries, Syst. Myc., 1: 160. 1821.

Illustrations:

Text fig. 7, nos. 4–5 (p. 91).

Fries, Icon. Sel. Hymen., 1, pl. 85, fig. 4.

Lange, Flora Agar. Dan., 2, pl. 54, H. 1936 (very good).

Pileus 2–5 mm. broad, subcylindric to obtusely conic with an appressed margin at first, becoming broadly conic to convex, surface glabrous, somewhat translucent striate, color bright, often delicate rose ("coral red") over all, gradually fading to grayish vinaceous—especially on the disc, the rose tints often persisting along the margin or the margin whitish; flesh membranous but not fragile, odor and taste not recorded; lamellae ascending, broadly adnate, distant, narrow to broad, pale rose at first, faces usually soon becoming whitish, the edges fading or remaining pale rose; stipe 2–4 cm. long, filiform or up to 0.5 mm. thick, toughish and flaccid, glabrous, concolorous with the pileus at first, becoming paler and hyaline, finally grayish brown, attached to fern debris by a thin mat or by groups of radiating white hairs, not readily separable from the tissue of the pileus.

Spores 8–10 (11) × 4–4.5 μ, narrowly ellipsoid to subcylindric, amyloid; basidia four-spored; pleurocystidia rare to scattered (sometimes apparently absent), similar to the cheilocystidia; cheilocystidia 20–32 × 9–14 μ, broadly ellipsoid or clavate to subglobose, apex studded with short rodlike projections, contents hyaline or pale pinkish; gill trama of enlarged cells, subhymenium very narrow and indistinct, all portions pale vinaceous red in iodine; pileus trama with a surface layer of broad hyphae (6–12 μ), the exposed surfaces of which are strongly echinulate, the tramal body made up of greatly enlarged cells (15–40 μ), subhymenium very thin, all parts vinaceous red in iodine; stipe vinaceous red in iodine, the surface cells covered with echinulations similar to those on the pileus.

Habit, habitat, and distribution.—Scattered to gregarious on decaying fern fronds; New York, Michigan, Washington, and Oregon in the United States and in Ontario in Canada. This species appears to be rare though very widely distributed, but on Mt. Baker, Washington, near Martin Lake, it was collected in great abundance. It is usually found during the late summer or fall.

Material studied.—Smith, 4476, 16573, 16746. Kauffman, Kentucky. Mains, 33–889, 34–49, 34–141, September 14, 1936, Michigan.

Observations.—The habitat and beautiful bright-rose color of the carpophores are the outstanding features of this species. The manner in which the gills fade is interesting but may cause some confusion to the taxonomist. At times both the faces and the edges become pallid, but sometimes the color persists in the edges, thus placing the fungus in the old Friesian section Calodontes if no other characters are considered. In this species the colored gill edge should not be considered an important diagnostic character. The pigment appears to be quite unstable.

From a study of tangential sections of the pileus it is evident that the trama has been reduced to a well-developed hypoderm and an almost nonexistent subhymenium. *M. pterigena* appears to be closely related to *M. juncicola.* The two are similar in habit, size, consistency, and, to a certain extent, in color as well as in their spores and cystidia. Macroscopically the habitat readily distinguishes them. Microscopically the gelatinous pellicle of *M. juncicola* distinguishes it at once.

36. Mycena juncicola (Fr.) Gillet

Champ. Fr., 1: 282. 1878

Agaricus juncicola Fries, Syst. Myc., 1: 160. 1821.

Illustrations:
Plate 1 A; Text fig. 7, nos. 3, 6 (p. 91).
Fries, Icon. Sel. Hymen., 1, pl. 85, fig. 6.
Smith, Mycologia, 28: 412, fig. 1, no. 4; p. 416, fig. 2, no. 2.

Pileus 1–3 mm. broad, conic to convex or broadly convex with or without a slight papilla, margin straight at first, surface uneven or rugulose around the disc, glabrous, moist and shining, opaque, sulcate-striate to the disc when fresh, disc tinged purplish to vinaceous or grayish vinaceous, the margin paler vinaceous to whitish; flesh membranous, fragile, odor and taste not distinctive; lamellae distant, broadly adnate, narrow or moderately broad in large caps, pale grayish vinaceous, edges whitish to gray; stipe 3–5 (25) mm. long, filiform, glabrous above, hyaline, grayish white or grayish vinaceous, inserted or attached by an inconspicuous flat vinaceous or brownish plate 0.5–1 mm. broad.

Spores 9–11 × 5–6 μ, narrowly ovoid, slightly amyloid; basidia four-spored; pleurocystidia rare, similar to the cheilocystidia; cheilocystidia numerous, 18–30 × 9–12 μ, broadly clavate to subglobose, verrucose or with a few short rodlike projections over the apex and verrucose over the remainder, hyaline; gill trama composed of enlarged cells, homogeneous, pale vinaceous in iodine; pileus trama with a thick gelatinous pellicle when revived in KOH (usually appearing only subgelatinous when in water mounts of fresh material), tramal body mostly of enlarged cells but a narrow band of filamentous tissue present next to the subhymenium, pale vinaceous in iodine except for the pellicle.

Habit, habitat, and distribution.—Scattered or in small clusters on sedge culms, spring and fall; Oregon and Washington.

Material studied.—Smith, 3512, 3545, 3613, 14000.

Observations.—This is a very distinctive little fungus that can readily be recognized by its dull pale-vinaceous color, size, cheilocystidia, and spores. The spores are typically amyloid, as Kühner has stated. My first study was made not long after the specimens had been dried and negative results were obtained. On checking this character later in the laboratory I obtained positive results. During the spring of 1939 the fungus was found in Washington. A similar series of tests was made, and the amyloid reaction was found to be very weak at first but readily demonstrable six months later. The plate of tissue at the base of the stipe is almost the color of the substratum and is very easy to overlook even in specimens on which it is well developed. The spores of herbarium specimens are apt to average smaller than those from deposits. This information is important to one who happens to have dried specimens sent to him for identification. The pileus is so small that the usual tests for viscidity are not reliable.

A closely related species is known from Europe—*M. tubarioides* (Maire) Kühner, Encyc. Myc., 10: 256. 1938. (*Omphalia tubarioides* R. Maire, Bull. Soc. Myc. Fr., 46: 218. 1930.) Kühner also suggests that *Mycena Typhae* is merely a variety of *M. tubarioides*. Since *M. Tubarioides* probably occurs in North America, I have included the following condensed translation of Kühner's description:

Pileus 2.5–5 mm. broad, convex, the margin a bit incurved at first, becoming convex-truncate and even a bit depressed on the disc, finally plane or with the margin elevated, obscurely grooved above the lamellae, clear brownish red, not hygrophanous or only very slightly so, pruinose-pubescent, then more or less glabrescent; flesh thin, thickish on the disc, concolorous, elastic, no odor, taste mild; lamellae 10–11 reach the stipe, one tier of lamellulae, very distant, concolorous with the pileus, the edge white-pruinose, broad, subhorizontal or finally subdecurrent, very broadly adnate, subtriangular, occasionally slightly ventricose, sometimes interveined; stipe 3–6 mm. long, 0.5–0.7 mm. thick, subequal, the base slightly thickened at times, white-villous, not radicating, concolorous with the pileus, the apex becoming purplish, covered over all by a white pruina, then glabrescent in the midportion, finely fistulose.

Spores very elongated and narrow, 11.5–14 $\times$ 3.5–5 μ, rounded at the apex, attenuated at the base, slightly comma-shaped, amyloid; basidia four-spored; cheilocystidia of the clavate-roughened type; pleurocystidia not differentiated.

The fungus occurs on the remains of Scirpus in wet places.

37. Mycena Smithiana Kühner

Encyc. Myc., 10: 252. 1938

Illustrations:

Text fig. 7, nos. 10–11 (p. 91).
Kühner, Encyc. Myc., 10: 253, fig. 80.
? Lange, Flora Agar. Dan., 2, pl. 57 H.

Pileus 1–5 (7) mm. broad, obtuse to convex at first with the margin connivent to the stipe, expanding and broadly convex or with a slightly depressed disc in age, faintly granulose or coarsely pruinose at first, soon glabrous, moist, hygrophanous, "vinaceous fawn" or "fawn color" when moist and with distant conspicuous grayish translucent striae, fading and becoming "light Congo pink" to "pale Congo pink" (dark grayish vinaceous to light grayish vinaceous and fading to pale pink), soon sulcate-striate, margin even or crenate; flesh very thin, membranous, and pliant, odor and taste not distinctive; lamellae distant, moderately broad, adnate or with a decurrent tooth, sordid flesh color at first or white, becoming concolorous with the faded pileus; stipe 3–5 cm. long, less than 1 mm. thick, equal, very weak and fairly pliant, evenly white-pruinose when young, base with scattered hairs, at first somewhat sordid grayish toward the apex, paler below, at maturity concolorous with the pileus.

Spores 8–11 × 4–5 μ, narrowly ellipsoid or abruptly tapered toward the point of attachment, smooth, very weakly amyloid; basidia four-spored; pleurocystidia not differentiated; cheilocystidia very abundant, 22–34 × 6–10 μ, clavate, the enlarged portion covered with obtuse short protuberances, the pedicels subgelatinous (the gill edge fanning out somewhat as in *M. vulgaris* when sections are mounted in KOH, but the gelatinization limited to the pedicels of the cells); gill trama homogeneous, hyaline in iodine, not appreciably gelatinizing or only very slightly so in the subhymenium; pileus trama with a pellicle and a vesiculose tramal body, scattered over the surface of the pellicle and variously arranged are small clavate to subcylindric cells 8–15 × 5–8 μ, the walls of which bear numerous projections, the main portion of the pellicle slightly gelatinous, vesiculose tramal body hyaline to pale brownish in iodine.

Habit, habitat, and distribution.—Scattered to gregarious on debris in wet places, late fall; Michigan and also in Europe. I have found it once. The description is from collection 1147, October 13, 1934, Mud Lake Bog, Cheboygan County, Michigan.

Observations.—As Kühner has pointed out, the fungus is very close to *M. capillaris.* It approaches *M. juncicola* somewhat in stature, color, and tendency of the pellicle to gelatinize. Kühner describes the amyloid reactions of the spores and tissues as weak, and I have been able to verify this observation.

Mycena Lohwagii Singer (*Bei. Bot. Centralbl.*, 46, Abt. 2: 93. 1930) apparently has duller colors than the preceding three, but falls in this general group and should be watched for in North America. The following is a summary of Singer's account:

Very delicate, densely cespitose; pileus white, brownish on the disc, truncate-campanulate, not umbilicate or papillate, glabrous, striate to the disc, 3–6 mm. broad; lamellae white, distant, adnate, or adnate-subdecurrent; stipe pallid yellowish brown, glabrous, smooth, 60 × 0.5 mm.; odorless; spores 7–10 × 4.5–6 μ, ellipsoid; cystidia lacking, gill edge homomorphic (lacking differentiated sterile cells). On fern debris, but most abundant on living ferns. Kühner (1938) gives the spores as weakly amyloid and the cheilocystidia as clavate and with roughened apices.

38. Mycena tenella (Fr.) Quélet

Champ. Jura et Vosges, p. 343. 1872

Agaricus tenellus Fries, Epicr. Syst. Myc., p. 111. 1836.

Illustrations: Plate 9; Text fig. 7, nos. 7–8 (p. 91).

Pileus 3–10 mm. broad, obtusely conic, remaining so in age, glabrous, moist, striate at first, somewhat sulcate in age, grayish to avellaneous or "light pinkish cinnamon" to "vinaceous buff" and fading to nearly white with a faint rosy tint or the disc finally creamy yellowish; flesh thin but cartilaginous and firm; lamellae narrow, close to subdistant, adnate, white or tinged with rose, edge at very first tinged with rose but soon white or concolorous with the faces; stipe 6–10 cm. long, 1 mm. or less thick, glabrous, dark drab gray with a paler apex, rather firm and elastic; spores 8–10 × 5–6 μ, smooth, hyaline, staining faintly bluish gray with iodine in chloral hydrate; cystidia abundant on edge and rare or scattered on the sides of the lamellae, their apices echinulate; basidia four-spored.

Habit, habitat, and distribution.—Cespitose to densely gregarious in low land; Washington and Oregon.

Material studied.—Smith, 1147, 2834, 4456, 17133, 17350, 17529. Kauffman, Mt. Hood, Oregon.

Observations.—This species is close to *M. mirata* Peck, from which it differs in its lighter and usually rosier color and cespitose habit. During October, 1941, it was found very abundantly in the vicinity of Mt. Angeles, Washington, and in most collections the pilei were merely pale gray.

Kühner has placed the fungus which Lange identified as *M. metata* under the name *M. vitrea* var. *tenella*. I have not followed this disposition of Lange's concept; in fact, I have followed Lange in his use of the name *M. metata*. In the abundant material studied of both *M. metata* of Lange and *M. vitrea* sensu Kühner (= *M. atroalboides* Pk.) the two have always been very easy to distinguish, and they did not impress me as being particularly closely related. The Washington material which has here been referred to *M. tenella* might possibly be better placed as a variety of *M. metata* than as a distinct species. However, two factors have influenced me in retaining it as distinct. First, a depauperate form of *M. metata* sensu Lange has been found in the West and, although it resembles *M. tenella* somewhat in stature, it is readily distinguished by its habit and different colors. Second, the consistency of *M. tenella* is more like that of *M. pterigena* than that of the fragile *M. metata*. This, I admit, is a relative character which is difficult to express in such a way that it can be accurately evaluated by an inexperienced collector, but it nevertheless deserves consideration.

Subsection Fuscae

39. Mycena minutissima Murrill

Mycologia, 8: 221. 1916

Prunulus minutissimus Murrill, North Am. Flora, 9: 326. 1926.

Illustrations: Text fig. 7, nos. 9, 12 (p. 91).

"Pileus conic to campanulate, not expanding, slightly umbonate, solitary, 2 mm. broad; surface dry, smooth, glabrous, griseous, murinous on the disk, margin pallid, striate: lamellae adnate, of medium breadth and distance, white: stipe filiform, smooth, glabrous, murinous, 1–2 cm. long, less than 1 mm. thick.

"Type collected among leaves in mixed woods at Unaka Springs, Tennessee, August 18–24, 1904, *W. A. Murrill 909* (herb. N. Y. Bot. Gard.).

"Distribution: Known only from the type locality."

The type has been examined and the following data have been obtained: The spores are smooth, narrowly ellipsoid, amyloid, and measure 5–6.5 × (2.5) 3–3.5 μ. The basidia are four-spored. Pleurocystidia are not differentiated. The cheilocystidia are scattered to abundant, 24–36 × 8–12 μ, broadly or narrowly clavate with the enlarged portion covered with short or elongated, often contorted or branched, protuberances. The gill trama is homogeneous and pale tawny brown in iodine. The pileus trama is characterized by a thin pellicle, the cells of which give off numerous rodlike projections. Beneath this is a fairly well differentiated hypoderm. The remainder is floccose. Both the tramal body and the hypoderm are pale brown in iodine.

I have occasionally collected specimens in New York, Michigan, and Oregon, and Mains has one collection from New Hampshire (4163), all of which appear to belong to this species, but I have never obtained enough fresh material to allow me to arrive at a reliable species concept. The specimens referred to here could perhaps just as well be classified as depauperate forms of the short-stiped form of *M. atroalboides*, in which the spores are slightly smaller than usual. The pilei in my collections measured 3–7 mm. broad. Because of the small spores the species is recognized provisionally. When it is finally collected in quantity, it may be very easy to characterize accurately.

40. Mycena capillaris (Fr.) Quélet

Champ. Jura et Vosges, p. 110. 1872

Agaricus capillaris Fries, Syst. Myc., 1: 160. 1821.
Pseudomycena capillaris Cejp, Publ. Fac. Sci. Univ. Charles, **104**: **141**. 1930.

Illustrations:
- Plate 2 B; Text figs. 7, no. 13 (p. 91); 8, nos. 1–2.
- Fries, Icon. Sel. Hymen., 1, pl. 84, fig. 6.
- Lange, Flora Agar. Dan., 2, pl. 56 B (good).
- Smith, Am. Journ. Bot., 22, pl. 3, fig. 5.

Pileus 2–6 (10) mm. broad, obtusely conic to convex, becoming broadly convex to nearly plane, sometimes remaining campanulate, margin appressed against the stipe at first, surface hoary but soon polished and moist, pale or dark gray fading to whitish, the disc often remaining darker or marked with a hyaline gray watery spot, translucent-striate, becoming sulcate-striate; flesh very thin or membranous, soft and fragile, whitish or grayish, odor and taste not

FIG. 8. *M. capillaris:* 1, cell from surface of cap; 2, cheilocystidia. *M. mirata:* 3, spores; 4, cheilocystidia. *M. supina:* 5, cheilocystidia; 6, spores. *M. oregonensis:* 7, spores; 8, hyphae and terminal cells from cap surface; 9, cheilocystidia; 10, pleurocystidia; 11, caulocystidia

distinctive; lamellae adnate, toothed or nearly free, moderately broad to very narrow, subdistant, 14–18 reach the stipe, one or two tiers of lamellulae, grayish or becoming white, edges even; stipe 3–7 cm. long, filiform, equal, flexuous, very fragile, inserted abruptly on leaves or sparsely strigose—particularly if decumbent—glabrous, apex pruinose at first, upper portion bluish black in young specimens, the color fading quickly and soon entirely pale gray or whitish.

Spores 8–10 × 4–5 μ (11–13 × 5–6 μ in two-spored forms), narrowly ellipsoid, smooth, hyaline; scarcely amyloid; basidia either four-spored or two-spored (both quite common in Michigan); pleurocystidia not differentiated; cheilocystidia fairly abundant, 25–37 × 7–11 μ, clavate to capitate, apex covered with 5–12 or more projections, sometimes the heads variously branched or irregular, hyaline; gill trama of vesiculose cells, homogeneous, vinaceous brown in iodine; pileus trama composed of enlarged cells, those near or at the surface with verrucose walls, vinaceous brown in iodine, pellicle poorly formed, of slender hyphae with obtuse protuberances covering most of their length.

Habit, habitat, and distribution.—Single to gregarious on fallen leaves of oak and beech during late fall. It is often very abundant locally in November. It comes up during warm rainy weather between severe frosts. I have examined specimens from New York, Michigan, Idaho, and Oregon in the United States and from Ontario in Canada. Because of its size it is difficult to make good herbarium specimens.

Material studied.—Smith, 32–664, 1175, 1252, 1270, 18207, October 1, November 12, 19, 1931, October 29, 1932 (two-spored), Michigan.

Observations.—The relatively long, narrow spores are characteristic of both the four-spored and the two-spored forms and are a distinguishing feature of the species. The color of the pileus is gray at first, either light or dark, and soon fades to white or retains a watery-gray disc and striations. I have not observed colors approaching those described for *M. Smithiana* by Kühner. As is frequently true of very small agarics, the thickness and width of the gills, their attachment to the stipe, and their spacing are all tremendously variable and must be used cautiously as taxonomic characters. The large granular or verrucose hyphae near or on the surface of the pileus that Kühner described and illustrated are also present in my material. The hyphae described as forming the poorly differentiated pellicle are very narrow and actually may be only the connecting hyphae which bind the larger cells together. The stipe, typically, is inserted on undecayed fallen leaves and is entirely glabrous. If a group of leaves is matted together, however, and the fruiting body arises within the mat, the portion of the stipe covered by the leaves will very likely be sparsely strigose. The size of the pileus as I have observed it varies greatly. Pilei over 6 mm. broad, however, are rare. The large form I illustrated previously (1935) is seldom found.

41. Mycena mirata (Pk.) Saccardo

Syll. Fung., 5: 290. 1887

Agaricus (*Mycena*) *mirata* Peck, Bull. Buffalo Soc. Nat. Sci., 1: 48. 1873.
Prunulus miratus Murrill, North Am. Flora, 9: 327. 1916.
Mycena supinoides Kühner, Bull. Soc. Linn. Lyon, 10: 124. 1931.

Illustrations:
Plate 8 C; Text fig. 8, nos. 3–4 (p. 111).
Smith, Am. Journ. Bot., 22, pl. 3, fig. 3.

Pileus 2–7 (10) mm. broad, evenly conic with an appressed margin or campanulate, sometimes convex and with or without a small papilla, surface even or slightly rugulose around the papilla, pruinose but soon naked and moist, translucent-striate, becoming sulcate, color dark grayish brown over the disc, margin gray or whitish, fading to drab or pallid in age, the disc remaining darker ("fuscous" on the disc, shading to "hair brown" or "avellaneous" toward the pallid margin); flesh very thin, grayish to pallid, odor and taste not distinctive, usually mild or nearly so; lamellae broad, narrowly adnate, usually toothed and often seceding, subdistant to distant, 9–13 reach the stipe, one to three tiers of lamellulae, white to grayish, edges even and pallid; stipe 2–6 cm. long, filiform, flaccid, flexuous, white or grayish, or at least pallid above, when young covered with a faint bloom, either inserted on the substratum with a characteristic strigose-echinulate base or rooting and white-strigose (the manner of attachment varies with the substratum).

Spores 9–12 × 5–7 μ, subellipsoid to ovoid, faintly amyloid; basidia two-spored, sterigmata 7–8 × 3–4.5 μ, pleurocystidia absent or rare, similar to cheilocystidia; cheilocystidia abundant, hyaline, clavate to globose with a short pedicel, the head studded with numerous short rodlike projections which sometimes become quite elongated and branched; gill trama homogeneous, composed of enlarged cells, vinaceous brown in iodine; pileus trama with a thin pellicle the hyphae of which bear numerous short projections; tramal body vinaceous brown in iodine and composed entirely of greatly enlarged hyphal cells; stipe tissue deep vinaceous brown in iodine.

Habit, habitat, and distribution.—Scattered to gregarious on bases of hardwood trees, old nut husks, pieces of pine bark, and other debris on the forest floor, usually late in the fall during cold weather, not abundant; New York and Michigan.

Material studied.—Smith, 32–490, 32–501, 32–553, 32–599, 32–618, 32–627, 32–631, 32–661, 33–890, 66, 66a, 80, 80b, 1260, 4440, September 18, 1931, Michigan. Atkinson, 13801, 13942. Mains, 32–272.

Observations.—*M. mirata* is one of the species very frequently neglected by the collector because of its small size and the lack of sufficient fruiting bodies to make a worth-while collection. Its cystidia and spores, along with the small size, are its most outstanding characters. The colors vary greatly, depending on the amount of available light. Carpophores in deep shade are nearly always paler than those in open areas. Sometimes the cap is white except for a fuscous central eyespot.

Some difficulty will very likely be experienced in distinguishing *M. capillaris* from *M. mirata.* If the two-spored forms are compared, the shape of the spores and particularly their width will serve to distinguish them at once. Since the four-spored form of *M. mirata* has not been reported, no comparison of it can be made with the four-spored form of *M. capillaris.* I have examined the type of *M. mirata* and found the microscopic characters to be as given in the description. When growing in the open, *M. mirata* has a very strong tendency to develop numerous white-strigose hairs at the base of the stipe; in fact, this is so pronounced that the base often appears decidedly echinulate. As pointed out under *M. capillaris,* that species, when fruiting in the open, is inserted onto the substratum by a naked base.

42. Mycena supina (Fr.) Gillet

Les Hymén., p. 279. 1874

Agaricus supinus Fries, Syst. Myc., 1: 142. 1821.

Illustrations:

Plate 10 B; Text fig. 8, nos. 5–6 (p. 111).
Lange, Flora Agar. Dan., 2, pl. 57 D.
Smith, Am. Journ. Bot., 22, pl. 3, fig. 4.

Pileus 2–7 mm. broad, conic, campanulate or convex, the margin appressed at first and often flaring at maturity, faintly pruinose at first, soon glabrous and moist, translucent-striate when moist, becoming sulcate when faded, surface even or slightly reticulate around the disc, the margin lobed at times, "fuscous" when young but with a hoary sheen, fading slowly to "army brown" or "wood brown" and finally a sordid yellowish gray or merely cinereous in age (dark fuliginous at first, developing a reddish cast as it fades, and finally the reddish giving way to yellow or ash color, generally with sordid colors); flesh cartilaginous and tough, grayish to brownish or pallid in age, sometimes staining reddish, taste not distinctive, odor faintly farinaceous; lamellae subdistant to close (or distant on very small caps), adnate, moderately broad, whitish or grayish, often staining a sordid vinaceous brown in age; stipe 1–3 cm. long, up to 1 mm. thick, terete, cartilaginous, equal, often flexuous and decumbent, cinereous to sordid vinaceous or whitish, the apex paler and pruinose in some, glabrous except for a somewhat strigose rooting base.

Spores 7–9 × 6–7.5 (8) μ or 8–10 × 7.5–8 μ, globose to broadly ellipsoid; basidia four-spored or two-spored; pleurocystidia scattered, similar to cheilocystidia; cheilocystidia 20–36 × 8–12 μ, capitate to clavate with numerous projections over the surface, hyaline; gill trama homogeneous, pale vinaceous brown in iodine; pileus with a thin, scarcely differentiated pellicle, the cells of which are relatively broad (5–8 μ) and have walls which are furnished with numerous short rodlike projections; beneath these hyphae are scattered slender individuals 2–4 μ thick, and below them is a fairly well differentiated rather broad hypoderm made up of cells only slightly broader than those comprising the remainder of the flesh, excretory hyphae present throughout the flesh, all but the pellicle pale vinaceous in iodine; stipe bright reddish in iodine.

Habit, habitat, and distribution.—Scattered to gregarious on debris and bark of hardwoods, usually rather rare and seldom found in any quantity. I have collected it during the late summer and fall in New York and Michigan, and Burke sent specimens from Alabama.

Material studied.—Smith, 32–374, 32–454, 32–463, 32–464, 32–496, 94, 443, 765, 943, 960, 7743, 16631. Burke, Alabama, 1942. Mains, 34–69.

Observations.—Kühner (1938) reports it as common on tree trunks around Paris, where it does not develop a very long stipe. Most of my collections have been from bark lying on or in the soil or from around the lower sides of prostrate trunks. In such habitats the stipe becomes more elongated, and the fruiting bodies are very apt to be decumbent. The distinguishing characters of the fungus are its small size, toughish consistency, somber colors, echinulate cheilocystidia, and nearly globose spores. As for other species, I do not consider the tendency to become reddish or the development of reddish stains to be a primary character.

One collection, Smith, 16632, from a balsam-fir log, near Noisy Creek, Baker Lake, Washington, appeared to be this species but lacked differentiated cheilocystidia. Its spores were subglobose and 6–7 × 5–6 μ in diameter. Inasmuch as there was no other character by which the collection could be distinguished from typical material, it appears best to recognize it as merely an intermediate form between *M. supina* and *M. subsupina.*

Because of the rather tough consistency and the tendency of the stipe to be about 1 mm. broad, *M. supina* could about as well be placed in the section Typicae.

Mycena Quercus-Ilicis Kühner (*Encyc. Myc.*, 10: 260. 1938) is a fungus which should be watched for in North America. It has spores similar to those of *M. supina* but should be readily distinguished by its other characters. A summary of Kühner's description follows:

In groups of fallen leaves of *Quercus Ilex.* Very common in August and September around Rayon. Pilcus 1–3 mm. broad, at first ovoid-hemispheric and concealing the stipe, becoming campanulate or convex, sometimes flattened, sometimes slightly depressed, rarely radially pellucid-striate but often grooved, white or a bit grayish, covered with mealy granules at first and usually remaining more or less powdery around the disc, the margin fringed; lamellae 6–12 reach the stipe, rarely forked, whitish or pallid grayish, always well developed and not plicate, broadly adnate, often subhorizontal, lamellulae lacking, stipe 1–10 mm. long, 0.1–0.25 mm. thick, flexuous, incurved, the base evenly thickened to a clavate small bulb, inserted on the substratum, grayish at first, then nearly hyaline, often slightly yellowish or brownish at the base, pruinose-powdery at first over all or more densely so toward the base.

Spores broadly ellipsoid to subglobose (6.5) 7–8 (8.7) × 5–6 (7) μ, strongly amyloid; basidia four-spored; pleurocystidia absent; cheilocystidia clavate-roughened, 8–16 μ in diameter, apices covered with short rodlike projections. The flesh of the pileus gelatinous under the disc, not gelatinous in contact with the backs of the lamellae, pellicle of rather broad hyphae, from which arise enlarged cells with verrucose walls (much as in *M. tenerrima*).

Kühner has also described var. *citrina*, which is pale sulphur yellow over all in button stages and has spores 8.5–9.5 × 5.5–6 μ. The particles (inflated cells) on the surface of the stipe and pileus are filled with droplets or granules of a bright citron-yellow color animated by Brownian movement.

Subsection Aciculae

43. Mycena oregonensis A. H. Smith

Mycologia, 28: 413. 1936

Mycena siskiyouensis A. H. Smith, Mycologia, 28: 414. 1936.

Illustrations:

Plate 7 B; Text fig. 8, nos. 7–11 (p. 111).
Smith, Mycologia, 28: 412, fig. 1, no. 3.

Pileus 2–10 mm. broad, obtusely conic to convex, often with a small papilla, margin appressed against the stipe when young, often flaring somewhat in age and becoming wavy, the extreme edge frequently broken, striatulate when moist, faintly hoary at first but soon polished, color "capucine yellow" on the disc, "deep chrome" toward the margin, opaque after losing moisture but the colors hardly fading; flesh thin, brittle, yellowish; odor and taste not distinctive; lamellae adnate but developing a rather distinct decurrent tooth in age, distant to subdistant or sometimes appearing close (depending on how well the lamellulae develop), 10–12 reach the stipe and up to three rows of lamellulae may develop, narrow, broadest at point of attachment, "massicot yellow" or appearing whitish, edge "deep chrome" or nearly white in old specimens; stipe 1–3 cm. long, 0.5–0.7 mm. thick, equal, not fragile, evenly covered with a faint yellowish pubescence or pruinosity, inserted on needles, base with scattered yellow hairs, evenly colored and concolorous with or paler than the pileus.

Spores subcylindric, tapered to a point at one end, often slightly curved, nonamyloid, 7–8 × 2.5–3 μ (four-spored), 8–10 × 3.5–4 μ (two spored); basidia 20–23 × 4–5 (6) μ; cheilocystidia abundant, 30–45 × 9–12 μ, subcylindric to fusoid-ventricose, smooth or rarely forked at the apex, outlines often wavy, filled with a bright sulphur-yellow substance; pleurocystidia similar, scattered to rare (often very difficult to locate on sections of revived material); gill trama yellowish in iodine; pileus trama with a thin adnate pellicle, its hyphae giving rise to various types of more or less upright protuberances or even to cells which give the surface its unpolished appearance; beneath the pellicle is a layer of enlarged cells filled with a bright-yellow substance, the central portion of the trama is of floccose hyaline hyphae, the subhymenium containing a few greatly enlarged cells with or without yellow contents, all parts yellowish in iodine; stipe tissue yellowish in iodine.

Habit, habitat, and distribution.—Scattered to gregarious on needles of Douglas fir and spruce; Washington, Oregon, and California. Kauffman reported it on oak leaves and pine needles.

Material studied.—Smith, 3018, 3586, 7805, 7810, 7842, 7865, 8025, 8130, 8240, 8625, 8816, 9279, 16751. Flett, 1940, Bremerton, Washington. Kauffman, Oregon.

Observations.—During the season of 1937 this species was collected in quantity in the Cascade Mountains around McKenzie Pass in Oregon and later in northern California. The protuberances which develop from the hyphae of the pellicle give the cap an unpolished appearance, but in carpophores in which they are only scattered over the surface the pilei are almost shiny. Neither the pileus nor the stipe has ever been found to be even subviscid to the touch (they feel dry when compressed between one's fingers), and no incrusting material has been seen, either on the caulocystidia or on the protuberances from the pellicle, which might lead one to describe these organs as glandular. The pleurocystidia are best demonstrated in fresh material. Their bright-yellow content makes them readily visible. As the fruiting body ages, the yellow fades, and in old caps the pleurocystidia are difficult to find. The cheilocystidia are very abundant and cause the gill edge to be margined with bright yellow. The color of the margin fades as the yellow content of its cystidia fades, and in old pilei the margins may be concolorous with the gill faces.

Mycena siskiyouensis was based on Kauffman's description (1930) of *M. aurantiidisca* Murrill. At the time I described both *M. oregonensis* and *M. siskiyouensis* the differences evident between my collections from Washington and Oregon and Kauffman's description appeared to be too great to be disregarded. In the light of the 1937 collections, however, the discrepancy in regard to the presence of pleurocystidia in one and their absence in the other is readily explained, as is also the reported difference in the color of the gills. Kauffman's description of glandular hairs over the pileus and stipe is more difficult to explain. I can only advance my studies of both Kauffman's and my own specimens as evidence for the conclusions given here. In Kauffman's type the protuberances from the pellicle are better developed than in my 1935 material, but the difference is one of degree. In the specimens collected in 1937 the same range of variation existing between Kauffman's and my earlier specimens was present, often in a single collection. Since no other differences of taxonomic importance were evident, all the material is referred to the one species.

44. Mycena acicula (Fr.) Quélet

Champ. Jura et Vosges, p. 107. 1872

Agaricus acicula Fries, Epicr. Syst. Myc., p. 114. 1838.

Illustrations:

- Plates 8 E, 11 E; Text fig. 9, nos. 1–4 (p. 120).
- Atkinson, Stud. Am. Fungi, fig. 98 (1900 edition).
- Bresadola, Icon. Mycol., 6, pl. 255 (excellent but large).
- Fries, Icon. Sel. Hymen, I, pl. 85, fig. 3.
- Lange, Flora Agar. Dan., 2 pls., 53 D (very good).
- Murrill, Mycologia, 5, pl. 92, fig. 8 (as *P. amabilissimus*).
- Ricken, Die Blätterpilze, 2, pl. 109, fig. 7.

Pileus 3–10 mm. broad, obtuse or convex when young, becoming broadly convex or campanulate, sometimes with a small abrupt umbo, margin appressed against the stipe when young, often flaring or slightly recurved as expansion takes place, a narrow sterile band which frequently becomes lobed or lacerated often forms the extreme margin, surface smooth, faintly translucent-striate when moist, at first hoary-pruinose, soon naked, color "coral red" when young, soon yellowish toward the margin, slowly fading to bright orange yellow, not hygrophanous; flesh thin, brittle, yellow, odor and taste not distinctive; lamellae adnate or slightly rounded next to the stipe, close to subdistant, 10–14 reach the stipe, two or three tiers of lamellulae, moderately broad, pale orange to whitish, often yellowish at the base and whitish along the edges; stipe 1–6 cm. long, filiform or up to 1 mm. thick, flexuous, brittle, base strigose with white hairs, surface densely white-pruinose at first but soon naked and orange-yellow or lemon yellow over all.

Spores subfusoid, 9–11 × 3.5–4.5 μ, nonamyloid; basidia four-spored, 20–22 × 5–6 μ; cheilocystidia and pleurocystidia similar,

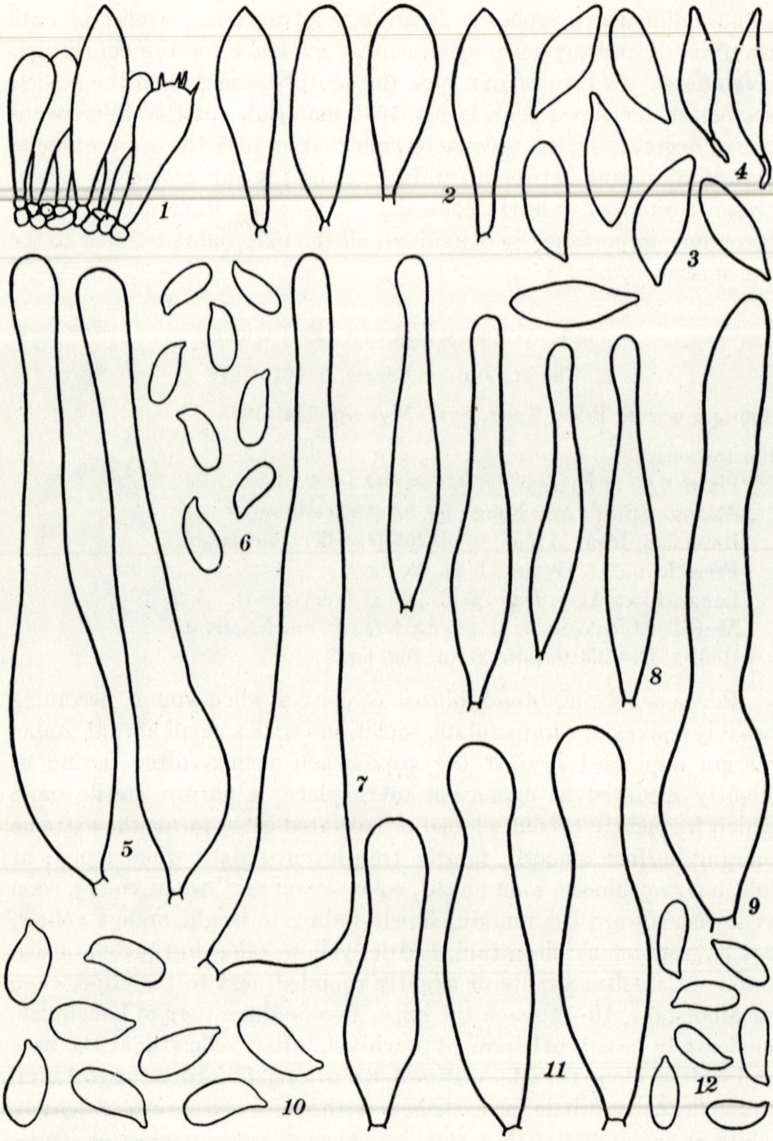

FIG. 9. *M. acicula:* 1, pleurocystidia and basidia; 2, cheilocystidia; 3, spores; 4, cells from surface of pileus. *M. fibula:* 5, pilocystidia; 6, spores; 7, caulocystidia; 8, pleurocystidia and cheilocystidia; 9, caulocystidia. *M. pseudogrisella:* 10, spores. *M. Swartzii:* 11, pleurocystidia and cheilocystidia, 12, spores

clavate to subfusoid or obovoid, apices often covered with a resinous secretion (in mounts revived in KOH), 25–32 × 6–9 μ, very inconspicuous on the gill edge and projecting only slightly from the faces; gill trama yellowish in iodine, pileus trama with a thin differentiated pellicle from which small scattered fusoid cystidia project, the region beneath the pellicle of slightly inflated cells, the remainder filamentous, yellowish in iodine; stipe tissue yellowish in iodine.

Habit, habitat, and distribution.—Single, gregarious or subcespitose on debris in wet places, particularly along streams or the borders of swamps. It is widely distributed throughout the eastern United States and Canada and occurs in Washington, Oregon, and California along the Pacific coast. I find it most frequently during the spring and early summer or during exceptionally wet weather in the fall.

Material studied.—Smith, 32–64, 33–22, 33–147, 33–209, 33–334, 33–490, 1312, 1373, 3969, 6308, 6336, 8324, 8472, 13445, 13702, 13884, 14139, 14176, 14240, 14584, 16416, 17557. Atkinson, 14666, 19851, 24079. Burke, 1942, Alabama. Hesler, 12439. Mains, 5063. Jackson, 1936, Ontario, Canada. Josserand, 1937, France. Kauffman, collections from North Carolina, Michigan, Washington, and Oregon.

Observations.—Davidson's report (1930) of this species from British Columbia is doubtful. She more than likely found *M. oregonensis*, a species of similar stature, but one which does occur in quantity on needle carpets. *M. acicula*, however, probably occurs in British Columbia. The brilliant colors, the nonmarginate gills, subfusoid rather large spores, slender stature, and cystidia distinguish it from other members of the genus. It is quite likely to be mistaken for a Hygrophorus by many.

45. Mycena fibula (Fr.) Kühner

Encyc. Myc., 10: 607. 1938

Agaricus (Omphalia) fibula Fries, Syst. Myc., 1: 163. 1821.
Omphalia fibula Quélet, Champ. Jura et Vosges, p. 101. 1872.
Omphalopsis fibula Murrill, North Am. Flora, 9: 314. 1916.

Illustrations:

Text fig. 9, nos. 5–9.
Bresadola, Icon. Mycol., 6, pl. 275, fig. 1 (as Omphalia).
Hard, The Mushroom, Edible and Otherwise, fig. 99.
Konrad et Maublanc, Icon. Sel. Fung., 3, pl. 235-I (as Omphalia).
Lange, Flora Agar. Dan., 2, pl. 61 G (as Omphalia).

Murrill, Mycologia, 10, pl. 8, fig. 6.
Thomas, Field Book of Common Gilled Mushrooms, pl. 12, fig. 94.

Pileus (3) 8–12 (15) mm. broad, slightly convex, flat or the disc faintly depressed at first, the margin straight or curved in slightly, in age the disc deeply depressed but the margin remaining decurved, surface moist when young and fresh but soon changing to sordid ochraceous or buff ("cinnamon buff" to "clay color"); flesh very thin, fragile, pale orange to buff, odor and taste not distinctive; lamellae subdistant to close, 17–20 reach the stipe, one or two tiers of lamellulae, long-decurrent, narrowed at the extremities but broad in the notch between pileus and stipe, white or whitish, often intervenose; stipe 1–4 cm. long, 1–1.5 (2) mm. thick, equal, fragile, hollow, finely pubescent over all from cystidia but glabrescent in age, concolorous with the pileus and fading in the same manner.

Spores 4.5–6 × 2.5 μ, narrowly ellipsoid, smooth, nonamyloid; basidia four-spored; pleurocystidia and pilocystidia scattered to abundant, 36–56 × 6–10 μ, subcyclindric to subfusoid, the apex tending to be more rounded than pointed, at times subcapitate, hyaline, thin-walled; gill trama homogeneous, yellowish in iodine; pileus trama homogeneous, yellowish in iodine, the surface covered with elongated pilocystidia 60–90 × 10–14 μ; caulocystidia abundant near the apex of the stipe, similar in size and shape to the pilocystidia.

Habit, habitat, and distribution.—Single, scattered, or gregarious on beds of moss throughout North America. Very common. I have examined material from North Carolina, Tennessee, Maryland, New York, Michigan, Idaho, Washington, Oregon, and California in the United States and from Nova Scotia and Ontario in Canada. It can be found during the spring, summer, and fall. Exceptionally fine specimens were collected in the vicinity of Mt. Baker, Washington, in 1941 (Smith, 16771).

Material studied.—Smith, 32–23, 33–607, 567, 3067, 4030, 6456, 6886, 13000, 13469, 16771, October 10, 1931, Michigan; May 28, 1932. Gruber, P-16. Hesler, 6509. Josserand, France. Kauffman, ten collections, North Carolina, New York, and Michigan. Kelly, one collection, Maryland. Mains, 32–146, 34–36. Rice, 5716. Wehmeyer, 567.

Observations.—The colors of the pileus vary greatly in intensity. Fresh vigorously developing pilei may be bright orange or ochraceous orange, but all intergradations to avellaneous occur. In some collections dull, pale-brown colors appear to be present from the first. I have not attempted to segregate any color forms.

46. Mycena Swartzii (Fr.), comb. nov.

Agaricus fibula var. *Swartzii* Fries, Syst. Myc., 1: 164. 1821.
Agaricus setipes var. *acrocyanea* Fries, Icon. Sel. Hymen, 1, pl. 75, fig. 4. 1873.
Omphalia Swartzi Quélet, Bull. Soc. Myc. Fr., 1: 27. 1885.
Omphalina fibula var. *Swartzii* Quélet, Enchir. Fung., p. 46. 1886.
Mycena fibula var. *Swartzii* Kühner, Encyc. Myc., 10: 608. 1938.

Illustrations:
Text fig. 9, nos. 11–12 (p. 120).
Bresadola, Icon. Mycol., 6, pl. 274 (as Omphalia).
Fries, *loc. cit.*
Lange, Flora Agar. Dan., 2, pl. 61 D (as Omphalia).

Pileus 6–12 mm. broad, flat when young, the margin decurved, at maturity the disc slightly depressed and the margin plane or slightly raised, surface moist and at first densely pruinose from cystidia, nearly glabrous in age, color "deep slaty brown" on disc and "vinaceous cinnamon" on the margin when young, the margin "avellaneous" at maturity and the disc "deep brownish drab" (disc violet brown to violet gray, the margin more or less cinnamon tinged with pink and fading to avellaneous); flesh thin and readily splitting radially; lamellae close, 17–20 reach the stipe, one or two tiers of lamellulae, long-decurrent, narrow at the extremities, but broad in the notch between pileus and stipe, whitish, densely pruinose under a lens; stipe 4–7 cm. long, 1 mm. thick, equal, cartilaginous, densely pruinose over all, "deep slaty brown" at apex, "pinkish cinnamon" over the lower half, finally fading to sordid violaceous gray.

Spores 4–5 $\times$ 2–2.5 μ, narrowly ellipsoid, smooth, nonamyloid; basidia four-spored; pleurocystidia and cheilocystidia scattered to abundant, hyaline, thin-walled, ventricose-subcapitate, 42–66 $\times$ 10–18 μ; gill trama homogeneous and yellowish in iodine; pileus trama homogeneous, the surface covered with numerous pilocystidia, 50–90 $\times$ 10–18 μ, cylindric above an inflated base or fusoid-ventricose with the neck greatly elongated; caulocystidia similar to pilocystidia and very abundant.

Habit, habitat, and distribution.—Usually it occurs scattered to gregarious in groups of six to a dozen carpophores. It fruits in the spring, summer, or fall, depending on the weather. On beds of moss or occasionally on debris; North Carolina, New York, Michigan, Washington, and California in the United States and Nova Scotia and Ontario in Canada. It is widely distributed throughout North America, but is seldom found in large quantities.

Material studied.—Smith, 566, 1477, 6385, 8413, 14207. Kauffman, North Carolina and Michigan. Oches, Michigan. Krieger, Ontario.

Observations.—The difference in color between *M. Swartzii* and *M. fibula* is very striking and, at least so far as the material examined is concerned, there is a distinct difference in the shape of the pleurocystidia, as is shown in the drawings. Because of these characters *M. Swartzii* has been given the rank of a species rather than of a variety.

47. **Mycena pseudogrisella,** sp. nov.

Illustrations: Text figs. 9, no. 10 (p. 120); 10, nos. 1–2.

Pileus 3–8 mm. latus, convexus vel subdepressus, hygrophanus, avellaneus; lamellae subdistantes, albidae, breviter decurrentes, ventricosae; stipes 10–18 mm. longus, 1.5 mm. crassus, fragilis, aequalis, pruinosus, avellaneus; sporae 7–8 × 3–3.5 μ, nonamyloideae; pleurocystidia et cheilocystidia (36) 50–80 × 8–15 (20) μ, subclavata. Specimen typicum in Herb. Univ. Mich. conservatum. Legerunt E. B. Mains et A. H. Smith, n. 33–690, prope Ontonagon, Mich., Sept. 14, 1933.

Pileus 3–8 mm. broad, convex or the disc slightly flattened, in age very slightly depressed, margin straight when young, becoming crenate in some at maturity, hygrophanous but unpolished, margin becoming sulcate-striate, surface pruinose when viewed under a lens, color pale avellaneous, whitish along the margin; lamellae subdistant, white, short-decurrent, often ventricose near the cap margin and narrow in the notch (the opposite of the triangular type), edges even; stipe 10–18 mm. long, 1.5 mm. thick, watery and fragile, equal, densely pruinose at first but becoming glabrous, concolorous with the pileus or more sordid.

Spores 7–8 × 3–3.5 μ, narrowly ellipsoid, smooth, nonamyloid; basidia 25–27 × 4–5 μ, four-spored; pleurocystidia and cheilocystidia abundant and similar (36) 50–80 × 8–15 (20) μ, subcapitate and ventricose to subclavate, obtuse, smooth, thin-walled, hyaline; gill trama yellowish in iodine, parallel to subparallel, of long-cylindric cells; pileus trama homogeneous beneath a turflike covering of pilocystidia similar to the cheilocystidia or more elongated (not a compact palisade), yellowish in iodine; stipe covered by caulocystidia at least near the apex (similar to the cheilocystidia).

FIG. 10. *M. pseudogrisella:* 1, pleurocystidia; 2, caulocystidia. *M. subsupina:* 3, spores; 5, cheilocystidia. *M. subcucullata:* 4, spores of two-spored basidia; 6, spores of four-spored basidia; 7, cheilocystidia; 8 and 9, pleurocystidia

Habit, habitat, and distribution.—Scattered on a wet bank among thalli of liverworts; September 14, 1933, Porcupine Mountains, Michigan (near Ontonagon). Known only from the type collection.

Material studied.—E. B. Mains & A. H. Smith, 33–960.

Observations.—Although known from only the one collection, the cystidia over the gills, pileus, and stipe make the species very easy to recognize in either Omphalia or Mycena. By virtue of its non-amyloid spores, homogeneous pileus, and cystidia it is placed in the same group with *M. fibula*. Its colors are very close to those of *O. grisella* as that species is illustrated by Lange on Plate 60 of the *Flora Agaricina Danica*, Vol. 2.

Subsection Cinereae

48. Mycena subcucullata A. H. Smith

Mycologia, 31: 279. 1939

Illustrations:

Text fig. 10, nos. 4, 6–9 (p. 125).

Smith, Mycologia, 31, fig. 1 O (spores).

Pileus 1–6 mm. broad, conic, campanulate or expanded, and with or without a conic umbo, frequently more or less cucullate, margin appressed against the stipe at first, prominently striate when moist and often splitting readily in age, colors fuscous on the disc or dark gray, margin paler and finally whitish, disc also fading and often sordid ochraceous in age, sometimes white except for a sordid yellowish umbo; flesh thin, grayish but becoming white, rather cartilaginous, odor and taste not distinctive; lamellae adnexed, subdistant to distant, broad, thickish, white, edges even and concolorous; stipe 10–20 mm. long, 0.5 mm. thick, filiform but rigid, white to pale gray, sordid below, rooting in mosses.

Spores broadly ovoid to subglobose, pointed at one end, 6–7 (8) × 5–6 μ, amyloid (the reaction very inconclusive); basidia four-spored, 15–17 × 7–8 μ; cheilocystidia smooth, saccate when young, subcylindric in age or remaining saccate, 25–30 × 9–14 μ; pleurocystidia scattered to rare, smooth, hyaline, saccate to cylindric, 26–34 × 8–12 μ; gill trama yellowish in iodine; pileus trama homogeneous beneath a thin adnate pellicle, the hyphal cells somewhat enlarged, yellowish in iodine; stipe tissue yellowish in iodine.

Habit, habitat, and distribution.—Gregarious on mossy logs during September; New York, Michigan, and Washington in the United

States and Manitoba in Canada. A two-spored form has been found in Washington. The species does not appear to be rare, but it is difficult to find sufficient specimens to justify a critical study.

Material studied.—Smith, 491, 779, 784, 913, 14170, 16954. Bisby, Winnipeg, Manitoba. Mains, 32–288, 34–150.

Observations.—The two-spored form was collected in the Olympic Mountains of Washington. The spores measure 8–10 × 4.5–5 μ, and no positive amyloid reaction was obtained when they were tested with iodine. The gills appeared to be fairly close, and 12–14 reached the stipe; they were 1.5–2 mm. broad.

The consistency is relatively tough for such a small species. The spores either remain hyaline or appear very faintly grayish in iodine. The latter reaction may be caused by the surrounding medium in the mounts. In view of the nonamyloid reaction of all other parts of the fruiting body, this character needs further study, preferably on spore prints which have been dried for some time.

49. Mycena subsupina A. H. Smith

Mycologia, 29: 340. 1937

Illustrations: Plate 10 D; Text fig. 10, nos. 3, 5 (p. 125).

Pileus 3–7 mm. broad, narrowly to broadly conic, not expanding, margin appressed against the stipe when young and flaring somewhat in age, surface covered with a delicate bloom but soon naked and polished, moist, hardly striate, becoming slightly sulcate in age, color near "mummy brown" when young, disc sometimes darker, "buffy brown" at maturity or with an ochraceous tint, margin pale avellaneous to whitish in age, sometimes pallid gray or watery grayish brown over all in age, hardly hygrophanous, fading slowly to sordid ashy brown; flesh thin, brownish, rather tough, odor and taste mild; lamellae ascending-adnate or with a tooth, distant, 10–13 reach the stipe, narrow to moderately broad (1 mm.), white to pale grayish, edges whitish; stipe 1–2 (3) cm. long, 1 mm. thick, equal, decumbent, rather tough and cartilaginous with a rather large hollow, glabrous and polished except for the white-strigose base, concolorous with the pileus, base not rooting.

Spores broadly ellipsoid, (8) 9–11 (12) × 5.5–7 (7.5) μ, strongly amyloid; basidia two- and four-spored; 18–20 × 7–8 μ; pleurocystidia rare to absent, if present similar to cheilocystidia; cheilocystidia abundant, fusoid-ventricose, the apices acute or obtuse and in some branched into two to several fingerlike prolongations above the inflated basal portion, 36–50 (80) × 9–14 μ, gill trama pale yellowish in iodine; pileus trama with a thin adnate pellicle, beneath it a region of slightly inflated cells, the remainder filamentous and floccose, all parts pale yellowish to yellowish brown in iodine.

Habit, habitat, and distribution.—More or less decumbent and densely gregarious on mossy logs, summer and fall. It was found on redwood logs in California and is known also from Tennessee, New York, Alabama, and Michigan, where it was collected on old logs of both hardwoods and conifers.

Material studied.—Smith, 33–375, 33–688, 33–719, 3782. Burke, Alabama, 1942. Hesler, 10154, Tennessee; no number, 1934.

Observations.—*M. supina* has the consistency and stature of *M. subsupina*, but the latter's cystidia distinguish it at once. In addition, the spores are ellipsoid rather than globose. Those of *M. supina* are globose to subglobose and vary to ellipsoid on two-spored forms. The spores of two-spored forms of *M. subsupina* vary to globose or subglobose. Their bluish reaction in iodine is very distinct.

Mycena subcucullata is readily distinguished from *M. subsupina* by its smaller cheilocystidia, thinner stipe, and weak amyloid reaction (if there is any) of its spores. There is also a distinct difference in spore size if only the four-spored forms of both are compared.

I have collected a number of specimens in Michigan and New York which seem to belong here but are characterized by rather abundant pleurocystidia. In some of these the pileus was quite rugose, but the character did not appear to be constant. Their basidia were mostly two-spored (33–719, 33–688 from Michigan), and the spores measured 9–12 × 6–7.5 μ. In 782, from Warrensburg, New York, the spores measured 7–9 × 4–5 (6) μ; they were produced on four-spored basidia. These forms need further study.

50. Mycena corticaticeps Kauffman & Smith

Pap. Mich. Acad. Sci., Arts and Letters, 17: 180. 1933

Illustrations: Text fig. 11, nos. 1–2.

Pileus 2–4 mm. broad, sharply conic to campanulate, broadly expanded and with a subpapillate umbo at maturity, margin appressed against the stipe when young, flaring in age, surface glabrous, moist, translucent-striate, becoming opaque, color grayish vinaceous

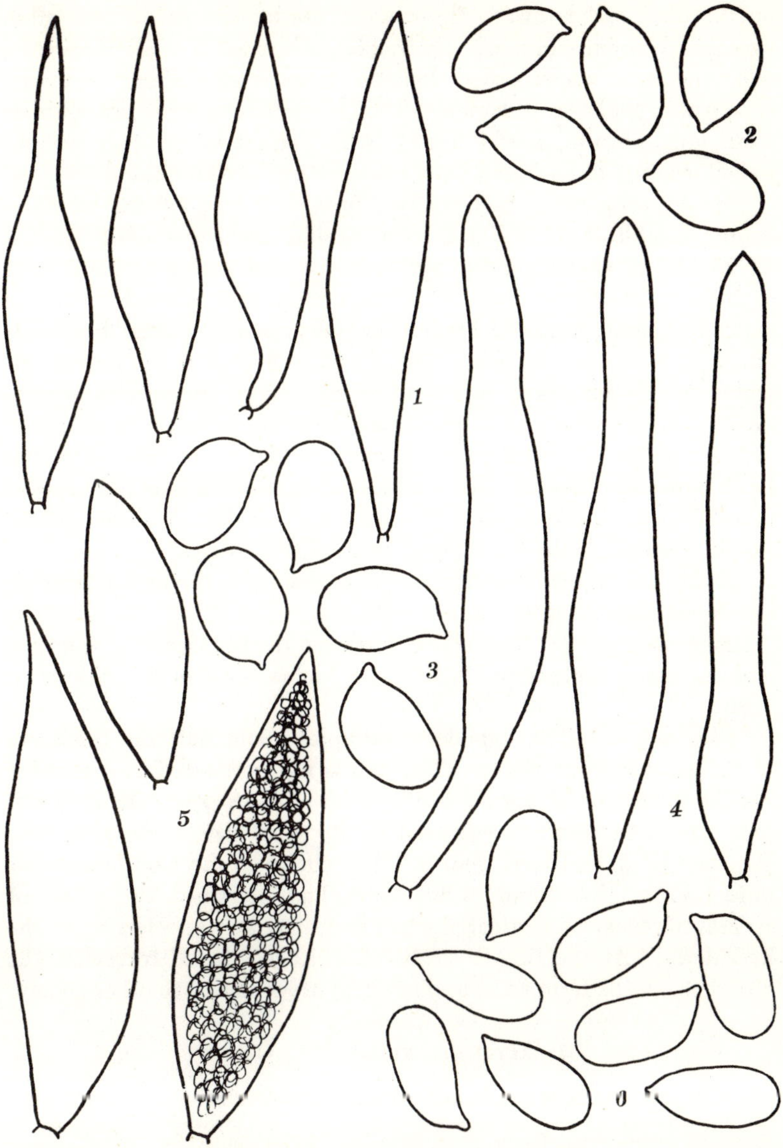

FIG. 11. *M. corticaticeps:* 1, pleurocystidia; 2, spores. *M. erubescens:* 3, spores; 5, pleurocystidia. *M. galopus:* 4, pleurocystidia; 6, spores

on the disc, pallid toward the margin, sordid and pallid over all in age; flesh thin but not markedly fragile (as is usual in small Mycenae), odor and taste not recorded; lamellae adnate, subdistant, very narrow, white, pruinose under a lens from projecting cystidia; stipe about 2 cm. long, filiform, flaccid, hyaline white, equal, terete, glabrous and naked, rooting in bark and debris, with strigose white hairs.

Spores subglobose to broadly ellipsoid, 7–9 × 4.5–5 μ on four-spored basidia, 9–11 × 7–9 μ when basidia are two-spored; basidia 23–25 × 7 μ; cheilocystidia and pleurocystidia similar and very abundant, broadly ventricose in the middle and tapered to a long aciculate point, smooth, hyaline or very faintly tinged lavender, 35–50 × 6–12 μ on the edge, (50) 60–80 × 9–15 μ on the faces; gill trama faintly vinaceous brown in iodine; pileus trama with a well-differentiated pellicle of very narrow hyphae, beneath this a well-defined region of pseudoparenchymatous tissue the cells of which have brownish contents, the remainder of narrow filamentous hyphae, in iodine pale vinaceous brown beneath the pellicle. Atkinson, 19802 (as *M. speirea*), is the four-spored form of this species.

Habit, habitat, and distribution.—Gregarious on the bark of rotting hardwood logs; Michigan and New York.

Material studied.—Two collections from the type locality, the type and one made September 11, 1932, by Mains and Smith. Atkinson, 19802.

Observations.—The hypoderm occupies about half the thickness of the trama of the pileus. The cap is not corticated in a manner similar to that of *M. rorida* or of species of Conocybe. *M. corticaticeps* should be readily recognized by its small size, large, abundant pleurocystidia, and large spores. The choice of the name was a bit unfortunate. Kauffman in his notes had described the pileus of several Mycenae as corticated when in reality he was referring to the hypoderm. At the time I studied his notes I failed to realize the difference in the term as he applied it to Mycena and to other genera.

51. Mycena debilis (Fr.) Quélet

Champ. Jura et Vosges, p. 107. 1872

Agaricus debilis Fries, Epicr. Syst. Myc., p. 112. 1838.

Illustrations:

Text fig. 21, nos. 4–5 (p. 201).

Bresadola, Icon. Mycol., 5, pl. 249, fig. 1.
Fries, Icon. Sel. Hymen., 1, pl. 82, fig. 4.

Pileus 4–8 mm. broad, obtusely conic to convex, soon broadly conic to campanulate, the margin straight at first but sometimes flared in age, surface glabrous and moist, translucent-striate to the disc, pale vinaceous brown, sordid brown, or grayish brown, paler grayish or sordid brown when faded, becoming sulcate; flesh very thin and delicate, odor and taste not distinctive; lamellae adnate, subdistant, moderately broad, whitish or tinged grayish, edges pallid or very faintly brownish; stipe 1.5–4 cm. long, 0.5–0.75 mm. thick, equal, tubular, very weak, glabrous or a few white fibrils around the base, concolorous with the pileus.

Spores 9–14 × 5–7 (8) μ, ovoid to ellipsoid, amyloid, smooth; basidia two-, three-, or four-spored; pleurocystidia not differentiated; cheilocystidia 34–68 × 9–15 μ, abundant, fusoid-ventricose to clavate, smooth, hyaline; gill trama with a thin pellicle, the remainder of vesiculose cells, vinaceous brown in iodine.

Habit, habitat, and distribution.—Scattered on sod in pastures or in open stands of conifers during the summer and fall; Michigan, Washington, and California. I have found it only after periods of heavy rain.

Material studied.—Smith, 33-596, 3273, 3753, 8493, 13996, 15984.

Observations.—The specimens of this species are very similar in stature to *M. sanguinolenta* as it occurs in sphagnum bogs in Michigan. As yet I have not found *M. debilis* in great quantity, and have not been able to answer certain questions in regard to some of its characters. In the specimens seen the spore size and shape were extremely variable. This was caused, no doubt, by the different number of spores borne on the basidia. It is quite apparent, however, that spores from four-spored forms will be found to measure 8–11 μ long. The colors of the pileus shade over into vinaceous, and cause the fruiting bodies to resemble closely those of *M. capillaripes*. The situation is further complicated by a form of *M. capillaripes* with large spores on two-spored basidia. Its gill edges are usually colored, but at times this coloring is very inconspicuous. Such forms differ from *M. debilis* in the presence of pleurocystidia and in having a sharp acidulous odor emitted by the crushed flesh. Observations on more abundant material should be made to determine whether pleurocystidia are ever present in *M. debilis* and whether forms developing vinaceous tints ever have faintly margined gills. Kauffman's report (1921) of this species is based on a slender form of *M. pectinata*. Kühner (1938) has placed this fungus in his Ciliatae–Calodontes alongside *M. capillaripes*, an arrangement which certainly appears to be a natural one.

SECTION LACTIPEDES

The species of the section Lactipedes are characterized by the presence of a milky or colored latex in the stipe and flesh of the pileus. The section is artificial, as is to be expected in view of the fact that it is based on a single character. It contains gray or brightly colored species. Among the gray species one is likely to encounter some difficulty in placing collections in the right section because the latex may be scanty and easily overlooked. Such forms, if that mistake is made, will key out in either the Deminutivae or the Typicae. In reality this section is not sharply separated from the Typicae. Many species in the latter have the lactifers in the flesh of the stipe and pileus but have no latex. Following Kühner, I have segregated a few species which have a copious watery juice as a section, Hydropus, of the Typicae. Singer recognizes Hydropus as a distinct genus.

KEY TO SPECIES

1. Latex blue .... *M. caerulea* Vog. (Eur.)
1. Latex white (milklike, often scanty) 2
1. Latex reddish to orange 5

2. Pleurocystidia absent .... 53. *M. cayugaensis*
2. Pleurocystidia present and conspicuous 3

3. Growing on petioles of palmetto palm, southern United States *M. Sabali*
3. Not with habitat given above 4

4. Taste mild; spores 6–7 × 4–5 μ .... 55. *M. copiosa*
4. Taste mild; spores 9–12 × 5–7 μ .... 52. *M. galopus*
4. Taste very bitter and lasting; spores 9–12 × 6–8 μ 54. *M. erubescens*

5. Cheilocystidia clavate-echinulate 6
5. Cheilocystidia fusoid-ventricose 7

6. Cuticle of pileus not hymeniform .... 56. *M. crocata*
6. Cuticle of pileus composed of a palisade of clavate to pear-shaped cells 57. *M. anomala*

7. Gill edges colored reddish or reddish brown 8
7. Gill edges not differently colored .... 58. *M. haematopus*

8. Stipe glabrous to densely pruinose but not fibrillose at apex 9
8. Stipe surface fibrillose to slightly scurfy because of dark reddish-brown fibrils .... 60. *M. Atkinsoniana*

9. Pileus with crenate margin .... 59. *M. haematopus* var. *marginata*
9. Margin of pileus not crenate 10

10. Spores 7–8.5 μ long; pleurocystidia absent ... 61. *M. subsanguinolenta*
10. Spores 8–10 (11) μ long; pleurocystidia present ..... *M. sanguinolenta*

52. Mycena galopus (Fr.) Quélet

Champ. Jura et Vosges, p. 107. 1872

Agaricus galopus Fries, Syst. Myc., 1: 148. 1821.
Mycena lactescens Schroeter, Pilze Schles., p. 632. 1889.
Agaricus leucogalus Cooke, Grevillea, 12: 41. 1883.
Agaricus lacticularius Britz., Revision der Diagnosen zu dem von M. Britzelmayr aufgestellten Hymenomyceten-Arten. Bot. Centralbl., 63: 19. 1898.

Illustrations:
Plate 10 A; Text fig. 11, nos. 4, 6 (p. 129).
Beardslee and Coker, Journ. Elisha Mitchell Sci. Soc., p. 40, pl. 9, lower fig.
Bresadola, Icon. Mycol., 5, pl. 248, fig. 1.
Konrad et Maublanc, Icon. Sel. Fung., 3, pl. 225.
Lange, Flora Agar. Dan., 2, pl. 50 E and pl. 51, figs. G, G[1].
Ricken, Die Blätterpilze, 2, pl. 109, fig. 11.

Pileus 5–25 mm. broad, ovoid when young, becoming conic or conic-campanulate, in age often with a recurved margin and a prominent umbo, surface with a hoary sheen at first, soon becoming naked and glabrous, margin entire and appressed against the stipe at first, translucent-striate when moist, somewhat sulcate when faded, color "fuscous black" over all except the whitish margin, fading to pale gray, the umbo remaining blackish or becoming dark gray, sometimes very pale ashy gray over all when moist, opaque and cinereous after losing moisture; flesh thin, soft, fragile, odor and taste not distinctive; lamellae subdistant, narrow, ascending-adnate, whitish to gray, usually darker in age, edges pallid or grayish; stipe 4–8 (12) cm. long, 1–2 mm. thick, equal, strict or flexuous, glabrous, fragile, dark blackish brown below or merely dark cinereous, apex pallid, base white-strigose, when broken exuding a white milklike juice.

Spores 9–13 × 5–6.5 μ, smooth, ellipsoid, occasionally somewhat pear-shaped, very weakly amyloid or at times apparently not at all amyloid; basidia four-spored; pleurocystidia and cheilocystidia similar and very abundant, 70–90 (110) × 9–15 μ, narrowly fusoid-ventricose and usually with abruptly pointed apices, sometimes forked or branched near the apex, hyaline, smooth; gill trama homogeneous, dark vinaceous brown in iodine; pileus trama with a thin but clearly differentiated pellicle, a well-developed hypoderm, and the remainder filamentous, all but the pellicle vinaceous brown in iodine.

Habit, habitat, and distribution.—Gregarious to scattered on humus under hardwoods or conifers. In the United States it is very abundant along the Pacific coast from Washington to California, and also in Tennessee and North Carolina. Kauffman failed to find it in Michigan, and as yet my efforts to discover it there have been equally unsuccessful.

Material studied.—Smith, 3151, 3209, 3215, 3335, 3454, 3471, 3668, 3677, 3724, 3737, 4475, 7541, 7795, 7942, 8012, 8207, 8370, 8522, 9027, 9113, 9149, 9177, 10855 16640, 16711. Atkinson, 1620, 20869, 24744, 15197 (as *M. leucogala*). Hesler, 6511, 7354, 9620, 12323, 12324, 14478. Overholts, Tennessee.

Observations.—*M. galopus* looks like a very ordinary slender gray or blackish Mycena and, if one does not take the trouble to ascertain the presence of a milklike fluid at the time the specimens are collected, he may never be able to identify the collection correctly. The latex is often scanty and not readily observed. The most accurate way to demonstrate its presence is to break the fruiting body from its point of attachment and carefully observe the broken base for a minute or two. If a milklike juice is present, at least a drop will exude slowly from the broken base. In robust specimens the latex will usually issue readily from any cut or broken portion of the pileus or stipe. In the laboratory one can usually identify dried material by the long pleurocystidia, relatively large spores, and strong reaction of the tissues of the gills and pileus to iodine. In California, where the species has been observed to be most abundant, variations from black to almost grayish white were seen and, at least in that region, there appears to be no justification for separating out forms or varieties on the basis of either size or color. I have not as yet seen any pure-white carpophores.

53. **Mycena cayugaensis,** sp. nov.

Pileus 5–10 mm. latus, saepe 10 mm. altus, campanulatus vel convexus, striatus, griseus; lamellae angustae, subdistantes, adnatae, pallidae; stipes 4–6 cm. longus, 1.5–2 mm. crassus, laevis, cavus, concolor vel pallidus; sucus albus, non copiosus; sporae 9.5–11 × 5–6 μ, ellipsoideae; cheilocystidia subventricosa, 40–50 × 10–12 μ. Specimen typicum in Atk. Herb., Cornell Univ., conservatum. Legit G. F. Atkinson, n. 15131, prope Ithaca, N. Y., June 24, 1903.

Pileus 5–10 mm. broad and nearly as high, campanulate to convex, thin, striate toward the margin, gray; lamellae narrow, subdistant, adnate, whitish; stipe 4–6 cm. long, 1.5–2 mm. thick, smooth, hollow, when broken exuding a whitish juice, concolorous with the pileus or paler above, base with spreading mycelium.

Spores 9.5–11 × 5–6 μ, ellipsoid; basidia 25–30 × 9–12 μ, four-spored; pleurocystidia not differentiated; cheilocystidia subventricose at the base, 40–50 × 10–12 μ, apices acute.

Habit, habitat, and distribution.—Gregarious to scattered on leaf mold. Known only from the region around Ithaca, New York, collected by Atkinson and Kauffman.

Observations.—The description has been compiled from Atkinson's notes on collection 15131 and from studies of its microscopic characters. Unfortunately, I did not obtain iodine reactions at the time I studied the specimens. The distinguishing characters are the lack of pleurocystidia and the presence of the white latex in the stipe. Atkinson had studied two collections (15131 and 15152) and had tentatively placed them in *M. galopus*. Since this species is universally interpreted as having large, conspicuous pleurocystidia and since the abundant material which I have examined has been constant in this character, I believe I am justified in describing the Ithaca collections as a distinct species.

54. Mycena erubescens von Höhnel

Fragm. zur Myk., XV, 794: 267. 1913

Mycena fellea Lange, Dansk Bot. Arkiv, 1 (5): 26. 1914.
Mycena cholea A. H. Smith, Mycologia, 26: 306. 1934.

Illustrations:

Plates 10 C, 12; Text fig. 11, nos. 3, 5 (p. 129).

Beardslee and Coker, Journ. Elisha Mitchell Sci. Soc., 40, pl. 26 (upper right as *M. fellea*).

Lange, Flora Agar. Dan., 2, pl. 50 B (as *M. fellea*).
Smith, Am. Journ. Bot., 22, pl. 4, fig. 2 (as *M. fellea*); fig. 3 (as *M. cholea*).

Pileus 2–12 (15) mm. broad, ovoid at first, becoming campanulate or obtusely conic, sometimes papillate and convex, surface smooth, moist to lubricous and polished, sometimes subviscid, color variable, "fuscous" to "clove brown" except for the abruptly whitish margin, fading through reddish browns to drab ("Mars brown," "wood brown," "warm buff," or yellowish casts often predominating), subhygrophanous and finally fading to grayish or to very sordid grayish brown, often appearing canescent and silvery in faded condition, opaque or translucent-striate when moist, sometimes sulcate when faded; flesh thin, rather pliant and cartilaginous, whitish, usually staining reddish when cut or bruised, no odor, taste distinctly bitter —like that of quinine; lamellae subcrowded to subdistant, ascending-adnate, narrow to moderately broad (somewhat ventricose in age), white to grayish or pallid vinaceous brown, sometimes turning bright pink when wounded, under a lens pruinose because of the numerous cystidia; stipe (1) 3–6 (10) cm. long, 1 mm. ± thick, flaccid-cartilaginous, equal, straight or flexuous, tubular, lower portion densely white-strigose when covered with leaves, white-pruinose over all when young, bluish black toward the apex at first, paler grayish brown below, fading through brownish grays and apex soon becoming paler than the lower portion, when broken exuding a few drops of a white milklike fluid.

Spores 9–12 × 6–8 μ, smooth, contents often granular, broadly ovoid, weakly amyloid; basidia two-spored; pleurocystidia and cheilocystidia abundant and similar, 54–72 × 10–16 μ, broadly fusoid with very sharply pointed apices or ventricose with two or more apical prolongations, contents usually coarsely granular when fresh; gill trama homogeneous, pale brownish in iodine; pileus trama with a moderately thick but not gelatinous pellicle and a sharply differentiated hypoderm, the remainder floccose-filamentous, vinaceous brown in iodine; stipe tissue reddish vinaceous in iodine.

Habit, habitat, and distribution.—Scattered to cespitose around logs and stumps or on chips in places where wood choppers have been working. September, October, and November; Michigan.

Material studied.—Smith, 32-335, 32-465, 32-481, 32-531, 32-548, 32-582, 32-595, 32-625, 5017, 14961, 15503, October 15, 16, 19, November 18, 1931, Michigan.

Observations.—*M. erubescens* varies greatly in stature, so much so that the character is best disregarded when identifying the species. When the plant is growing in the open, the fruiting bodies may be short-stiped and rather strict in their bearing. When it is growing in piles of fallen leaves around logs or stumps, the pilei are often small and the stipes exceptionally long and flexuous. The latex is extremely scanty and easily overlooked, particularly in small specimens. Consequently one is likely to place a collection in the wrong section. The color change to red is also unreliable. The most useful character is the very bitter taste. It develops slowly but remains for a long time and causes a distinctly unpleasant sensation. The cystidia, as Kühner (1938) has pointed out, are also characteristic because of their guttulate content. However, I have found specimens in which the content of the cystidia was homogeneous and hyaline. These two characters will usually enable one to identify any of the growth forms he may collect in either dry or wet weather.

Only the two-spored form has been found in North America. Both it and the four-spored form occur in Europe. According to Kühner's interpretation, the four-spored form is usually more robust and accounts for von Höhnel's attempt to compare the species to the larger fragile gray Mycenae, with which it has little in common. I have followed Kühner (1938) in considering *M. fellea* and *M. cholea* synonyms of *M. erubescens*. The form I previously referred to *M. fellea* has not been found since and is best regarded as a variation in which the milky juice was lacking. In conversation when he visited North America in 1939 Dr. Lange stated that *M. fellea* does have a slight amount of a milklike juice. In a way the entire situation in regard to naming this fungus is a "comedy of errors." Von Höhnel neglected to mention the taste. This, in addition to other discrepancies in his original description, caused me to regard my specimens as different from his. The assumption was that, if taste was not recorded, it was not specific. Lange overlooked the milky juice or neglected to mention it, and placed *M. fellea* in the Filipedes. Because of this I did not refer my specimens, which possessed a distinct latex, to his species, and ended by describing them as "new." After I had talked with Lange all doubts about differences between *M. cholea* and *M. fellea* were cleared up. Kühner has apparently proceeded on the assumption that there is only one species of Mycena with a milklike latex which has the cystidia filled with highly refractive granules or droplets. I believe his assumption is justified until it is definitely disproved by the discovery of additional species with these characters. *M. Sabali*, unfortunately, is not sufficiently well characterized at present to be considered here. It is the only other species known to have similar cystidia. Its habitat is peculiar and should aid materially in its recognition.

55. Mycena copiosa Cejp

Publ. Fac. Sci. Univ. Charles, 104: 103. 1930

Illustration: Beardslee and Coker, Journ. Elisha Mitchell Sci. Soc., 40, pl. 8 (upper right, as *M. erubescens*).

"Cap 3–8 mm. broad, conic-campanulate, becoming broadly campanulate with age, dull and opaque in appearance, somewhat hygrophanous, yellow-brown, shading to deep brown at the center when moist, pale gray-brown with the center darker when dry, margin incurved and almost fluted when young, striate almost to the center with age, glabrous, thin and membranous. No odor or taste.

"Gills narrow, not ventricose, somewhat distant, adnate, white, becoming grayish with age, margin minutely flocculose.

"Stem 5–7 cm. long, scarcely 2 mm. thick, glabrous, colored like the cap below, paler above, attached by white mycelial fibers to old leaves, etc. Wounds of the cap and stem exude a copious milky juice and soon become stained red.

"Spores ellipsoid, $4–4.5 \times 6–7\ \mu$. Cystidia numerous, $7–12\ \mu$ thick at the gills, extending $30–40\ \mu$ to a slender point. Basidia 4-spored.

"This is smaller than von Hohnel's species is described and has somewhat different spores, but it seems too close for separation from his species, of which it seems the American representative. It is unusual in its milky juice and in the red color which is soon assumed when the cap is cut. So far it has been detected in only one station."

This is the original account of a fungus Beardslee identified as *M. erubescens*. Kühner placed *M. copiosa* Cejp in synonymy with *M. galopus* in spite of the decided discrepancy in spore size. Since Beardslee and Cejp agree on the spore size, I am putting the Ohio collection in Cejp's species but recognize at the same time that, because of the red stains developed by Beardslee's material, it may be different from the European form. However, in view of the situation in regard to *M. cholea*, *M. fellea*, and *M. erubescens*, I hesitate to name it as a new species.

56. Mycena crocata (Fr.) Gillet

Les Hymén., p. 260. 1874

Agaricus crocatus Fries, Syst. Myc., 1: 148. 1821.

Illustrations:

Konrad et Maublanc, Icon. Sel. Fung., 3, pl. 226, II.
Lange, Flora Agar. Dan., 2, pl. 55 D.

Pileus 1–2 cm. broad, campanulate, subhemispheric or submammillate, color variable, olivaceous, cinereous, whitish, or tinged reddish or orange around the disc, surface glabrous, appearing silky when faded, margin striate at first; flesh thin, odorless, taste mild; lamellae adnate or attenuated at the stipe, subdistant to moderately close, broadest in front, subventricose, whitish, edges not differently colored; stipe (2) 4–12 cm. long, 1–2 mm. thick, equal or slightly attenuated upward, hollow, base fibrillose and rooting somewhat, fragile, glabrous above, color variable, yellowish orange to dark reddish, darker below and paler above, when broken exuding a bright ochraceous orange latex, or latex darker and even purplish at times.

Spores 9–11 $\times$ (5) 6–7 μ, ellipsoid, smooth, amyloid; basidia four-spored; pleurocystidia scattered, 29 $\times$ 9–12 μ, clavate with an abrupt needlelike apical projection; cheilocystidia clavate with the apices roughened with short projections; pileus trama covered with a nonviscid but slightly gelatinous pellicle.

Habit, habitat, and distribution.—On leaf mold, particularly of beech; very rare in North America. Beardslee and Coker report it from North Carolina.

Observations.—I have not studied either fresh or dried specimens and have drawn the description largely from Kühner. The colored juice, nonmarginate gills, and cheilocystidia are sufficient to distinguish it from other species.

57. Mycena anomala Beardslee

Journ. Elisha Mitchell Sci. Soc., 40: 58. 1924

"Cap 4–8 mm. broad, thin and membranous, at first campanulate, becoming expanded and plane or even depressed with age, sulcate striate with forking furrows, mahogany-red to brownish red, very minutely downy.

"Gills distant, few in number, broad, almost triangular, adnate decurrent, colored like the cap, margin darker, deep red-brown.

"Stem slender, 4–6 cm. long, scarcely 1 mm. thick, pale, but with a slight red tint, very minutely downy, attached by a circle of fibers to pine needles at the base, exuding a dark red juice when wounded.

"Spores narrowly ellipsoid, 4 × 11–12 μ. Cystidia abundant on the margins of the gills, globose to rounded obovate with pointed, rod-like projections.

"This interesting species was found once growing in abundance on the fallen needles under pines. Unfortunately, the photograph which was made was not successful and there was no other opportunity to make one. The drawing made at the time will show its peculiarities and it will be readily recognized by its unusual characters. In many ways this species is suggestive of *Omphalia*. The broad gills which seem quite decurrent as the cap becomes expanded would almost indicate its reference to that genus. In its bleeding stem, margined gills, and setulose cystidia it finds, however, close relatives in *Mycena* and it has seemed best to place it with these and ignore the somewhat decurrent gills."

This is the original description and comments. Mr. Beardslee has shown me the drawings of the microscopic characters as he made them from the study of the fresh specimens. The pileus is corticated by a palisade of pear-shaped cells. This is an additional important character which, along with the others, justifies the specific name. The species is as peculiar in the Lactipedes as *M. rorida* is in the Glutinipes.

58. Mycena haematopus (Fr.) Quélet

Champ. Jura et Vosges, p. 244. 1872

Agaricus haematopus Fries, Syst. Myc., 1: 149. 1821.

? *Agaricus haematopus* var. *umbilicatus* Berkeley & Montagne, Syll. Crypt., p. 111. 1856.

Galactopus haematopus Earle, Bull. New York Bot. Gard., 5: 426. 1909.

Illustrations:

Plates 13, 14 A; Text fig. 12, nos. 1, 7 (p. 142).

Atkinson, Mushrooms, Edible, Poisonous, etc. (1900), fig. 100.

Beardslee and Coker, Journ. Elisha Mitchell Sci. Soc., 40, pl. 6.

Bresadola, Icon. Mycol., 5, pl. 247 (lower figures).

Fries, Icon. Sel. Hymen., 1, pl. 83, fig. 1.

Güssow and Odell, Mushrooms and Toadstools, pl. 56, fig. 2.

Hard, The Mushroom, Edible and Otherwise, fig. 90.

Konrad et Maublanc, Icon. Sel. Fung., 3, pl. 226 I.

Krieger, A Popular Guide to the Higher Fungi (Mushrooms) of New York State, fig. 108.
Lange, Flora Agar. Dan., 2, pl. 50 G (rather pale).
Marshall, The Mushroom Book, plate facing page 93, upper figure.
White, Conn. State Geol. and Nat. Hist. Surv., Bull. 15, pl. 10.

Pileus 1–3.5 (5) cm. broad, buttons ellipsoid to egg-shaped, expanding to obtusely campanulate or broadly conic, sometimes convex, sometimes sharply conic, in age plane or with a recurved margin and an acute or obtuse umbo, margin appressed against the stipe and usually characterized by a sterile band, which becomes lacerated or merely crenate, surface at first appearing dry and covered with a dense pruinose coating, soon becoming polished and moist, translucent-striate at maturity but often opaque at first, sometimes becoming sulcate in age, color dark reddish brown ("carob brown") on the disc and paler grayish vinaceous toward the margin, margin whitish in some and the disc pale sordid reddish brown, often stained dark sordid reddish brown in age; flesh thin, fragile, grayish vinaceous or pallid, exuding a dark blood-red latex when cut, odor not distinctive, taste mild or bitterish; lamellae close to subdistant, 20–30 reach the stipe, two or three tiers of lamellulae, narrowly adnate, ascending, narrow to moderately broad, whitish or grayish vinaceous, soon stained sordid reddish brown, edges flocculose, pallid or whitish; stipe (3) 4–8 (14) cm. long, 1–2 (3) mm. thick, equal, rigid, fragile, hollow, base strigose, upper portion covered with a dense coating of pallid to pale cinnamon drab pruina, becoming polished in age, exuding a dark blood-red latex when broken.

Spores 8–11 $\times$ 5–7 μ, broadly ellipsoid, amyloid; basidia four-spored; pleurocystidia and cheilocystidia similar, rare, scattered or abundant on faces, very numerous on the edges, 33–60 (80) $\times$ 9–12 μ, fusoid-ventricose, apices acute, sometimes forked, hyaline; gill trama with numerous lactifers, otherwise homogeneous, bright vinaceous red in iodine; pileus trama with a well-differentiated pellicle and hypoderm, the remainder floccose, vinaceous red in iodine (all parts except the pellicle), lactifers numerous.

Habit, habitat, and distribution.—Single to cespitose on decaying wood throughout the United States and Canada. I have examined material from Alabama, North Carolina, Pennsylvania, New York, Ohio, Michigan, Wisconsin, Illinois, Missouri, Idaho, Washington, Oregon, and California in the United States and from Nova Scotia, Ontario, Manitoba, and British Columbia in Canada. This species

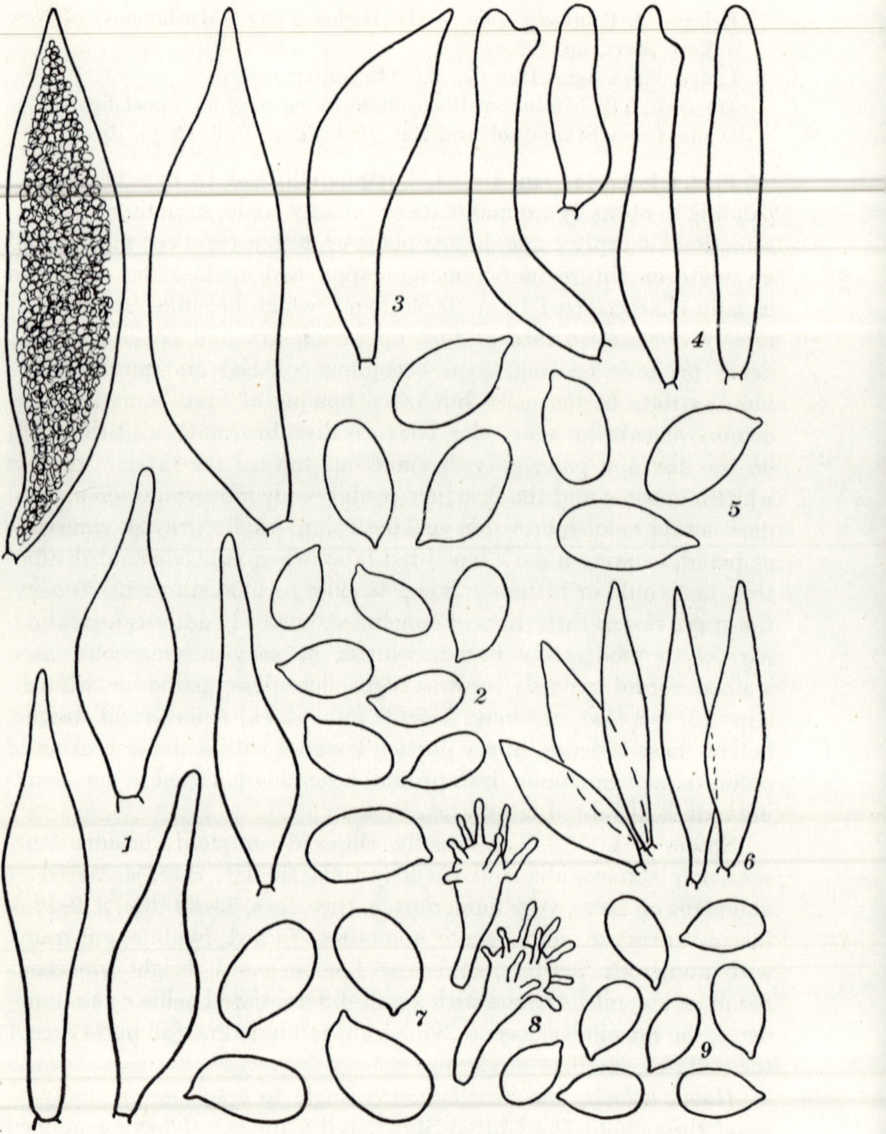

FIG. 12. *M. Sabali:* 3, pleurocystidia; 4, cells from cap surface; 5, spores. *M. Atkinsoniana:* 2, spores; 6, cheilocystidia. *M. haematopus:* 1, cheilocystidia; 7, spores. *M. lineata:* 8, cheilocystidia; 9, spores

is the commonest and the most easily recognized one in the genus. It can be found during the spring, summer, and fall months in the northern areas and very likely throughout the winter where weather conditions are favorable.

Material studied.—Smith, 32–226, 32–283, 32–314, 131, 228, 482, 514, 1032, 2568, 4383, 4674, 6309, 6390, 7016, 7071, 7249, 8194, 8249, 8317, 8373, 8486, 8821, 10869, 10937, 13262, 13397, 13609, 13706, 13770, 13773, 16574, 17308. Atkinson, numerous collections, numbers not recorded. Bisby, July 6, 1932. Brown, October 1, 1925. Flett, June 3, 1935, and 1941, Washington. Gruber, Oregon. Hesler, 2486, 12321. Kauffman, New York, Michigan, Idaho, Washington, and Oregon. Kelly, 1093, Ontario. Mains, 34–139, 32–433, 5028. Morse, August 28, 1930, California. Overholts, 800, 895, 1532, 1533, 4253. Povah, 7, Florida. Routien, 1059. Wehmeyer, 726, 726a, 726b.

Observations.—The dull-red to reddish-brown colors, habitat on wood, crenate margin of the pileus, fusoid cystidia, and densely grayish-pruinose cap and stipe (when young), as well as dark blood-red juice, readily distinguish this species. Although very easily recognized, it contains many intergrading forms. Its stature varies with the substratum, the nutrient, and the climatic conditions. Along our Pacific coast it often attains gigantic proportions for a Mycena. Such characters as the density and color of the pruinosity and the colors of all parts of the carpophore also depend on the growing conditions, being generally much more striking in robust individuals. The spore size likewise is variable. Large specimens with spores 9–12 × 5–7 μ have been found as well as specimens with spores 7–8 × 4–5 μ. The latter are rare. Were it not for the fact that the great majority of specimens have a spore range intermediate between these two, both might be considered worthy of taxonomic recognition. The pleurocystidia vary in number in much the same manner as those of *M. alcalina.* I have always been able to find at least a few on any specimens examined. In some collections they are abundant.

59. Mycena haematopus var. marginata Lange

Dansk Bot. Arkiv., 1 (5): 20. 1914

Illustrations:

Plate 15.

Lange, Flora Agar. Dan., 2, pl. 50 D.

The variety is distinguished from typical material by the reddish content of the cystidia. The gill edges are usually bordered with dull reddish as a result. I have observed specimens in which the gill margins were very dark. The variation in color is caused by a variation either in the number of cystidia present or in the intensity of the color of their contents. The commonest form of the variety is a tall (10–15 cm.), robust form which is particularly abundant in the Olympic Mountains of Washington, but which also occurs in Michigan and Nova Scotia. Its spores measure 9–12 × 6–7 μ. There appear, however, to be about as many forms of the marginate variety as there are of var. *typicus*, and they are distinguished by the same characters—spore size, color, abundance of pleurocystidia, and stature. Variety *marginata* is to be expected throughout the range of the species.

Material studied.—Smith, 374, 795, 826, 3741, 17308. Hesler, 11469. Overholts, 3917. Stuntz, F 603. Wehmeyer, 726c, 726e, 726f, 826.

60. **Mycena Atkinsoniana,** nom. nov.

Mycena fagicola A. H. Smith, Mycologia, 27: 587. 1935 (not *M. fagicola* Grog., Fung. Gall., 3104).

Illustrations: Plate 14 B; Text fig. 12, nos. 2, 6 (p. 142).

Pileus (5) 10–30 mm. broad, obtuse to convex at first, becoming obtusely umbonate, plane or at times the disc slightly depressed, margin connivent to the stipe at first, expanded in age and somewhat wavy, extreme margin striatulate when moist, sulcate-striate in age, hoary-pruinose at first, glabrous, disc "claret brown" when young, soon "mahogany red" to "Morocco red," in age "bay," margin "Sanford's brown" and becoming "Isabella color"; flesh buff, exuding an orange-yellow juice when cut, odor and taste not distinctive; lamellae bluntly adnate, close to subcrowded, 23–26 reach the stipe, narrow to moderately broad, "chamois" (pale yellow) when young, somewhat darker in age, edges near "Victoria lake" or "maroon," crenulate or even; stipe 2–4 cm. long, 2–3 mm. thick, equal, often compressed, rooting among the leaves and debris, base strigose, surface scantily fibrillose with "claret brown" fibrils, dull purplish brown over all but color ceasing abruptly at the apex, exuding a dull reddish-brown juice when cut or juice dull orange in old specimens.

Spores narrowly to broadly ellipsoid, 7–9 × (3) 4–5 (6) μ, amyloid;

basidia four-spored, 28–30 × 6–7 μ; cheilocystidia numerous, forming a sterile band, narrowly fusoid, smooth, contents dark reddish, 30–40 × 6–8 μ; pleurocystidia absent or rare, similar to cheilocystidia; gill trama yellowish to very faintly vinaceous brown in iodine; pileus trama consisting of a thin pellicle of narrow hyphae filled with a dark-reddish substance, a region of vesiculose cells beneath the pellicle and the remainder of narrower floccose hyphae, both regions beneath the pellicle becoming very faintly vinaceous brown in iodine; stipe tissue dark vinaceous brown in iodine.

Habit, habitat, and distribution.—Gregarious on leaves or subcespitose in beech and beech-hemlock forests during the summer and fall; Connecticut, New York, Ohio, and Michigan.

Material studied.—Smith, 936, 955, 972, 2381. Atkinson, 3933, 5029. Stuntz, 831. Walters, November, 1941, Cleveland, Ohio.

Observations.—The color of the pileus varies from dark lateritious to testaceous, and in age the sordid yellow margin is quite distinctive. The dull-yellow gills with their dark borders are also unusual. The stipe may become nearly glabrous, and in such specimens there is a slight resemblance to large forms of *M. subsanguinolenta.*

61. Mycena subsanguinolenta A. H. Smith

Mycologia, 31: 280. 1939

Illustrations:

Plate 16 A; Text fig. 13, no. 11 (p. 148).
Smith, Mycologia, 31, fig. 1 E (spores).

Pileus 10–25 mm. broad, conic, becoming obtusely conic-campanulate, in age sometimes nearly plane but always with an abrupt obtuse umbo, margin appressed against the stipe when young, surface hoary at first, soon naked, moist, translucent-striate, in age sulcate to the disc, color "burnt umber" near and on the disc, paler and near "vinaceous buff" on the margin, in age a strong yellowish cast is evident throughout; flesh thin, yellowish or sometimes reddish under the disc, pliant, when cut exuding a watery orange-yellow juice, odor and taste mild; lamellae ascending-adnate with a decurrent tooth, distant to subdistant, 17–20 reach the stipe, moderately broad (4 mm.), faces pale incarnate, edges dark reddish brown; stipe 3–8 cm. long, 1–2 mm. thick, equal, fragile, hollow, base sparsely strigose with whitish hairs, apex hoary-pruinose but soon naked, color pallid incarnate or concolorous with pileus margin, in age with a yellowish cast, when broken exuding a drop of a blackish-red juice but the remaining drops dull orange.

Spores subovoid, 7–8.5 × 3.5–4 μ, amyloid; basidia four-spored, 17–19 × 7–8 μ; cheilocystidia numerous, forming a sterile band along the gill edge, somewhat fusoid with subacute apices, smooth, contents sordid reddish brown, 28–33 × 6–9 μ; pleurocystidia not differentiated; gill trama dark vinaceous brown in iodine; pileus trama with a thin adnate pellicle, beneath it a region of somewhat inflated cells, the remainder floccose, becoming dark vinaceous brown in iodine below the pellicle; stipe tissue dark vinaceous brown in iodine.

Habit, habitat, and distribution.—Densely gregarious on needle beds under Douglas fir and Sitka spruce; Washington, Oregon, and California.

Material studied.—Smith, 7775, 7809, 7975, 8020, 8272, 8329, 8499, 8618, 9078, 9085, 13100, 13336, 14077, 14827, 17064. Flett, Bremerton, Washington.

Observations.—This species is separated from *M. sanguinolenta* by the smaller spores, lack of pleurocystidia, and the more pronounced yellowish colors of the latex as well as those of both the pileus and the stipe. Both species were observed growing in great abundance in northern California and were readily recognizable. *M. subsanguinolenta* can nearly always be distinguished from *M. Atkinsoniana* by its glabrous stipe. The two species are also readily distinguished by their habitats and general appearance. The appearance of *M. Atkinsoniana* is similar to that of *M. haematopus*, whereas that of *M. subsanguinolenta* is similar to that of *M. sanguinolenta*. From the collections of both which have been examined there appears to be a distinct difference in color, also, but since there is likely to be considerable variation in the shades of red, reddish brown, or purplish brown, it seems best not to emphasize it.

62. Mycena sanguinolenta (Fr.) Quélet

Champ. Jura et Vosges, p. 244. 1872

Agaricus (*Mycena*) *sanguinolentus* Fries, Syst. Myc., 1: 149. 1821.
Galactopus sanguinolentus Murrill, North Am. Flora, 9: 319. 1916.

Illustrations:

Plate 16 B; Text fig. 13, nos. 1–2 (p. 148).
? Beardslee and Coker, Journ. Elisha Mitchell Sci. Soc., 40, pl. 7.
Fries, Icon. Sel. Hymen., 1, pl. 83, fig. 3.
Lange, Flora Agar. Dan., 2, pl. 50 A.
Ricken, Die Blätterpilze, 2, pl. 110, fig. 7.

Pileus 3–10 (25) mm. broad, either convex or conic when young, becoming broadly convex or campanulate, not expanding completely, margin appressed when young, surface at first covered with a dense grayish hoary-pruinose coating, becoming naked and glabrous, moist, margin opaque or slightly striatulate, soon becoming sulcate, color variable but always some shade of bright or dull reddish brown with a more or less vinaceous to avellaneous margin ("brick red" becoming "dragon's-blood red" on the disc and "avellaneous" to "light vinaceous-fawn" on the margin, sometimes "cameo brown" to "Prussian red" on the disc or "liver brown" to "Hessian brown"); flesh thin, not very fragile, sordid reddish, exuding a reddish juice when cut, odor and taste not distinctive; lamellae adnate or slightly toothed, subdistant to distant, narrow to moderately broad, sordid reddish to grayish, the edges very dark reddish brown and even; stipe 2–6 (7) cm. long, 1–1.5 mm. thick, equal, tubular, fragile, white-strigose at the base, the remainder covered with a drab pruina, soon polished and more or less concolorous with the pileus, exuding a bright or dull-red juice when cut or broken.

Spores 8–10 (11) × 4–5 (6) μ subellipsoid, only weakly amyloid; basidia four-spored (occasionally two- or three-spored); pleurocystidia rare to scattered or sometimes quite abundant, narrowly to broadly fusoid-ventricose, 36–54 × 8–13 μ, filled with a sordid-reddish substance; cheilocystidia similar to pleurocystidia or shorter and more obese, very abundant; gill trama of broad hyphae the cells of which are often vesiculose in age, pale reddish brown in iodine; pileus trama covered with a thin adnate pellicle, hypoderm moderately well differentiated, the remainder floccose and filamentous, all except the pellicle pale vinaceous brown in iodine, lactiferous hyphae abundant.

Habit, habitat, and distribution.—Gregarious on leaf mold, moss beds, or needle carpets during the spring and fall; from Maine to Washington and south to North Carolina and California in the United States, and from Nova Scotia to British Columbia in Canada. Common and very widely distributed.

Material studied.—Smith, 32–29, 32–98, 33–97, 33–186, 33–344, 33–377, 33–403, 33–914, 11, 88, 331, 505, 906, 1381, 3307, 4345, 4450, 6254, 6515, 8619, 9670, 15077. Atkinson, numerous collections from Cayuga Lake Basin, New York. Davidson, August 15, 1932, British Columbia. Flett, Washington. Hesler, 906, 4345, 6515, 9670, 11484, 14187, 14226. Kauffman, 8 collections, Tennessee, Maryland, Michigan, Washington, Oregon. Mains, 32–576, 33–120, 33–620, 5073. Morse, January 27, 1931. Wehmeyer, 505.

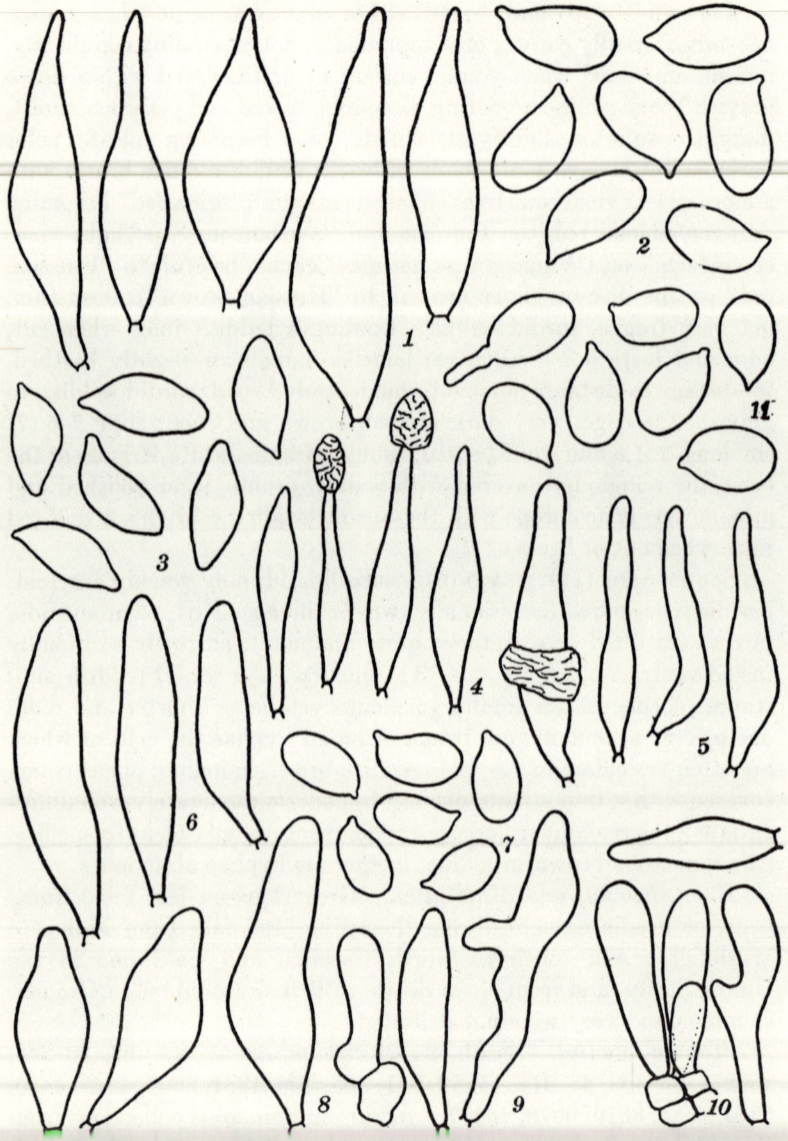

FIG. 13. *M. sanguinolenta:* 1, pleurocystidia and cheilocystidia; 2, spores. *M. gypsea:* 3, spores; 4, pleurocystidia; 5, cheilocystidia. *M. olida* var. *americana:* 6, pleurocystidia; 7, spores; 8, cheilocystidia; 9, caulocystidia; 10, cells from surface of pileus. *M. subsanguinolenta:* 11, spores.

Observations.—The larger spores and the presence of pleurocystidia distinguish this fungus from *M. subsanguinolenta.* Along the Pacific coast it occurs in troops of hundreds of carpophores under second-growth spruce. Here it reaches its maximum size of around 2.5 cm. across the base of the pileus. In Michigan the common form occurs in bogs and seldom has pilei more than 1 cm. broad. The species occurs sparingly in hardwood forests. In such locations isolated fruiting bodies are most commonly found.

SECTION ADONIDAE

In this section are grouped the medium-sized white species and those generally with bright colors but with the gills not margined. Two species with marginate gills are grouped in the last subsection. A large number of very small bright-colored species are grouped in the section Deminutivae. Certain of the species in the section Typicae are nearly white or fade to white, and one, *M. incarnatifolia*, is rather delicately colored. It is not included here because it obviously belongs in the group which is cespitose on wood and which centers around *M. alcalina.* Many members of the section Typicae may develop rather delicate pink tints in the gills. *M. niveipes*, *M. incarnatifolia*, *M. radicatella*, and occasionally *M. rubrotincta* of the section Typicae may be sought here. They are rather large cartilaginous species with amyloid spores, and all grow on old logs, stumps, or lignicolous debris.

KEY TO SUBSECTIONS

1\. Fruiting body relatively fleshy; colors dull to bright and variable (white in *M. subaquosa*), usually with tint of lilac or purplish or pink; odor and taste of radish .... Subsection Ianthinae, p. 185
1\. Fruiting body relatively thin and membranous 2

2\. Fruiting body white over all .... Subsection Albidae, p. 149
2\. Fruiting body yellow, pink or red (or some other bright color) Subsection Euadonidae, p. 163

SUBSECTION ALBIDAE

KEY TO SPECIES

(Large pale lignicolous species should be sought for in the Typicae)

1\. Cheilocystidia clavate-roughened .... 63. *M. lineata*

1. Cheilocystidia with one or more thornlike projections .. 204. *M. insignis*
1. Cheilocystidia of the smooth fusoid-ventricose type 2

2. Spores globose to ellipsoid 3
2. Spores elongated to fusoid 6

3. Spores globose 4
3. Spores ellipsoid 5

4. Gills adnate; carpophores scattered on needle beds under conifers 68. *M. Rickenii*
4. Gills short-decurrent: cespitose on wood (but delicate) 66. *M. translucentipes*

5. On old acorns and nuts of hickory or walnut (white form) 75. *M. luteopallens*
5. Odor nitrous; on humus and debris under conifers (*M. subaquosa* might key out here also) .... 67. *M. delectabilis*
5. Odor none; carpophores gregarious on hardwood logs and debris 65. *M. olida* var. *americana*

6. Spores 8–10 (11) × 4–5 μ; subventricose .... 64. *M. gypsea*
6. Spores nearly cylindric, seldom over 3 μ wide 7

7. Pleurocystidia scattered to quite abundant, 36–60 × 9–12 μ 69. *M. pseudolactea*
7. Pleurocystidia, if present, embedded and difficult to locate, 18–26 × 4–6 μ .... 70. *M. delicatella*

63. Mycena lineata (Fr.) Quélet sensu Lange

Flora Agar. Dan., 2: 46. 1936

Agaricus lineatus Fries, Syst. Myc., 1: 152. 1821.

Illustrations:

Text fig. 12, nos. 8–9 (p. 142).

Beardslee and Coker, Journ. Elisha Mitchell Sci. Soc., 40, pl. 15 (upper right).

Fries, Icon. Sel. Hymen., 1, pl. 78, fig. 5.

Lange, Flora Agar. Dan., 2, pl. 55, figs. A, A[1].

Pileus 5–10 mm. broad, ovoid, becoming convex, in age broadly conic to slightly umbonate, apex obtuse, surface moist, glabrous, with conspicuous broad translucent striations extending to the disc, color "pale smoke gray" or the gray tinged with olive over the disc and striations, the remainder watery white, fading, becoming opaque in age and then chalky white; flesh rather thin and fragile, whitish or grayish under the disc, odor and taste not distinctive; lamellae broadly adnate or slightly decurrent, narrow to moderately broad, soon seceding, white, edges even and pallid; stipe 2–4 cm. long, 1 mm. thick, equal above a slightly enlarged glabrous base or at times the base slightly strigose, tubular, rather firm and pliant, not pruinose, pure hyaline white or tinged watery grayish.

Spores 6–8 × 3.5–5 μ, ovoid to ellipsoid, smooth, faintly amyloid, basidia four-spored; pleurocystidia not differentiated; cheilocystidia embedded in the gill edge, clavate, the apices studded with contorted fingerlike prolongations; gill trama homogeneous, yellowish brown in iodine; pileus trama with a thin pellicle, a differentiated hypoderm, and a filamentous tramal body, sordid yellowish brown beneath the pellicle in iodine.

Habit, habitat, and distribution.—Scattered to gregarious on sticks and humus under conifers in late summer and fall; California and Ontario.

Material studied.—Smith, 3622, 4729.

Observations.—Although I have only two collections of this species, it is recognized here because it appears to be distinct from other Mycenae and to fit into Lange's concept of *M. lineata* remarkably well. Lange's form *pumila*, however, has not come to my attention. The spores of my material are faintly but distinctly amyloid, and the hypoderm is very distinct; these characters remove the fungus from the vicinity of the brightly colored forms and indicate a relationship to species of the Typicae. The cheilocystidia, very pale watery gray but conspicuous striations of the pileus, white stipe, and smaller spores distinguish it from *M. filopes*, with which one would not be likely to confuse it. The olive tints were obscure in my material, and no tendency to yellow was noted. Kühner has placed this fungus among the forms of *M. vitilis*.

64. Mycena gypsea (Fr.) Quélet

Champ. Jura et Vosges, p. 342. 1873

Agaricus gypseus Fries, Epicr. Syst. Myc., p. 104. 1838.

Illustrations:

Plate 17, Text fig. 10, nos. 3–5 (p. 148).

Beardslee and Coker, Journ. Elisha Mitchell Sci. Soc., 40, pl. 18 (upper figs.).

Pileus 8–20 mm. broad, convex to slightly umbonate at first, the margin incurved, becoming broadly convex to plane in age, the margin becoming wavy, surface pruinose at first, moist and shining at maturity, translucent-striate along the margin, chalk white to creamy white at first or the disc more yellowish, fading to chalk white or retaining yellowish tints; flesh thin, pliant, and cartilaginous, white, odor and taste not distinctive; lamellae bluntly adnate, horizontal or only slightly ascending at first, horizontal in age and then slightly decurrent, narrow, close, pure white, edges white and somewhat eroded; stipe 2–4 (8) cm. long, 1–2 mm. thick, base rooting in the decayed wood (3 cm. ±), strigose and with adhering debris, equal, cartilaginous, tubular, pure white, covered at first with a dense white pruinose coating, somewhat glabrescent in age.

Spores 8–10 (11) × 4–5 μ, smooth, subventricose to subfusoid, yellowish in iodine; basidia two- and four-spored; pleurocystidia and cheilocystidia similar, very abundant, 28–42 × 5–8 μ, narrowly fusoid-ventricose with rather acute apices, the tips frequently incrusted with a globular mass of a resinous substance; gill trama of slender branched hyphae, yellowish in iodine; pileus trama homogeneous, of slender hyphae, yellowish in iodine, surface covered with a turflike cuticle of slender cystidia 28–60 × 5–8 μ, which are subfusoid to cylindric and occasionally have a resinous incrustation; caulocystidia similar to the pilocystidia.

Habit, habitat, and distribution.—Gregarious to subcespitose on decaying wood. I have found it in Michigan during June. According to my experience, it is rare, although Kühner reports it as common around Paris.

Material studied.—Smith, 15061, 18659, June 24, 1935, Michigan.

Observations.—Kühner's interpretation (1938) of this species has been followed. The fungus is very similar to *M. delicatella* in many of its characters, but is easily distinguished because of its broader spores. In my specimens the gills were close and narrow, and the margin of the pileus was distinctly incurved. In these respects they differ slightly from European species. In *M. delicatella* these characters are rather variable, so that little emphasis has been placed on them here. *M. olida* var. *americana* has much the same consistency and color but both its spores and cystidia should readily distinguish it. Both grow in the same habitats and fruit at the same time. I have not found a fungus of this type with the fragrant odor mentioned by Coker in his notes (Beardslee and Coker [1924]), and Kühner has not admitted the specimens of Coker in his concept.

65. Mycena olida var. **americana**, var. nov.

Illustrations:

Plates 11 C, 18 B; Text figs. 13, nos. 6–10 (p. 148); 14, nos. 1–4 (two-spored form) (p. 154).

Beardslee and Coker, Journ. Elisha Mitchell Sci. Soc., 40, pl. 15 (lower right).

Pileus 3–20 mm. latus, obtuse conicus demum campanulatus, impolitus, glabrescens, striatus, lacteo-albus; sapor mitis; lamellae adnatae, confertae vel distantes, latae, albidae; stipes 1–8 cm. longus, 1–2 mm. crassus, deorsum strigosus, sursum dense pruinosus, albidus; sporae 7–8 × 3.5–4.5 μ; pleurocystidia 34–68 × 8–16 μ, subcylindrica vel fusoide ventricosa, levia; cheilocystidia 26–42 × 8–12 μ. Specimen typicum in Herb. Univ. Mich. conservatum. Legit A. H. Smith, n. 15244, prope Dexter, Mich., Sept. 2, 1940.

Pileus 3–20 mm. broad, obtusely conic with the margin straight at first, becoming broadly conic, campanulate or nearly plane with a wavy or scalloped margin in age, sometimes merely convex and becoming plane, surface pruinose-pubescent at first (under a lens) but soon glabrous and polished, lubricous when moist, hygrophanous, striate to the disc before fading, pure white to milky white and soon sordid creamy yellowish over the disc, often becoming pale tan in age; flesh thin but quite tough and cartilaginous, white or becoming yellowish in age, no odor, taste mild or slightly farinaceous; lamellae ascending-adnate to uncinate, close (large specimens) to distant (small individuals), moderately broad and somewhat ventricose, white but becoming creamy yellowish, edges even and whitish; stipe 1–8 cm. long, (0.5) 1–2 mm. thick, equal or tapered below, base often rooting in the debris and densely strigose over the lower portion, upper part densely pruinose at first but soon glabrous and polished, white, sometimes becoming yellowish or sordid tan in age.

Spores 7–8 × 3.5–4.5 μ, ellipsoid, smooth, hyaline, apiculus rather prominent, spores from two-spored basidia 7–8.5 × (5) 6–7.5 μ and broadly ellipsoid to globose, yellowish in iodine; basidia usually four-spored, two-spored forms rare; pleurocystidia rare to scattered, 34–68 × 8–16 μ; gill trama homogeneous, yellowish in iodine; pileus trama with a pseudoparenchymatous layer of enlarged cells covering the surface, the remainder filamentous, yellowish in iodine, surface hyphae at first with a rather dense covering of upright pilocystidia or

FIG. 14. *M. olida* var. *americana* (two-spored form): 1, caulocystidia; 2, spores; 3 and 4, cheilocystidia. *M. hiemalis:* 5, spores; 6, cheilocystidia. *M. Rickenii:* 8, cheilocystidia; 9, spores. *M. delectabilis:* 7, spores; 10, pleurocystidia

differentiated hyphal tips 28–35 × 6–9 μ, which may be clavate, subfusoid, or branched; caulocystidia similar to pilocystidia.

Habit, habitat, and distribution.—Gregarious to subcespitose on old logs and rich humus; North Carolina, New York, Michigan, and Missouri in the United States and Ontario and Manitoba in Canada. It usually occurs in lower Michigan on elm logs or debris, but farther north I have found it on pine logs.

Material studied.—Smith, 32–281, 32–293, 32–518, 32–521, 33–730, 787, 871, 1368, 1404, 4751, 7146, 7626, 7673, 7690, 11036, 15244. Bisby, Manitoba. Kanouse, Michigan. Kauffman, four collections (as *M. nivea* Quélet), Michigan; one collection (as *M. minutula*), North Carolina. Overholts, 1049.

Observations.—The cystidia on the pileus are not always demonstrable on revived material, but have been noted on several of my collections. The American material differs in some respects from European specimens, but most of the differences do not impress one as being constant. I found small pleurocystidia on my four-spored specimens from Ann Arbor, but observed that on specimens from farther north the pleurocystidia were rather abundant and within the size range given by Kühner (1938). The two-spored specimens from Ann Arbor had the largest cystidia in the collections, but the difference was only in size. The pilocystidia apparently collapse more readily than do the caulocystidia, with the result that the pileus often has a polished appearance when the stipe is still pruinose. The specimens previously referred to *M. lactea* var. *pithya* (Smith, 1935b) belong here. Bisby's report of *M. collariata* may also be referred to a form of this variety. Kühner does not recognize *M. collariata.*

Varietal ranking has been given to the American material because it has never been found to have a farinaceous taste pronounced enough to be distinct. In fact, all my Ann Arbor collections have been perfectly mild. Bisby (September 18, 1935, St. Vitae Park, Manitoba) also described it as being without any odor or taste. The basidia in his collection are nearly all bispored, and the spores measure 7–9 × 6–7 μ. The cystidia measure 50–80 × 9–12 μ.

For a time I regarded this variety as *M. lactea* Fries because of the very striking resemblance of the fruiting bodies to the illustration (*Icones Selectae Hymenomycetum*) of that species. Kühner, however, has followed Lange and used the name "*lactea*" in a very different sense. I have abandoned this name for any species because of the possibilities of confusion. In my estimation Fries's descriptions of

M. lactea cannot apply to the fungus Peck named *M. crystallina*, and there is no doubt whatever in my mind that Fries's illustration does not apply to *M. crystallina* (*M. delicatella* of this work; see page 161). For additional comments see *M. Rickenii* also (p. 158).

66. Mycena translucentipes (Murr.), comb. nov.

Omphalopsis translucentipes Murrill, North Am. Flora, 9: 312. 1916.
Omphalia translucentipes Murrill, Mycologia, 8: 220. 1916.

"Pileus very thin and delicate, convex, depressed at the center, gregarious to subcespitose, scarcely reaching 1 cm. broad; surface glabrous, not striate, pure-white, margin entire, concolorous: lamellae simple, short-decurrent, subdistant, rather narrow, white: stipe very slender, cylindric, smooth, glabrous, white, subtranslucent, 2 cm. long, less than 1 mm. thick.

"Type collected on an old stump in wet woods in City Park, New Orleans, Louisiana, September 6, 1908, *F. S. Earle 72* (herb. N. Y. Bot. Gard.).

"Habitat: Logs or stumps in wet woods.

"Distribution: Vicinity of New Orleans, Louisiana."

The type consists of a good collection. The spores measure 4–5 × 3.5–4 μ and are broadly ovoid, hyaline, smooth, and amyloid. The basidia are apparently four-spored, but, because of the very fine sterigmata and the failure of most of the hymenium to revive, the point could not be determined positively. No pleurocystidia were seen. The cheilocystidia measured 26–37 × 8–12 μ and had obtuse apices. They were smooth, hyaline, and fusoid-ventricose. The tramae of the pileus and gills each appeared to be homogeneous and turned yellowish in iodine. Neither revived very well. One might question whether or not this species is identical with *M. papillata*. The spore sizes of the two are about the same. The cespitose habit of *M. translucentipes*, its southern distribution, and its truly diaphanous character seem sufficient to characterize it.

67. Mycena delectabilis (Pk.) Saccardo

Syll. Fung., 5: 262. 1887

Agaricus (*Mycena*) *delectabilis* Peck, Ann. Rep. New York State Mus., 27: 93. 1875.
Prunulus delectabilis Murrill, North Am. Flora, 9: 323. 1916.

Omphalia nitrosa Kauffman, Pap. Mich. Acad. Sci., Arts, and Letters, 11: 202. 1930.
Omphalia delectabilis Smith, Bull. Torr. Bot. Club, 64: 482. 1937.
Prunulus leucophaeus Murrill, North Am. Flora, 9: 336. 1916.
Mycena leucophaea Murrill, Mycologia, 8: 220. 1916.

Illustrations:
Plate 8 A; Text fig. 14, nos. 7, 10 (p. 154).
Peck, Ann. Rep. New York State Mus., 27, pl. 1, figs. 22–25.

Pileus (3) 8–20 mm. broad, obtusely conic with an appressed margin at first, sometimes campanulate, becoming broadly conic to broadly campanulate, sometimes papillate, the margin usually spreading in age, watery white at first or the disc with a faint watery-gray cast, almost chalk white when faded or in age a bit yellowish around the disc; flesh very thin and very fragile, white, taste hardly distinctive, odor strongly nitrous or nearly lacking (sometimes evident only if the flesh is crushed); lamellae arcuate at first and soon becoming long-decurrent, subdistant to distant, narrow, white, edges even or fimbriate; stipe 2–4 cm. long, 0.5–1.5 (2) mm. thick, equal, nearly filiform at times, fragile-cartilaginous, tubular, base white-strigose and apex faintly pruinose, otherwise glabrous and naked, white.

Spores 5–7 × 3.5–4 μ, subellipsoid, smooth, nonamyloid; basidia four-spored; pleurocystidia scattered to abundant, 33–58 × 7–12 μ, fusoid-ventricose, elongating to subcylindric or nearly filamentous at times (usually in age), hyaline, smooth; cheilocystidia similar to the pleurocystidia, numerous; gill trama homogeneous, pale yellow in iodine; pileus trama homogeneous beneath a thin pellicle, the hypoderm not differentiated, all parts yellowish in iodine.

Habit, habitat, and distribution.—Scattered to gregarious on needle beds under conifers, particularly around and under fallen treetops or heaps of dead branches, not uncommon during the late summer and fall in wet weather. It has been found in New York, Idaho, Washington, Oregon, and California in the United States and in Nova Scotia and Ontario in Canada.

Material studied.—Smith, 788, 904, 1065, 2784, 3391, 4562, 7770, 7931, 8021, 8497, 8020, 8016, 8091, 14464, 14497, 14510, 16752. Atkinson, 13491 (as *M. leptocephala*). Gruber, Oregon. Kauffman, two collections from Oregon (as *Omphalia nitrosa*). Slipp, UIFP: 2601, 2674.

Observations.—The small, moderately broad spores, cystidia on both faces and edges of the gills, the nitrous odor, and obtuse pileus characterize the species. The stipe is not particularly fragile, but the pileus breaks up very easily, with the result that it is difficult to get the specimen back to the laboratory in perfect condition. The taste may seem nitrous, but this is a false impression since taste and odor cannot be readily distinguished when testing a fungus of this type. If one chews a pileus for a minute the nitrous effect vanishes, and then there is no appreciable taste. In one of my collections the odor was absent (14464).

The type of *M. leucophaea* is well preserved, but I was unable to demonstrate the presence of any gelatinous layers on either the pileus or the stipe, and hence doubt whether they were viscid. The spores measure 5–6 × 2 μ, but only a few were found. Fusoid-ventricose cystidia were noted on both the sides and the edges of the gills. They measured 32–40 × 7–11 μ, and are similar in shape to those of *M. delectabilis*. An occasional individual cystidium was observed with a slight incrustation over the apex, apparently formed by a dried mucilaginous substance. Iodine reactions were not obtained. In all its aspects the type resembles a specimen of *M. delectabilis* so closely that I have no hesitation in referring *M. leucophaea* to it as a synonym. *M. delectabilis* is very common in western Washington, and Murrill could hardly have failed to collect it.

68. **Mycena Rickenii,** sp. nov.

Illustrations: Text fig. 14, nos. 8–9 (p. 154).

Pileus 5–8 (15) mm. latus, obtuse conicus dein umbonatus, pruinosus, albidus; lamellae confertae, angustae, adnatae, albidae; stipes 1–3 cm. longus, 1.5–2 mm. crassus; aequalis, cartilagineus, pruinosus, deorsum strigosus; sporae 5–6 × 4–5 μ; cystidia 30–60 × 9–18 μ, subcylindrica vel clavata; basidia bispora. Specimen typicum in Herb. Univ. Mich. conservatum. Legit A. H. Smith, n. 33–996, prope Ives Lake, Huron Mts., Mich., Sept. 16, 1933.

Pileus 5–8 (15) mm. broad, obtusely conic, becoming expanded umbonate, pruinose at first, soon glabrous, translucent-striate toward the margin, soon opaque, pure white; flesh thin, white, fragile, odor and taste not distinctive; lamellae close to crowded, adnate, narrow but becoming slightly ventricose, white, edges even and concolorous with the sides; stipe 1–3 cm. long, 1.5–2 mm. thick, equal, tubular, pliant-cartilaginous, white, base strigose, remainder densely pruinose, somewhat glabrescent in age.

Spores 5–6 × 4–5 μ, subglobose to globose, hyaline, smooth, nonamyloid; basidia two-spored, 20–24 × 5–6 μ; pleurocystidia and cheilocystidia abundant, 30–60 × 9–18 μ, ventricose with broad apices or narrowly to broadly clavate, not incrusted, smooth; gill trama homogeneous, surface area compact but no true hypoderm differentiated, surface hyphae radially arranged in a rudimentary pellicle, the hyphae having smooth walls.

Habit, habitat, and distribution.—Scattered on needles under species of Picea and Abies; Huron Mountains, Michigan. Also known from Germany.

Material studied.—Smith, 33–996.

Observations.—This species is named in honor of Ricken, who first collected and adequately described it (under the name *M. lactea*). Apparently it is very rare. The small somewhat globular spores and large clavate to subcylindric cystidia distinguish it. In the field it is almost indistinguishable from small forms of *M. olida* if one disregards habitat. The two-spored form of the latter, however, has larger spores and a more rooting stipe. To me there is a very strong possibility that this *is* the *M. lactea* of Fries; but since that name has been rather freely interpreted, as previously pointed out under *M. olida* var. *americana* (p. 153), it seems best to disregard it in order to avoid more confusion.

69. Mycena pseudolactea Kühner

Encyc. Myc., 10: 632. 1938

Illustrations: Plate 11 A; Text fig. 15, nos. 1–2 (p. 160).

Pileus (3) 5–15 mm. broad, broadly conic to convex, margin decurved or slightly incurved at first, becoming broadly campanulate or plane, the margin wavy or slightly crenate in age, surface pruinose and dull but becoming moist and faintly translucent-striate, chalk white and opaque when faded, becoming yellowish ("light buff," "cream color," or "Naples yellow") in age; flesh thin, pliant, white, odor and taste not distinctive; lamellae close to subdistant, 18–25 reach the stipe, two or three tiers of lamellulae, narrow but becoming ventricose in age, ascending-adnate at first, horizontally adnate in age, white or cream-colored, sometimes intervenose, edges even; stipe (1) 3–7 cm. long, (0.5) 1–1.5 mm. thick, equal or with either or both the apex and the base slightly enlarged, either with a small tubule or solid, cartilaginous but pliant, base conspicuously strigose, remainder

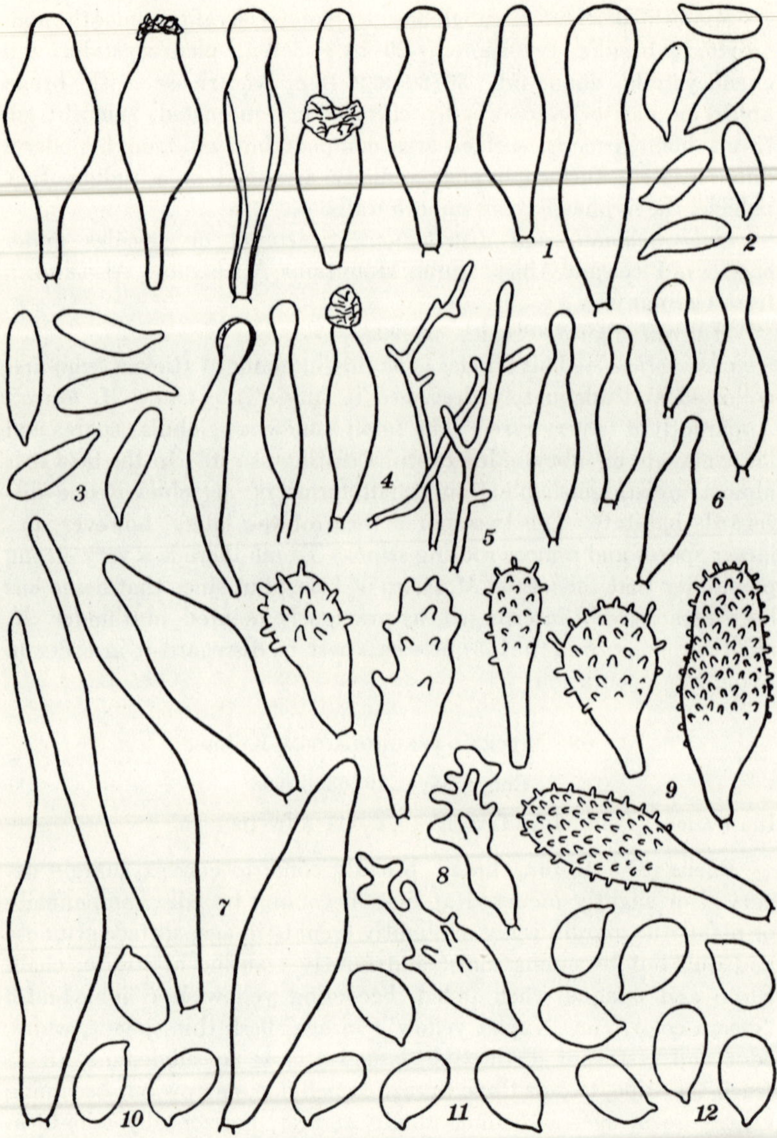

FIG. 15. *M. pseudolactea:* 1, pleurocystidia and cheilocystidia; 2, spores. *M. delicatella:* 3, spores (of type); 4–5, cystidia and hairs from surface of pileus; 6, pleurocystidia and cheilocystidia. *M. carolinensis:* 7, pleurocystidia; 8, cheilocystidia; 10 and 11, spores. *M. flavifolia:* 9, cheilocystidia; 12, spores

finely pruinose under a lens, appearing glabrous to the naked eye, somewhat translucent white at first, soon chalk white.

Spores 6–8 × 2.5–3 μ, subcylindric with a long oblique apiculus, yellowish in iodine (nonamyloid); basidia four-spored; pleurocystidia scattered to quite abundant, 36–60 × 9–12 μ, obtusely fusoid-ventricose to capitate or merely with a broad rounded apex, hyaline, with or without an amorphous incrustation; cheilocystidia similar to pleurocystidia or smaller, more or less incrusted; gill trama homogeneous, yellowish in iodine; pileus trama homogeneous, yellowish in iodine, surface with numerous projecting filaments or with subcapitate pilocystidia 12–20 × 3–9 μ, often with incrusted amorphous globules of a resinous substance around the apex; caulocystidia similar to the pilocystidia, rather abundant.

Habit, habitat, and distribution.—Gregarious to scattered on conifer debris and in sphagnum bogs during the summer and fall; Michigan in the United States and Ontario in Canada. Apparently this species is not so abundant as *M. delicatella.*

Material studied.—Smith, 4323, 4407, 15750, 15855.

Observations.—*M. pseudolactea* is most likely to be confused with *M. delicatella.* The large, rounded, conspicuous pleurocystidia, however, distinguish it at once. In addition, it characteristically has a longer stipe, broader pileus, and, at least in my specimens, not so distinct a resinous feel when a cap or a stipe is rubbed between the fingers. I have found them growing together and must admit that it would be very easy to get a mixed collection.

70. **Mycena delicatella** (Pk.), comb. nov.

Agaricus (*Collybia*) *delicatellus* Peck, Ann. Rep. New York State Mus., 30: 39. 1878.
Collybia delicatella Saccardo, Syll. Fung., 5: 224. 1887.
Gymnopus delicatellus Murrill, North Am. Flora, 9: 354. 1916.
Mycena crystallina Peck, Ann. Rep. New York State Mus., 41: 63. 1888.
Prunulus crystallinus Murrill, North Am. Flora, 9: 322. 1916.

Illustrations: Plate 11 D; Text fig. 15, nos. 3–6.

Pileus 3–10 mm. broad, obtusely conic to convex, the margin usually incurved at first, soon becoming plane or with a slight umbo, surface pruinose to minutely pubescent, remaining unpolished or becoming glabrous only in age, very faintly translucent-striate (appearing hygrophanous if water-soaked), color watery milk white at first, opaque and chalk white when faded or the disc becoming slightly creamy in age; flesh thin, equal, flexible to somewhat fragile, white, odor and taste not distinctive; lamellae close to subdistant, narrow, free or narrowly adnate, pure white, edges even or concolorous with the surface; stipe 1–3 cm. long, 1 ± mm. thick, equal, solid or with a tubule, pliant but cartilaginous, base not at all bulbous and very densely strigose, the hairs radiating to form a rather distinct mat, densely pruinose-pubescent over all at first, appearing somewhat glabrous in age, not changing color or becoming only slightly yellowish.

Spores 7–9 × 2.5–3 μ (four-spored), 9–12 × 3–3.5 μ (two-spored); subcylindric to slightly ventricose with a tapered apiculus, nonamyloid; basidia two- or four-spored; pleurocystidia and cheilocystidia similar and very abundant but nevertheless difficult to demonstrate, 18–26 × 4–6 (9) μ, nearly cylindric or somewhat ventricose, the apex obtuse or rounded into a small head, often with a conspicuous resinous deposit over the head; gill trama homogeneous, yellowish in iodine, of narrow hyphae; pileus trama compact, homogeneous, yellowish in iodine, the surface covered with a dense turf-like growth of cystidia and filamentous hyphae 15–28 × 3–8 μ, and these often with contorted or rounded incrusted apices, sometimes capitate; caulocystidia very numerous and similar to the pilocystidia.

Habit, habitat, and distribution.—Gregarious to scattered on fallen twigs and needles of conifers; New York, Michigan, Idaho, Washington, Oregon, and California. It is fairly common during the late summer and fall.

Material studied.—Smith, 33–726, 33–783, 33–819, 34–104, 277, 627, 727, 925, 998, 3134, 3226, 3443, 3510, 3542, 3619, 3671, 3702, 3709, 4322, 4407, 5072, 8036, 8041, 8748, 8820, 9398, 14979, 17846, 17964, 18148, August 30, 1934, New York. Kanouse, September 2, 1923. Kauffman (as *M. ludia* sensu Ricken), Wyoming; Mt. Hood and Takilma, Oregon. Slipp, UIFP: 2595, 3131, 3680. Stuntz, 810.

Observations.—This species is best characterized by its long, cylindric spores, the cystidia of the gills, pileus, and stipe, and its generally robust stature. The resinous incrustation of the cystidia (best demonstrated in revived material) causes both the pileus and the stipe to feel slightly sticky to the touch. Because of this character one must be careful not to confuse *Marasmius resinosus* with *Mycena delicatella.* The stipe is not readily separable from the tissue of the pileus, and for this reason the species cannot be placed in the subgenus Pseudomycena. I overlooked the cystidia in my first examination of the type of *Collybia delicatella.* In order to demonstrate them clearly one needs to use an oil-immersion lens. A reëxamination of the type revealed that it did possess the embedded cystidia which are characteristic of *M. crystallina* and that the pileus and stipe were covered with similar incrusted cystidia. In fact, the microscopic characters of both species are identical in every respect, including the reactions to iodine. Peck obviously was led to place the first material he collected in Collybia because of the incurved margin of the pileus. All stages of variation from an inrolled pileus margin to one which was perfectly straight have been observed, and in this species one should not regard the character as taxonomically significant. In view of the additional data obtained on *C. delicatella* I have no hesitation in referring it to Mycena and in making *M. crystallina* a synonym of it.

Kühner has used the name "*M. lactea*" for this fungus. There is no question in my mind whether or not this species occurs in Europe, but the use of the name "*M. lactea*" for it does have certain disadvantages, the most serious one being that the fungus as I know it can by no stretch of the imagination be reconciled with Fries's illustration of *M. lactea* in *Icones Selectae Hymenomycetum*. In his *Monographia Hymenomycetum Sueciae* Fries described what he apparently considered to be two forms of *M. lactea*—A and B. Later (1872) he apparently referred form B to *Agaricus lacteus* var. *pithyus*, thus indicating that he considered them at least somewhat distinct. This variety apparently approaches the present concept of the species in Europe more closely than does the form Fries considered typical. In addition, European workers have not been consistent in their interpretations of the microscopical characters of the Friesian species. With this situation existing, it seems best to drop the name "*M. lactea*" and to use a name based on an existing type. Most reports of *M. echinipes* in North America very likely are based on *M. delicatella*. Kauffman's report of *M. ludia* (1926) also is based on *M. delicatella*.

Subsection Euadonidae

KEY TO SPECIES

1. Pileus yellow, orange to flame-scarlet 2
1. Pileus pink to coral red 9

2. Cheilocystidia clavate with roughened apices or at least this type mixed with the others 3
2. Cheilocystidia fusoid-ventricose, smooth or sometimes incrusted (when revived in KOH) 4

3. Pileus smoky yellow; pleurocystidia absent .... 71. *M. flavifolia*
3. Pileus yellow; pleurocystidia abundant .... 72. *M. carolinensis*

4. Pileus pale yellow to cadmium yellow 5
4. Pileus reddish to reddish yellow, at least when young 6

5. Growing singly from buried nuts and acorns or from hulls of these fruits .... 75. *M. luteopallens*
5. Growing gregarious on humus, usually under conifers in mountains
   (*a*) Spores 7–9 μ long .... 73. *M. flavoalba*
   (*b*) Spores 5–6 μ long .... 74. *M. flavoalba* var. *microspora*

6. Pleurocystidia absent or rare and usually occurring near gill edge .. 7
6. Pleurocystidia abundant and conspicuous 8

7. Spores 5–6.5 × 5–6 μ (4-spored) .... 76. *M. leptophylla*
7. Spores 7–9 × 5–6 μ (4-spored) .... 77. *M. roseipallens*

8. Pileus scarlet when fresh, fading to orange; cystidia pointed 80. *M. adonis*
8. Pileus as above; cystidia obtuse .... 78. *M. fusipes*
8. Pileus orange when young .... 79. *M. aurantiidisca*

9. Cheilocystidia roughened (but smooth type may also be present) ... 10
9. Cheilocystidia smooth and fusoid-ventricose 11

10. Pleurocystidia absent .... 84. *M. monticola*
10. Pleurocystidia present (fusoid-ventricose) .... 83. *M. subincarnata*

11. Gregarious to subcespitose on leaf mold under hardwoods, occasionally on bark of trees or on debris .... 81. *M. roseocandida*
11. Gregarious to scattered on beds of moss, particularly sphagnum, and on needle carpets under conifers (a form of *M. flavoalba* with a pinkish disc keys out here also) .... 82. *M. amabilissima*

71. Mycena flavifolia Peck

Bull. New York State Mus., 167: 28. 1913

Prunulus flavifolius Murrill, North Am. Flora, 9: 324. 1916.

Illustrations: Text fig. 15, nos. 9, 12 (p. 160).

"Pileus thin, slightly submembranaceous, conic or convex, sulcate striate, somewhat plicate-crenate on the margin, glabrous, pale smoky yellow becoming pale pinkish brown or subalutaceous in drying, sometimes slightly umbonate; lamellae thin, close, broad at the outer extremity, narrowed toward the stem, pale yellow, becoming pallid in drying; stem slender, equal, glabrous, hollow, chestnut colored; spores ellipsoid or subovoid, 6–8 × 4–5 μ.

"Gregarious. Under balsam fir trees. North Elba. September. The center of the pileus is often more highly colored than the rest."

The original description is here quoted. I have not seen fresh specimens. Sections from the type showed that the basidia were four-spored; the spores smooth, 7–8 × 4–5.5 μ, and broadly ellipsoid. Pleurocystidia were not found. The cheilocystidia measured 12–26 × 8–12 μ and were clavate, and the enlarged portions were covered with numerous short rodlike protuberances. The fungus is related to *M. elegans*, as that name is used in this work, but apparently differs in not having a yellow gill edge and in lacking pleurocystidia. Both apparently have a tendency, rather common among Mycenae, to stain reddish brown in age.

Mains (5183, August 24, 1940, near Estes Park, Colorado) has found one collection. It is the only authentic material I have seen besides the type. The spores, apparently, are nonamyloid, but the reaction was not convincing.

72. Mycena carolinensis Smith and Hesler

Journ. Elisha Mitchell Sci. Soc., 56: 319. 1940

Illustrations:

Text fig. 15, nos. 7–8, 10–11 (p. 160).

Smith and Hesler, Journ. Elisha Mitchell Sci. Soc., 56: Fig. I, nos. 8–10. 1940.

Pileus 5–10 mm. broad, convex, remaining broadly convex or becoming plane, margin appressed against the stipe at first, surface subviscid to viscid, glistening when moist, translucent-striate and soon sulcate to the flattened disc, glabrous, color "maize yellow" when young, fading on the margin and the disc becoming tinged with brown in age, sometimes sordid brownish over all; flesh thin, pliant (but not reviving), whitish, odor and taste mild; lamellae bluntly adnate, moderately broad (up to 2 mm.), subdistant, 10–13 reach the stipe, one tier of lamellulae, yellowish white, edges pallid, under a lens the faces appearing pruinose from the projecting cystidia; stipe 2–3 cm. long, 0.5–1 mm. thick, strict, equal, pliant, pale yellow when young, becoming sordid brownish in age, dull and unpolished (pubescence hardly visible under a lens), inserted on the substratum.

Spores 7–8 × 3.5–4 μ, amyloid, ellipsoid; basidia four-spored; cheilocystidia embedded or projecting, clavate to subcapitate with echinulate apices or more or less contorted, some similar to pleurocystidia; pleurocystidia abundant, smooth, with pale-yellowish content, 60–70 × 8–12 μ, somewhat ventricose with an elongated neck; caulocystidia abundant, some resembling pleurocystidia and some the roughened cheilocystidia, some variously contorted or elongated and branched; gill trama homogeneous, dark red in iodine, pileus trama composed of a subgelatinous pellicle made up of yellow hyphae; tramal body dark vinaceous red in iodine.

Habit, habitat, and distribution.—Not uncommon, single or in twos and threes on dead twigs of Rhododendron and dead stems of *Leucothoë Catesbaei* (Walt.) Gray; North Carolina and Tennessee.

Material studied.—Smith, 7359, 10413, 10871, 14879, 14885. Hesler, 9624, 11479, 14186, 14228.

Observations.—The stipe of *M. carolinensis* revives somewhat, and at first it seemed questionable whether or not the species should be placed in Marasmius. Dried specimens do not revive so well as those of *M. corticola*, however. This, in view of the nature of the trama of the pileus, its reactions to iodine, and the two types of cystidia on the gill edge, is more indicative of Mycena than of Marasmius. Dr. Jakob E. Lange, who saw fresh material from Cades Cove, Tennessee, pronounced it a Mycena. Although differing sharply in color, *M. carolinensis* has a superficial resemblance to *M. corticola*. The consistency, the manner in which the stipe is attached to the substratum, and the shape of the pileus are all suggestive of that species. It is entirely possible that *M. carolinensis* at times has gills margined with pale yellow. This would happen if the fusoid-ventricose cystidia were more abundant than usual along the gill edge. The relationships of this species are a bit obscure, to say the least. It may possibly be related to *M. viridimarginata* or to *M. avenacea*, but because of its consistency, a gelatinous pellicle over the pileus, and the two distinct types of cystidia such a relationship does not seem likely.

73. Mycena flavoalba (Fr.) Quélet

Champ. Jura et Vosges, p. 103. 1872

Agaricus (Mycena) flavoalba Fries, Epicr. Syst. Myc., p. 103. 1838.

Illustrations:

- Plate 19 A; Text fig. 16, nos. 4–5 (p. 170).
- Bresadola, Icon. Mycol., 5, pl. 231.
- Fries, Icon. Sel. Hymen., I, pl. 79, fig. 5.

Konrad et Maublanc, Icon. Sel. Fung., 3, pl. 230, I.
Lange, Flora Agar. Dan., 2, pl. 53 G (very good).
Ricken, Die Blätterpilze, 2, pl. 110, fig. 8.

Pileus 1–2 (2.5) cm. broad, obtusely to rather sharply conic when young, with an appressed margin, becoming obtusely campanulate, broadly conic or at times nearly convex and then occasionally with an abrupt papilla on the disc, the disc sometimes flattened, the margin flaring at maturity or slightly recurved, surface smooth, glabrous, moist, translucent-striate, hygrophanous, "cream-buff" or a clearer yellow at first, the margin paler and almost white, fading to "cartridge buff" on the disc and a dead yellowish white along the margin, buttons pink in one form; flesh yellowish to white, thick under the disc, otherwise thin, moderately fragile, odor and taste not distinctive; lamellae ascending and somewhat uncinate or toothed, narrow at first but becoming rather broad (2.5 mm. and becoming 3–4 mm.), nearly equal or slightly ventricose in age, close to subdistant, 18–24 reach the stipe, two tiers of lamellulae, intervenose at times, white to creamy white, edges even and whitish, waxy in appearance and consistency; stipe 3–8 cm. long, 1–2.5 mm. thick, equal, tubular, somewhat elastic, cartilaginous, and not particularly fragile, base white-strigose or surrounded with a matted white mycelium, glabrous above the base, pruinose toward the apex, when moist translucent and slightly transversely undulate or uneven, white to pale yellow.

Spores 7–9 × 3–4.5 μ, ellipsoid, nonamyloid; basidia four-spored; pleurocystidia and cheilocystidia similar and abundant, fusoid-ventricose with long, rather narrow necks, 46–62 × 9–14 (18) μ, the neck often incrusted with a mucilaginous substance, otherwise smooth, hyaline; gill trama homogeneous, pale yellow in iodine; pileus trama with a thin, poorly differentiated pellicle, a somewhat differentiated hypoderm (most pronounced in old caps) and the remainder made up of somewhat enlarged cells and fairly compact, pale yellow in iodine.

Habit, habitat, and distribution.—Scattered to densely gregarious on needle beds under conifers and on humus in oak woods during the fall months; North Carolina, Michigan, Colorado, Wyoming, Idaho, Washington, and Oregon. The species is rare, but occurs in large quantities in certain localities.

Material studied.—Smith, 7744, 7840, 8024, 8026, 8035, 11073, 15859, 17012, 17049. Gruber, Oregon. Hesler, 14689. Kauffman, three collections, Oregon, Idaho, Wyoming. Mains, 5024. Overholts, 1932; July 20, 1914, Colorado. Slipp, UIFP: 1528, 3089.

Observations.—When growing under conifers the base of the stipe is hardly curved and is covered with a thin mat of white mycelium or scattered hairs; when growing in oak woods the stipe is often decidedly curved at the base and has a rootlike strigose portion which may extend for some distance under the leaves. I regard this difference as one caused by the nature of the habitat. Only one collection from under oak was made in Michigan, but the species was very abundant at Lost Creek, Oregon (just west of McKenzie Pass). Two forms were found there, each growing in troops of several hundred carpophores. One had a yellow cap at all stages. The other was characterized by buttons with pinkish pilei, which by the time they matured faded to pale yellow or retained a slight pinkish tinge around the disc. In age the two forms were indistinguishable. Josserand (1930) has found a somewhat similar situation in Europe. Kühner (1938) has used the name *M. floridula* for this reddish form, and I must admit the difference in color appears to be constant. The impression one gets from collecting these two together, however, is that at best one is just a variety of the other. *M. amabilissima* Peck is very distinct from the pink variety of *M. flavoalba* not only in its brighter colors but also in the manner in which they fade from red to white—not yellow.

My specimens of the pink variety of *M. flavoalba* did not have colored gills at first, and they differ in this respect from *M. floridula* of Josserand and Kühner.

74. Mycena flavoalba var. **microspora**, var. nov.

Pileus 5–15 mm. latus; stipes 2–4 cm. longus, 1–1.5 mm. crassus; sporae 5–6 × 3 μ.

In all other respects the collection resembled typical material. The specimens were found September 8, 1941, at Park Creek, 1000 feet elevation, Mt. Baker, Washington. A. H. Smith 16753—type. The specimens were growing gregariously under conifers (pine, hemlock, and Douglas fir). Since the spore size of *M. flavoalba* has been remarkably constant in all my other collections, it appeared advisable to segregate this material. It occurred in considerable quantity, and the spore size was constant. Slipp, UIFP: 3565, has collected it in Idaho.

75. Mycena luteopallens (Pk.) Saccardo

Syll. Fung., 9: 37. 1891

Agaricus (*Mycena*) *luteopallens* Peck, Ann. Rep. New York State Mus., 32: 27. 1880.

Prunulus luteopallens Murrill, North Am. Flora, 9: 325. 1916.

Marasmius nucicola McDougal, Trans. Ill. State Acad. Sci., 17: 84. 1925.

Illustrations:

Text fig. 16, nos. 1–3, 6 (p. 170).

Beardslee and Coker, Journ. Elisha Mitchell Sci. Soc., 40, pl. 20 (upper fig.).

McDougal, Trans. Ill. State Acad. Sci., 17: 84, fig. 1 (as *Marasmius nucicola*).

Pileus 8–15 mm. broad, ovoid with an appressed margin at first, becoming broadly conic, campanulate, convex, or in age plane or slightly umbonate, the margin often becoming wavy or irregular but sometimes remaining decurved, surface canescent but soon naked and glabrous, translucent-striate when moist, somewhat tuberculate-striate in age, hygrophanous, "cadmium yellow" to "buff yellow" (brilliant orange to rich yellow), fading to whitish; flesh thin and somewhat pliant, yellowish to pallid, odor and taste not recorded; lamellae depressed-adnate to adnexed, rather broad, sometimes ventricose in age, subdistant, pruinose under a lens from cystidia, margins pallid, faces yellowish to tinged incarnate; stipe 5–9 cm. long, 1–2 mm. thick, equal, flexuous or strict, terete, base prolonged and coarsely strigose (but not forming a typical pseudorhiza), tubular, concolorous with the pileus toward the apex, base paler, upper portion pruinose at first.

Spores 7–9 × 4–5.5 μ, ellipsoid to subovoid, immature individuals slightly but distinctly amyloid and very slightly roughened (use oil immersion), mature spores smooth, nonamyloid; basidia two- or four-spored, 26–30 × 6–8 μ; pleurocystidia abundant, 50–80 (110) × 8–18 μ, ventricose with a long or a short neck and a rather distinct and usually curved pedicel, smooth, hyaline, content homogeneous or with a few granules in water mounts; cheilocystidia abundant, hyaline, 38–52 × 9–10 μ; gill trama interwoven, homogeneous, pileus trama with a distinct pellicle, a sharply differentiated hypoderm, and a filamentous tramal body, amyloid reaction faint and not conclusive (a tinge of vinaceous red appeared in a few places); stipe tissue deep vinaceous red in iodine.

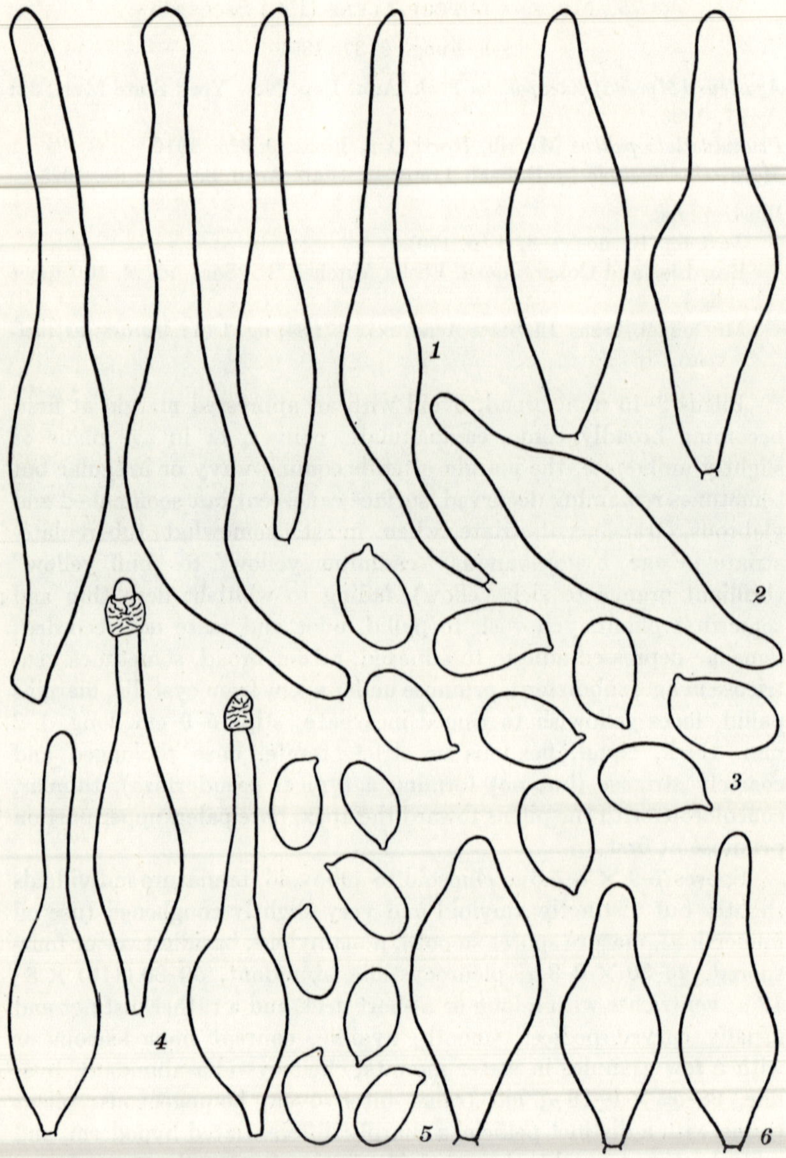

FIG. 16. *M. luteopallens:* 1 and 2, pleurocystidia; 3, spores; 6, cheilocystidia. *M. flavoalba:* 4, pleurocystidia; 5, spores

Habit, habitat, and distribution.—Scattered or in groups of two to four, usually on buried remains of hickory nuts and walnuts in September, October, or November; North Carolina, Tennessee, New York, Michigan, and Illinois.

Material studied.—Smith, 32–479, 32–519, 15501 (white form), 15515. Hesler, 5770, 6512.

Observations.—This agaric can be recognized at sight by the curious habitat, brilliant color, and almost Hygrophorus-like appearance. It is one of the very brightly colored species, but does not appear to be closely related to such Mycenae as *M. roseipallens* and *M. amabilissima.* Its conspicuous cystidia, the well-differentiated hypoderm of the pileus, and the amyloid immature spores are its outstanding microscopic characters. As Kühner noted, the spores often appear to be roughened by thin plates of adhering material which give a very strong amyloid reaction. The mature spores which I have examined do not appear to be roughened (under oil immersion) and do not give an amyloid reaction, but I have had only one spore deposit from which to study this reaction.

A single fruiting body of a fungus identical with *M. luteopallens* in all respects except color was found near Ann Arbor in 1940 (15501). It was white in all parts. It grew from an old walnut husk in the locality where *M. luteopallens* usually occurs. It was fertile and had smooth spores, 7–9 × 4–5 μ, and the usual cystidia and hypoderm. There was no possibility of its being a faded carpophore. If this form is ever found in any quantity, it will bring up an interesting question concerning its taxonomic position and also the emphasis that should be placed on color as a specific character. With the information at hand it hardly appears justifiable to consider it anything more than an albino form.

76. Mycena leptophylla (Pk.) Saccardo

Syll. Fung., 5: 304. 1887

Agaricus (*Mycena*) *leptophyllus* Peck, Ann. Rep. New York State Mus., 24: 63. 1872.

Prunulus leptophyllus Murrill, North Am. Flora, 9: 325. 1916.

Illustration: Text fig. 17, no. 3 (p. 173).

"Pileus thin, campanulate or convex, subpapillate, smooth, striatulate when moist, pale reddish-yellow, the disk brighter colored;

lamellae close, narrow, widest at the middle, pointed at the outer extremity, sharply uncinate at the inner, whitish or yellow with a flesh-colored tint; stem slender, tough, hollow, smooth, whitish.

"Plant 1′–1.5′ high, pileus 3″–5″ broad, stem .5″ thick.

"Old mossy logs and rotten wood in woods. Greig. September.

"The papilla of the pileus is sometimes absent."

I have seen fresh material of one collection of this species from near Ann Arbor. It was collected by a member of a mycology class in 1930, but the dried specimens were lost. The pilei measured 10–15 mm. broad, and the discs were obtuse. Neither color notes nor a good description was obtained. However, the spores measured 6–8 μ and were subglobose, with a conspicuous oblique apiculus. Only cheilocystidia were observed. Among Kauffman's collections I found one (B. B. Kanouse, October 23, 1926, at Ann Arbor) which also belongs here. Kauffman had placed it in *M. adonis.* Sections of Peck's type have been made, and the basidia were found to be four-spored. The spores were subglobose with an oblique apiculus and measured 5–6.7 × 5–6 μ. The cheilocystidia were smooth, obtuse, and fusoid-ventricose. A few were present on the sides a short distance from the gill edge, but they could hardly be classed as pleurocystidia. No others were found on the faces of the gills. The species is best separated from *M. roseipallens* by its spores.

77. Mycena roseipallens Murrill

Mycologia, 8: 221. 1916

Prunulus roseipallens Murrill, North Am. Flora, 9: 324. 1916.

Illustrations:

Plate 20; Text fig. 17, nos. 1–2.

Bresadola, Icon. Mycol., 5, pl. 229 (as *M. floridula*).

Smith, Am. Journ. Bot., 22, pl. 2, fig. 1.

Pileus 5–18 (20) mm. broad, obtusely conic when young, the margin either appressed against the stipe or slightly incurved, becoming campanulate to expanded-umbonate, the margin sometimes flaring and sometimes recurved in age and often becoming lobed, the umbo sometimes disappearing in age, pruinose at first but soon naked and moist, translucent-striate, color dark to light brick red at first, becoming yellowish incarnate or yellowish tan in age, the margin sometimes whitish ("ferruginous" to "rufous" at first, paler in age); flesh thin, not markedly fragile, pallid incarnate to yellowish or

FIG. 17. *M. roseipallens:* 1, cheilocystidia; 2, spores. *M. leptophylla:* 3, spores. *M. fusipes:* 4, spores; 5, cheilocystidia. *M. aurantiidisca:* 6, spores; 8, pleurocystidia. *M. adonis:* 7, pleurocystidia; 9, spores. *M. subincarnata:* 10, cheilocystidia

whitish, odor faintly spermatic at times (Kühner reports it as faintly fragrant), taste not distinctive; lamellae narrow to moderately broad, somewhat ascending, close, 16–24 reach the stipe, two or three tiers of lamellulae, rather thick, whitish to pale incarnate when young, often sordid pale rufous orange in age, edges even and concolorous with the faces; stipe 2–7 cm. long, 1–2 mm. thick, tubular, moderately fragile, equal, the base frequently prolonged and densely strigose (sometimes forming a slight pseudorhiza), surface white, glabrous or very finely pruinose-pubescent, becoming yellowish in age, finally discoloring at times and becoming dark sordid brown (near "mummy brown").

Spores (6) 7–9 × 5–6.7 μ, ovoid to somewhat ellipsoid, non-amyloid; basidia four-spored or two-spored (then the spores 8–10 × 6–8 μ); pleurocystidia very rare or apparently absent, similar to cheilocystidia; cheilocystidia numerous, 28–47 × 7–12 μ, subcylindric, subclavate, or somewhat fusoid-ventricose, hyaline, smooth; gill trama homogeneous, the cells rather broad and interwoven, yellowish in iodine; pileus trama with a distinct pellicle, a moderately well-differentiated hypoderm and the remainder filamentous, yellowish in iodine.

Habit, habitat, and distribution.—Scattered to gregarious on debris of elm and ash, and on alder debris in Washington, particularly on the bark of partly decayed logs, late June and early fall; not uncommon in New York, Michigan, Missouri, and Washington.

Material studied.—Smith, 32–227, 32–244, 32–259, 32–282, 32–309, 32–332, 32–342, 32–619, 33–722, 776, 1465, 6832, 7628, 7674, 11081, 15239, 17256, five unnumbered collections, Michigan. Atkinson, 23409, 23147, 23369 (as *M. adonis*). Kauffman, two collections (as *M. subincarnata*), Michigan. Mains, 1932, Michigan.

Observations.—The habit and consistency of *M. roseipallens* reminds one very much of *M. olida* var. *americana*. Both occur in similar habitats at the same time of year. The colors readily separate the two if young carpophores are compared, but in age either one may become sordid yellowish tan. The almost complete absence of pleurocystidia distinguishes *M. roseipallens* from the confusing group of rosy-colored species centering around *M. amabilissima*.

78. Mycena fusipes Murrill

Mycologia, 8: 220. 1916

Prunulus fusipes Murrill, North Am. Flora, 9: 337. 1916.

Illustrations: Text fig. 17, nos. 4–5 (p. 173).

"Pileus conic to convex, becoming nearly expanded, solitary or gregarious, about 1 cm. broad; surface moist, glabrous, striate, miniatous or incarnate, the margin paler, entire, appressed when young: lamellae nearly free, crowded, inserted, ventricose, white with a yellowish tint: spores ellipsoid, smooth, hyaline, 7–8 × 4–5 μ: stipe unusually large at the center and tapering at both ends, smooth, glabrous, translucent, pale-yellow, 6 cm. long, 2–3 mm. thick.

"Type collected on the ground, probably on dead wood, in woods near Seattle, Washington, October 20–November 1, 1911, *W. A. Murrill 468* (herb. N. Y. Bot. Gard.).

"Distribution: Known only from the type locality."

I have looked for this species in vain. A study of the type yielded the following data: The spores measure 7–8 × 4–5 μ and are ellipsoid, hyaline, smooth, and nonamyloid. The basidia seen were all two-spored, but "tetrads" and "diads" of spores were both noted in mounts from the type, which indicates that four-spored basidia were present. There appears to be a copious secretion of a mucilaginous substance in this species; most of the spores adhere in groups when the sections are revived in KOH. Pleurocystidia and cheilocystidia are similar and quite abundant, and have their apices incrusted with an amorphous substance in revived mounts. They are decidedly ventricose, 10–15 μ broad, and have short necks terminating in obtuse apices. Many smaller cystidia are also present in the hymenium. These are of the same shape as the others but measure 22–36 × 7–10 μ. The larger individuals are 30–46 μ long and project up to 10 μ beyond the basidia. The gill trama is yellowish in iodine. The pileus trama is very compact and is characterized by a well-developed pellicle, the hyphae of which are covered with numerous short rodlike projections. All parts were yellowish in iodine, and there did not appear to be a distinct hypoderm.

If one had fresh material in hand it might be very easy to characterize this species accurately. I feel almost certain that the character of the stipe, which Murrill emphasized, is merely a chance variation. The species clearly falls in the very confusing group of reddish Mycenae centered around *M. adonis*. At first I thought it was identical with *M. roseocandida* because of its obtuse cystidia. After studying the types again, however, I am inclined to keep the two separate. Since dried specimens of all species in this group look about alike, not much emphasis can be placed on gross characters in herbarium material. Because of its obtuse cystidia it is apparently distinguishable from *M. adonis*.

79. Mycena aurantiidisca Murrill

Mycologia, 8: 220. 1916

Prunulus aurantiidiscus Murrill, North Am. Flora, 9: 336. 1916.

Illustrations: Text fig. 17, nos. 6, 8 (p. 173).

Pileus 7–20 mm. broad, obtusely conic when young, not expanding or becoming merely broadly conic, margin appressed against the stipe when young, flaring somewhat in age, striate when moist, glabrous, "mikado orange" when young and fresh, fading to "mustard yellow" on the disc and whitish along the margin; flesh thin and fragile, orange or yellow, odor and taste mild; lamellae bluntly adnate to slightly hooked, close to subdistant, 20–24 reach the stipe, narrow to subventricose, white at first, becoming tinged yellowish in age, margin concolorous with the faces; stipe 2–3 cm. long, 1 mm. thick, equal, strict, hollow, fragile, surface faintly pruinose above, scarcely fibrillose at the base, upper portion white, basal portion yellowish.

Spores ellipsoid, 7–8 × 3.5–4 μ, nonamyloid; basidia four-spored, 20 × 6 μ; cheilocystidia and pleurocystidia similar, 38–47 × 5–14 μ, fusoid with a long narrow neck, smooth in mounts of fresh material, frequently having the apex or the neck incrusted with an amorphous mass in mounts of revived material, hyaline; pileus trama with a poorly differentiated pellicle, the tissue beneath it homogeneous or the upper portion composed of only slightly enlarged hyphae, faintly vinaceous brown in iodine.

Habit, habitat, and distribution.—Gregarious under Douglas fir and pine during the spring and fall; Idaho, Washington, and Oregon. Not common.

Material studied.—Smith, 7750, 7778, 7880, 7932, 8239, 8415, 8928, 9305, 14641, 16463, 16507. Slipp, UIFP: 3249.

Observations.—It remains an open question whether *M. aurantiidisca* is a distinct species or merely a form of *M. adonis.* The difference in the color of the young stages is striking, but no more so than that of the two forms of *M. flavoalba.* In fact, there actually is less difference. The young stages of *M. adonis* are scarlet, whereas those of *M. aurantiidisca* are brilliant orange. The latter fades to whitish, at least on the margin, and the former to yellow, but both are characterized by dominantly yellow colors when first reaching maturity.

80. MYCENA ADONIS (Fr.) S. F. Gray

Nat. Arr. Brit. Plants, 1: 620. 1821

Agaricus adonis Fries, Syst. Myc., 1: 152. 1821.

Illustrations:

Plate 19 B; Text fig. 17, nos. 7, 9 (p. 173).
Bresadola, Icon. Mycol., 5, pl. 224, fig. 1.
Lange, Flora Agar. Dan., 2, pl. 53 A (pileus too pink).
Ricken, Die Blätterpilze, 2, pl. 109, fig. 6.

Pileus 5–12 (15) mm. broad, either sharply or obtusely conic when young, becoming broadly conic in age or narrowly campanulate, not expanding, margin appressed against the stipe at first, sometimes moist, opaque or nearly so at first, "scarlet" when fresh and moist, becoming orange or yellowish orange before losing moisture, hygrophanous, fading to "orange buff"; flesh thin, concolorous with the surface, fragile, odor and taste not distinctive; lamellae ascending-adnate or attached by a tooth, subdistant to close, 14–16 reach the stipe, two or three tiers of lamellulae, narrow, yellowish or tinged incarnate at first, margin paler and concolorous with the faces; stipe 2–4 cm. long, 1–2 mm. thick, equal, tubular, fragile, base hardly strigose, pruinose at first, polished and glabrous in age, pale yellow, becoming whitish, base often sordid yellow or brownish (not bister).

Spores narrowly ellipsoid, 6–7 × 3–3.5 μ, nonamyloid; basidia four-spored, 20–22 × 6–7 μ; cheilocystidia and pleurocystidia abundant and similar in shape and markings, (36) 40–58 × (8) 10–15 μ, fusoid and usually with a long aciculate neck (which is branched in some), smooth (but when dried material is revived in KOH an amorphous substance apparently holds spores and debris around the neck or apex, making them appear incrusted); gill trama very faintly vinaceous brown in iodine; pileus trama with a thin, poorly differentiated pellicle, beneath it a region of slightly enlarged cells, the remainder filamentous, the filamentous portion vinaceous brown in iodine.

Habit, habitat, and distribution.—Scattered to gregarious on needle beds under spruce and hemlock in the wet coastal conifer forests, or in the higher mountains, not uncommon in the spring and fall; Washington, Oregon, and California.

Material studied.—Smith, 2490, 2597, 3332, 3711, 13102, 13655.

Observations.—The occurrence of this fungus in the higher mountains and along the coast in our western states is in line with the distribution of a rather large number of alpine plants in the Olympic Peninsula of Washington. As the species is recognized here, it differs from the *M. amabilissima* group in becoming bright yellow as it fades instead of merely a very pale weak yellow or white. In their brilliance the colors of *M. adonis* remind one of those of *M. acicula*, *M. oregonensis*, and *M. strobilinoides*, but there appears to be no close relationship to any of these. Kühner lists *M. rubella* Quélet sensu Boudier, *Icones*, pl. 68, as a synonym of *M. adonis*. Boudier's fungus had two-spored basidia and spores 10–12 × 3.7–5.7 μ. Kühner's concept of *M. adonis* thus applies to rose-colored forms here treated under the names *M. roseocandida* and *M. amabilissima*.

81. Mycena roseocandida (Pk.) Bull. Saccardo

Syll. Fung., 5: 262. 1887

Agaricus (*Mycena*) *roseocandidus* Peck, Bull. Buffalo Soc. Nat. Sci., 1:47. 1873.

Prunulus roseocandida Murrill, North Am. Flora, 9: 323. 1916.

Illustrations:

Plate 11 B.

Beardslee and Coker, Journ. Elisha Mitchell Sci. Soc., 40, pl. 19 (as *M. amabillissima*).

"Pileus convex or broadly campanulate, subpapillate, striate nearly to the apex, white or rosy-red; lamellae close, uncinate, colored like the pileus; stem slender, smooth, white.

"Plant 2″ high, pileus 4‴–6‴ broad.

"Among mosses in woods. Adirondack Mountains. July.

"Usually the whole plant is pure white, but sometimes the pileus has a delicate rosy hue except on the apex and the margin. The striations of the pileus remain in the dried specimens. The papilla is sometimes very prominent, sometimes wanting."

Peck's original description is quoted. I have examined the type but found very few spores. Those seen were attached to sterigmata, measured 4.5–5.5 × 3–4 μ, and were broadly ellipsoid. Other spores scattered in the mount measured 6–7 × 4 μ. Because of their scarcity not much reliance can be placed on the spore size as obtained from the type. More than likely, spores from a deposit would have been larger than any of those found in my mounts. Pleurocystidia and cheilocystidia are similar and fairly abundant. On the faces of the gills two kinds were observed: (1) In one they were typically fusoid-ventricose, with the apex sometimes embedded in an amorphous substance; (2) In the other they were smooth, clavate, and with the incrustations over the broad apex. The former measure 48–54 × 8–10 μ; the latter, 32–46 × 10–18 μ. The clavate individuals were not abundant. Only the typcially fusoid-ventricose variety was found on the gill edge. Iodine reactions were not obtained.

I have never found fresh material with the clavate pleurocystidia, and as a result have never been certain of Peck's species. For the last ten years, however, I have been collecting in the vicinity of Ann Arbor a fungus which could easily pass for *M. roseocandida* if one overlooks the character mentioned above. This was previously referred to *M. amabilissima*, but it is as distinct as *M. flavescens*. The following is a description of the Michigan material:

Pileus 5–15 (25) mm. broad, obtusely conic or the disc flattened slightly, the margin incurved slightly at first, expanding and at maturity broadly umbonate with a plane or a recurved margin, surface smooth, glabrous, moist, and translucent, the margin translucent-striate and frequently incised, hygrophanous or only subhygrophanous, becoming opaque when faded, color pale pink ("seashell pink") over all, soon fading (while still moist) to whitish or yellowish ("ivory yellow"), the margin usually paler and the disc darker, the color sometimes persisting along the margin, often white when still apparently young and moist; flesh thin and fragile, pinkish but soon white, no odor, taste bitterish but soon mild; lamellae very narrow (1 mm. ±), bluntly adnate or becoming slightly adnexed, close, 15–20 reach the stipe, one to three tiers of lamellulae, pinkish like the pileus and fading to white or yellowish, intervenose, the edges slightly pruinose; stipe short or long, depending on the habitat, 10–15 × 1–2.5 mm. when growing on bark of trees or in other exposed places, 3–5 cm. long, 1–2 mm. thick when growing on humus or fallen leaves, equal or nearly so, solid but soon hollow and rather fragile, watery, terete or compressed, base white-strigose (if growing from leaf mats often curved and conspicuously strigose), pruinose above but soon polished and translucent-white to yellowish toward the base, pinkish above when young but usually fading quickly to white.

Spores 6–7 (8) × 3–3.5 μ, narrowly ovoid, smooth, hyaline, nonamyloid; basidia four-spored; pleurocystidia and cheilocystidia similar and abundant to scattered, 33–46 × 7–11 (13) μ, narrowly fusoid-ventricose with long, slender necks and acute apices, smooth; gill trama homogeneous, yellowish in iodine; pileus trama homogeneous beneath a thin pellicle, the cells of which are covered with short rod-like projections, yellowish in iodine.

Habit, habitat, and distribution.—Gregarious to subcespitose on bark, debris, or fallen leaves of deciduous trees in September and October; vicinity of Ann Arbor, Michigan.

Material studied.—Smith, 32–471, 32–483, 32–492, 32–517, 32–616, 33–1059, 978, 11077, 15195.

Observations.—During the season of 1940, which was a very unusual year for agarics in the vicinity of Ann Arbor, specimens of this species were found growing over the bark of a hop-hornbeam tree in much the same manner as *M. corticola.* Ordinarily, fruiting bodies occur in groups of two to five on debris under the fallen leaves, or among mosses in oak woods.

This material has now been tentatively placed in *M. roseocandida*, but it may be just a local variation of *M. amabilissima.* From field experience it appears to me that in various parts of North America there are many variations of these pink to rosy-red Mycenae with fusoid pleurocystidia and that they do not differ enough from one another to be classed as species. I have adhered to the traditional concepts of the various species, however, in order to keep the present treatment as consistent as possible with the existing literature, and to avoid the confusion in the use of names which would certainly arise from wholesale rearrangements. No matter how one treats the species of this group, the same difficulties are always encountered.

82. Mycena amabilissima (Pk.) Saccardo

Syll. Fung., 9: 37. 1891

Agaricus (*Mycena*) *amabillissimus* Peck, Ann. Rep. New York State Mus., 39. 1887.

Prunulus amabilissimus Murrill, North Am. Flora, 9: 324. 1916.

Illustrations:

Plate 21 A, B; Text fig. 18, nos. 3, 5 (p. 183).

Smith, Mycologia, 27: 594, figs. a, b.

Pileus 3–20 mm. broad, obtuse, or conic at first, becoming broadly conic or campanulate, sometimes expanded-umbonate, margin appressed against the stipe at first and in age often flaring or recurved, surface glabrous, polished, lubricous when moist, color "light coral red" (brilliant pinkish red) when young, fading to white in age or the disc pale creamy, opaque or distinctly translucent-striate; flesh thin, fragile, pinkish, becoming white, odor and taste not distinctive; lamellae adnate or toothed, usually ascending, narrow or in age slightly ventricose, subdistant to distant, 15–21 reach the stipe, two or three tiers of lamellulae, intervenose, white or tinted coral red, edges concolorous with the faces; stipe 3–5 cm. long, (0.5) 1–2 mm. thick, equal, hollow, watery and fragile, tinged coral red or entirely white, if colored at first soon fading, base becoming slightly yellowish in age, surface glabrous but pruinose and unpolished at first, soon polished and appearing moist.

Spores 7–9 × 3–4 μ, narrowly ellipsoid, pointed at one end, hyaline, smooth, yellowish in iodine; basidia four-spored; pleurocystidia and cheilocystidia similar, moderately abundant, fusoid-ventricose, with long, slender, almost filamentous necks, 40–65 × 8–15 μ, hyaline (in mounts of revived material spores often adhere to the neck as if held by glue); gill trama homogeneous, yellowish in iodine.

Habit, habitat, and distribution.—Gregarious on moss or on needle beds under conifers, often widely scattered. It is to be expected wherever sphagnum bogs are found. It fruits during the spring, late summer, and fall, but is most abundant in the fall. I have examined material from New York, Michigan, Washington, Oregon, and California in the United States and from Ontario in Canada.

Material studied.—Smith, 32–515, 924, 2551, 3131, 3280, 3388, 4820, 4923, 7590, 7841, 7880, 7936, 8017, 8107, 8629, 8742, 8917, 9280, 13654, 13998, 14496, 17014, 17068, September 5, 1929. Atkinson, 23590. Flett, Bremerton, Washington. Mains, 5023. Slipp, UIFP: 2644.

Observations.—This species is readily recognized by its clear rose color without yellow tints. In very vigorously developing carpophores the stipe, gills, and pileus are all colored bright rose. The stipe soon becomes entirely white, and the gills very pale. The gills, however, often retain a pale-rose tint until the pileus has faded to white. The cap may fade to pure white or retain splashes of rose around the disc or along the margin. Microscopically the spores and cystidia are quite similar to those of other species of this series and therefore are not very helpful. The lack of yellow in the color of the fruiting body is the most outstanding character. After the red has faded the pileus and stipe may assume slightly yellowish tints, but these are not present until the fruiting bodies are obviously old.

Previously I referred my material to *M. clavus.* Kühner tentatively placed *M. clavus* sensu Smith under *M. floridula* sensu Ricken.

In order to avoid confusion I have gone back to a name based on a type concerning which, in the restricted sense employed here, there is no confusion. For a discussion and arrangement of the European forms see Kühner (1938), pp. 553–562.

83. Mycena subincarnata (Pk.) Saccardo

Syll. Fung., 5: 262. 1887

Agaricus subincarnatus Peck, Ann. Rep. New York State Cab., 23: 83. 1872.
Prunulus subincarnatus Murrill, North Am. Flora, 9: 325. 1916.

Illustrations: Text figs. 17, no. 10 (p. 173); 18, nos. 1–2, 4.

Pileus 8–12 mm. broad, conic to convex, with the margin appressed at first, becoming campanulate to convex, sometimes papillate on the disc, or the disc depressed, surface glabrous and naked, moist, smooth, translucent on the margin, pale bright or dull incarnate over all at first, soon becoming paler and delicately yellowish around the disc, whitish along the margin, the intermediate area remaining delicate pink; flesh very thin and pinkish to white, pliant, odor and taste not distinctive; lamellae ascending-adnate with a slight tooth, becoming notched, moderately broad, subdistant, delicately rose color at first, soon whitish, margin not differently colored; stipe 4–5 cm. long, 1–1.5 mm. thick, equal, not fragile, glabrous above, base white-strigose, tubular, white- or rose-tinted above, more or less grayish incarnate below or sometimes grayish incarnate over all.

Spores 7–9 (10) × 4–5 μ, ellipsoid, slightly but distinctly amyloid; basidia four-spored; pleurocystidia abundant, fusoid-ventricose, with acute apices, smooth, 40–62 × 7–12 μ, hyaline; cheilocystidia either similar to the pleurocystidia or somewhat clavate and with the enlarged portion smooth or more or less covered with short obtuse projections, 28–37 × 6–13 (15) μ, the contents of some with a smoky tinge when fresh; gill trama homogeneous; hyaline to faintly brownish in iodine; pileus trama with a distinct pellicle, the hyphae of which bear numerous short projections, hypoderm differentiated, the remainder of fairly large cells, excretory hyphae present, 3–4.5 μ thick, iodine reactions not convincing (all parts hyaline to dull brownish).

Habit, habitat, and distribution.—Scattered on needle beds of white pine during the fall months; New York and Ontario. Apparently it is a very rare species.

Material studied.—Smith, 700, 710, 1107, 4250.

Observations.—Kühner has arranged this species in the same series

FIG. 18. *M. subincarnata:* 1, pleurocystidia; 2, cheilocystidia; 4, spores. *M. amabilissima:* 3, spores; 5, cheilocystidia. *M. monticola:* 6, spores; 7, cheilocystidia. *M. strobilinoides:* 8, pleurocystidia; 9, spores

as *M. amabilissima* and, in fact, the two look almost alike. From Peck's comments and material it appears that he had collected the species in quantity and knew it well. My collections in New York and Ontario have been confined, unfortunately, to only a few scattered fruiting bodies. From limited observations on fresh specimens, and from a study of the type, I believe that the fungus is more closely related to *M. rosella* than to any other Mycena. The consistency and color of the carpophores and the two kinds of cheilocystidia strikingly emphasize this relationship. The greatest difference appears to be that in *M. subincarnata* the gill edges are not differently colored. This point was checked carefully in the field against fresh material of *M. rosella*. In my collections the colors of *M. subincarnata* were even more delicate than those of *M. rosella*, with the result that the former appeared rather waxy and Hygrophorus-like.

The drawings were made from collection 700, Warrensburg, New York. The microscopic details of the type are those given in the description above, but my drawings of its spores and cystidia were not made with the camera lucida. Kühner, after a study of some specimens which I sent him, reported the spores as nonamyloid, an observation I have not been able to confirm.

84. Mycena monticola A. H. Smith

Mycologia, 31: 273. 1939

Illustrations:

Plate 18 A; Text fig. 18, nos. 6–7 (p. 183).
Smith, Mycologia, 31, fig. 1 F (spores).

Pileus 1–3 cm. broad, conic to obtusely campanulate, becoming either plane or umbonate, the margin at times appressed against the stipe in young specimens and sometimes connivent with it, frequently with a wavy uplifted margin in age, translucent-striate when moist, becoming slightly sulcate when faded, glabrous, moist, and hygrophanous, "Pompeian red" on the disc and "light jasper red" to "coral pink" toward the margin, sometimes the disc not darker, fading to "flesh color"; flesh thin, incarnate, brittle, odor and taste not distinctive; lamellae ascending-adnate, becoming horizontal and slightly toothed, close, 23–28 reach the stipe, moderately broad (3 mm. ±), edges even and whitish or tinged "flesh pink" to "coral pink" when the faces are similarly tinted; stipe 4–7 cm. long, 1–3.5 mm. thick, equal above a narrowed crooked base, hollow, juice watery and scanty, base sparsely fibrillose, apex faintly frosted, soon naked and polished above, "coral pink" over all when fresh, soon becoming sordid brown from the base upward and finally "bister" over the lower portion.

Spores ellipsoid or pointed at one end, 7–10 × 4–5 μ, nonamyloid; basidia four-spored, 26–30 × 7–8 μ; cheilocystidia embedded, hyaline, clavate, 28–35 × 9–12 μ, the enlarged portion covered with short rodlike projections; no pleurocystidia; gill trama vinaceous brown in iodine; pileus trama homogeneous beneath a thickish subgelatinous (in KOH) pellicle, becoming vinaceous brown in iodine below the pellicle.

Habit, habitat, and distribution.—Gregarious under pine; Washington and Oregon (3500 to 4500 feet elevation). It is one of the species that occur in great quantity on beds of pine needles during the fall. It was second only to *M. rosella* in abundance in the type locality, McKenzie Pass, Oregon.

Material studied.—Smith, 8014, 8104, 17063, 18134.

Observations.—The beautiful pinkish-red color makes this one of the most attractive species in the genus. A tinge of this color is preserved in the dried specimens, so that the species can be readily recognized from herbarium material. The change in the color of the stipe reminds one of *M. roseipallens* and furnishes another valuable macroscopic character. The gill edges are not margined, and there is no evidence to indicate that the species is closely related to members of the Calodontes.

Subsection Ianthinae

KEY TO SPECIES

1. Pileus and stipe white .... 85. *M. subaquosa*
1. Pileus and stipe colored vinaceous, purplish, lilac, etc. 2

2. Gills broadly adnate to arcuate, subdecurrent, not marginate 87. *M. Kuehneriana*
2. Gills sinuate to adnexed, not marginate .... 86. *M. pura*
2. Gills adnate to adnexed, marginate 3

3. Spores 5.5–7 × 3–3.5 μ .... 88. *M. pelianthina*
3. Spores 7–9 (10) × (3.5) 4–5 μ .... 89. *M. rutilantiformis*

85. Mycena subaquosa A. H. Smith

Mycologia, 31: 278. 1939

Illustration: Plate 22.

Pileus 2–3.5 cm. broad, obtuse to convex, becoming broadly convex to nearly plane, margin appressed against the stipe when young but soon becoming connivent with it and somewhat recurved in age, translucent-striate to the disc when moist, glabrous, hygrophanous, watery and dull white except for the milky-white disc, at maturity the disc becoming tinged with watery gray, fading and becoming shining whitish; flesh thickish, watery white, very soft and fragile, odor and taste very pronounced, resembling those of radish or more pungent; lamellae adnate but soon becoming adnexed, close, 26–32 reach the stipe, in three to four tiers, broad and ventricose (3–4 mm.), concolorous with the pileus, edges even and whitish; stipe 4–9 cm. long, 2–3 mm. thick, hollow, equal, very fragile, glabrous except for sparse white hairs at the base, apex naked or faintly frosted, white and translucent over all.

Spores narrowly ellipsoid, 5–6.5 (7) × 2.5–3 μ, amyloid; basidia four-spored, 20–23 × 5–6 μ; cheilocystidia abundant, broadly fusoid with obtuse apices or the necks somewhat elongated, smooth, hyaline, 30–45 × 9–18 μ; pleurocystidia scattered to numerous, fusoid-ventricose with rounded apices, hyaline, 40–60 × 10–16 μ; gill trama vinaceous brown in iodine; pileus trama with a scarcely differentiated pellicle, below this a region of compact radially arranged hyphae, the remainder of floccose filamentous tissue, in iodine vinaceous brown except for the pellicle.

Habit, habitat, and distribution.—Gregarious under western red cedar on moss and debris during October; Washington and Oregon.

Material studied.—Smith, 7813, 17146, 17251.

Observations.—This species is similar to *M. pura* in its spores, cystidia, and radishlike odor. Although *M. pura* was found everywhere in the conifer forests around McKenzie Pass during the 1937 season, it was consistently different from the specimens described above. According to my experience, in the white form of *M. pura* the odor is not exceptionally strong, the colors are usually faintly pinkish or lilac on the disc of the cap and at the apex of the stipe, and the stature and consistency are the same as for the other color forms of the species. In stature *M. subaquosa* is more like *M. polygramma*. This difference cannot be considered a variation caused by the habitat because typical specimens of *M. pura* were collected in the same moss beds. The glabrous translucent stipe of *M. subaquosa* also aids in distinguishing it.

86. Mycena pura (Fr.) Quélet

Champ. Jura et Vosges, p. 103. 1872

Agaricus purus Fries, Syst. Myc., 1: 151. 1821.
Prunulus purus Murrill, North Am. Flora, 9: 332. 1916.

Illustrations:

Plates 23, 24 A; Text fig. 19, nos. 1–2 (p. 188).
Atkinson, Mushrooms, Edible, Poisonous, etc., fig. 95 (1900).
Beardslee and Coker, Journ. Elisha Mitchell Sci. Soc., 40, pl. 8 (upper left and below).
Bresadola, Icon. Mycol., 5, pls. 226, 227 (as var. *multicolor*).
Hard, The Mushroom, Edible and Otherwise, fig. 95.
Murrill, Mycologia, 7, pl. 158, fig. 3; 9, pl. 11, fig. 3.
Thomas, Field Book of Common Gilled Mushrooms, pl. 12, no. 91.

Pileus 2–4 (6.5) cm. broad, obtuse when young with the margin straight or slightly incurved, becoming obtusely umbonate, convex, plane, or with the margin elevated, disc rarely slightly depressed, surface naked, moist, surface translucent-striate, the margin thin and splitting in broadly expanded forms, hygrophanous, opaque and frequently wrinkled in age, color extremely variable, bright rosy red, purplish, lilac gray, yellowish, or white with a faint bluish or purplish tinge on the disc ("pale lobelia violet" to "pale lilac," "deep hellebore red" to "livid brown," often fading to "light purplish vinaceous"); flesh moderately thick, usually abruptly tapered halfway to the margin, purplish, livid bluish to sordid lilac, becoming pallid or whitish, odor and taste resembling those of radish; lamellae close to subdistant, adnate, adnexed, or uncinate, broad, becoming ventricose and 5–8 mm. broad, intervenose, color variable, tinged purplish lilac or bluish, often shaded more or less with gray, sometimes white, edges whitish; stipe (3) 4–10 cm. long, 2–6 mm. thick, terete or compressed, equal, sometimes enlarged below, hollow, toughish, sometimes more or less twisted-striate, sometimes even, glabrous, or scabrous-pruinose, sometimes subscaly owing to the lacerated cuticle, base slightly myceloid, whitish or concolorous with the pileus, often paler or with only slightly different tones.

Spores 6–9 (10) × 3–3.5 μ, subcylindric to narrowly ellipsoid, amyloid; basidia four-spored; pleurocystidia scattered, rare, or abun-

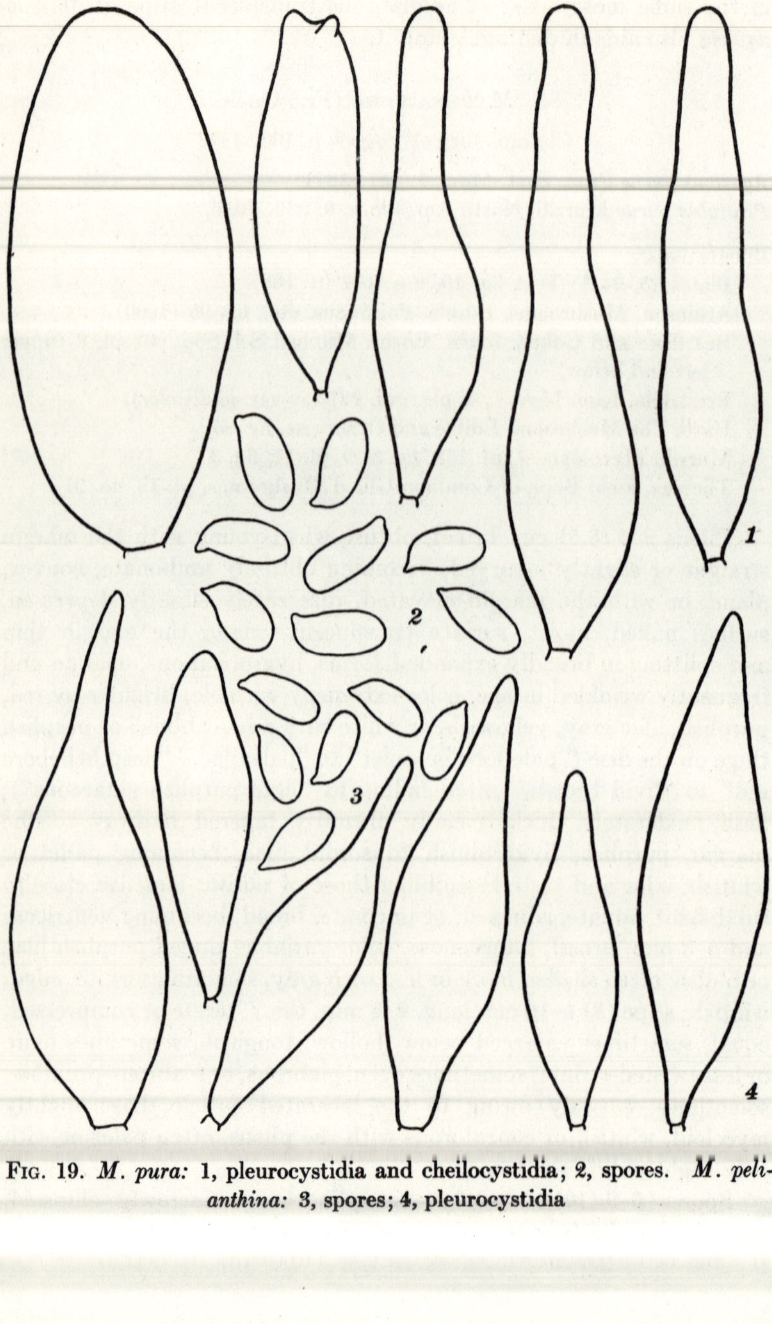

FIG. 19. *M. pura:* 1, pleurocystidia and cheilocystidia; 2, spores. *M. pelianthina:* 3, spores; 4, pleurocystidia

dant, 40–70 (100) × 10–20 (25) μ, ventricose with obtuse apices or elongated into a tapered neck, occasionally with one or two obtuse protuberances, sometimes saccate above a narrow pedicel; cheilocystidia abundant, similar to pleurocystidia or saccate; gill trama homogeneous, of long, moderately broad cells, vinaceous brown in iodine; pileus trama with a poorly differentiated pellicle, a distinct hypoderm, and a rather thick body of filamentous hyphae, vinaceous brown in iodine.

Habit, habitat, and distribution.—Scattered to gregarious on humus in both coniferous and hardwood forests, common throughout the United States and Canada in the spring, early summer, and fall. I have examined material from Alabama, North Carolina, Tennessee, New Jersey, Massachusetts, New York, Ohio, Michigan, Wisconsin, Colorado, Wyoming, Montana, Idaho, Washington, Oregon, and California. In Canada it is known from Nova Scotia to British Columbia.

Material studied.—Smith, 32–384, 32–467, 33–1089, 487, 495, 529, 918, 953, 2678, 2986, 3379, 3585, 3691, 4354, 4937, 6393, 6795, 7177, 7275, 7534, 7807, 8022, 8073, 8108, 8263, 8586, 8766, 10172, 11110, 13848, 13864, 14049, 14540, 15081, 15256, 15856, 15912, 16249. Atkinson, 2992, 3100, 5456, 20102, 22711. Burke, Alabama, 1942. Ellis and Everhardt, North Am. Fungi, 2005. Epling, 109. Flett, Zenith and Bremerton, Washington. Gruber, Oregon. Hesler, 6522, 12693, 12956, 14085. Kelly, 1000, 1203. Kauffman, six collections from New York, Michigan, Montana, Idaho, Oregon. Mains, 1914, New York. Morse, California. Overholts, 1935, 1962. Slipp, UIFP: 1592, 2651, 3139, 3230, 3551. Wehmeyer, 561, 561a.

Observations.—*Mycena pura* is very easily recognized but is an exceedingly variable agaric, and a long list of color varieties have been described. These range from white, yellow, and purplish tan to lilac, bluish gray, rosy, and purple. In addition to these so-called varieties one usually finds numerous intergradations between them, so that, although one quickly learns to recognize the species, he soon gives up trying to distinguish any subdivisions of it. I have not been able to correlate any slight differences in spore size with shape and abundance of pleurocystidia, but it may be possible to recognize certain forms or varieties on this basis. Collections with spores 6–7 × 3 μ have been observed, and a few in which the spores measure 8–10 × 3 μ, all on four-spored basidia, were also found. Such comparisons must be made from spore deposits. A study of herbarium specimens has not verified any distinct spore differences. In dried material the spores usually measure 5–7 (8) × 3 μ. This is true even of those collections in which the spores from deposits were larger. So far as my experience goes, each collection shows some difference in the abundance and distribution as well as in the shape of the pleurocystidia.

87. **Mycena Kuehneriana,** sp. nov.

Illustration: Plate 24 B.

Pileus 1–2.5 cm. latus, convexus, glaber, striatus, griseo-vel brunneo-incarnatus, saepe sublilaceus; lamellae pallidae, late adnatae vel decurrentes; stipes 3–6 cm. longus, 1.2–2.5 mm. crassus, subincarnatus, nitens, cavus; sporae 5–7.5 × 2.5–4.5 μ, amyloideae. Specimen typicum in Herb. Univ. Mich. conservatum. Legit A. H. Smith, n. 11035, prope Ann Arbor, Mich., Sept. 8, 1938.

Pileus 1–2 cm. broad, convex, obtuse in age, fleshy, translucent-striate when moist, color pale avellaneous with a tinge of a mixture of rose and lilac, almost white when faded, odor and taste of radish; lamellae close, broad, arcuate, short-decurrent or very broadly adnate, pallid or tinged rose-lilac; stipe 2–3 cm. long, 3–4 mm. thick above, 1–2 mm. below, glabrous, fragile, concolorous with the pileus.

Spores 5–6 × 2.5–3 μ, amyloid; basidia four-spored; pleurocystidia not seen; cheilocystidia scattered to rare, 50–60 × 10–14 μ, subfusoid to nearly cylindric, smooth.

Habit, habitat, and distribution.—Gregarious on mats of Polytrichum in oak woods; Michigan, and Manitoba (collected by Bisby).

Observations.—*Mycena pseudopura* Cke. sensu Kühner is very similar to *M. Kühneriana*, which, however, differs in having distinctly amyloid spores. Kühner emphasized that the basidia and trama were strongly amyloid, but that the spores were negative. I am inclined to believe that Kühner's *pseudopura* is the same as *M. Kühneriana* and that the iodine reaction in his material was aberrant. The following account is taken from Kühner:

Pileus 1–2.5 cm. broad, convex, sometimes submammillate, striatulate halfway or more to the disc, colored grayish or clear brown mixed with flesh color or lilac, margin entire, sometimes whitish, hygrophanous, fading from the disc outward, glabrous, not viscid; flesh thin, fragile, pallid when dry, odor and taste slightly radishlike; lamellae 24–25 reach the stipe, one to three tiers of lamellulae, sordid whitish, fairly crowded with a faint reflection of lilac, horizontal and decurrent by a tooth, the edge arcuate-concave and not ventricose, always broadly adnate, slightly interveined; stipe 3–6 cm. long, 1.2–2.5 mm. thick, subequal or weakly attenuated from the base upward, with long white rhizoids penetrating the humus, concolorous with the pileus, shining, polished, glabrous (the apex may be faintly pruinose), hollow.

Spores elliptic, 5–7.5 × 2.5–4.5 μ, nonamyloid; basidia four-spored; cheilocystidia abundant, 50–70 × 9–15 μ, very obtuse, clavate to cylindric, smooth, hyaline; gill trama amyloid; pileus trama with a thin pellicle, a hypoderm and a filamentous tramal body, all but the pellicle strongly amyloid.

Although I have seen about all the variations described for *M. pura* and have observed many more, forms with the gills attached, as in my 11035, have not been among them. In their bearing the fruiting bodies, with their turbinate caps and decurrent gills, remind one of small carpophores of *Hygrophorus pratensis*.

Bisby's collection from Lac du Bonnel in Manitoba, September 9, 1935, is the same as my Michigan collection. The iodine reactions are the same, with the spores very distinctly amyloid. In his notes Bisby described the pileus as watery-brown-striate, then yellowish, and with a trace of pink. The lamellae were lavender pink and broadly adnate or decurrent by lines. The stipe was translucent whitish.

Kühner's interpretation of *M. pseudopura* does not appear justified. There is nothing in Cooke's original description to indicate that his species had broadly arcuate to subdecurrent gills. He described them as merely adnate. Rea and others have placed Cooke's fungus in synonymy with *M. pura*, considering it a color variation.

88. Mycena pelianthina (Fr.) Quélet

Champ. Jura et Vosges, p. 102. 1872

Agaricus pelianthinus Fries, Syst. Myc., 1: 112. 1821.
Prunulus denticulatus S. F. Gray, Nat. Arr. Brit. Plants, 1: 630. 1821 (see Murrill, North Am. Flora, 9: 332. 1916)

Illustrations:
Plate 25; Text fig. 19, nos. 3–4 (p. 188).
Beardslee and Coker, Journ. Elisha Mitchell Sci. Soc., 40, pl. 13.
Bresadola, Icon. Mycol., 5, pl. 220.
Konrad et Maublanc, Icon. Sel. Fung., 3, pl. 223.

Lange, Flora Agar. Dan., 2, pl. 49 c.
Ricken, Die Blätterpilze, 2, pl. 110, fig. 3.

Pileus 1.5–3.5 cm. broad, obtuse to convex, in age broadly convex to plane, margin sometimes uplifted, glabrous, moist, hygrophanous, "Sayal brown" to "clay color" when moist and the margin translucent-striate, fading to "tilleul buff" or "pale olive buff," sometimes darker and "avellaneous" (dull sordid brown fading to pallid or avellaneous), sometimes a sordid-purplish shade pervading both the moist and the faded carpophores, in age frequently becoming livid gray; flesh thin, pallid, fragile, unchanging when cut or bruised, odor and taste distinctly of radish; lamellae horizontally adnate or toothed, broad (up to 1 cm.), broadest at point of attachment, moderately close, 26–32 reach the stipe, two to three tiers of lamellulae, sordid grayish vinaceous or flushed with purplish, the edges dark purplish; stipe 3–5 cm. long, 2–5 mm. thick, equal, sometimes enlarged at either end, hollow, fragile, innately appressed longitudinally fibrillose-striate, apex glabrous or with fine fibrils or fibrillose points under a lens, apex sometimes tinged purplish from the fibrils, otherwise pallid (not yellow), base white-strigose.

Spores 5.5–7 × 3–3.5 μ, smooth, amyloid, pointed at one end, obtuse at the other; basidia four-spored; pleurocystidia and cheilocystidia similar and very abundant, with a dull purplish-brown content, smooth, fusoid-ventricose, apices acute, 46–64 × 9–15 μ; gill trama homogeneous, hyphae of long and moderately broad cells, parallel to subparallel, bright vinaceous red in iodine; pileus trama homogeneous beneath a thin pellicle, bright vinaceous red in iodine.

Habit, habitat, and distribution.—Single to gregarious on humus in oak-hickory forests, late summer and fall; Michigan, Washington, and Oregon in the United States and Nova Scotia and Manitoba in Canada. This species has been widely reported throughout the United States, but there is no way to distinguish *M. rutilantiformis* from *M. pelianthina* in most accounts. The description was drawn from Michigan material.

Material studied.—Smith, 33–1053, 506, 4578, 5012, 6862, 7256, 8040, 8197, 8458, 8598, 10966, 11004, 11064, 11087, 14036, 14539, 14577, 16430, 16798, 16894, 17310, 17349, 18558, August 10, 1931, Nova Scotia. Kauffman, October 2, 1922, Mt. Hood, Oregon.

Observations.—The distinguishing features of *M. pelianthina* are its narrow spores and the lack of yellow in the apex of the stipe. *M. rutilantiformis* has spores 4 μ or more broad and bright yellow tints in the apex of the stipe. There is a great deal of variation in size and color of the pileus in both, and the length of the spores varies, so that it alone is not an accurate distinction. One's first impression of these two is very likely to lead to the conclusion that a single variable species is all that should be recognized. I was of this opinion until 1938, when both fruited in great abundance in the vicinity of Ann Arbor. They were constant in the characters as given above. *M. rutilantiformis* appears to be much more common in North America than *M. pelianthina*, most reports of which in the literature very likely apply to the former species. I know from field experience with Kauffman that he did not distinguish between them, but that he was dissatisfied with the species concept he had at the time. The report of *M. pelianthina* by Kauffman and Smith from Rock River, Michigan, is based on specimens of *M. rutilantiformis*.

89. Mycena rutilantiformis Murrill

Mycologia, 8: 221. 1916

Mycena denticulata Peck, Bull. Torr. Bot. Club, 32: 77. 1905 (not *M. denticulata* Quélet, 1888).
Prunulus rutilantiformis Murrill, North Am. Flora, 9: 334. 1916.
Mycena pseudopelianthina Lange, Mycologia, 26: 9. 1934.

Illustrations:
Text fig. 20, nos. 1–2 (p. 194).
Smith, Mycologia, 31, fig. 1 p.

Pileus (1.2) 2–7 cm. broad, convex, becoming broadly convex or in age at times with an elevated wavy margin, margin connivent with the stipe in young stages and frequently splitting in age, glabrous, striatulate when moist, surface moist and lubricous, hygrophanous, "Natal brown" to "deep brownish drab" and fading to near "avellaneous" or "vinaceous buff," often paler and with a sordid yellowish cast, sordid purplish tints often persistent; flesh moderately thick, yellowish to whitish, usually whitish in age, cuticle vinaceous in section, odor resembling that of radishes, taste similar or bitter and scarcely radishlike; lamellae adnate but becoming sinuate to adnexed, seceding, close to subdistant, 30–38 reach the stipe, broad (1 cm. ⊥), intervenose, edges eroded and sordid reddish purple, faces "vinaceous fawn" or paler; stipe 3–8 cm. long, (3) 5–10 mm. thick, hollow, equal or base enlarged, somewhat longitudinally sulcate-striate, with scattered appressed purplish fibrils above, sometimes lacerate-scaly

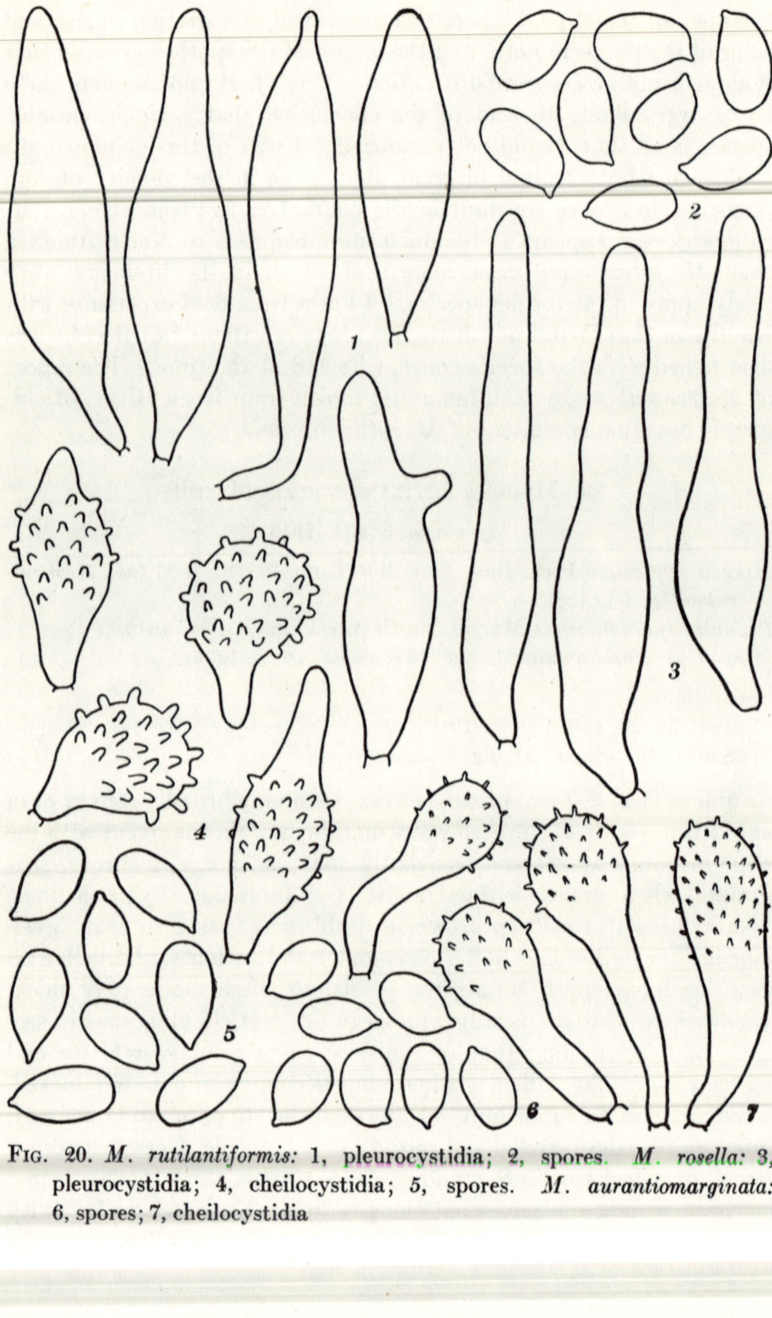

FIG. 20. *M. rutilantiformis:* 1, pleurocystidia; 2, spores. *M. rosella:* 3, pleurocystidia; 4, cheilocystidia; 5, spores. *M. aurantiomarginata:* 6, spores; 7, cheilocystidia

from the broken cuticle, pallid grayish over all or the apex bright to sordid yellow beneath the purple fibrils, flesh grayish below, yellowish in the apex.

Spores subovoid to ellipsoid, pointed at one end, 8–10 × (3.5) 4–5 μ, amyloid; basidia four-spored, 26–28 × 5–6 μ, cheilocystidia abundant, smooth, narrowly ovoid to somewhat fusoid and with obtuse apices, contents dark reddish brown, 40–60 × 9–15 μ; pleurocystidia abundant, more elongated than the cheilocystidia, 60–80 × 9–15 μ, contents reddish brown, gill trama vinaceous brown in iodine; pileus trama with a thin subgelatinous pellicle, the cells of which may have reddish contents, an indefinite region of enlarged cells beneath it, the remainder floccose-filamentous (the pellicle may become washed or worn away in old pilei), all tissue beneath the pellicle vinaceous brown in iodine.

Habit, habitat, and distribution.—Gregarious to scattered on humus and debris, usually under oak and hickory during the spring and fall; Maryland, Pennsylvania, New York, Ohio, Michigan, Missouri, and Washington in the United States and Ontario in Canada.

Material studied.—Smith, 32–500, 1095, 2565, 3389, 3980, 5013, 5039, 6431, 6943, 9569, 10933, 11065, 11086, 15089, 15296, three collections in 1929, one in 1932 and one in 1940, Michigan. Kauffman, Pennsylvania. Kelly, 492, 709. Walters, 1941, Cleveland, Ohio.

Observations.—The pellicle is somewhat separable and gelatinous but not enough so to be of very much aid in identifying the fungus. The purplish-brown colors of the cap vary greatly and are quite bright at times. The yellow at the apex of the stipe may be completely obscured by the dense covering of bright-purplish fibrils.

The specimens Kauffman (1918) referred to *M. denticulata* belong in *M. Kauffmanii* Smith. Both *M. pelianthina* and *M. rutilantiformis* are typically large fungi, but the type collection of the latter happened to consist of exceptionally small individuals. I have seen specimens which were well within the range in size originally given for the species. For a comparison of these two see *M. pelianthina*.

SECTION CALODONTES

Mycena Kauffmanii, *M. marginella* var. *rugosodisca*, *M. rutilantiformis*, *M. pelianthina*, and *M. oregonensis*, in addition to all members of the section Lactipedes and the marginate species in the subgenus Glutinipes, are excluded from this section.

Within the Calodontes the first subsection, the Granulatae, is formed by a group of very closely related species intermediate between the Adonidae and members of the section Typicae. In the subsection Ciliatae two lines are present, one characterized by a more or less lignicolous habitat and the other by a terrestrial one. Their relationships to other sections are the same as those of the Granulatae; they join the brightly colored forms of the Adonidae with smooth cystidia to the gray forms of the Typicae with similar sterile organs.

KEY TO SUBSECTIONS OF CALODONTES

1. Gill edges with roughened cheilocystidia .... Granulatae
1. Gill edges with smooth cheilocystidia or cheilocystidia variously branched at times .... Ciliatae, p. 206

Subsection Granulatae

KEY TO SPECIES

1. Pleurocystidia abundant, fusoid-ventricose, enlarged portion smooth or roughened; gills edged sordid-rose color .... 94. *M. rosella*
1. Pleurocystidia clavate-echinulate (when present) 2
2. Subcespitose to scattered, growing among oak leaves; base of stipe densely strigose .... 93. *M. flavescens*
2. Gregarious under conifers on needle beds 3
3. Pileus dark fuscous with tinge of sulphur yellow, at least along margin; gills edged with pale sulphur yellow .... 92. *M. elegans*
3. Pileus, if fuscous, with strong orange cast; gills edged with "cadmium orange," at least in young carpophores ... 91. *M. aurantiomarginata*
3. Pileus flame-scarlet to brilliant orange, occasionally fading to yellow or whitish; gills edged with flame-scarlet, at least when young 90. *M. strobilinoides*

90. Mycena strobilinoides Peck

Ann. Rep. New York State Mus., 45: 23. 1893

Prunulus strobilinoides Murrill, North Am. Flora, 9: 337. 1916.
Prunulus aurantiacus Murrill, ibid., p. 336.
Mycena aurantiaca Murrill, Mycologia, 8: 220. 1916.

Illustrations:

Plate 26 B; Text fig. 18, nos. 8–9 (p. 183).

Kühner, Encyc. Myc., 10, fig. 89.

Pileus 1–2 cm. broad, acutely or obtusely conic when young, either not expanding or becoming campanulate, margin appressed against the stipe when young, becoming slightly crenate and often flaring somewhat in age, surface moist and smooth, lubricous, translucent-striatulate near the margin when moist, somewhat sulcate at maturity, color "grenadine red" to "flame scarlet" when young and fresh, fading slowly to "cadmium yellow" or "capucine yellow," sometimes whitish, not hygrophanous; flesh thin and pliant, yellowish, odor and taste not distinctive; lamellae ascending-adnate and with a slight decurrent tooth, subdistant, 15–20 reach the stipe, narrow (2–3 mm.), yellow to "light salmon orange" with a "flame scarlet" edge at least when young; stipe 3–4 cm. long, 1–2 mm. thick, equal, terete, solid or with a very small tubule, pliant and cartilaginous, strigose with orange fibrils at the base, covered with an orange pruinose coating toward the apex, soon naked and glabrous except at the base, evenly "orange chrome" to "orange" or paler.

Spores ellipsoid, 7–9 × 4–5 (5.5) μ, amyloid (reaction strong); basidia four-spored, 26–30 × 6–7.5 μ; cheilocystidia and pleurocystidia similar, scattered to abundant on the sides and very abundant on the edges, clavate to subfusoid, the upper portion covered with rod-like projections, contents bright to pale orange; gill trama vinaceous brown in iodine; pileus trama with a thin poorly differentiated pellicle, the cells immediately beneath it filled with a bright-orange content, the hypoderm scarcely differentiated and grading evenly into the tramal body, all tissue beneath the pellicle vinaceous brown in iodine.

Habit, habitat, and distribution.—Densely gregarious, often forming extensive carpets on needle beds, usually under pine, during late summer and fall; Massachusetts, Michigan, Oregon, and California in the United States and British Columbia in Canada. It is common in the West, but rare in eastern North America.

Material studied.—Smith, 2747, 3152, 3184, 7890, 7927, 16462, 18206. Davidson, 219, Mt. Garibaldi, British Columbia. Stuntz, F 482. Kauffman, 1925, Mt. Hood, Oregon. Mains, 34–180, Michigan.

Observations.—The microscopic characters of the type of *M. aurantiaca* are the same as those of Peck's type. The cystidia on the sides of the gills in both species have the characteristic orange content, and in both the cystidia on the gill edge have brighter contents than those on the sides, which indicates that the gill edges were brightly colored. This, along with Murrill's description of the pileus, suggests that he had in hand faded specimens of *M. strobilinoides* when he described *M. aurantiaca.* One of the outstanding features of *M. strobilinoides* is that it does not change color noticeably in drying, a character Murrill also observed in his *M. aurantiaca.* The bright-red color of the young pilei fades slowly, and in age the entire plant is usually a brilliant yellow. In the vicinity of Chain Lakes, Mt. Baker, Washington, however, carpophores were found which had faded to almost pure white. In such faded specimens the gill edges are bright or pale yellow rather than red.

Mycena strobilinoides is most abundant at elevations of 2,500 feet or more in the mountains. It fruits more abundantly under pine than under other conifers but is not restricted in its habitat. Mains found the most luxuriant specimens I have yet seen under red pine on the south shore of Lake Superior near Marquette, Michigan.

Kauffman (1926) did not regard the echinulate-sterile structures in the hymenium of this species as cystidia and confused them with paraphyses. As is now known, there is no sharp distinction between the smooth and the roughened types of cystidia in the genus as a whole. Nearly all possible combinations of both are represented, and both are correctly termed cystidia.

91. Mycena aurantiomarginata (Fr.) Quélet

Champ. Jura et Vosges, p. 240. 1872

Agaricus (*Mycena*) *aurantio-marginatus* Fries, Syst. Myc., 1: 113. 1821.

Illustrations:

Plates 26 A, 27; Text fig. 20, nos. 6–7 (p. 194).
Bresadola, Icon. Mycol., 5, 221, fig. 1.
Konrad et Maublanc, Icon. Sel. Fung., 3, pl. 222.
Lange, Flora Agar. Dan., 2, pl. 54 G (as *M. elegans*).
Ricken, Die Blätterpilze, 2, pl. 110, fig. 5.

Pileus 8–20 mm. broad, obtusely conic to campanulate, becoming nearly plane in age, faintly hoary-pruinose at first but soon polished, lubricous, color when moist dark olive fuscous on the disc and the marginal area bright orange shaded with fuscous, sometimes the disc "orange citrine" to "medal bronze" (yellowish olive) and the margin "cadmium orange" (brilliant orange), striate, not hygrophanous, fading slowly to pallid olive brown with an orange tinge or near "Naples yellow," becoming somewhat plicate in age; flesh thin, rather pliant, odor and taste not distinctive; lamellae narrow, becoming broad in age, close, bluntly adnate or with a decurrent tooth, pallid to grayish olive on the faces, edges bright "cadmium orange"; stipe 3–6 cm. long, 1.5–2 (3) mm. thick, rigid, and cartilaginous, equal, terete or compressed, tubular to hollow, "buffy brown" to "medal bronze" (olive or grayish olive), sometimes tinged orange, glabrous, with an orange pruina toward the apex, base strigose with "cadmium-orange" hairs.

Spores 7–9 × 4–5 μ, ellipsoid, smooth, amyloid; basidia four-spored; pleurocystidia and cheilocystidia abundant and similar, 28–36 × 7–12 μ, clavate to subcapitate, the apices sparsely or densely echinulate, filled with a bright-orange pigment; gill trama homogeneous, pale vinaceous brown in iodine; pileus trama with a thin pellicle over the surface upon which are found scattered pilocystidia similar to the pleurocystidia and cheilocystidia; beneath this is a differentiated hypoderm of enlarged cells, the remainder floccose-filamentous and of moderately broad hyphae.

Habit, habitat, and distribution.—Widely scattered, gregarious or subcespitose under conifers on moss and needle carpets. I have found it in Washington, Oregon, and California. During the fall of 1937 it was very abundant under spruce around Crescent City, California.

Material studied.—Smith, 2595, 3080, 3283, 3393, 3435, 3455, 3725, 7771, 7971, 8015, 8271, 18081. Kauffman, 1925, Washington.

Observations.—The species is best recognized by the brilliant orange margins of the gills and the echinulate pleurocystidia and cheilocystidia. The color of the pileus varies greatly from olive fuscous to orange shaded with gray or dull yellowish olive, depending on the proportions of the various pigments present. It is readily separated from *M. Leaiana* by its habitat, manner of growth, and the lack of a gelatinous layer over the stipe.

Kühner (1938) uses the name *M. elegans* (Fr. ex Pers.) sensu Schroeter for this fungus. As I (1936) have previously pointed out, Fries made a definite distinction between the color of the gill edges of *M. aurantiomarginata* and that of *M. elegans*, so that, by following Wharton's interpretation of Fries's color names, the former can be recognized by the orange edges of the gills, whereas the gills of the latter should have yellow margins. Regardless of the various interpretations of the color term "croceus"—whether it is merely a full pale yellow or an orange—I believe that the important point to consider is that Fries was trying to distinguish between two closely related species with differently colored gill edges. If we are to use the Friesian names at all, they must be accepted on the basis of his descriptions, since there are no specimens to shed more light on the question. Kühner places the fungus I have referred to *M. elegans* in *M. chlorantha* Fries sensu Oort. I do not know *M. chlorantha.* Fries, who described it, placed it in the section Adonidae, among the brightly colored species—those without fuscous or cinereous shades and in which *the margins of the gills are not differently colored.* The concept of Oort is entirely out of line with the Friesian description and, in my estimation, should not be maintained. In view of this situation I am adhering to my original disposition of the two species, with the exception that the forms from oak woods previously placed in *M. elegans* are now referred to *M. flavescens.*

Kühner, in support of his opinion, points out that Fries admitted var. *hyperborea* to *M. elegans* and that the latter had gill edges "subcinnabarini." In considering these additional facts two things should be kept in mind:

(1) Kalchbrenner was apparently responsible for the variety, and very likely Fries included it under *elegans* because that was where Kalchbrenner described it. To me the significant thing is that the color of the gill edge was one of the distinguishing characters of the variety, a fact which supports my contention that Fries regarded a Mycena with a pale-yellow gill edge as typical of the species.

(2) It must be remembered that our concepts of species do not go back to Persoon, and, consequently, in establishing a Friesian concept we are not concerned with whether those of Fries and Persoon agree. One of the reasons for selecting Fries's *Systema Mycologicum* as the starting point in agaric classification and nomenclature was to avoid the confusion caused by conflicting species concepts among the earlier investigators.

92. Mycena elegans (Fr.) Quélet

Champ. Jura et Vosges, p. 241. 1872

Agaricus elegans Fries, Syst. Myc., 1: 149. 1821.

Illustrations:

Plate 28; Text fig. 21, nos. 1–2.

Smith, Mycologia, 28: 416, fig. 2, no. 4.

FIG. 21. *M. elegans:* 1, cheilocystidia; 2, spores. *M. flavescens:* 3, cheilocystidia; 6, spores. *M. debilis:* 4, cheilocystidia; 5, spores

Pileus 1–3.5 cm. broad, convex to obtusely conic, becoming broadly campanulate or developing a low obtuse umbo in age, surface moist, at first densely white-pruinose, soon polished, subhygrophanous, "dark olive" (blackish olive) when fresh, the margin "light chalcedony yellow" (pale clear yellow), fading to sordid olive gray with a whitish margin before losing moisture, pale "avellaneous" to pale olive gray when faded, striate to the disc when moist, margin entire; flesh watery gray, fragile, odor and taste not distinctive; lamellae narrow, moderately close, becoming subdistant, ascending-adnate, dull pale olive gray, margin "pale green yellow," usually staining sordid purplish brown in age; stipe 4–12 (15) cm. long, 1.5–3 mm. thick, tubular, fragile, equal or base a bit enlarged and strigose with "pale green-yellow" hairs, faintly pruinose above, polished in age, pale or dark olive gray with a decided yellowish tinge, apex pale yellow at times, in age often staining dull reddish brown from the base upward.

Spores 7.5–9 × 4–5 μ, ellipsoid, smooth, amyloid; basidia four-spored; pleurocystidia and cheilocystidia similar, 28–33 × 7–10 μ, abundant to scattered, clavate to subcapitate, the enlarged portion echinulate, contents pale yellow to nearly hyaline; gill trama homogeneous, pale vinaceous brown in iodine; pileus trama with a thin pellicle, a differentiated hypoderm, and a tramal body of filamentous but moderately broad cells, all parts below the pellicle vinaceous brown in iodine.

Habit, habitat, and distribution.—Gregarious to scattered under conifers, *Thuja plicata* in particular; Michigan, Washington, Oregon, and California. It appears to be very abundant in Oregon.

Material studied.—Smith, 977, 3453, 3508, 3546, 3714, 3729, 3730, 3757, 5056, 5091, 7966, 8248, 17122, 17553. Mains, 32–258.

Observations.—The colors vary greatly, but an olive-fuscous shade tinged with pale yellow is its outstanding character. *M. flavescens* has the same combination of colors but grows in hardwood forests and has a different habit. The habit, however, could easily be caused by the growth conditions imposed by the broad flat oak leaves which cover the forest floor. The reddish-brown stains of the gills and stipe of *M. elegans* are not given much emphasis here, although in the specimens I have examined their presence seemed constant. They develop best in wet, foggy weather of the type which prevails in western Oregon and Washington in October and November. They were not present in the Michigan collection, which was found under pine in a local plantation. Kühner uses the name *M. chlorantha* for this species. For further comments see *M. aurantiomarginata.*

93. Mycena flavescens Velenovský

České Houby, p. 323. 1920

Illustrations:

Plate 29 A; Text fig. 21, nos. 3, 6 (p. 201).

Lange, Flora Agar. Dan., 2, pl. 54 E (as *M. luteoalba* var. *sulphureomarginata*). 1936.

Pileus 10–20 mm. broad, up to 10–12 mm. high, obtusely conic or in age somewhat campanulate, sometimes nearly flat, the margin straight at first and frequently becoming recurved, surface delicately white-pruinose at first, soon polished and smooth, moist, color when moist "fuscous" on the umbo and grayish toward the whitish or faintly yellowish margin, sometimes an olive tint is evident, becoming paler while still moist, the disc and striae pale gray, the margin and portions between the striae whitish, fading to yellowish gray, grayish, or whitish and becoming somewhat sulcate; flesh thin, brittle, white or grayish, odor and taste not distinctive; lamellae ascending-adnate or almost horizontal in age, often seceding, narrow, becoming slightly ventricose, close to subdistant, whitish at first, becoming grayish, especially toward the base, the margin usually pale yellow (but merely pallid in some); stipe 3–6 cm. long, 1–2 mm. thick, equal, tubular, fragile, usually flexuous, base white-strigose and rooting among the fallen leaves, apex faintly pruinose, smooth or somewhat longitudinally striate, pallid or drab gray, the apex pallid, the base dark, when young the apex sometimes fuscous and the base a lighter grayish brown.

Spores 8–10 × 4–5.5 μ, ellipsoid, very strongly amyloid; basidia four-spored; pleurocystidia and cheilocystidia similar, scattered to abundant, ventricose to clavate, sessile or with a long pedicel, apices obtuse and covered with thin, short, hairlike projections or somewhat echinulate, contents hyaline or yellowish, 36–78 × 9–12 (18) μ; gill trama homogeneous, weakly amyloid; pileus trama with a thin pellicle, a well-differentiated hypoderm, and a tramal body of somewhat enlarged hyphae, amyloid except for the pellicle.

Habit, habitat, and distribution.—Gregarious to subcespitose among oak leaves; Michigan. To date it has appeared during October and November, and has been found abundantly in only one locality.

Material studied.—Smith, 32–430, 32–474, 32–491, 32–578, 33–1048, 5038, October 7, 1931, Michigan.

Observations.—Previously (1936) I regarded this species as merely a form of *M. elegans.* Its cystidia, however, constantly average larger than those of *M. elegans* and have, in addition, a rather distinctive shape. Failure to find the fungus in quantity in more than one locality precludes drawing conclusions about the effect of locality on the habit. As I have observed the fungus to date, the lack of reddish stains on the gills and stipe, the subcespitose manner of growth, flexuous stipe, and more or less rooting base, along with the very pale-yellow gill edges, characterize it macroscopically. Fruiting bodies, particularly old pilei, have been found in which there was no yellow on the gill edges. Sometimes, however, the yellow colors become more pronounced as the specimens age. During the season of 1939 a group of old carpophores was found which were similar in color to old specimens of *M. citrinomarginata.* The echinulate cystidia of *M. flavescens* readily distinguish such forms.

94. Mycena rosella (Fr.) Quélet

Champ. Jura et Vosges, p. 241. 1872

Agaricus rosellus Fries, Syst. Myc., 1: 151. 1821.
Omphalia rosella S. F. Gray, Nat. Arr. Brit. Plants, p. 615. 1821.
Prunulus roseolus Murrill, North Am. Flora, 9: 324. 1916.

Illustrations:

Plate 26 C, 30; Text fig. 20, nos. 3–5 (p. 194).
Konrad et Maublanc, Icon. Sel. Fung., 3, pl. 224 I.
Lange, Flora Agar. Dan., 2, pl. 54 F, F[1] (good).
Ricken, Die Blätterpilze, 2, pl. 109, fig. 5.

Pileus (5) 10–20 (30) mm. broad, obtusely conic with a slightly incurved margin at first, occasionally sharply conic, becoming campanulate to convex and in age often becoming plane, occasionally with a small conic umbo (almost a papilla), surface moist to lubricous, translucent-striate, "light russet vinaceous" to "flesh pink" at first, becoming paler and developing ochraceous tints in age (bright pink tinged with gray on the disc, the margin a clearer pink at first or evenly colored over all, fading and becoming more yellowish in age);

flesh thin, pliant, sordid pinkish to whitish, odor and taste not distinctive; lamellae moderately close to subdistant, 15–24 reach the stipe, one or two tiers of lamellulae, usually horizontal and broadly adnate in mature pilei, sometimes arcuate, sometimes slightly notched, narrow to moderately broad (2.5 mm. ±), intervenose, pale to rather bright rose color, the edges a darker sordid-reddish color; stipe 2.5–7 cm. long, 1–1.5 (2.5) mm. thick, equal, terete, strict or flexuous, tubular, rather flexible, base white-strigose or white-mycelioid, glabrous above, color pale rose or grayish rose, clearer above, more sordid below, translucent, lubricous but not viscid.

Spores 7–9 × 4–5 μ, narrowly ellipsoid, amyloid; basidia four-spored; pleurocystidia abundant, 60–80 × 10–14 μ, arising from the gill trama and projecting 15–30 μ beyond the basidia, narrowly fusoid-ventricose, smooth, with reddish contents when fresh; cheilocystidia clavate to subfusoid, 21–36 × 9–15 μ, covered over all with short obtuse projections, the apex occasionally elongated and smooth and then only the ventricose portion roughened, contents dark reddish; gill trama homogeneous, not appreciably amyloid; pileus trama with a thin somewhat gelatinous pellicle, a well-differentiated hypoderm and the remainder filamentous, not appreciably amyloid.

Habit, habitat, and distribution.—Gregarious in large troops or scattered on needle beds under conifers, common and widely distributed. One may expect to find it in conifer regions throughout North America. I have collected it in such regions in Michigan, Idaho, Washington, Oregon, and California in the United States and in Nova Scotia and Ontario in Canada. It is often abundant in open stands of pine.

Material studied.—Smith, 34–187, 723, 806, 3207, 3270, 3333, 3434, 3479, 4329, 5034, 7926, 8119, 8344, 8764, 8922, 9084, 16512, 16770, 16796, 18176. Atkinson, 22725, 24242, 18491 (as *M. rubromarginata*), 13996 as *M. subincarnata*. Ellis & Everhardt, North Am. Fungi, 2006. Kauffman, Pennsylvania, Michigan, Montana, Wyoming, Oregon. Mains, 34–79, 1935, Michigan. Overholts, 18945. Slipp, UIFP: 3248, 3581. Wehmeyer, 794, 794a.

Observations.—Kühner described the tramae of the pileus and gills as amyloid, a character which did not appear very distinctive in my sections. In dried specimens the pellicle of the pileus gelatinizes more in weak KOH than it does in water mounts of fresh material. Thus one is likely to obtain a false impression of the viscidity of the cap if he is limited to a study of herbarium specimens. The flexible nature of the entire fruiting body, the pale-rose colors, bordered gills, and two different types of cystidia on the lamellae are its most distinctive characters. I have not noted the nitrous odor mentioned by Kauffman, and I believe that he confused *M. rosella* with *M. capillaripes*. *M. rosella* is usually brighter in color than Peck's species and is by no means as fragile.

Subsection Ciliatae

Mycena Kauffmanii and *M. rosella* may be sought for in this group. For the former see page 384; for the latter, page 204.

KEY TO SPECIES

1. Terrestrial species 2
1. Lignicolous species (on dead twigs above ground or on stumps, logs, or branches) 7

2. Rather fleshy species; stipe 2–5 mm. thick 3
2. Slender, membranaceous, and fragile species 4

3. Spores 5.5–7 × 3–3.5 μ .... 88. *M. pelianthina*
3. Spores 7–9 (10) × (3.5) 4–5 μ .... 89. *M. rutilantiformis*

4. Gills edged sordid fuliginous (*M. Kauffmanii* may key out here also) 99. *M. atromarginata*
4. Gills not as above 5

5. Odor nitrous, spores 7–10 × 4–5 μ (4-spored) .... 104. *M. capillaripes*
5. Odor not nitrous 6

6. Gill edges reddish brown; spores 11–14 × 6–8 μ (4-spored); on sphagnum .... 103. *M. cheboyganensis*
6. Gill edges reddish brown; spores 4–5.5 μ wide; not on sphagnum 102. *M. avenacea*
6. Gill edges yellowish; spores 8–11 × 4–5.5 μ .. 100. *M. citrinomarginata*
6. Gill edges yellowish; spores 6–8 × 4.5 μ ...... 101. *M. olivaceobrunnea*

7. Pileus gray to grayish brown 8
7. Pileus deep reddish brown, purplish lilac, or pinkish 9

8. Gill edges dark reddish to purplish brown ..... 97. *M. rubromarginata*
8. Gill edges dark olive brown .... 98. *M. viridimarginata*
8. Gill edges grayish to fuliginous (*M. marginella* also keys out here) 99. *M. atromarginata*

9. Pileus reddish to pinkish brown; edges of gills pale pink to sordid pale vinaceous .... 96. *M. elegantula*
9. Pileus purplish, fading, and with lilac cast when faded; gill edges at first dark slate purple .... 95. *M. purpureofusca*

95. Mycena purpureofusca (Pk.) Saccardo

Syll. Fung., 5: 255. 1887

Agaricus (*Mycena*) *purpureofuscus* Peck, Ann. Rep. New York State Mus., 38: 85. 1885.

Prunulus purpureofuscus Murrill, North Am. Flora, 9: 333. 1916.

Illustrations:

Plate 31; Text fig. 22, nos. 1–2 (p. 208).

Beardslee and Coker, Journ. Elisha Mitchell Sci. Soc., 40, pl. 12 (above as *M. rubromarginata*).

Pileus 5–25 mm. broad, obtusely conic, the margin often bent in slightly at first, becoming broadly conic, sometimes campanulate, occasionally nearly plane in age, the margin usually remaining bent in slightly, seldom flaring, surface hoary at first but soon naked and moist, slightly hygrophanous, translucent-striate to the apex when mature and moist, usually more opaque when young, color dark purplish over the disc to pale lilac toward the margin, fading to purplish gray ("slate purple" to "dull Indian purple" on the disc at first, toward the margin "vinaceous lavender," "pale vinaceous lilac," or "light vinaceous lilac," fading on losing moisture); flesh thin but rather cartilaginous and pliant, purplish gray becoming pallid to white, odor and taste not distinctive; lamellae ascending, narrowly adnate, equal, narrow, moderately close, pallid to grayish, edges "dark slate purple" (very dark grayish purple) and slightly fimbriate; stipe 3–10 cm. long, 1–2 mm. thick, equal, terete, tubular, rather cartilaginous and tough, base white-strigose and often prolonged into a pseudorhiza, glabrous, more or less concolorous with the pileus or paler above.

Spores 8–10 × 6–7 μ (10–14 × 6.7–8.5 μ, two-spored), broadly ellipsoid, amyloid; basidia two- or four-spored; pleurocystidia not differentiated; cheilocystidia abundant and conspicuous, 30–50 (64) × 7–12 (15) μ, more or less fusoid-ventricose, the apices often becoming forked in age, filled with a dull-purplish sap, content granular or amorphous and dark sordid brown when revived in chloral-hydrate iodine solution; gill trama homogeneous, vinaceous in iodine; pileus trama with a well-differentiated pellicle, a distinct hypoderm, and a filamentous tramal body, all but the pellicle vinaceous brown in iodine.

Habit, habitat, and distribution.—Single to cespitose on conifer wood and debris; North Carolina, Tennessee, New York, Michigan, Montana, Idaho, Washington, Oregon, and California in the United

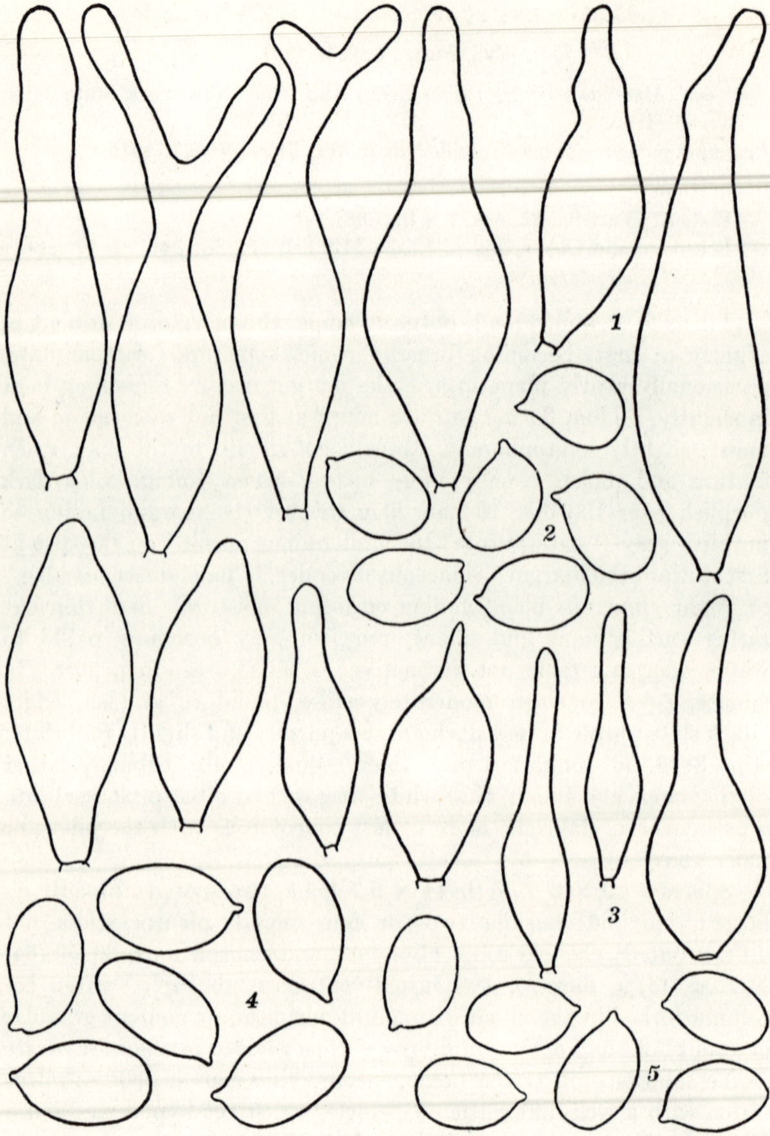

FIG. 22. *M. purpureofusca.* 1, cheilocystidia; 2, spores. *M. elegantula:* 3, cheilocystidia; 4, spores of two-spored form; 5, spores of four-spored form

States and Ontario in Canada. In Michigan it often occurs on old hemlock knots lying buried in the soil and then is usually found growing singly. On logs and stumps it is more likely to be cespitose.

Material studied.—Smith, 32–52, 32–54, 32–55, 32–75, 33–160, 33–370, 34–68, 774, 906, 1302, 2511, 2546, 3143, 3437, 3448, 3722, 4209, 8027, 8482, 9515, 14826, 14889. Atkinson, 24129, 33464 (two-spored). Hesler, 14206. Kauffman, five collections, Michigan and Montana. Slipp, UIFP: 3116.

Observations.—*Mycena purpureofusca* is well characterized by its dark-purplish rather than pale-rosy gill edges, the deep-purplish fuscous disc of the pileus, from which it derives its name, its rather tenacious consistency, and the very broadly ellipsoid spores. One will experience the most difficulty in distinguishing it from *M. elegantula*, which usually has rosy to vinaceous-brown gill edges, and the dominant color of the pileus is vinaceous brown.

96. Mycena elegantula Peck

Bull. Torr. Bot. Club, 22: 199. 1895

Prunulus elegantulus Murrill, North Am. Flora, 9: 337. 1916.
Mycena rubromarginata var. *laricis* Smith, Mycologia, 26: 307. 1934.

Illustrations:
Plates 29 B, 32; Text fig. 22, nos. 3–5.
Smith, Am. Journ. Bot., 22, pl. 1, fig. 4.

Pileus (5) 10–25 (50) mm. broad, obtusely conic with a straight margin at first, remaining conic or becoming convex-umbonate to campanulate, surface glabrous and moist, opaque, but becoming translucent-striate to the disc at times before fading, color variable, usually bright or dark vinaceous brown at first, the disc darker and the margin brighter, fading to near avellaneous or retaining a strong pinkish tint ("hellebore red" on the disc, "pale rhodonite pink" along the margin, or "warm blackish brown" to "dark vinaceous brown" on the disc and "livid brown" to "brownish vinaceous" on the margin, sometimes "fawn color" or "wood brown" in age), becoming sulcate-striate on the margin; flesh moderately thick and rather firm, pallid grayish to vinaceous, odor and taste not distinctive; lamellae close to subdistant, 16–22 reach the stipe, two tiers of lamellulae, bluntly adnate or hooked, narrow to moderately broad, white to grayish, edges pale pink to sordid vinaceous, hymenial tissue forming a collar around the stipe in some; stipe 2–6 (8) cm. long, 1–3 mm. thick, equal, fragile, cartilaginous, tubular, glabrous except for the white-strigose base, polished and translucent, variable in color but usually tinted more or less like the pileus, sometimes nearly white in age, sometimes with a strong pinkish tint.

Spores 8–10 (11) × 5–6.5 μ (four-spored), 10–12 × 5.5–7 μ (two-spored), broadly ellipsoid, smooth, amyloid; basidia two- or four-spored; pleurocystidia very rare or absent, similar to cheilocystidia; cheilocystidia scattered to fairly abundant, 40–75 × (5) 8–15 μ, smooth or occasionally with two or three fingerlike prolongations, clavate to fusoid-ventricose, contents usually pinkish; gill trama homogeneous, vinaceous brown in iodine; pileus trama with a well-differentiated pellicle, the hyphae of which give off short, slender projections, the projections sometimes forming a turflike covering over the surface, hypoderm well differentiated, the remainder filamentous, all except the pellicle vinaceous brown in iodine.

Habit, habitat, and distribution.—Usually single or gregarious but at times cespitose. On decayed conifer wood in the spring and fall; Nova Scotia, New York, and Michigan in eastern North America and Idaho, Washington, Oregon, and California in the West. Seldom found in large quantities. The material collected in Michigan has all been two-spored; that from the Pacific Coast, four-spored.

Material studied.—Smith, 32–229, 32–234, 32–236, 32–245, 33–102, 33–162, 33–525, 635, 690, 690a, 1323, 1363, 2545, 2580, 3262, 3295, 3528, 3606, 6287, 6313, 6389, 7805, 7894, 8065, 8416, 8483, 8923, 9279, 9297, 14850, 15092, 15926, 17065, 17553, 18121, 18132, 18140, 18197. Atkinson, 24109, 24235. Gruber, 43, Idaho. Morse, January 21, 1934, California. Wehmeyer, 625.

Observations.—Whenever a turflike covering is formed over the pileus by the short, slender branches arising from the hyphae of the cuticle, the pileus appears glaucous or very slightly pruinose. According to my observations, the character is usually found on relatively old carpophores. The variations in color of this species are great but follow a definite pattern, always remaining in the series of pinkish or vinaceous browns. The amount of gray present determines the degree of brightness. In some collections the gills separate from the stipe but adhere to each other to form a collar. This appears to be merely a chance variation. The habit of this species is more like that of *M. hemisphaerica* than that of *M. rubromarginata*, and its consistency also relates it to the cespitose cartilaginous gray species.

97. Mycena rubromarginata (Fr.) Quélet

Champ. Jura et Vosges, p. 435. 1875

Agaricus rubromarginatus Fries, Syst. Myc., 1: 150. 1821.

Illustrations:

Plate 33; Text fig. 23, nos. 1–2 (p. 212).
? Bresadola, Icon. Mycol., 5, pl. 222, fig. 1.
Fries, Icon. Sel. Hymen., 1, pl. 78, fig. 4.

Pileus 7–20 mm. broad, obtusely conic to nearly convex, becoming obtusely campanulate to broadly convex, the margin appressed against the stipe at first and sometimes flaring slightly at maturity, surface densely pruinose in buttons, soon naked, moist, color "fuscous" on the disc with a faint tinge of vinaceous at times, soon becoming "chaetura drab" or "hair brown," the margin near "ecru drab" or paler (dark gray with a vinaceous tinge, soon becoming paler gray, the disc remaining darker than the margin, sometimes the margin tinged reddish at first), sometimes very pale when faded; flesh thin, fragile, watery, grayish to pallid, odor and taste not distinctive; lamellae ascending but broadly adnate and with only a slight tooth, moderately broad (3 mm. ±), subdistant, 12–17 reach the stipe, one or two tiers of shorter individuals, intervenose, pallid to dull grayish but with bright reddish-brown edges which soon become sordid reddish brown (as seen under a lens); stipe 2–4 (7) cm. long, 1–3.5 mm. thick, equal, fragile, hollow, terete or compressed, pale watery gray, translucent, glabrous, base naked or nearly so.

Spores 10–12 × 5–7 μ, broadly ellipsoid, germinating readily (one often finds germinated spores in a deposit), amyloid; basidia four-spored; pleurocystidia very rare (usually only near the margin), similar to the cheilocystidia; cheilocystidia 28–42 × 8–12 μ, broadly fusoid-ventricose when young, elongating and in age somewhat irregular or narrowly fusoid-ventricose to subcylindric, apices sometimes forked; gill trama homogeneous, of broad hyphae, faintly vinaceous brown in iodine; pileus trama with a well-differentiated pellicle and hypoderm, remainder floccose and of broad hyphae, pale vinaceous brown in iodine; stipe tissue bright vinaceous in iodine.

Habit, habitat, and distribution.—Scattered on branches of fir and spruce, either on the naked limbs of dead trees or on the lower dead branches of living trees, often abundant in slashings. Fairly common in the fall. In the northern regions it is not uncommon from New York to Oregon, and is known from British Columbia. In the eastern

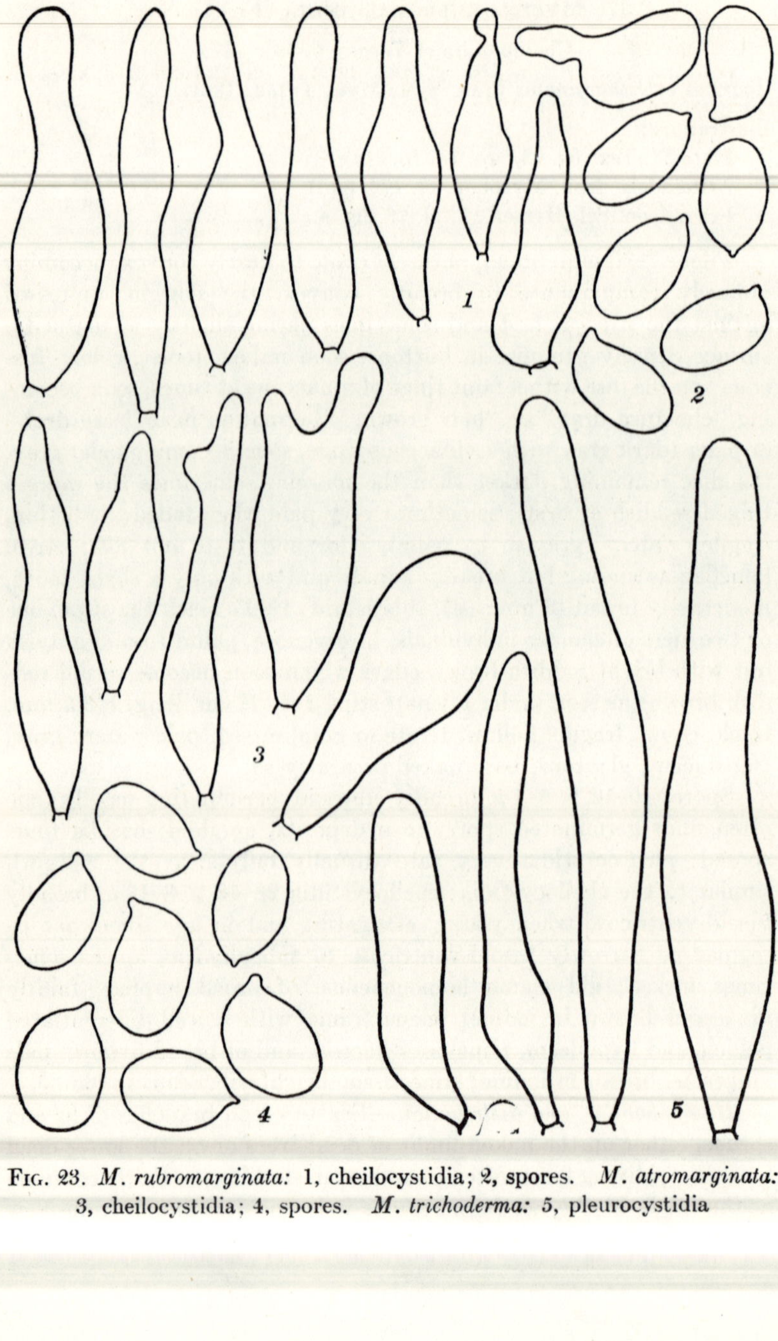

FIG. 23. *M. rubromarginata:* 1, cheilocystidia; 2, spores. *M. atromarginata:* 3, cheilocystidia; 4, spores. *M. trichoderma:* 5, pleurocystidia

United States it extends south in the mountains to North Carolina and Tennessee.

Material studied.—Smith, 33–195, 33–342, 33–439, 33–916, 33–973, 712, 725, 749, 756, 915, 1063, 1136, 2492, 4815, 4864, 4906, 7370, 7824, 8773, 9760, 14887, 16509. Davidson, 210. Kauffman, New York, Michigan, Oregon.

Observations.—This is typically a gray Mycena that may occasionally become whitish. Its habitat, the bright to dull reddish-brown gill edges, and the large spores on four-spored basidia, as well as the fusoid cheilocystidia, distinguish it. Kühner groups it along with *M. luteoalcalina*, *M. atromarginata*, *M. purpureofusca*, and *M. viridimarginata*. *M. rubromarginata* differs from the species named above by the characters used in the key. There is a wide range of variation in the color of the gill edges, but reddish brown is always dominant in contrast to rose, purplish, or olive tones in the other species.

In northern Michigan I found on an alder stick near the Big Garlic River a perfect carpophore (33–439, June 13, 1933) which appears to be intermediate between *M. rubromarginata* and *M. viridimarginata*. The gills were thick, distinctly veined, and glaucous grayish green. After standing overnight they had changed to brownish olive, but the edges were still dull red. The pileus was 1 cm. broad, bister in color, and scarcely faded. The spores measured 10–12 × 5–6 μ, the basidia were four-spored, and the cheilocystidia were fusoid-ventricose. They measured 38–43 × 9–14 μ and had a dark-red homogeneous content. I have not found more like it, and I include it here as a chance variation which appears to be intermediate between *M. rubromarginata* and *M. viridimarginata*.

98. Mycena viridimarginata Karsten

Hedwigia, 31: 218. 1892

Pileus 8–14 mm. broad, broadly conic to convex, the margin appressed at first, broadly umbonate in age, surface glabrous and moist, hygrophanous, dark watery gray to avellaneous or with a sordid-olivaceous cast on the disc, fading to sordid gray or with an obscure vinaceous tint near the margin, sometimes yellowish gray when faded, striate when moist, becoming slightly sulcate; flesh soft and fragile, odor and taste none; lamellae adnate, at times with a spurious collar, subdistant to distant, narrow to moderately broad, concolorous with the pileus or pallid, edges dull brown to olivaceous brown, rarely somewhat reddish brown in young specimens, intervenose; stipe 1–3 (5) cm. long, 1.5–2.5 mm. thick, equal or enlarged slightly at the base, usually curved, fragile and hollow, pale gray to avellaneous over all, apex pruinose, glabrous toward the base, which is furnished with scattered white mycelial hairs.

Spores 10–12 × 5–6 μ (four-spored), 10–14 × 6–8 μ (two-, three-spored), ellipsoid, smooth, amyloid; basidia two-, three-, four-spored; no pleurocystidia; cheilocystidia 33–65 × 8–14 μ, fusoid-ventricose, smooth, occasionally forked, contents homogeneous and dull brown; gill trama homogeneous, of broad interwoven cells which are vinaceous brown in iodine; pileus trama with a well-formed nongelatinous pellicle, a broad hypoderm, and the remainder floccose, all but the pellicle vinaceous brown in iodine.

Habit, habitat, and distribution.—Scattered on dead branches and trunks of alder in August and September; rare; Northern Michigan and Washington in the United States and Ontario in Canada.

Material studied.—Smith, 33-440, 33-471, 1359, 3073, 4156.

Observations.—This fungus was first reported for North America under the name *M. olivaceomarginata* Massee sensu Kühner. In Kühner's later work (1938), however, it is referred, as above, to Karsten's species. Massee described *M. olivaceomarginata* as having small spores, so that it is hardly logical to place a large-spored fungus under this name. The species is almost the counterpart of *M. rubromarginata* in habit, color, stature, and consistency but differs in the sordid olive-gray to yellow-gray colors that develop, especially along the gill edges.

99. Mycena atromarginata (Lasch) Saccardo sensu Kühner

Encyc. Myc., 10: 424. 1938

Agaricus atromarginatus Lasch, Linnaea, 3: 387. 1828.

Illustrations:

Text fig. 23, nos. 3–4 (p. 212).

Fries, Icon. Sel. Hymen., 1, pl. 78, fig. 3.

Pileus 22–35 mm. high, obtusely conic, 14–32 mm. across the base, the margin tending to recurve, long-striate or somewhat sulcate, grayish brown on the disc, the margin pale and sordid but translucent, very slightly wrinkled-fibrillose under a lens. Flesh thin, concolorous, almost inodorous (not with a nitrous odor), taste mild. Gills 19–27 reach the stipe, one to three tiers of lamellulae, sordid whitish or white with a grayish reflection, with the edge finely bordered fuliginous or grayish black (especially under a lens), ascending, hardly broad, a bit ventricose, adnate and not uncinate. Stipe 4–5 cm. long, 1.5–3 mm. thick, equal, sometimes prolonged into a root about 13 mm. long, grayish brown or merely laved with gray, paler above, polished and glabrous, almost shining, fistulose.

Spores ellipsoid to pruniform-ovoid, 7–13 × 5.2–7.2 μ; basidia four-spored (exceptionally three-, five-, six-spored), clavate, 27–35 × 8–10 μ; gill edge heteromorphic from numerous obtuse fusiform cystidia 40–60 × 11–14 μ, with their apices sometimes forked, with grayish-fuliginous or sordid-brownish contents (never yellow or reddish violet); pleurocystidia lacking; subhymenium indistinct; spores, pileus, flesh, and trama of gills distinctly amyloid.

Habit, habitat, and distribution.—Rare in North America, known only from one collection in Colorado, on debris on the forest floor. In Europe it fruits on decaying wood.

Material studied.—Mains, 5208.

Observations.—Mains's collection was made on August 26, 1940, at Bear Lake in the Rocky Mountain National Park. The spores measure 8–11 × 5–7 μ and the cheilocystidia 36–54 × 9–13 μ. The iodine reactions were as Kühner described them. The description is adapted from that of Kühner. The relationships of the species are very likely with *M. rubromarginata* and *M. viridimarginata.*

100. Mycena citrinomarginata Gillet

Les Hymén., p. 266. 1874

Mycena flavicitrina Murrill, Mycologia, 8: 220. 1916.
Prunulus flavicitrinus Murrill, North Am. Flora, 9: 336. 1916.

Illustrations:

Plate 34; Text figs. 24, nos. 1–2 (p. 216); 25, nos. 7–8 (p. 223).
Bresadola, Icon. Mycol., 5, pl. 222, fig. 2 (as *M. elegans*).
Lange, Flora Agar. Dan., 2, pl. 50 F (very good).

Pileus (5) 10–35 (40) mm. broad, evenly and obtusely conic when young, the margin appressed against the stipe, soon broadly conic or obtuse and the margin frequently incurved, at maturity conic-campanulate, obtusely umbonate, or broadly convex, the margin frequently flaring, surface moist, pruinose but soon naked and pol-

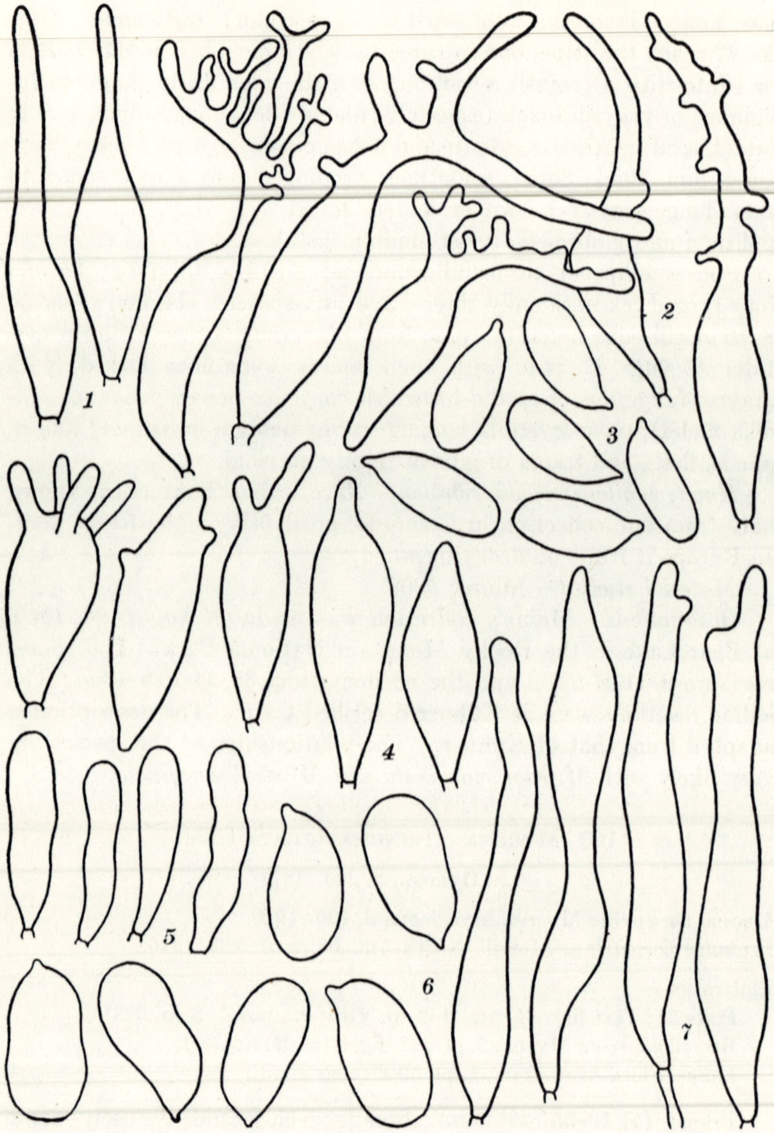

FIG. 24. *M. citrinomarginata:* 1 and 2, cheilocystidia. *M. arenacea:* 3, spores; 4, cheilocystidia; 5, cells from surface of pileus. *M. cheboyganensis:* 6, spores; 7, cheilocystidia

ished, translucent-striate, hygrophanous, color extremely variable, usually some shade of grayish or brownish yellow but sometimes bright yellow or nearly white (buttons "baryta yellow" or "citron yellow," changing to "olive lake" in age), fading to yellowish or olive gray, sulcate in age; flesh thick on the disc in large caps, otherwise thin, grayish yellow to whitish, fragile, odor and taste not distinctive; lamellae ascending-adnate, equal, distant to subdistant, 14–20 reach the stipe, two or three tiers of lamellulae, narrow (2–3 mm. $\pm$), intervenose, whitish or cinereous, edges pale yellow or sordid yellowish brown (the color often very inconspicuous in faded specimens); stipe 3–8 cm. long, (0.5) 1–2.5 mm. thick, equal, fragile, hollow, terete or compressed, base somewhat mycelioid, either glabrous or pubescent (depending on moisture conditions at the time of development), the apex pruinose at first, either smooth or somewhat striate, color olivaceous gray or yellowish, usually paler and brighter above.

Spores 8–11 $\times$ 4–5.5 (6) μ (12–14 $\times$ 5–6 μ, two-spored), narrowly ellipsoid to subcylindric, distinctly bluish in iodine; basidia two-, three-, four-spored; pleurocystidia absent or very rare to occasional, fusoid-ventricose, 36–52 $\times$ 9–15 μ; cheilocystidia abundant and variable, awl-shaped, fusoid-ventricose or clavate (all of these types either smooth, variously branched, or with fingerlike prolongations over the upper portion); gill trama homogeneous, vinaceous brown in iodine; pileus trama characterized by a thin nongelatinous pellicle, a well-differentiated hypoderm, and a filamentous tramal body, all but the pellicle vinaceous in iodine.

Habit, habitat, and distribution.—Single, scattered, or gregarious among mosses or on needle carpets under conifers, also on humus and among fallen leaves in groves of oak and beech. I have examined specimens from New York, Michigan, Washington, Oregon, and California in the United States and from Ontario, Manitoba, and British Columbia in Canada. It fruits during the spring and fall months, but is quite sporadic.

Material studied.—Smith, 32–1, 32–61, 32–520, 32–644, 707, 734, 1376, 1610, 1747, 2408, 2549, 3289, 3882, 4320, 5055, 6235, 6997, 8588, 8663, 8722, 8916, 13846, 15038, 15857, 16008, 16624. Atkinson, 15122, 19023, 24106. Bisby, Winnipeg, Manitoba. Davidson, 201. Mains, 31–678, 32–61.

Observations.—In the course of my study of this species I have examined many more collections than those cited above, particularly from localities around Ann Arbor. It is a most variable fungus in all four of its distinguishing characters: color of cap, type of cheilocystidia, color of gill edge, and spore size. Nevertheless, it is one of the easiest Mycenae to recognize both in the field and in the herbarium.

The cheilocystidia are extremely variable. The drawings in text figures 24, nos. 1–2 (p. 216), and 25, no. 8 (p. 223), were made from sections of a single pileus. In some collections the smooth fusoid-ventricose type is most common or is the only kind present, whereas in other collections a generous mixture of all types is present. The age of the fruiting bodies must be taken into account when one is studying the cystidia of this species. In mature or old pilei the cheilocystidia are more irregular than they are in young individuals. My comments on the variation in the shape of the cystidia are based on a study of freshly matured caps in order to avoid the possible error of comparing merely the different stages of development. The pleurocystidia are not sufficiently numerous to be taxonomically significant.

A distinct correlation between the color of the carpophores and the amount of light in the habitat has been found. Out in the open the fruiting bodies are very dark brown and have sordid-brown gill edges. In the deep shade of conifer thickets a very pale delicate yellow to whitish form is found. The majority of the fruiting bodies collected in moderately open stands are intermediate between the extremes noted above and form the bulk of the collections examined.

The color of the gill edges varies with the color of the pileus. In dark forms they are sordid yellowish brown, whereas in pale forms they are scarcely margined at all. If one uses a hand lens on the pale form, one can nearly always find some gills with colored edges.

Bisby and others (1938) have briefly commented on a form (under the name *M. chlorantha*) which appears to be very close to *M. citrinomarginata*. Bisby's notes are as follows: "In deciduous woods . . . . Pileus very small, greenish to yellowish, translucent striate, umbonate; gills white; stem 1 mm. wide, pale green above, darker green below, mycelioid at base; spores 12–14 × 5–6 μ; pointed sterile cells at edges of gills." This collection approaches very closely the classical concept of *M. chlorantha*. Since the name *M. chlorantha* has been used for a species very different from *M. citrinomarginata* by Kühner (see "*M. elegans*," p. 199), it is doubtful whether the name should be continued in use. If we continue to use it, it should be applied to a fungus of this type. If the fungus Bisby collected is found to have slightly colored gill edges, it should be considered a variety of *M. citrinomarginata*; otherwise, it should be placed in the Adonidae. This point needs clarification.

101. Mycena olivaceobrunnea A. H. Smith

Mycologia, 29: 339. 1937

Illustrations;
Plate 49 D; Text fig. 25, nos. 5–6 (p. 223).
Smith, Mycologia, 29: 353, fig. 3 C.

Pileus 5–10 mm. broad, obtusely conic, becoming obtusely campanulate but seldom expanding, margin appressed against the stipe when young, surface glabrous, moist, striate to disc with broad dark lines, color when moist "buffy brown" to "dark olive buff" with a pallid margin, fading to sordid yellowish gray; flesh very thin and membranous, pallid, odor and taste not distinctive; lamellae ascending-adnate, close to subdistant, 14–16 reach the stipe, narrow to moderately broad at maturity (1–1.5 mm. $\pm$), sordid yellowish gray, edges bright lemon yellow; stipe 2–6 cm. long, (0.5) 1–1.5 mm. thick, equal, tubular, rather weak and very fragile, glabrous and polished above, base sparsely white-strigose, color "olive brown" toward the base, "deep olive buff" or sordid yellowish gray above.

Spores 6.5–8 $\times$ 4–4.5 μ (8–10 $\times$ 4–5 μ, two-spored), ellipsoid, amyloid; basidia two- or four-spored; pleurocystidia not differentiated; cheilocystidia 30–44 $\times$ 8–18 μ, fusoid-ventricose to clavate, with or without a narrowed neck, the neck sometimes branched or occasionally with scattered blunt projections over the enlarged portion; pileus trama with a thin pellicle, a well-differentiated hypoderm, and a narrow region of filamentous tissue; both the gill and pileus trama sordid vinaceous brown in iodine.

Habit, habitat, and distribution.—Densely gregarious on humus, moss carpets, and needle beds under second-growth Douglas fir during May, June, September, and October; Washington to California. Often scattered in deep moss and very inconspicuous.

Material studied.—Smith, 2507, 3279, 8277, 13992, 14000, 10710.

Observations.—This fungus is distinguished from *M. citrinomarginata* by its consistently smaller size and smaller spores. From *M. flavifolia* it differs by having a colored gill edge and smoother cystidia. *M. olivaceomarginata* Massee appears to be close but is apparently not well understood in Europe. Rea reduces it to a variety of *M. avenacea*, saying that it differs only in its smaller spores (6–7 × 4–5 μ). It is described as having the gill edge sordid olivaceous, a rather striking contrast to the lemon-yellow gill edges of my Washington collections. However, this does not exclude the possibility that the two are synonymous. The situation in regard to the color of the gill edges may be the same as that discussed under *M. citrinomarginata*.

102. Mycena avenacea (Fr.) Quélet, sensu Kühner

Encyc. Myc., 10: 413. 1938

Agaricus avenaceus Fries, Syst. Myc., 1: 150. 1821.

Illustrations:

Text fig. 24, nos. 3–5 (p. 216).
Konrad et Maublanc, Icon. Sel. Fung., pl. 223, fig. 2 (very good).
Lange, Flora Agar. Dan., 2, pl. 49 B.

Pileus (5) 10–20 mm. broad, obtusely conic with an appressed margin when young, becoming broadly conic to campanulate, pruinose but soon naked and glabrous, striate to the disc with dark-colored striations when moist, soon sulcate-striate and with a flaring margin (which has a faint incarnate tint), hygrophanous, fading through "olive brown" to sordid olive or yellowish gray; flesh thin, fragile, odor and taste not distinctive; lamellae ascending and adnate with a slight tooth, distant to subdistant, moderately broad, pallid to grayish, edges even and concolorous or very faintly reddish brown; stipe 2–4 cm. long, 1–1.5 mm. thick, strict, equal, hollow, fragile, pruinose above, glabrous downward or the base sparsely mycelioid, concolorous with the pileus and fading with it.

Spores 9–12 (13) × 4–5.5 μ, subcylindric to narrowly ellipsoid, weakly amyloid; basidia four-spored; pleurocystidia not differentiated; cheilocystidia 32–40 × 9–14 μ, broadly fusoid to ventricose, the neck tapered to the obtuse apex or variously branched, the enlarged portion often giving off fingerlike branches, content reddish brown; gill trama weakly amyloid; pileus trama with a thin adnate pellicle from which numerous pilocystidia often arise (the cystidia 15–20 × 5–7 μ, cylindric and with obtuse apices or merely filamentous), hypoderm well differentiated, all but the pellicle pale sordid brown in iodine.

Habit, habitat, and distribution.—Single to scattered. I have one collection found under white pine, September 8, 1934; Warrensburg, New York.

Material studied.—Smith, 733.

Observations.—The dried specimens have a vinaceous rather than a yellowish cast, and in this respect the species resembles *M. capillaripes* more closely than *M. citrinomarginata.* The lack of pleurocystidia and the larger spores readily separated it from the former, and the reddish gill edges from the latter, although I am inclined to believe it intergrades with *M. citrinomarginata.*

103. **Mycena cheboyganensis,** sp. nov.

Illustrations: Text fig. 24, nos. 6-7 (p. 216)

Pileus 5–15 mm. latus, late conico-campanulatus vel convexo-umbonatus, sulcatus, fuscus demum griseus; lamellae latae, subdistantes, late adnatae, albidae vel griseae, rubro-marginatae; stipes 3–8 cm. longus, 1–2 mm. crassus, aequalis, fragillissimus, glaber, griseus; sporae (10) 11–14 × 6–8 μ; basidia tetraspora; pleurocystidia et cheilocystidia similia, subventricosa vel subcylindrica, 40–65 × 6–9 μ. Specimen typicum in Herb. Univ. Mich. conservatum. Legit A. H. Smith, n. 32–92, prope Rees's Bog, Cheboygan Co., Mich., June 6, 1932.

Pileus 5–15 mm. broad, broadly conico-campanulate to convex or umbonate, umbo blunt but abrupt, margin appressed at first, surface moist, soon dry, and in the faded condition appearing finely appressed-fibrillose under a lens (10×), sulcate-striate to the umbo, fuscous on the disc and along the striations, paler toward the margin and in the interspaces, pale grayish over all when faded except for the reddish-brown margin, margin not incised or lobed; flesh thin, pallid, fragile, no color change noted when bruised, odor and taste not recorded; lamellae broad, subdistant to distant, broadly adnate and decurrent by a line, white to grayish with dark reddish-brown edges; stipe 3–8 cm. long, 1–2 mm. thick, equal, fragile, hollow, apex pruinose but soon polished, base white-mycelioid, pale gray to pallid.

Spores (10) 11–14 × 6–8 μ, smooth, ovoid to subpyriform, amyloid; basidia four-spored; pleurocystidia rare to scattered, 40–60 × 6–9 μ, subventricose to subfiliform, the apices occasionally forked, hyaline; cheilocystidia very abundant and similar to pleurocystidia, gill edge with the coloring matter apparently not located in the cystidia (its position was not ascertained at the time the fresh specimens were in hand); gill trama amyloid; pileus trama with a thin non-gelatinous pellicle, the cells of which give off more or less upright branches 6–10 × 2.5–3 μ, hypoderm sharply differentiated, cells 20–40 μ thick, filamentous tramal body very narrow and irregular, all except the pellicle vinaceous brown in iodine.

Habit, habitat, and distribution.—Scattered on sphagnum along with *M. praelonga* in June; Michigan. Known only from the type locality.

Observations.—The spores prevent this species from being referred to *M. capillaripes*, but it also differs in lacking an odor and in the scarcity of pleurocystidia. It appears to be closest to *M. rubromarginata*, but is distinguished by its terrestrial habitat, narrow cystidia, and long stipe. It has the habit of *M. praelonga*, and appears to be a very rare fungus. I have searched bogs for it repeatedly since 1932 without finding it again.

104. MYCENA CAPILLARIPES Peck

Ann. Rep. New York State Mus., 41: 63. 1888

Prunulus capillaripes Murrill, North Am. Flora, 9: 328. 1916.
Mycena plicosa var. *marginata* Lange, Stud. Agar. Dan., 1: 18. 1914.
Mycena Langei R. Maire, Bull. Soc. Myc. Fr., 44: 39. 1928.

Illustrations:

Plate 35 A, B, C, D; Text fig. 25, nos. 1–4.
Lange, Flora Agar. Dan., 2, pl. 49, fig. A (as *M. plicosa* var. *marginata*).
Smith, Mycologia, 28, fig. 1, no. 2.

Pileus (5) 10–20 (25) mm. broad, oval to convex when young, becoming obtusely conic to campanulate, in age remaining campanulate or becoming plane, frequently obtusely umbonate, margin appressed against the stipe when young, often flaring somewhat at maturity or sometimes recurved, occasionally crenate or lobed, surface smooth and with a hoary bloom at first, soon naked and appearing moist, obscurely to conspicuously translucent-striate, sulcate-striate when faded, color generally some shade of vinaceous gray, at times nearly fuscous and only the margin reddish, occasionally dark grayish red over all ("light seal brown" to nearly "fuscous" on the disc and fading to "cinnamon drab" or "pale vinaceous fawn" in one form, "Prussian red" to "dark Indian red" at first and fading to "vinaceous fawn" or pale gray with a tinge of vinaceous in age in another form); flesh thin, fragile, pallid grayish or vinaceous, taste

FIG. 25. *M. capillaripes:* 1, cheilocystidia; 2, pleurocystidia; 3, spores from two-spored basidia; 4, spores from four-spored basidia. *M. olivaceo-brunnea:* 5, cheilocystidia; 6, spores. *M. citrinomarginata:* 7, spores; 8, cheilocystidia

slightly acidulous, odor weakly or strongly nitrous if fresh material is crushed, occasionally lacking or somewhat radishlike; lamellae ascending-adnate and usually toothed, subdistant to distant (sometimes appearing close in young pilei), 12–18 reach the stipe, one to three tiers of lamellulae, narrow to moderately broad, faces pale or dark grayish vinaceous, edges colored pale pink or rose color at least on the larger gills, the edges of the lamellulae sometimes concolorous with the faces; stipe (2) 3–6 (8) cm. long, 1–1.5 (2.5) mm. thick, equal, hollow, very fragile, with a faint bloom at first, soon naked and polished, translucent, base sparsely white-strigose, color pale or dark grayish vinaceous, apex sometimes pallid and base sordid.

Spores ellipsoid to slightly obovoid, 7–9 (10) × 4–5 μ (10–13 [14] × 4.5–6 μ in two- and three-spored forms), amyloid; basidia two-, three-, and four-spored; cheilocystidia and pleurocystidia similar in size and shape but variable in distribution, usually abundant though at times rather rare on either the sides or the edges or both, fusoid-ventricose with obtuse apices or elongating and subcylindric with slightly tapered apices, smooth to rarely forked, contents pale or dark pinkish brown; gill trama vinaceous brown in iodine; pileus trama with a thin adnate surface pellicle, beneath it a broad region of vesiculose cells half the trama in thickness (the hypoderm), the remainder filamentous, dark vinaceous brown beneath the pellicle in iodine, contorted lactiferous hyphae present in both cap and gill trama.

Habit, habitat, and distribution.—Densely gregarious to scattered on needles under conifers, summer and fall, often common. Occasionally on humus or moss in oak woods; Tennessee, New York, Michigan, Idaho, Washington, Oregon, and California in the United States and Ontario in Canada.

Material studied.—Smith, 33–1021, 691, 701, 713, 724, 834, 1270, 1345, 1418, 1616, 1649, 1741, 1742, 2509, 2532, 2548, 2743, 4041, 4060, 4378, 5037, 7001, 7976, 8042, 8111, 8414, 8765, 8768, 8771, 8826, 13772, 13953, 14202, 14631, 18115. Atkinson, 18381. Kauffman, September, 1915, Michigan; September 12, Idaho, 1922. Mains, 34–116, 34–191, 34–194. Morse, California. Stuntz, 589.

Observations.—The two color forms were collected in the same plantation at the same time and appeared to be genetically distinct. Both were two-spored. The dark-red form has been found only in the one locality.

The color of the gill edges varies directly with the number of cystidia present, and in forms with only scattered cheilocystidia the gills may not be margined. This, however, is an extreme variation. In the specimens I have seen from oak woods the larger gills were distinctly margined, whereas the lamellulae were not. The pale grayish-vinaceous or reddish colors, marginate gills, smooth cystidia, and alkaline odor serve to distinguish *M. capillaripes* from other marginate Mycenae.

In the Smith River Canyon in northern California an odorless form (8765) was found in which the stipe was bluish black. The consistency was more like that of *M. rosella* than that of typical *M. capillaripes*, but the cystidia were all smooth. This collection is placed here tentatively.

SECTION TYPICAE

The species of this section form the center of the genus, and it is in relation to these that all the others are grouped. Although the section is a large one, it is not divided into subsections because of the lack of practical divisions. In general, one might say that the old sections Filipedes, Fragilipedes, and Rigidipedes of Fries have been brought together here. The divisions used in the key serve to group the obviously closely related species. The section represents a most monotonous series of blackish, brown, gray, bluish-gray, or brownish-gray species mostly with ascending gills and generally large to moderate stature. The very small species are placed in the Deminutivae, even though their closest relatives may be here. The stipes are usually smooth, but a few species with the combination of fibrillose stems and amyloid spores are also included. The section Floccipedes (those with fibrillose stipes and nonamyloid spores) will not cause much trouble for the beginner because it is a very small section, which can be eliminated from consideration by glancing through the characters of the several species. The section Omphaliariae will cause more difficulty at first. In general, species growing cespitose on wood, even if their gills are broadly adnate, arcuate, or subdecurrent, are placed in the typicae. *M. semivestipes* and *M. laevigata* are the two commonest examples. Certain terrestrial species have been keyed out in both sections to facilitate their identification.

KEY TO SPECIES

1. Fragile species: usually gregarious on humus or scattered on pieces of debris on ground, frequently found in large troops under conifers (see no. 33 for fragile lignicolous species) 2

1. Cartilaginous-pliant species: usually on decaying wood and often cespitose (if on humus, cespitose, or not fragile; if fragile, then lignicolous). The large cartilaginous *whitish* or *very pale species* often cespitose on wood are keyed out here rather than in the Adonidae . . 33

2. Cheilocystidia roughened (at least over ventricose portion if they are fusoid-ventricose), with rodlike projections or numerous contorted fingerlike projections or merely verrucose 3
2. Cheilocystidia, if present, smooth, fusoid-ventricose, or with two to four or, occasionally, more obtuse prolongations 16

FRAGILE SPECIES WITH ROUGH CHEILOCYSTIDIA

3. Pleurocystidia fusoid-ventricose, 54–90 × 8–12 μ; ventricose portion often roughened . . . . 124. *M. latifolia*
3. Pleurocystidia, if present, clavate and with roughened apices 4

4. Pellicle gelatinous to subgelatinous or broad gelatinous layer present just beneath pellicle 5
4. No gelatinous layers present in pileus 6

5. Odor alkaline or lacking; subpellicular layer gelatinous
125. *M. constans*
5. Pellicle gelatinous . . . . *M. pseudoclavicularis* (p. 374)
5. Odor none; pellicle subgelatinous, thickish 5a
   (*a*) Stipe breaking readily at base when collected; pileus usually plicate-striate to scalloped . . . . 133. *M. plicosa*
   (*a*) Stipe not as above; pileus slightly sulcate only in age
126. *M. pusilla*

6. Pileus russet, tawny, or vinaceous fawn tinged with gray at times . . . 7
6. Pileus dominantly gray, fuscous, bluish gray, or grayish brown 8

7. Fruiting body Galerina-like (russet to tawny); pleurocystidia absent
129. *M. psammicola*
7. Fruiting body vinaceous fawn when fresh, fading to ashy or avellaneous in age; pleurocystidia present . . . . 130. *M. metata*

8. Stipe long in relation to width of cap (cap 5–25 mm., stipe 6–12 cm.) . . 9
8. Stipe shorter in relation to width of cap (cap 1–4 cm., stipe 3–8 cm.) . . 11

9. A strong odor of iodoform developing after fruiting bodies have been gathered . . . . 137. *M. iodiolens*
9. Odor not distinctive 10

10. Growing in sphagnum, hence long stipe . . . . 135. *M. atroalboides*
10. Growing in open on humus or needle beds or on moss . . 136. *M. filopes*

11. Pileus typically large (2–3.5–5 cm.) 12
11. Pileus 5–10 (20) mm. 13

12. Spores 8–11 × 5–6 μ; pleurocystidia present ..... 132. *M. hudsoniana*
12. Spores 6–8 (9) × 4–5 μ; pleurocystidia absent ...... 131. *M. piceicola*

13. Cheilocystidia quite irregular in outline, projections (4–7) fingerlike and often contorted; odor fragrant .... 127. *M. alcaliniformis*
13. Cheilocystidia more nearly verrucose to echinulate, only in age with contorted fingerlike projections 14

14. Pileus decidedly dark bluish drab or lilac gray ..... 128. *M. urania*
14. Pileus drab gray to fuscous 15

15. Pleurocystidia rare to absent; in wet weather fruiting bodies sometimes staining sordid reddish in all parts (see *M. plicosa* and the odorless form of *M. constans*, which may key out here also) 135. *M. atroalboides*
15. Pleurocystidia scattered, margin of cap becoming sordid ochraceous in age; no other color change noted............. 134. *M. subplicosa*

FRAGILE SPECIES WITH SMOOTH CHEILOCYSTIDIA

16. Pleurocystidia present and usually very prominent 17
16. Pleurocystidia, if present, not with greatly elongated necks; cheilocystidia inclined to be branched 23

17. Pleurocystidia with highly refractive granular contents; spores 8–11 × 5–6 μ .... 106. *M. Sabali*
17. Pleurocystidia not as above 18

18. Spores 10–12 × 5–6 μ (4-spored) (see *M. galopus* also) 120. *M. fusco-ocula*
18. Spores of 4-spored basidia smaller 19

19. Pleurocystidia 36–50 × 6–11 μ 20
19. Pleurocystidia 60–90 μ long 21

20. Odor alkaline .... 153. *M. alcalina*
20. Odor not alkaline; cap trama composed of contorted "metallic" hyphae (lactifers ?) .... 200. *M. brevipes*

21. Pileus without distinct pellicle, surface more or less covered with irregularly disposed pilocystidia 15–60 × 9–16 μ and with obtuse apices .... 108. *M. trichoderma*
21. Not as above .... a
    (*a*) Stipe glabrous; spores 6–8 μ long .... b
    (*a*) Stipe furfuraceous, at least when young 22
    (*b*) Stipe 3–6 cm. long; on needle beds under conifers or on humus under brush .... 107. *M. pseudotenax*
    (*b*) Stipe (4) 6–12 cm. long; on decaying wood of hardwoods 121. *M. Kauffmaniana*

22. Spores ellipsoid; pileus 1–3 cm. broad .... 109. *M. scabripes*
22. Spores globose to subglobose, 5–6 μ .... 190. *M. floccipes*

23. Odor alkaline when flesh is crushed .... 111. *M. leptocephala*
23. Odor not as above 24

24. Spores subcylindric, 9–12 (13–15) × 4.5–5 (6) μ; pileus dark to pale gray .... 110. *M. Abramsii*
24. Spores more nearly ellipsoid; if 9–11 μ long, then 6–7 μ wide 25

25. Pileus and stipe blackish, hardly fading, not staining reddish in age (*M. praelonga* may key out here also) .... 118. *M. atrocyanea*
25. Pileus blackish but very hygrophanous; stipe whitish, staining reddish where bruised .... 112. *M. subvitrea*
25. Pileus gray to avellaneous, or, if blackish at first, gills pallid and stipe not staining reddish 26

26. Cheilocystidia absent, cap turning lilac when dried 105. *M. syringescens*
26. Cheilocystidia typically obovate-mucronate (in age becoming more or less branched); young pilei faintly tinged vinaceous along margin (see *M. plumbea* also) .... 117. *M. pectinata*
26. Cheilocystidia usually fusoid-ventricose to variously branched; pileus not with vinaceous margin 27

27. Stipe 9–16 (20) cm. long 28
27. Stipe 3–6 (8) cm. long 30

28. Exceedingly watery-fragile; on fern debris .... 116. *M. fragillima*
28. Not as above 29

29. Cheilocystidia obovoid-mucronate; on needle carpets, stipe very brittle 29a
    (*a*) Pleurocystidia absent .... 114. *M. plumbea*
    (*a*) Pleurocystidia present .... 115. *M. plumbea* var. *robusta*
29. Cheilocystidia typically fusoid-ventricose, occasionally branched; stipe tenacious, growing on sphagnum .... 119. *M. praelonga*

30. Stipe almost filiform, about 1 mm. thick 31
30. Stipe 2–3 mm. thick 32

31. With bearing and consistency of *M. filopes* (tall and slender), but cheilocystidia smooth .... a
    (*a*) Pleurocystidia absent .... 122. *M. subfusca*
    (*a*) Pleurocystidia present and distinctive ...... 123. *M. subfumosa*
31. Very fragile and watery, growing on fern debris, western United States .... 116. *M. fragillima*

32. Pileus long, remaining decidedly glaucous owing to persistent hoary-pruinose coating .... 114. *M. plumbea*
32. Pileus pruinose but soon naked, hygrophanous, with stannous sheen when faded .... 113. *M. stannea*

CARTILAGINOUS OR FRAGIL LIGNICOLOUS SPECIES

33. Cheilocystidia fusoid-ventricose type; if branched, see no. 43 (apex echinulate in *M. borealis*) 34
33. Cheilocystidia clavate-roughened or, if fusoid-ventricose, sometimes giving rise to two or more fingerlike prolongations, sometimes merely irregular in outline (see *M. semivestipes*, *M. laevigata*, and *M. ochraceicinerea*, which may have typically fusoid cystidia but are keyed out here) 43

34. Pleurocystidia very conspicuous 35
34. Pleurocystidia rare to absent and not projecting much if present 41

35. Odor nitrous 36
35. Odor none 37

36. Stipe smooth, color remaining dark; pleurocystidia 60–90 μ long 152. *M. macrocystidia*
36. Stipe fibrillose-striate at least when young, white, pileus pale in age; cystidia 60–110 μ long ..... 151. *M. niveipes*
36. Stipe smooth; pileus dark; pleurocystidia up to 50–54 μ long 153. *M. alcalina*

37. Pleurocystidia with echinulate apices ..... 147. *M. borealis*
37. Pleurocystidia smooth 38

38. Stipe soon smooth and polished 39
38. Stipe fibrillose to somewhat striate, at least when young 40

39. Gills broadly and deeply adnexed; pellicle gelatinous in KOH 148. *M. excisa*
39. Gills hooked or ascending-adnate (see also *M. floccipes* and *M. Kauffmaniana*) ..... 153. *M. alcalina*

40. Cespitose; pileus blackish to dull sordid brown and scarcely fading; gills close to crowded ..... 150. *M. algeriensis*
40. Solitary to gregarious; gills subdistant; pileus dark; with appearance of Collybia ..... 149. *M. tenuiceps*
40. Pileus fuscous on disc, whitish toward margin, fading to whitish; stipe very soon almost snow white ..... 151. *M. niveipes*

41. Pileus grayish, sometimes fading to white 42
41. Pileus whitish; gills pale incarnate (see also *M. laevigata*) 154. *M. incarnatifolia*

42. Spores globose, 3.5–4 μ, nonamyloid ..... 191. *M. ulmicola*
42. Spores broadly ellipsoid to subglobose, 5–6 (7) × 4–5 μ 155. *M. collybiiformis*
42. Spores ellipsoid, larger than above; cheilocystidia smooth, 23–30 × 5–7 μ; carpophore similar to *M. inclinata* in appearance 159. *M. pseudoinclinata*

42. Spores ellipsoid; cheilocystidia 26–40 × 9–15 μ; carpophore similar to *M. parabolica* in appearance .... a
(*a*) Odor none .... 160. *M. alnicola* var. *typica*
(*b*) Odor of radish .... 161. *M. alnicola* var. *odora*

CHEILOCYSTIDIA ROUGHENED

43. Terrestrial and cartilaginous-pliant; stipe usually 8–12 cm. long and sometimes rooting 44
43. Typically lignicolous or usually cespitose 52

44. Pileus 5–15 (20) mm. broad 45
44. Pileus 15–30 (50) mm. broad 47

45. Stipe with strong purplish to vinaceous tinge in apex at least; cap more or less dark vinaceous brown .... 140. *M. pullata*
45. Vinaceous or purplish tints absent 46

46. Gills very broadly adnate, nearly horizontal in age ... 138. *M. peltata*
46. Gills ascending, adnate by slight tooth .... 139. *M. vitilis*

47. Spores 7–10 (11) μ long 48
47. Spores 9–13 (17) μ long 51

48. Stipe longitudinally striate, at least when young .. 141. *M. polygramma*
48. Stipe smooth and polished 49

49. Spores 8–11 × 3–4.5 μ .... 146. *M. fagetorum*
49. Spores 7–9 (10) × 5–7.5 μ 50

50. Spores 7–9 × 6–7.5 μ; pleurocystidia rare .... 142. *M. magna*
50. Spores 7–10 × 5–6.5 μ; pleurocystidia absent .... 143. *M. longipes*

51. Spores 12–17 × 6–8 μ; stipe with long pseudorhiza 144. *M. megaspora*
51. Spores 9–13 × 4–5.5 (6) μ; pseudorhiza lacking .. 145. *M. griseiconica*

52. Cheilocystidia fusoid-ventricose, with tendency to branch or develop short projections, or basidium-like with somewhat irregular outlines, or not differentiated 53
52. Cheilocystidia broad (10–18 μ) and mucronate or with several rather long and thick projections 54
52. Cheilocystidia clavate, broad apices covered with numerous rodlike projections or projections fewer and larger as well as irregular in outline and manner of branching 55

53. Cheilocystidia not differentiated; spores 7–9 × 6–7 μ 158. *M. ochraceicinerea*
53. Cheilocystidia clavate to basidium-like, outlines often irregular; spores 4–5 × 2.5–3 μ .... 157. *M. semivestipes*
53. Cheilocystidia narrowly fusoid-ventricose, neck often with one or more rodlike projections; spores 6–8 × 3–4 μ.... 156. *M. laevigata*

54. Pileus blackish when fresh, becoming pale gray, margin conspicuously sulcate-striate; spores 9–11 μ long .... 162. *M. parabolica*
54. As above, but spores 6–9 μ .... 160. *M. alnicola*
54. Pileus, stipe, or gills or all parts developing strong pinkish to reddish tints or stains in age or when bruised; on hardwood logs and debris 163. *M. rubrotincta*
54. Pileus, gills, or stipe and hairs at base of stipe developing tawny to sordid-tan colors in age; on wood of pine ...... 164. *M. intertexta*

55. Stipe covered (sparsely in some) with white-fibrillose flecks, at least when young .... 165. *M. inclinata*
55. Stipe glabrous from beginning (apex may be pruinose) 56

56. Spores 3–5 μ wide 57
56. Spores 5–7 μ wide 59

57. Gills soon stained with sordid-reddish stains .... 167. *M. maculata*
57. Gills not stained as above 58

58. Cheilocystidia very irregular, varying from smooth to covered with rodlike projections, apices frequently elongated and contorted 168. *M. occidentalis*
58. Cheilocystidia clavate, apices covered with short rodlike projections; spores 7–9 × 4–5 μ .... 169. *M. hemisphaerica*
58. Cheilocystidia as above; spores 5–7 × 4–5.5 μ .. 166. *M. subinclinata*

59. Pileus very pale ("deep olive-buff") often appearing whitish (but young caps distinctly pale gray); spores 6.5–8 × 5.5–7 μ or 8–9 × 5–7 μ (2-, 3-, 4-spored) .... 172. *M. radicatella*
59. Pileus darker, fuscous when young; spores 8–10 (11) × 5–7 μ 60

60. Odor and taste farinaceous to raphanoid; often with pseudorhiza 170. *M. galericulata*
60. Odor and taste none; pseudorhiza, if present, very short 171. *M. rugulosiceps*

Species of Uncertain Position

105. Mycena syringescens Murrill

Bull. Torr. Bot. Club, 67: 235. 1940

Prunulus syringescens Murrill, *ibid.*, p. 234.

"Pileus convex, not expanding, solitary, 1 cm. broad; surface dry, smooth, glabrous, opaque, pale-avellaneous, margin even, entire; context membranous; lamellae sinuate with decurrent tooth, broad, inserted, distant, pallid, toothed; spores ovoid, smooth, hyaline, 1-guttulate, about 5 × 3 μ; cystidia none; stipe equal, smooth, glabrous, shining, white, 2.5 cm. long, scarcely 1 mm. thick. The entire hymenophore changes to pale-lilac on drying.

"Type collected by W. A. Murrill on the ground in mixed woods just east of Gainesville, Fla., Oct. 29, 1938 (*F 18379*). A rare species, springing a surprise by its entire change of color."

Through the kindness of Dr. Murrill I have been able to examine a small portion of the type. The details of the tramae of the pileus and gills could not be made out clearly, but neither of the tissues was amyloid. No pleurocystidia or cheilocystidia were observed. The basidia were four-spored. Spores were abundant, measured 5–6 × 3–3.5 μ, were narrowly ellipsoid and very distinctly amyloid. The systematic position of this species is a puzzle to me. Because of the amyloid spores and the pale-avellaneous pileus it appears to be close to *M. subfusca*, but it is readily distinguished from all in this group by the change in color in drying. This change may indicate a relationship to the Adonidae or the Calodontes.

106. Mycena Sabali Murrill

Mycologia, 8: 221. 1916

Prunulus Sabali Murrill, North Am. Flora, 9: 325. 1916.

Illustrations: Text fig. 12, nos. 3–5 (p. 142).

"Pileus convex, scarcely umbonate, not fully expanding, solitary, 1 cm. broad; surface glabrous, rugose, striate, dry, pale-luteous, slightly darker on the disk: lamellae adnate, subdistant, narrow, white: stipe cylindric, glabrous, smooth, white, pubescent at the base, 3 cm. long, 1 mm. thick.

"Type collected on a dead petiole of a dwarf palmetto in City Park, New Orleans, Louisiana, September 6, 1908, *F. S. Earle 77* (herb. N. Y. Bot. Gard.).

"Distribution: Known only from the type locality."

Observations.—I have examined the type and made the following observations: The spores measure 8–11 × 5–6 μ, are smooth, subellipsoid, and amyloid. Both two- and four-spored basidia were seen. The pleurocystidia and cheilocystidia, which are abundant, are broadly fusoid with acute apices and smooth. Many have a highly refractive granular-appearing content similar to those of *M. erubescens*. The surface of the pileus is characterized by the presence of thin-walled pilocystidia 30–100 × 8–10 μ. These are hyaline and apparently quite delicate (many were collapsed). The trama of the pileus did not revive well, and few details could be ascertained. The tissue of the gills and pileus was yellowish in iodine. Murrill did not mention a milky juice, and I observed no lactifers in the flesh of either the pileus or the gills. The stipe was not examined.

Material studied.—The type and one collection by C. A. Brown; December 31, 1934, Louisiana.

107. Mycena trichoderma Josserand

In Kühner, Encyc. Myc., 10: 523. 1938

Illustrations: Text figs. 23, no. 5 (p. 212); 26, nos. 1–2 (p. 234).

Pileus 3–6 mm. broad, conic, the margin straight, dark fuscous on the disc, sordid brownish near the margin, surface somewhat reticulate around the disc, opaque, becoming slightly sulcate-striate; flesh rather cartilaginous, very thin, sordid brownish, odor and taste not distinctive; lamellae hooked, rather thick, narrow, subdistant, sordid grayish, edges grayish or pallid; stipe 10–15 mm. long, 1.5 mm. thick, equal, cartilaginous, tubular, dark sordid grayish brown, somewhat scabrous (glabrous in one old fruiting body), base with a few white hairs.

Spores 8–10 × 5–6.5 μ (10–12 × 6–7 μ), ovoid to ellipsoid, very weakly amyloid; basidia two-, three-, and four-spored; pleurocystidia and cheilocystidia very abundant, smooth, hyaline, subcylindric to slightly ventricose, the apices rounded, 60–90 (110) × 9–15 μ; gill trama homogeneous, yellowish in iodine; surface of the pileus covered with appressed, decumbent, or nearly upright cells 9–16 μ thick and 15–60 μ long, not forming a true pellicle and not arranged in a palisade, hypoderm very well differentiated—the cells vesiculose, the remainder of the trama filamentous, yellowish in iodine.

Habit, habitat, and distribution.—Known in North America from a single collection of three fruiting bodies, made September 2, 1933, by Mains and Smith in a mixed forest near Emerson, Michigan.

Observations.—The following is a description summarized from that of Josserand: Pileus 12–24 mm. across the base, conic-campanulate, slightly mammillate, sometimes a bit irregular, not truly hygrophanous (but the flesh is), dry, dark blackish brown (*bistre noirâtre*), glabrous, unpolished, radially plicate-striate at maturity, margin straight, finally slightly recurved, slightly sinuate, not striate. Cuti-

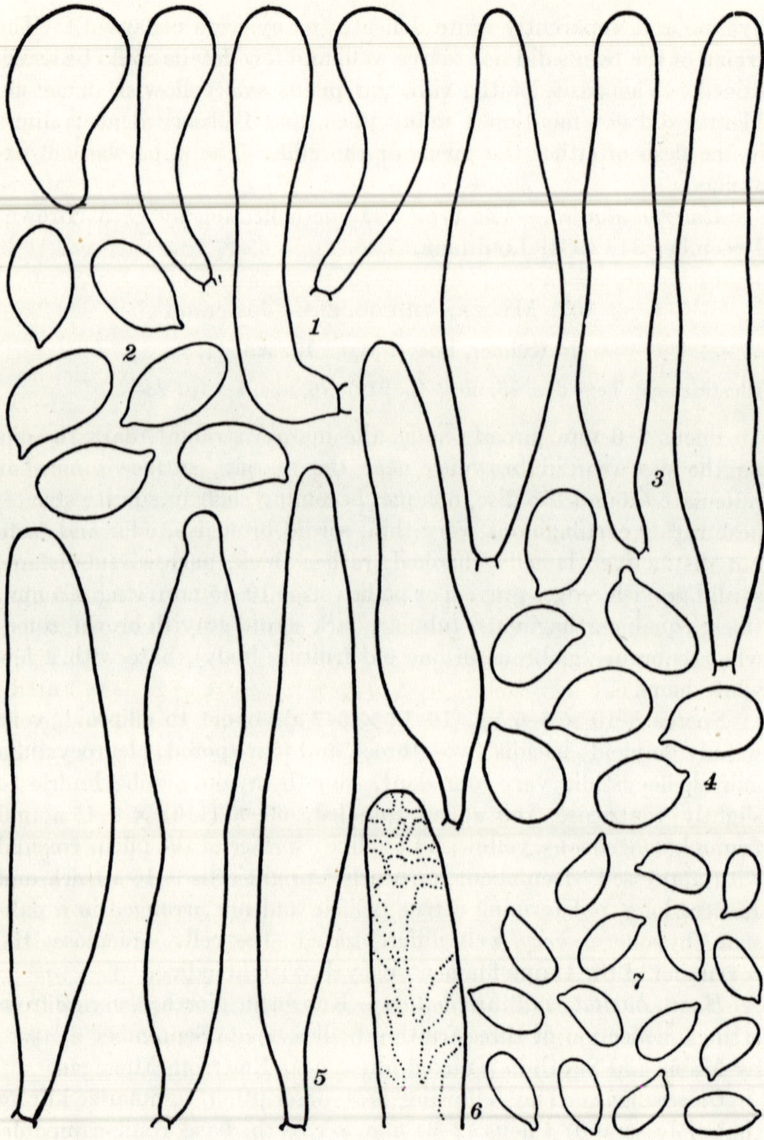

FIG. 26. *M. trichoderma:* 1, cells from surface of pileus; 2, spores. *M. scabripes:* 3, pleurocystidia; 4, spores. *M. pseudotenax:* 5, pleurocystidia; 6, pleurocystidia revived in KOH showing refractive content; 7, spores

cle poorly defined, not separable, in other respects similar to the flesh. Flesh very thin, blackish brown, fading to pallid grayish. Lamellae not very close, unequal, three to five (seven) tiers of lamellulae, simple, moderately broad, thickish, a bit ventricose, rounded, slightly sinuate-adnate, white or grayish, especially toward the base. Edge entire, somewhat obtuse, quite pale. Stipe moderately tenacious, 35–50 × 2–3 mm., equal, not bulbous, tubular or hollow, dark sordid grayish brown, slightly furfuraceous-punctate, the particles rather dark but stipe as a whole lighter than the pileus, dry. Practically no odor or taste.

Basidia four-spored, 40–48 × 7.6–9 μ; spores 7.5–9.4 × 5–5.5 (6) μ, ellipsoid, amyloid; pleurocystidia and cheilocystidia similar, numerous on the edge, scattered on the faces, large, 60–90 × 9–15 μ, irregularly cylindric or cylindric-fusoid, obtuse at the apex, hyaline; gill trama with a broad filamentous central strand; covering of the pileus distinctive. On the well-developed cellular hypoderm are decumbent epicuticular, separate, very loose filaments which do not mask the hypoderm. These hyphae have differentiated clavate extremities which at times are almost decumbent, subdecumbent, or almost upright and then simulate a hymeniform layer, the cells of which are free from each other. These hairs are 25–60 (120) × 6–18 μ. They are packed with a dark-brownish pigment located in the vacuoles; the epicuticular hyphae from which these arise are furnished with a second pigment located in peripheral incrusting plates. The hairs on the stipe are numerous, are more or less grouped in tufts, and resemble the pilocystidia except that they are more irregular. They measure 35–80 × 6–10 μ and have the pigment in the vacuoles. The hyphae of the pileus lamellae and stipe are not amyloid.

I did not note the pigments in the pilocystidia and caulocystidia of the Michigan collection. On revived material of the stipe the contents of the apical cells still appeared sordid brownish, indicating that a dark pigment was present in the fresh condition. In my specimens the caulocystidia were of two types; toward the apex they were like those on the gills, but over the remainder they resembled those on the pileus There is a discrepancy in spore shape between Josserand's material and mine, but, considering the variable number of sterigmata produced on basidia in the Michigan collection, the discrepancy is best disregarded. Singer (1942) places this species in Hydropus.

108. Mycena scabripes Murrill

Mycologia, 8: 221. 1916

Prunulus scabripes Murrill, North Am. Flora, 9: 331. 1916.

Illustrations: Plate 36; Text fig. 26, nos. 3–4 (p. 234).

Pileus 1.5–3.5 (5) cm. broad, obtusely conic, papillate or almost convex at first, generally becoming broadly campanulate, umbonate, papillate or plane, the margin appressed or sometimes slightly incurved at first, surface hoary-pruinose but soon naked and polished, moist to lubricous, even, hygrophanous to merely subhygrophanous, opaque when faded, color "mummy brown" to "clove brown" on the disc, "olive brown" to "buffy brown" toward the margin, "drab" when faded or the margin "pale smoke gray" (colors generally dark blackish brown with an obscure olive tinge or dark gray to blackish on the disc when moist, the margin slightly paler, fading to deep gray on the center and pale gray tward the edge); flesh thick on the disc, thin toward the margin, dark brownish to grayish, rather firm, odor and taste not distinctive; lamellae close to subdistant, 21–25 reach the stipe, two or three tiers of lamellulae, narrow at first but soon becoming broad and ventricose (3.5–6 mm. broad), ascending-adnate or becoming depressed-adnate, sometimes broadly adnexed or attached only by a slight tooth, seceding readily and often forming a collar around the stipe, thickish, at times whitish or pale grayish, sometimes spotted sordid reddish brown in age, pruinose under a lens from the projecting cystidia, edges even; stipe (2) 4–9 (12) cm. long, 2–3.5 (5) mm. thick, equal or the base slightly enlarged, hollow, fragile, surface fibrillose-furfuraceous to coarsely pruinose at first and sordid grayish from the fibrils, more or less glabrescent and sometimes merely pruinose or silky-fibrillose from a thin coating of whitish fibrils, dry, concolorous with the gills toward the apex, olivaceous gray or yellowish toward the base.

Spores 7–9 (10) × 4–5 (6) μ, amyloid, broadly ellipsoid; basidia four-spored, rarely two-spored; pleurocystidia and cheilocystidia similar, abundant, 60–90 (100) × 9–16 μ, slightly fusoid-ventricose to cylindric, the apices obtuse, smooth, hyaline; gill trama homogeneous, hyaline in iodine or becoming pale yellowish; pileus trama with a thin pellicle, from which a few cystidia project, hypoderm narrow but fairly well differentiated; the remainder of the trama floccose and filamentous, pale yellowish brown in iodine.

Habit, habitat, and distribution. Gregarious to scattered, occasion ally in clusters of two or three fruiting bodies, on humus under hardwoods or conifers; Tennessee, New York, Michigan, Montana, Washington, Oregon, and California in the United States and Ontario in Canada. It fruits most abundantly in the fall and is rather common under both oak and redwood My best collections have been made under second-growth oak and young redwoods.

Material studied.—Smith, 32–515, 32–539, 32–546, 32–555, 32–584, 32–632, 32-647, 33–1073, 33–1122, 3736, 3752, 3785, 3812, 3871, 4558, 4854, 6168, 7882, 8351, 8551, 8895, 8919, 9079, 9184, 9300, 9380, 9470, 15364, 17147, 17877, 17950. Flett, Washington. Hesler, 9619, 10918. Mains, November 20, 1931, Michigan; 6092, Glacier National Park.

Observations.—The entire fruiting body is often of a curious soft, fragile consistency. As in *M. floccipes*, the stipe may be practically glabrous at times, and such specimens are, of course, rather difficult to identify. In none of the numerous collections which I have observed was the stipe polished and translucent, as it is in *M. pseudotenax*. The pileus of *M. quinaultensis* is practically identical in color with that of *M. scabripes*, and specimens of the former which have dried out somewhat and in which the stipe is not so viscid as usual may easily be mistaken for it. Kühner reports the two-spored form as rare in Europe, but, as in North America, both it and the four-spored form have been collected He also gives a detailed account of the anatomical characters and iodine reactions of the various tissues. I did not obtain positive amyloid reactions from either the gill trama or the flesh of the pileus. Kühner observed a greater range of variation in spore size on his four-spored forms than I have found over here, but in the two-spored forms from both continents the spores measure about the same. In one of my collections a faint fragrance was present.

Fragile Terrestrial Species

109. Mycena pseudotenax A. H. Smith

Mycologia, 31: 275. 1939

Illustrations:

Plate 37; Text fig. 26, nos. 5–7 (p. 234).
Smith, Mycologia, 31, figs. 1, I; 2 A; 4.

Pileus 1–3 cm. broad, obtusely conic or the apex rounded at first, becoming expanded and umbonate or nearly plane, the umbo usually broad and somewhat flattened, seldom with a minute papilla in the center, margin appressed against the stipe at first and in age often flaring—particularly in unexpanded individuals—surface glabrous, lubricous when wet, translucent-striate to disc when moist, the margin often creased or crenate in age, subhygrophanous, "fuscous" to pale watery gray with a pallid margin in young stages, usually pale watery gray over all at maturity or sometimes the margin remaining paler, fading to ash-gray and appearing as if pruinose; flesh gray, thin but distinctly cartilaginous, and thus causing the pileus to be very rigid, odor and taste mild; lamellae adnate, becoming slightly sinuate, moderately close, 18–23 reach the stipe, narrow to moderately broad and slightly ventricose at times (3–4 mm.), white, becoming grayish in age but not developing reddish spots, edges even and concolorous with the faces; stipe 3–5 (6) cm. long, 1.5–2 (3) mm. thick, equal, strict, cartilaginous, glabrous, apex frosted from projecting cystidia, soon polished, concolorous with the pileus or paler, apex often whitish, base of stipe and surrounding fibrils whitish but soon becoming sordid yellow.

Spores subellipsoid, pointed at one end, 5.5–7 (8) × 3.5–4 μ, nonamyloid; basidia four-spored, 22–24 × 5–6 μ; cheilocystidia and pleurocystidia similar, abundant, 50–60 (90) × 10–12 (15) μ, subcylindric with more or less rounded apices, thin-walled, hyaline, arising in the subhymenium; gill trama nonamyloid; pileus trama with a very thin nongelatinous adnate pellicle, beneath it a region composed of rather compactly arranged hyphae with vesiculose cells, the remainder floccose-filamentous, nonamyloid; stipe tissue nonamyloid.

Habit, habitat, and distribution.—Gregarious in troops on needle beds under Douglas fir and western red cedar and under scattered conifers in pastures; rarely around redwood. It has been found in the Cascade Mountains of Washington and Oregon and along the coast in southern Oregon and northern California.

Material studied.—Smith, 7835, 7931, 7967, 8025, 8038, 8049, 8105, 8763, 8823, 8835, 9071, 9152, 14038, 14561, 17069, 17426, 18152.

Observations.—This species is easily recognized macroscopically by the cartilaginous, tough consistency, the mild taste, the lutescent hairs at the base of the stipe, the slippery feel, and the appearance of the faded pilei. Here, as in many other gray Mycenae, the colors are distinctive to one who has seen abundant material, even though all the descriptions read very much alike.

The spores were first described as bluish in iodine (Smith, 1939). These observations were made on freshly dried material. On reexamining material from the type and other collections it was found that a positive amyloid reaction was not obtained on either the spores or any part of the tissue of the pileus or stipe. The error in the previous observation was very likely caused by certain spores remaining hyaline in the yellowish iodine solution and, consequently, appearing very faintly bluish in contrast to the surrounding field.

The nonamyloid reaction of the spores is unique for a Mycena of this kind and furnishes an excellent character by which dried material can be identified. *M. plumbeibrunnea* Murrill (see "Excluded Species," p. 487) has amyloid spores, but since it could be placed in any one of several of the gray species, the name has been discarded.

110. Mycena Abramsii Murrill

Mycologia, 8: 220. 1916

Prunulus Abramsii Murrill, North Am. Flora, 9: 338. 1916.

Illustrations: Plate 38; Text fig. 27, nos. 1–4 (p. 240).

Pileus 1–3 cm. broad, 1–2 (3) cm. high, nearly cylindric with an obtuse apex when young, becoming broadly conic to conic-campanulate, broadly expanded and umbonate in age or remaining unexpanded, margin appressed against the stipe when young, often flaring in age, surface pruinose and with a stannous sheen at first, glabrous and soon naked, somewhat translucent-striate when moist, usually sulcate in age, colors "Chaetura black" when young and fresh, soon "fuscous" to "drab," very hygrophanous and soon fading to pale gray, the margin frequently paler than the disc throughout its development; flesh very fragile in large caps, pallid grayish, odor and taste not distinctive; lamellae ascending-adnate, subdistant to distant, 18–20 reach the stipe, narrow to moderately broad (2.5–3 mm.), equal to subventricose; pallid gray, edges even and pallid; stipe 3–8 cm. long, 1–3 mm. thick, equal, very fragile, watery, hollow, base white-strigose, lower portion sometimes with scattered appressed white fibrils, upper porton at first covered with a conspicuous bloom which soon disappears, in age glabrous and polished, concolorous with the pileus or merely pale gray with a whitish apex.

Spores subcylindric (10) 11–13 (15) × 4.5–5.5 (6) μ, amyloid; basidia four-spored, 26–30 × 6–7 μ; cheilocystidia and pleurocystidia similar, rare to scattered on sides, abundant on edges of gills, varying greatly in shape from club-shaped to vesiculose and sometimes elongated and fusoid-ventricose with acute apices, smooth or with a few fingerlike protuberances developing and present on the elon-

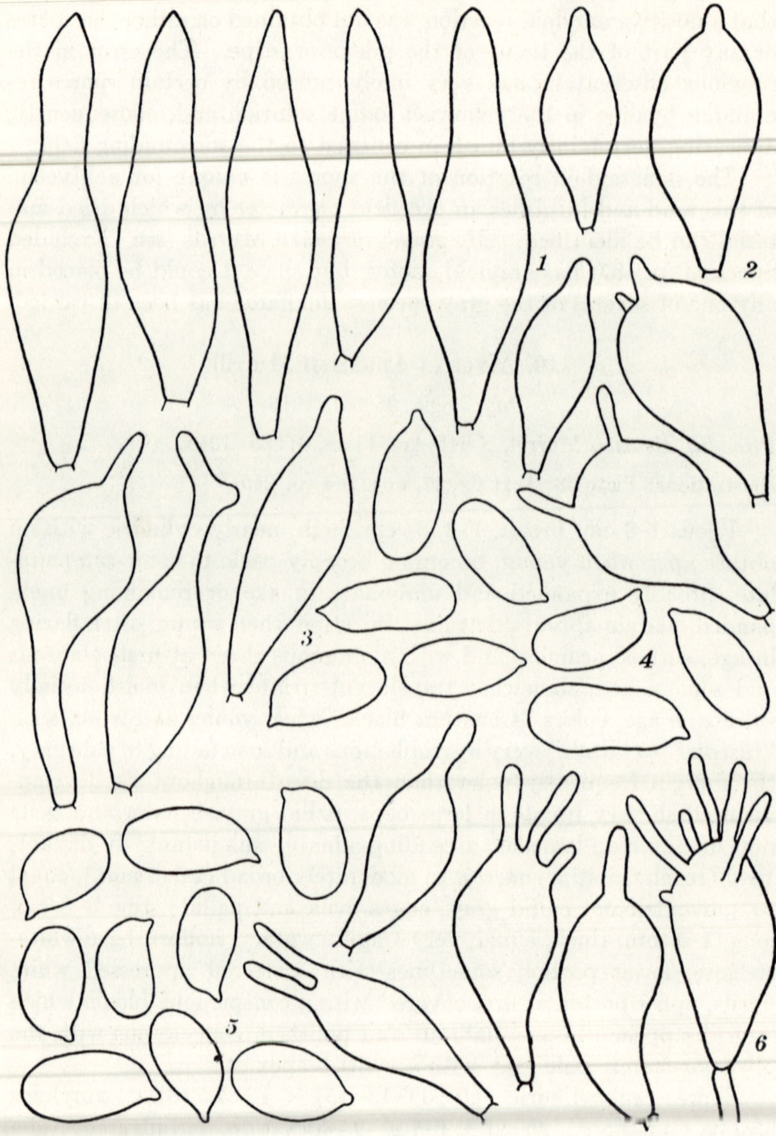

FIG. 27. *M. Abramsii:* 1, 3, pleurocystidia; 2, cheilocystidia; 4, spores. *M. griseiconica:* 5, spores; 6, cheilocystidia

gated cystidia as well as on the vesiculose individuals, 36–52 × 9–16 μ; gill trama vinaceous brown in iodine; pileus trama with a thin pellicle, a broad hypoderm, and a moderately broad filamentous portion, all but the pellicle vinaceous brown in iodine; contorted metallic-appearing lactifers present in the flesh of the pileus and in the gill trama.

Habit, habitat, and distribution.—Gregarious to cespitose on leaves and debris in hardwood forests; Tennessee, New York, Michigan, and California in the United States and Manitoba in Canada. Very common in Michigan in the spring on the forest-covered sand dunes along Lake Michigan.

Material studied.—Smith, 32-43, 32-50, 32-113, 32-522, 32-563, 32-622, 33-210, 33-212, 33-369, 49, 6344, 9514, 15022, 15034, 15047. Atkinson, 6029, 23394, 24092, 24132L, Cayuga Lake Basin, New York. Hesler, 9625. Mains, 5075.

Observations.—The long, narrow, almost cylindric spores are the outstanding character of the species. Its somewhat cespitose manner of growth and its cylindric buttons, which at times remind one of those of many species of Coprinus, are also distinctive. The cystidia are fairly constant in size and shape in each collection, but vary considerably among collections. Nos. 6344 and 9514 show the variability found in the Michigan material.

111. Mycena leptocephala (Fr.) Gillet

Les Hymén., p. 267. 1874

Agaricus alcalinus leptocephalus Fries, Syst. Myc., 1: 143. 1821.
Agaricus leptocephalus Fries, Epicr. Syst. Myc., p. 109. 1838.
Agaricus ammoniacus Fries, *ibid.*
Mycena ammoniaca Quélet, Champ. Jura et Vosges, p. 106. 1872.

Illustrations:

Plates 39, 40 C; Text fig. 28, nos. 1–2 (p. 242).

Beardslee and Coker, Journ. Elisha Mitchell Sci. Soc., 40, pl. 22, lower left figures (small specimens).

Lange, Flora Agar. Dan., 2, pl. 51, fig. A (as *M. ammoniaca*, normal form), fig. D (as *M. alcalina* f. *chlorinella*); fig. F, robust form (very good).

Ricken, Die Blätterpilze, 2, pl. 111, fig. 2.

Smith, Am. Journ. Bot., 22, pl. 1, fig. 2.

Pileus 1–3 (4) cm. broad, obtusely conic with the margin appressed at first, becoming broadly conic to convex, sometimes broadly cam-

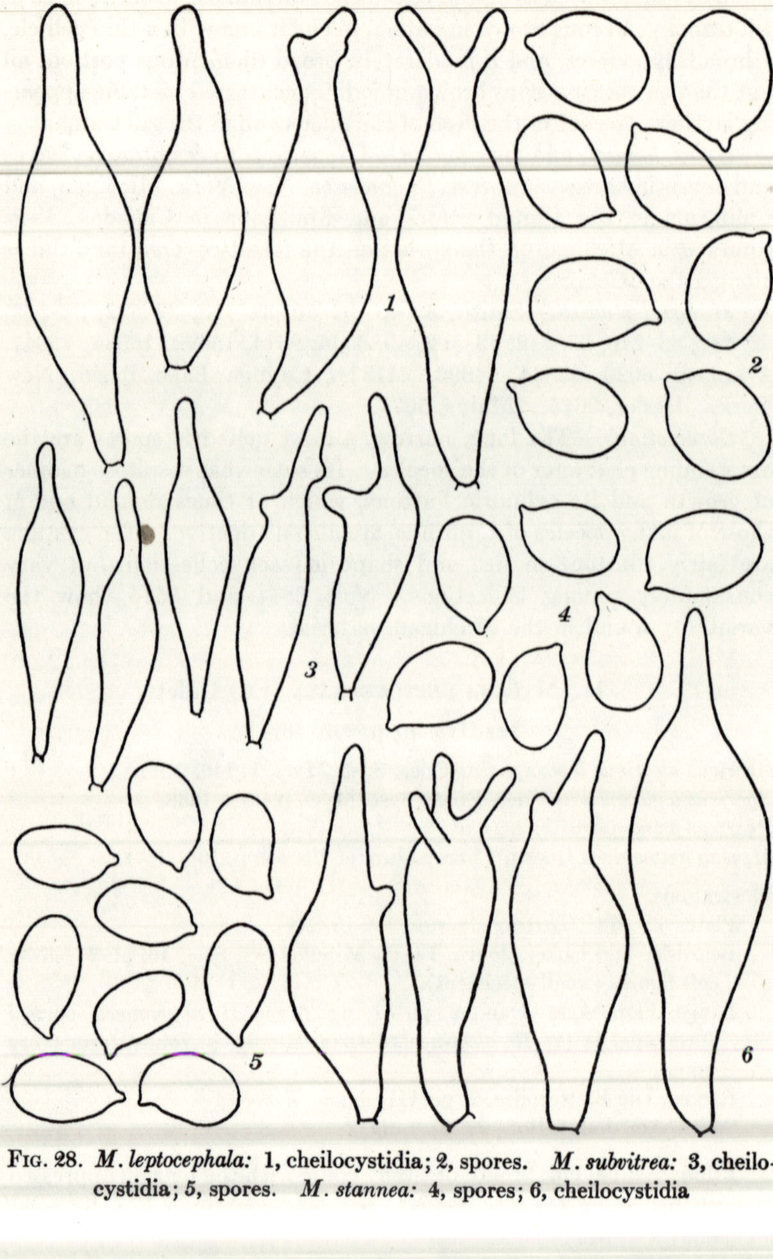

FIG. 28. *M. leptocephala:* 1, cheilocystidia; 2, spores. *M. subvitrea:* 3, cheilocystidia; 5, spores. *M. stannea:* 4, spores; 6, cheilocystidia

panulate, sometimes convex with a flaring margin, hoary owing to a pruinose coating, slowly becoming naked and moist, glabrous, translucent-striate, becoming sulcate, the surface becoming uneven at times, fuscous to blackish at first (after removal of pruinose coating), soon fading from dark to light gray and finally cinereous; flesh thin and fragile, grayish, taste acidulous, odor weakly alkaline (rather strong if flesh is crushed); lamellae narrow, equal, ascending-adnate and toothed, subdistant, 18–27 reach the stipe, one or two tiers of lamellulae, pallid or cinereous, edges pallid and even; stipe (2) 4–6 (12) cm. long, 1–2 mm. thick, equal, hollow, very fragile, usually bluish black at first (darker than the pileus) but soon sordid brownish gray, finally fading to pallid or cinereous, densely white-pruinose over all but soon polished and translucent, base nearly glabrous to rather densely white-strigose.

Spores 7–10 × 4–6 μ, broadly ellipsoid, amyloid; basidia usually four-spored (two- and three-spored forms have been found on which the spores measured 11–12 × 6–6.5 μ or 8–10 × 3.5–4.5 μ); pleurocystidia scattered, rare or absent, 30–44 × 9–13 μ, variable in shape, fusoid-ventricose to clavate, the apex forked in some, the clavate individuals occasionally bearing two or three fingerlike prolongations; cheilocystidia numerous, similar to pleurocystidia; gill trama homogeneous, composed of enlarged hyphae, vinaceous brown in iodine; pileus trama with a well-differentiated pellicle, the cells of which bear numerous rodlike prolongations, hypoderm well formed, the remaining tissue floccose, all but the pellicle vinaceous brown in iodine.

Habit, habitat, and distribution.—Scattered to gregarious on fallen sticks and on needle carpets under conifers, rather common during early summer and again in the fall; Nova Scotia, Manitoba, to Washington and south to California and North Carolina.

Material studied.—Smith, 32–383, 32–466, 32–519, 32–598, 32–603, 32–604, 32–605, 32–654, 33–879, 780, 876, 3079, 3515, 5092, 6189, 7774, 8037, 8045, 8052, 8498, 8751, 9077, 14112, 14260, 14564, 14957, 15039, 16208, 16418, 16511, 16583, October 31 and November 21, 1931. Drayton, Ottawa, Canada. Flett, Washington, February 10, 1940. Kauffman, Maryland, New York, Michigan, Washington, Oregon (as *M. ammoniaca*). Mains, 34–75, 5078. Parlin, 15329, Maine. Slipp, UIFP: 2598, Idaho. Stuntz F 418, 420 Wehmeyer, 876.

Observations.—In most respects *M. leptocephala* is just another gray Mycena of fragile consistency. Its alkaline odor, medium size, typically fusoid cystidia, which are rarely abundant on the faces of the gills, and the terrestrial habitat are its chief distinguishing characters. I have never seen the pileus or gills stain brown. The presence of the alkaline odor is about the only character which sharply distinguishes it from *M. stannea.* Numerous variations of *M. leptocephala* are found, the more slender of which have been referred by many workers to *M. ammoniaca.*

112. Mycena subvitrea A. H. Smith

Mycologia, 31: 281. 1939

Illustrations:

Plate 41; Text fig. 28, nos. 3, 5 (p. 242).
Smith, Mycologia, 31, fig. 1 H (spores).

Pileus 1–3 cm. broad, obtusely conic, or becoming campanulate, the disc somewhat flattened at times, margin appressed against the stipe when young and flaring somewhat in age, surface moist, conspicuously translucent-striate to disc, sulcate in age, glabrous, very hygrophanous, black to "fuscous" on the disc, watery gray toward the margin, fading to ashy or blackish gray over all; flesh very thin and fragile, dark watery gray, odor and taste not distinctive; lamellae bluntly adnate, rather distant, 19–21 reach the stipe, narrow (2.5 mm. ±), dark gray and staining reddish brown in age or where bruised, edges even and pallid; stipe 5–8 cm. long, 1–2 mm. thick, equal, very fragile and watery, hollow, glabrous, apex frosted, base scarcely strigose, color pale grayish white, readily staining reddish brown when bruised or in age.

Spores ovoid, 8–10 (11) × (4.5) 5–6.5 μ, amyloid; basidia four-spored, 26–28 × 7–8 μ; cheilocystidia inconspicuous, somewhat fusoid or with an abruptly narrowed neck, smooth, hyaline, 30–38 × 9–12 μ; pleurocystidia not differentiated; gill trama vinaceous brown in iodine; pileus trama with a thin adnate pellicle, beneath which is a well-differentiated hypoderm with its cells having a brown content, the remaining tissue of filamentous hyaline hyphae, all but the pellicle becoming vinaceous brown in iodine.

Habit, habitat, and distribution.—Gregarious in small groups under Douglas fir in Oregon during October. There were from three or four to about two dozen fruiting bodies in the groups. It usually appeared in places where *M. tenax* was very abundant.

Material studied.—Smith, 7811, 8050, 8160, 18124.

Observations.—The fragile, conspicuously striate dark-gray pileus, pale fragile stipe, and tendency to stain reddish brown when bruised or in age characterize it among the gray Mycenae. The dark-ashy to blackish-gray evenly colored faded pilei are also distinctive. Most species become appreciably lighter in color as moisture escapes.

This species differs from *M. stannea* by its conspicuous broad blackish striations, which extend to the disc, by the tendency of all parts to stain reddish brown, and by its more watery-fragile consistency, as well as by its lack of pleurocystidia. It appears to be very close to *M. strobilicola* Favre et Kühner but lacks the distinctive odor of *M. strobilicola* and has no pleurocystidia. Kühner has commented on some interesting observations which Vandendries made upon cultures of *M. strobilicola.* In single-spore cultures there is apparently a segregation of two types of mycelium; one type remains white and the other soon turns reddish brown in the older parts. Thus it would not be surprising to find two strains of the species in nature, one producing fruiting bodies which soon stain reddish brown and the other producing carpophores that do not stain.

113. Mycena stannea (Fr.) Quélet

Champ. Jura et Vosges, p. 242. 1872

Agaricus stanneus Fries, Epicr. Syst. Myc., p. 111. 1836.
Prunulus murinus Murrill, North Am. Flora, 9: 331. 1916.
Mycena murina Murrill, Mycologia, 8: 221. 1916.
Mycena dealbata Velenovský, České Houby, p. 327. 1920.

Illustrations:

Plates 42, 43; Text fig. 28, nos. 4, 6 (p. 242).
Bresadola, Icon. Mycol., 5, pl. 244 (as *M. aetites*).
Fries, Icon. Sel. Hymen., 1, pl. 81, fig. 5.
Smith, Am. Journ. Bot., 22, pl. 4, fig. 1 (as *M. murina*).
Smith, Mycologia, 27, fig. 2 a (as *M. aetites*).

Pileus 1–3 (4) cm. broad, obtusely conic at first, becoming broadly conic, conic-campanulate, convex-umbonate, or nearly plane and with an obtuse or sharply conic umbo, the margin appressed at first, often recurved in age, surface even and pruinose when young, soon naked and sometimes cracking radially, glabrous, translucent-striate when moist or remaining opaque, colors variable, blackish to dark gray except for the whitish margin when young (sometimes the entire pileus evenly colored dull gray), becoming paler in age, usually some shade of watery gray over the disc and striations, hygrophanous and opaque when faded and then often with a stannous sheen, sometimes whitish and canescent when faded, at other times dull lead color over all, including the extreme margin, margin sometimes splitting radially or becoming slightly crenate; flesh thickish under the disc but tapering abruptly and thin over the remainder of the pileus, grayish, odor and taste not distinctive; lamellae ascending-adnate, close to subdistant, narrow but sometimes slightly ventricose, at maturity pale or dark gray with a pallid margin; stipe 3–8 cm. long, 1–4 mm. thick, hollow, terete or compressed, fragile, equal or nearly so, glabrous except for the white-strigose base, apex pruinose at first, glabrous, watery and translucent, or opaque and then frequently twisted-striate (particularly in forms with compressed stipes), dark to light gray, usually more or less concolorous with the pileus, the apex usually pallid.

Spores 8–11 × 5–6 (7) μ, broadly ellipsoid to subovoid, smooth, amyloid; basidia usually four-spored (two-spored forms are known); pleurocystidia rare to scattered, sometimes hard to find, fusoid-ventricose or with one to several short or long obtuse protuberances over the part which projects, 36–64 × 9–16 μ; cheilocystidia similar but more inclined to have obtuse projections over the enlarged portion; gill trama vinaceous brown in iodine; pileus trama with a thin pellicle, a well-differentiated hypoderm; and the remainder of broad hyphae loosely interwoven, all but the pellicle vinaceous brown in iodine.

Habit, habitat, and distribution.—Gregarious on humus in hardwood and coniferous forests; very abundant during cool, wet fall seasons throughout the United States and Canada. I have examined material from New York, Michigan, Colorado, Idaho, Washington, Oregon, and California in the United States and from Ontario and Manitoba in Canada.

Material studied.—Smith, 32–596, 32–655, 33–764, 33–1137, 1750, 3547, 3719, 3833, 3844, 3935, 4377, 6160, 6162, 6163, 6280, 7742, 7754, 7779, 8177, 9072, 9185, 9254, 14470, 15033, 16232, 18139. Beausignour, November 29, 1935. Bisby, September 27, 1933, and November 29, 1935, Manitoba, Canada. Mains, 5221. Overholts, 1992. Slipp: UIFP, 3042.

Observations.—Previously I considered *M. aetites* and *M. murina* distinct species. Continued studies have convinced me that these are merely variations of one species, for which I am recognizing the name *M. stannea* instead of *M. aetites*. Kühner used the latter name and included the North American material under it. However, certain discrepancies in his description of the European fungus are apparent when it is applied to my specimens. I have never detected an odor or a taste as noted by Kühner, and my material was characteristically hygrophanous and very frequently had a glistening appearance after fading. In addition, the pleurocystidia are not very abundant. For these reasons I venture to place the American and European material in separate categories, although I readily admit that in the dried specimens, aside from the abundance of pleurocystidia, one might not find any differences by which to distinguish them. The characters I have used form the classical distinction between *M. aetites* and *M. stannea* of Fries, and it seems to me desirable to keep the Friesian name. *M. murina*, as I previously interpreted it, is a simple growth form of *M. stannea* in which the striations are more conspicuous than usual.

In my present conception of *M. stannea* I include any medium-sized fragile gray Mycena growing on the ground, with spores 8–11 × 5–7 μ, and with scattered to rare ventricose pleurocystidia (if any are present), which may or may not be furnished with obuse protuberances. The pileus is hygrophanous, and no distinctive odor or taste is evident. The very fine radial striations which are sometimes quite noticeable and which are formed by the splitting of the cuticle are a secondary character of little taxonomic significance. The variation in the shape of the cystidia is similar to that found in *M. citrinomarginata*.

114. Mycena plumbea (Fr.) Saccardo

Syll. Fung., 5: 284. 1887

Agaricus (*Mycena*) *plumbeus* Fries, Hymen. Eur., p. 144. 1874.

Illustrations: Plates 44, 45 A; Text fig. 30, nos. 3, 5–6 (p. 253).

Pileus 1–3 cm. broad, subellipsoid, nearly cylindric in button stages, margin appressed, becoming conic to obtusely campanulate, finally convex, plane or umbonate, the margin often flaring in age, surface hoary when young, slowly becoming naked, glabrous, moist, translucent-striate, hygrophanous, opaque, and frequently sulcate when faded, "dark mouse gray" at first (dark bluish black), becoming "hair brown," "drab," and finally "drab gray" (or paler when faded); flesh thin, grayish, very rigid and fragile, no odor or taste; lamellae ascending-adnate at first, horizontally adnate in age or sometimes toothed, subdistant, 23–28 reach the stipe, two or three tiers of lamellulae, white to nearly concolorous with the pileus, edges pallid and even; stipe (4) 8–15 cm. long, 2–3 mm. thick, strict, base white-strigose, equal, hollow, very fragile, pruinose above at first, soon naked, pallid above, concolorous or paler than the pileus below (bluish black at very first but quickly fading).

Spores 9–11 × 5–6.5 μ, broadly ellipsoid, amyloid; basidia four-spored; pleurocystidia not differentiated; cheilocystidia rather abundant, clavate to ventricose and mucronate, the apex at times with 2–3 fingerlike projections; gill trama homogeneous, pale vinaceous in iodine; pileus trama with a thin pellicle, the hyphae of which bear numerous rodlike projections, hypoderm well differentiated, the cells with brownish contents, filamentous portion rather narrow.

Habit, habitat, and distribution.—Single to scattered on spruce needles, June, and again late in the fall; New York, Michigan, and Colorado in the United States and Manitoba in Canada. I have collected it for ten years in a local plantation, but have never found it elsewhere.

Material studied.—Smith, 33–939, 33–1045, 587, 741, 757, 774, 907, 1105, 1374, 1621, 1622, 6233, 6298, 7867, 15080. Atkinson, 23750, Cayuga Lake Basin, New York. Bisby, July 18–24, 1935, Winnipeg, Canada. Mains, 34–72, 5223.

Observations.—The cystidia resemble those of *M. pectinata*, and both species fruit most abundantly during the same period in the spring. The vinaceous tinge present in the pileus of *M. pectinata* distinguishes it from *M. plumbea*, which is also close to *M. stannea*. The short-stiped forms of both are likely to be confused by the casual observer. There is a difference in general bearing and appearance which quickly separates them in the field, but it is difficult to describe. It may possibly be expressed by stating that *M. plumbea* is more glaucous. The cystidia of both species intergrade to some extent, but a consistently higher percentage of the cheilocystidia of *M. plumbea* are mucronate rather than fusoid-ventricose. The opposite is true in *M. stannea*. Sometimes the short-stiped form is very abundant on needle carpets under spruce, and I have frequently found from six to twelve fruiting bodies on a single old cone. The name "*M. plumbea*" has been used for these collections because Fries placed it in the Filipedes and emphasized the dense pruinose covering of the cap as well as the fragile stipe. I have long had misgivings over separating this species from *M. stannea*, but, in view of the difficulties in establishing the limits of many of the Friesian species, the course followed here seemed best. For the last ten years this species, along with ***M. stannea***, has been recognizable in the same spruce plantation.

115. Mycena plumbea var. **robusta**, var. nov.

Illustration: Plate 46.

Pileus 1–3.5 cm. latus, conicus demum campanulatus, glaber, lubricus, striatus demum sulcatus, fuscus demum cinereus; lamellae confertae vel subdistantes, angustae, adnatae, pallidae vel cinereae; stipes 9–13 cm. longus, 2–3 mm. crassus, cinereus, glaber; sporae 7–9 (10.5) × 4–5.5 μ; pleurocystidia et cheilocystidia 36–52 × 9–15 μ, ventricosa vel obovata. Specimen typicum in Herb. Univ. Mich. conservatum. Legit A. H. Smith, n. 18154, prope Lost Creek, Ore., Oct. 30, 1941.

Pileus 1–3.5 cm. broad, obtusely conic when young, becoming obtusely conic-campanulate and finally expanded-umbonate, surface glabrous and polished or when young very faintly pruinose, lubricous to subviscid to the touch when wet but pellicle not separable, conspicuously translucent-striate to the disc when moist, sulcate to subplicate over the marginal half when faded, "fuscous" when young, soon "drab" or "hair brown," at times very pale watery gray, cinereous when faded (blackish to pale brownish gray, fading to ash color); flesh very thin, pliant and not markedly fragile, grayish, no odor, taste mild, no color change when bruised; lamellae close to subdistant (26–29), two or three tiers of lamellulae, narrow (2.5–4 mm.), equal, ascending-adnate, pallid to dark cinereous, sometimes concolorous with the pileus, edges even, no color change when bruised; stipe 9–13 cm. long, 2–3 (4) mm. thick, equal or slightly enlarged below, hollow but not markedly fragile, pale gray, darker below, pallid at apex, glabrous and polished, moist but not viscid, base only slightly white-strigose.

Spores 7–9 (10.5) × 5–5.5 μ, ellipsoid, smooth, weakly amyloid; basidia four-spored, dark yellow in iodine; pleurocystidia scattered, fusoid-ventricose, apices simple or branched, with one or several fingerlike projections; cheilocystidia frequently obovate-mucronate but also fusoid-ventricose, often with one or several fingerlike prolongations; gill trama strongly amyloid; pileus trama with a thin pellicle of narrow hyphae covered with short rodlike projections, hypoderm well differentiated, all but the pellicle amyloid.

Habit, habitat, and distribution.—Densely gregarious to scattered on humus and debris in a mixed woods of *Alnus rubra* and *Thuja plicata;* A. H. Smith, 18154, October 30, 1941, near McKenzie Pass, Lost Creek, Oregon. It was also found in great abundance near Detroit, Oregon, along the Santiam River in a similar plant association. I have also seen material from Warrensburg, New York, but the specimens were not preserved.

Material studied.—Smith, 18154, 18178, 18179, 18180, 18181, 18182, 18183, 18184, 18210.

Observations.—The variety differs from typical material in lacking a pronounced bluish tint in the gray color, in its more cartilaginous consistency, and in the presence of pleurocystidia. There appears to be a slight difference in spore size if spores from deposits of both are measured, but it is not always evident when one compares only the dried material. In stature the variety is more like *M. longipes* than *M. stannea*, and in its consistency it also approaches the cartilaginous species.

116. Mycena fragillima A. H. Smith

Mycologia, 31: 269. 1939

Illustrations: Text fig. 29, nos. 1–2.

Pileus (0.5) 1.5–3.5 cm. broad, obtusely conic, becoming broadly campanulate or nearly plane in age, margin appressed against the stipe at first and in age frequently flaring somewhat, surface covered with a faint bloom but soon polished, moist to watery, dark watery gray and then translucent-striate to the disc, becoming pale watery gray or "hair brown" when still moist, hygrophanous and fading to very pale cinereous; flesh very thin, watery and fragile, grayish to pallid, odor and taste not distinctive; lamellae adnate or slightly hooked, close in large caps, subdistant to distant in small individuals, narrow, pallid grayish with even, whitish edges; stipe variable but usually long and slender (3–7) 9–15 cm. long, (1) 1.5–2 (3) mm. thick, usually decumbent, very fragile, pale watery gray and minutely pubescent over all at first, soon polished and translucent, base white-strigose and sometimes slightly bulbous.

Spores subovoid, pointed at one end, 7–9 (10) × 4–5 (5.5) μ, amyloid; basidia four-spored, 16–18 × 10–20 μ, hyaline, smooth; cheilocystidia forming a broad sterile band along the gill edge, broadly fusoid with acute apices, which become drawn out into long narrow necks (15–25 μ long) in age; pleurocystidia rare to absent, when present similar to cheilocystidia; gill trama vinaceous brown in iodine; pileus

FIG. 29. *M. fragillima:* 1, cheilocystidia; 2, spores. *M. subfumosa:* 3, spores; 4, pleurocystidia. *M. subfusca:* 5, cheilocystidia; 6, spores. *M. latifolia:* 7, spores; 8, cheilocystidia; 9, pleurocystidia

trama with a very thin adnate pellicle over the surface, the tramal body made up entirely of inflated hyphal cells (the hypoderm), vinaceous brown in iodine below the pellicle.

Habit, habitat, and distribution.—Gregarious in and around clumps of ferns (*Polystichum munitum*), on fern debris, or on the soil; Idaho Washington, Oregon, and California. It was abundant in November and December, 1935, after severe cold weather, but was found only once during the warm season of 1937.

Material studied.—Smith, 3620, 3670, 3716, 3802, 9397, 16679. Kauffman, September 8, 1922, Idaho. Stuntz, F 342. Gruber, Oregon.

Observations.—The fragile nature of this species should be emphasized. It is very difficult to get it into the laboratory undamaged. The delicate pubescent covering of the stipe, along with the fragility, makes it readily recognizable. The hairs of the stipe are narrow, elongated, delicate hyphae. The variability in the stature of the carpophores is such as to prevent this species from being accurately classified in either Filipedes or Fragilipedes, the old Friesian sections.

Small forms of *M. fragillima* may resemble specimens of *M. debilis* or, if they have longer stems, *M. subfusca*. The spore size distinguishes it from the former, and the lack of a layer of filamentous hyphae in the flesh of the pileus separates it from the latter. The iodine reactions of the flesh also appear distinctive, but not much emphasis should be placed upon this character. Actually, *M. subfusca* differs in many small characters. Its cheilocystidia do not become elongated into a long, narrow neck, its stipe is glabrous even under very moist conditions, its habitat is quite different, and the pileus is not evenly colored.

Large forms of *M. fragillima* closely resemble the fruiting bodies with long stipes which I have referred to *M. plumbea*, but differ in being usually decumbent, and in having paler colors, slightly smaller spores, pubescent stipes, and in the structure of the pileus. Apparently there is no pigment in the vacuoles.

117. Mycena pectinata Murrill

Mycologia, 8: 221. 1916

Prunulus pectinatus Murrill, North Am. Flora, 9: 333. 1916.

Illustrations: Plates 40 B, 45 B; Text fig. 30, nos. 1–2.

Pileus (5) 10–20 (30) mm. broad, obtusely conic with an appressed margin, becoming broadly conic and the margin often flaring, finally

FIG. 30. *M. pectinata:* 1, cheilocystidia; 2, spores. *M. plumbea:* 3, 6, cheilocystidia; 5, spores. *M. atrocyanea:* 4, cheilocystidia; 7, spores

campanulate or nearly convex, with a conspicuous bloom when young, soon naked, glabrous, translucent-striate to near the apex when fresh, becoming sulcate in age, disc "fuscous" to "benzo brown" (dark grayish brown), "drab" toward the whitish margin, the margin at first darker grayish brown and usually tinged pale vinaceous on buttons, soon becoming paler, in age pallid over all except the disc; flesh very thin and papery in mature pilei, grayish to pallid, odor and taste not distinctive; lamellae adnate, usually somewhat ascending, often slightly toothed but readily seceding, close to subdistant, 17–23 reach the stipe, one or two tiers of lamellulae, narrow at first, in age ventricose and moderately broad, intervenose, white or sometimes yellowish in age, edges pallid and even; (1) 3–7 cm. long, 1–2.5 mm. thick, tubular, very fragile, equal or slightly thickened at the base, covered with a conspicuous bloom at first, the base white-strigose, soon naked over the upper portion, "hair brown" over all or the apex pallid, in age pallid over all or the base dingy.

Spores (7) 8–10 × 4–5.5 μ, ellipsoid, smooth, amyloid; basidia four-spored; pleurocystidia rare to scattered, 38–46 × 9–12 (14) μ, broadly fusoid-ventricose with the apices usually forked or branched (unbranched in young specimens); cheilocystidia numerous, 40–62 × 10–20 μ, ventricose and mucronate or the apex with several fingerlike prolongations, occasionally merely fusoid-ventricose; gill trama homogeneous, vinaceous brown in iodine; pileus trama with a thin pellicle, beneath that a well-differentiated hypoderm, the remainder composed of a relatively narrow band of filamentous tissue, vinaceous brown in iodine except for the pellicle.

Habit, habitat, and distribution.—Scattered to gregarious or in clusters of two or three individuals. Usually found on sticks of *Vaccinium corymbosum* L. and the soft maple (*Acer rubrum*), occasionally in spruce bogs, where it may be densely gregarious on the needle carpet (*V. corymbosum* has always been found in the locality, however). Most abundant in the spring (early June) but occasionally fruiting in the fall as well. It is known from New York and Michigan in the United States and from Ontario and Manitoba in Canada.

Material studied.—Smith, 32–4, 32–31, 32–558, 32–559, 33–98, 33–183, 33–380, 63, 1410, 1617, 6232, 15035, October 12, 1929; June 13, 1934; July 13, 1938, Michigan. Bisby, July 8, 1931, Manitoba. Kauffman, Michigan (as *M. debilis*).

Observations.—Although it was originally described as cespitose, I have seldom found fruiting bodies in clusters. The thin, sulcate (in age) brownish pileus with its whitish margin and the broad mucronate cheilocystidia distinguish it. The cheilocystidia are its most distinctive character. The faint vinaceous tint along the edge of the pileus is a fairly reliable character, but must be determined from immature pilei. *M. plumbea* has somewhat the same type of cheilocystidia, though it has a distinctly different appearance as well as slightly larger spores.

118. Mycena atrocyanea (Fr.) Gillet

Les Hymén., p. 271. 1874

Agaricus atrocyaneus Fries, Syst. Myc., 1: 147. 1821.
Mycena nigricans Bresadola, Fungi Trid., 1: 33. 1882.

Illustrations:
Plate 47 C; Text fig. 30, nos. 4, 7 (p. 253).
Bresadola, Icon. Mycol., 5, pl. 245.

Pileus 5–15 mm. broad, obtusely conic to convex, the margin appressed, becoming broadly convex to slightly umbonate, surface at first hoary because of the presence of a glaucous bloom, slowly becoming naked and moist, opaque at first, at times becoming faintly striate when mature, sometimes becoming rugulose, scarcely fading, color "Chaetura black" and fading to "hair brown" or "drab" near the edge (black to bluish black when fresh and fading to dark sordid gray); flesh thin but rather cartilaginous, two-layered, upper layer dark blackish brown, lower layer pallid (under a hand lens), no odor, taste not recorded; lamellae adnate or toothed, distant, narrow to moderately broad, pale gray, edges even and dark grayish; stipe 2–5 cm. long, about 1 mm. thick, extending a short distance into the soil but hardly rooting, equal, tubular, rigid, and cartilaginous, base white-strigose, apex hoary, concolorous with the pileus (apex seldom paler).

Spores 8–10 × 5–6 µ, ellipsoid, amyloid; basidia four-spored; pleurocystidia rare to scattered, 38–54 × 8–12 µ, fusoid-ventricose, subclavate or sometimes irregular but seldom branched; cheilocystidia abundant, similar to pleurocystidia; gill trama with a thin surface pellicle, the hyphae of which bear numerous short projections, beneath the pellicle a broad hypoderm of greatly enlarged cells filled with a dark-brown sap, the remainder made up of broad but filamentous hyphae.

Habit, habitat, and distribution.—Gregarious on humus under conifers or in mixed conifer and hardwood forests, summer and fall; New York and Michigan in the United States and Nova Scotia and Manitoba in Canada.

Material studied.—Smith, 33-373, 551, 1315, 1356, 1746, 3975, 6054, 6299, 6968. Atkinson, 22677. Bisby, July 28, 1935. Wehmeyer, 551.

Observations.—When the pileus fades it often assumes a glistening or silvery-streaked appearance which may cause it to appear silky to the naked eye. The consistently very dark color is a constant feature of this species, as is also the strict appearance. Small forms which appear bluish because of their glaucous coating may lead one to confuse *M. atrocyanea* with *M. urania*. A microscopic examination will readily separate the two since their cystidia are very different.

119. Mycena praelonga (Pk.) Saccardo

Syll. Fung., 5: 282. 1887

Agaricus (Mycena) praelongus Peck, Ann. Rep. New York State Cab., 23: 81. 1872.

Prunulus praelongus Murrill, North Am. Flora, 9: 330. 1916.

Illustrations: Plate 40 A; Text fig. 31, nos. 1–2.

Pileus 5–15 (20) mm. broad, obtusely conic at first, the margin appressed or bent in slightly, becoming obtusely campanulate to plane with an abrupt sharp or obtuse umbo, surface hoary but soon polished and naked, glabrous, near "bone brown" when fresh or even more blackish, becoming paler grayish brown ("hair brown" to "pale drab gray") in age, the margin sometimes pallid; flesh thin, sordid grayish brown, rather cartilaginous, pliant, odor and taste not distinctive; lamellae subdistant, narrow, adnate or toothed, pallid gray, sometimes stained sordid reddish brown, pallid or whitish along the edges; stipe 8–16 (20) cm. long, 1–1.5 (2) mm. thick, equal, tubular, straight or flexuous, terete, cartilaginous, base sometimes white-strigose, glabrous above, the apex sometimes pruinose at first, concolorous with the pileus or paler, especially above, in age sometimes spotted sordid reddish brown near the base.

Spores 8–9 (10) $\times$ 5–6 μ, broadly ellipsoid, amyloid; basidia four-spored; pleurocystidia rare to scattered, similar to the cheilocystidia, cheilocystidia abundant, 30–62 $\times$ 8–12 (14) μ, fusoid-ventricose with acute to subacute apices; gill trama of enlarged cells and scattered

FIG. 31. *M. praelonga:* 1, cheilocystidia; 2, spores. *M. fusco-ocula:* 3, pleurocystidia; 4, cheilocystidia; 6, spores. *M. Kauffmaniana:* 5, pleurocystidia; 7, spores

lactifers, pale brownish in iodine; pileus trama with a distinct pellicle of narrow hyphae, a well-differentiated hypoderm the cells of which are filled with a dark-brown pigment, the remainder floccose and the hyphae rather narrow, all but the pellicle sordid brownish to yellowish brown in iodine.

Habit, habitat, and distribution.—Scattered to gregarious on sphagnum in bogs during late spring and early summer; Northb Carolina, New York, and Michigan. Probably common in bogs throughout northeastern North America.

Material studied.—Smith, 32–94, 33–198, 33–532, 33–539, 33–564, 1275, 1285, 1330, 6256. Atkinson, 24128 (not typical). Hesler, June 4, 1942.

Observations.—The best way to characterize this species is to say that it is a long-stemmed odorless *M. alcalina*. Without question it is a segregate of *M. alcalina*, but one which seems distinct enough to be readily recognized. It often fruits in great quantity.

Mycena plumbea often has the stature of *M. praelonga* and both have rather dark colors. The former, however, has a more brittle stipe and a very conspicuous pruinose covering over the young pileus. The spore size of the two is about the same, but the cheilocystidia of *M. plumbea* are generally more mucronate. I have never seen it stain reddish brown, but the stains are not always present on *M. praelonga*, so that the character is not a reliable distinction.

120. Mycena fusco-ocula A. H. Smith

Mycologia, 29: 338. 1937

Illustrations:

Text fig. 31, nos. 3–4, 6 (p. 257).
Smith, Mycologia, 29, figs. 1 a, b, c.

Pileus 5–15 (20) mm. broad, obtusely conic at first, becoming more or less campanulate in age, sometimes convex, margin appressed against the stipe when young, becoming sulcate-striate in age, surface glabrous, moist, not viscid, color "fuscous" on the disc and "avellaneous" or "cartridge buff" elsewhere, in age drab to "pinkish buff," with a paler margin which may become tinged with pale tan; flesh pallid, thin and pliant, odor and taste mild; lamellae ascending and narrowly adnate, close to subdistant, 17–19 reach the stipe, narrow, 1.5–2 mm., whitish, densely pruinose from projecting cystidia; stipe 4–7 (10) cm. long, 1–1.5 mm. thick, tubular, equal, fragile, glabrous above the white-strigose base, moist to subviscid to the touch but no gelatinous surface layer demonstrable in dried specimens, concolorous with the pileus or paler above, often "avellaneous" except for the whitish apex.

Spores narrowly ellipsoid, 10–12 × 5–6 μ, amyloid; basidia four-spored, 28–30 × 7–8 μ; cheilocystidia very numerous, ventricose with a rather abruptly tapered neck, the elongated portion occasionally forked, the ventricose portion smooth, 36–54 × 9–14 μ; pleurocystidia abundant, 65–95 × 10–14 μ, ventricose to nearly equal, with a long, cylindric neck usually tapered abruptly to a point, smooth, the apices sometimes forked or branched; gill trama dark vinaceous brown in iodine; pileus trama with a thin adnate pellicle over the surface, the tramal body composed entirely of inflated hyphal cells, subhymenium very narrow, becoming vinaceous brown beneath the pellicle in iodine; stipe tissue dark vinaceous brown in iodine.

Habit, habitat, and distribution.—Gregarious on needle beds under hemlock in Washington during October and under redwood in California during November.

Material studied.—Smith, 9044. Kauffman, 1925, Lake Quinault, Washington.

Observations.—The only specimens seen in the fresh condition were those from California. They were mature and, although they had endured a heavy rain, were still in good condition. No indications of a white milklike juice or latex were noted when the stipes were fractured. The stems were slightly sticky to the touch, but no gelatinous layer was observed in sections. If originally present, the gelatinous layer must have been washed off by the rain, as sometimes happens to carpophores of *M. quinaultensis*. The question of the subgeneric position of the species still remains unsettled. Kühner (1938) suggests a relationship with *M. galopus*. I have examined the stipes of some specimens of the type collection and find that they possess scattered hyphae with granular contents similar to those found in *M. galopus*. This, when considered in connection with the spores and cystidia, might lead one to place *M. fusco-ocula* in synonymy with *M. galopus*. Because of certain discrepancies I do not believe that such a conclusion is justified at present. Kauffman's notes describe the stem as subviscid and elastic, and he distinctly mentioned young and old carpophores. He did not mention a milklike juice, and it is reasonable to assume, since he was working on a monograph of Mycena at the time he collected his material, that he checked that character. The stipe of *M. galopus*, according to my experience, is fragile, not elastic. Cystidia of the type found in *M. fusco-ocula* also occur in *M. tenax*. In view of the situation which developed in regard to the viscidity of the stipe of *M. quinaultensis* (see p. 439), Kauffman's observation on the character of the stipe cannot be dismissed without further study. The species is placed here because the only fresh material I have seen indicates such an arrangement. *M. atroalba* sensu Ricken is apparently very close to *M. fusco-ocula*, and the two may not be distinct. The color differences which appear to distinguish them may be mere variation. Kühner lists *M. atroalba* as one of the poorly known European species.

121. **Mycena Kauffmaniana,** sp. nov.

Illustrations: Text fig. 31, nos. 5, 7 (p. 257).

Pileus 5–15 mm. latus, obtusus demum conico-campanulatus, fuscus dein cinereus, striatus, udus, glaber; lamellae confertae demum subdistantes, angustae, adnatae, pallidae demum cinereae; stipes 4–10 (15) cm. longus, 1–1.5 mm. crassus, glaber, cinereus; sporae 7–9 × 4.5–5.5 μ; pleurocystidia et cheilocystidia 60–90 × 5–9 μ. Specimen typicum in Herb. Univ. Mich. conservatum. Legit A. H. Smith, n. 16622, prope Baker Lake, Wash., Sept. 4, 1941.

Pileus 5–15 mm. broad, ovoid to somewhat acorn-shaped at the very first, the margin connivent to the stipe, soon obtusely conic to conic-campanulate, in age umbonate with a recurved margin, striatulate at maturity, fuscous to bluish black when young but soon fading to "hair brown" and finally cinereous, subsulcate in age; flesh thin, pliant, white or grayish, odor and taste not distinctive; lamellae close to subdistant, narrow, ascending, hooked ("uncinate-adnexed"—Kauffman), faintly grayish at maturity, whitish at first, edges sometimes fimbriate (under a lens); stipe 4–10 (15) cm. long, 1–1.5 mm. thick, equal or attenuated slightly toward the apex, hoary when young but soon polished, hollow, glabrous, strict or flexuous, terete, rarely compressed, concolorous with the pileus and fading like it, white-strigose at the base.

Spores 7–9 × 4.5–5.5 μ, ellipsoid, smooth, amyloid; basidia four-spored; pleurocystidia and cheilocystidia similar and abundant, 60–90 × 5–9 μ, greatly elongated above a slightly inflated base, almost filamentous in age, smooth or the outline merely wavy or irregular; gill trama slightly amyloid; pileus trama with a thin adnate pellicle, a well-differentiated hypoderm, and a thin region of floccose interwoven narrower cells.

Habit, habitat, and distribution.—Scattered to gregarious ("subcespitose"—Kauffman) on old logs, sticks, and debris of hardwoods; Tennessee, Michigan, and Washington in the United States and Manitoba in Canada.

Material studied.—Smith, 32–540, 32–666, 16622 (type). Bisby and others reported it as *M. atroalboides*. Hesler, 3816. Kauffman, August 28, 1910, October 11, 1913, Michigan.

Observations.—According to my experience, this is a rare species, which does not fruit in great abundance. It is highly probable, however, that under certain conditions it, like other Mycenae, will be found in quantity. It is readily distinguished from *M. fusco-ocula* by its smaller spores and flexuous cystidia. In Kauffman's *Agaricaceae of Michigan* it is described under the name "*M. atroalboides*." Since Peck's type of *M. atroalboides* has clavate-echinulate cheilocystidia, Kauffman's fungus cannot be placed there, even though a very striking macroscopic similarity does exist. Forms of *M. fragillima* and *M. subfusca* may also resemble *M. Kauffmaniana* in stature, but are easily separated by their lack of abundant conspicuous flexuous pleurocystidia. The smaller spores distinguish it from *M. atroalba* sensu Ricken.

122. **Mycena subfusca,** sp. nov.

Illustrations: Text fig. 29, nos. 5–6 (p. 251).

Pileus 5–10 (14) mm. latus, conicus demum campanulatus, glaber, subhygrophanus, pallide fuscus, striatus; lamellae adnatae, latae, subdistantes, albidae; stipes 3–6 cm. longus, 1 mm. ± crassus, aequalis, deorsum albo-strigosus, sursum pruinosus, albidus vel pallide griseus; sporae 7–9 (10) × 5–6 μ; basidia bispora; cheilocystidia laeva, subcylindrica vel ventricosa, 30–46 × 10–18 μ. Specimen typicum in Herb. Univ. Mich. conservatum. Legit A. H. Smith, n. 15258, prope Ann Arbor, Mich., Sept. 6, 1940.

Pileus 5–10 (14) mm. broad, obtusely conic at first, becoming campanulate or in age with a flaring or recurved margin and an obtusely conic umbo, margin appressed against the stipe at first, surface faintly pruinose, glabrous, moist, subhygrophanous, color pale fuscous to grayish brown over the disc, the margin whitish or very pale gray ("fuscous" to "hair brown" on the disc), both sulcate- and translucent-striate when young, opaque when faded, flesh very thin but only moderately fragile, no odor, taste perfectly mild; lamellae narrowly adnate, ascending, moderately broad (1.5–2 mm.), subdistant. 13–18 reach the stipe, one or two tiers of lamellulae, white, edges even and concolorous with the surface; stipe 3–6 cm. long, 1 mm. or less thick, equal, fragile, solid, base white-strigose and somewhat rooting, juice watery and scant, pale gray to hyaline white over all, no color change when bruised.

Spores 7–8 (10) × 5–6 μ, broadly ovoid to subellipsoid, amyloid; basidia two-spored; no pleurocystidia, cheilocystidia 30–40 × 10–18 μ, saccate to subcylindric or fusoid-ventricose with broad obtuse apices, smooth, hyaline; gill trama homogeneous, not colored in iodine; pileus trama with an adnate pellicle, the hyphae of which give off numerous short projections, hypoderm hardly differentiated (the layer visible more because of the compact arrangement of the cells rather than their great volume), the remainder floccose and filamentous, not colored appreciably in iodine.

Habit, habitat, and distribution.—Scattered on debris in wet places, September to December; Alabama and Michigan.

Material studied.—Smith, 32–452, 32–516, 6307, 15258 (type). Burke, Alabama.

Observations.—This species was collected in 1932, 1937, and 1940. It has the stature of a long-stiped *M. atroalboides* or *M. filopes*, but is smaller in all its proportions. The cheilocystidia are its most distinctive character and separate it from the species mentioned above. The structure of the pileus is peculiar for such a small Mycena. Most species of this size have the hypoderm very well differentiated; it forms the greater part of the flesh of the pileus. The spores from four-spored basidia should be somewhat smaller than the measurements given in the description and should furnish an additional character to separate the species from *M. fragillima*. For additional comments on the relationship to that species see page 252.

123. Mycena subfumosa Murrill

Mycologia, 8: 221. 1916

Prunulus subfumosus Murrill, North Am. Flora, 9: 328. 1916.

Illustrations: Text fig. 29, nos. 3–4 (p. 251).

"Pileus convex to subexpanded, thin, scarcely umbonate, scattered, 6–10 mm. broad; surface minutely silky to subglabrous, striate, dry, pale-fumosous, margin concolorous, slightly crenate: lamellae adnate, rather broad, subdistant, white: stipe filiform, pruinose to glabrous, subconcolorous below, white above, 4–5 cm. long, 0.5 mm. thick.

"Type collected on fallen dead wood in a wet thicket at New Orleans, Louisiana, September 8, 1908, *F. S. Earle 126* (herb. N. Y. Bot. Gard.)."

Distribution.—Louisiana and possibly Tennessee.

Material studied.—Earle, 126. Hesler, 6519.

Observations.—An examination of the type revealed that the spores were hyaline, smooth, narrowly ellipsoid, 7–9 × 3.5–4.5 μ, and weakly amyloid. The basidia were four-spored. The pleurocystidia and cheilocystidia, which were similar, were numerous, smooth, 36–48 × 6–11 μ, hyaline, subfusoid, and with a homogeneous content. The gill trama was not distinctive beyond being yellowish in iodine. The pileus trama possessed a well-developed nongelatinous pellicle, the hyphae of which gave off numerous short rodlike projections. The hypoderm was well differentiated, the large cells being filled with a smoky-brown content. The remaining portion was made up of narrow filamentous interwoven hyphae. All parts were yellowish in iodine.

In stature the dried specimens somewhat resemble those of *M. debilis*, but the resemblance is only superficial. Hesler's notes on the Tennessee collection follow:

"Pileus 8–15 mm. in diameter, convex-campanulate, obtusely umbonate, grayish-brown, umbo dark brown, sulcate, not viscid, glabrous. Flesh whitish, membranous. Gills adnate then uncinate, white and remaining white, subdistant, intervenose, alternately long and short, edge white-fimbriate. Stipe 6–7.5 cm. × 0.5–1.0 mm., hollow, apex white and pruinose, darker below, livid at the base, which is white-strigose, often twisted, not viscid. Spores variable: some 9–9.5 × 4.5–5.5 μ; others 4 × 3 μ; apiculate, elliptical. Cystidia lanceolate, base enlarged, extending 35–40 μ beyond the hymenium; about 50–55 × 9.5 (base), 3.2 μ (apex). Odor and taste none."

Species with More or Less Roughened Cystidia; Gregarious

124. Mycena latifolia (Pk.) Saccardo

Syll. Fung., 5: 268. 1887

Agaricus latifolius Peck, Ann. Rep. New York State Cab., 23: 81. 1872.
Mycena pinetorum Lange, Dansk Bot. Arkiv, 1 (5): 30. 1914.

Prunulus latifolius Murrill, North Am. Flora, 9: 327. 1916.
Omphalopsis Bakeri Murrill, *ibid.*, p. 315.
Omphalia Bakeri Murrill, Mycologia, 8: 220. 1916.

Illustrations:

Plate 48 A, B; Text fig. 29, nos. 7–9 (p. 251).
Lange, Flora Agar. Dan., 2, pl. 55 B (good).
Peck, Ann. Rep. New York State Cab., 23, pl. 6, figs. 8–14.
Smith and Wehmeyer, Pap. Mich. Acad. Sci., Arts, and Letters, 21, pl. 26, fig. 2.

Pileus 8–25 mm. broad, obtusely campanulate when young, the margin straight, becoming broadly campanulate, broadly conic, or broadly convex, sometimes expanded and with a low obtuse umbo, faintly hoary but soon moist and lubricous, sometimes subviscid (large specimens), color blackish to pale gray, the margin pallid, "olive gray" on the disc or pale cinereous over all in age, translucent-striate when moist, at times somewhat sulcate after fading; flesh thin but tough or somewhat fragile in small carpophores, grayish to pallid, odor slight, taste farinaceous; lamellae broadly adnate, often with a decurrent tooth, close to subdistant, rather broad, white or grayish, edges even and pallid; stipe 2–5 (7) cm. long, 1–2 mm. thick, equal or tapered at the base, cartilaginous and fragile, tubular, terete, or compressed, base covered with a dense white tomentum or coarsely white-strigose, glabrous above, the apex pruinose at first, concolorous with the pileus or nearly so, usually pallid toward the apex.

Spores 7–9 $\times$ 3.5–4 μ, narrowly ellipsoid, strongly amyloid; basidia four-spored; pleurocystidia scattered to abundant, 54–90 $\times$ 8–12 μ, fusoid-ventricose to subcylindric, with long slightly tapered necks, the ventricose portion verrucose or smooth; cheilocystidia variously shaped, clavate to fusoid-ventricose, the enlarged portion covered with short or long obtuse protuberances (occasionally contorted or branched and the protuberances also branched in some); gill trama homogeneous, faintly vinaceous brown in iodine; pileus trama with a well-differentiated pellicle, the hyphae of which give off numerous rodlike projections, the hypoderm well differentiated, the remainder of rather broad hyphae, pale vinaceous brown except for the pellicle.

Habit, habitat, and distribution.—Gregarious on needle beds under conifers, often in great quantity, from Nova Scotia, Canada, across the continent to Cape Flattery in the state of Washington and south to California in the West and North Carolina and Tennessee in the East; generally most abundant during late summer or early fall.

Material studied.—Smith, 745, 754, 985, 2491, 3132, 3907, 4137, 4748, 6517, 7509, 7525, 8043, 8818, 9047, 10838, 14729, 17015. Hesler, 9623, 10459, 11492, 12387, 12828. Kauffman, August, 1924 (two collections as *M. albogrisea*), North Carolina. Mains, 34–168, 34–172, 34–183. Parlin, Maine. Slipp, UIFP: 3225. Wehmeyer, 643, 645, 755, 755a.

Observations.—The broad gills are a rather reliable macroscopic character, but the cystidia are the most significant microscopic feature. Aside from these two characters, this species could be regarded as just another gray Mycena. Its colors are not distinctive, and its stature and consistency vary a great deal, depending on the location. Fruiting bodies from dense mossy forests are more fragile than those found in the open. The lubricous to subviscid feel of the pileus is caused by the well-developed pellicle and is most noticeable in the carpophores with a cartilaginous consistency. The pellicle does not gelatinize in either water or KOH. My first observations were based on rather cartilaginous specimens. Further collecting revealed the differences mentioned above. The species in no way reminds one of *M. galericulata*. *M. clavicularis* has the same stature, appearance, and broadly adnate gills but is readily separated by its viscid stipe and different pleurocystidia.

125. Mycena constans (Pk.) Saccardo

Syll. Fung., 5: 296. 1887

Agaricus (*Mycena*) *constans* Peck, Ann. Rep. New York State Mus., 27: 93. 1875.

Prunulus constans Murrill, North Am. Flora, 9: 327. 1916.

"Pileus submembranaceous, campanulate or convex, striate, palecinereous; lamellae close, ascending, uncinate, white; stem slender, equal, smooth, colored like the pileus, with hairy filaments at the base; odor alkaline.

"Plant about 2′ high, pileus 1″–3″ broad.

"Among mosses in woods. Forestburgh. Sept.

"In general appearance it bears some resemblance to small forms of *A. vulgaris*, but it is easily separated by its dry stem and its odor of hartshorn, which is very distinct when the plant is first gathered."

The following data were obtained from an examination of the type: The spores measure 6–8 × 3–3.5 μ and are narrowly ellipsoid, smooth, and hyaline. The basidia are four-spored. Pleurocystidia were not seen. The cheilocystidia, which are quite numerous, measure 26–34 × 8–12 μ and are clavate; their apices are covered with scattered short rodlike projections. The gill trama is homogeneous. The pileus trama is very distinctive. The surface of the pileus is covered with a very thin layer (one hypha ± in thickness) of nongelatinous hyphae, the walls of which give off numerous short rodlike projections. Beneath this is a broad layer of gelatinous hyphae almost half the thickness of the cap. Below the gelatinous layer is a region of enlarged cells which have smoky-brown contents (this layer would ordinarily be the hypoderm), the trama next to the subhymenium is filamentous, but only a thin layer is present, and the subhymenium is not gelatinous. The gill edges do not gelatinize, and there is no gelatinous layer over the stipe. Iodine reactions were not obtained.

I have one small collection from Lake Timagami, Ontario, which may be this species. However, it lacked an alkaline odor. Mains (32–121) has one collection from Rock River, Michigan, which also appears to belong here, but he did not note the odor. Numerous collections of a small Mycena from around Mt. Baker, Washington, also apparently should be referred here, but they did not have a very distinct gelatinous layer beneath the pellicle. In fact, it could be demonstrated only from dried material revived in KOH. These collections also lacked a distinctive odor. Overholts, 1872, found what appears to be the typical form of this species in Colorado. The following are his notes:

"Pileus 3–8 mm. broad, campanulate, 'neutral gray' to pallid neutral gray, pruinose to glabrous, dry, striate almost to apex, flesh tough, thin, whitish, somewhat alkaline; gills adnexed or free, distant, 1 mm. broad, unequal, white; stipe central, terete, equal, white (at least at apex), glabrous above, with a few white hairs, hollow, 3–4 cm. long, 0.5–1 mm. thick; spores white, 6–7 × 3–4 μ."

The collection was meager, but the caps I sectioned had a gelatinous layer like those from around Mt. Baker. The cheilocystidia were typical. *M. pusilla* is closely related to *M. constans*, but differs in having a rather thick pellicle that in wet weather causes the cap to be subviscid. This is quite a contrast to the thick gelatinous layer beneath the pellicle of *M. constans*. The gills of the latter were described as uncinate, whereas those of *M. pusilla* are rather broadly adnate and, though more ascending, somewhat resemble those of *M. latifolia*.

126. Mycena pusilla A. H. Smith

Mycologia, 31: 277. 1939

Illustrations: Plate 49 B; Text fig. 32, nos. 1, 3 (p. 268).

Pileus 5–10 mm. broad, convex to obtuse or with a slightly flattened disc, in age more or less expanded, plane or umbonate, margin appressed against the stipe at first, often flaring and faintly scalloped, striate to the disc when moist, surface hoary at first but soon polished, lubricous when wet, glabrous, color pale watery gray with a whitish margin when young, hardly paler at maturity, fading slowly to pallid cinereous over all; flesh thin and membranous, rather pliant, pallid to grayish, odor and taste not distinctive; lamellae ascending-adnate, close, 18–20 reach the stipe, moderately broad (3 mm. ±) or somewhat ventricose and quite broad for such a small fungus, the white edges even and concolorous with the faces; stipe 2.5–4 cm. long, 1 mm. thick, equal, tubular. hoary when young, soon polished and lubricous but not viscid, tough, whitish above, darker grayish brown toward the base, which is scarcely strigose.

Spores broadly ovoid to ellipsoid, occasionally slightly pear-shaped, 7–10 × (4) 5–6 μ, amyloid; basidia four-spored, 26–29 × 6–7 μ; cheilocystidia clavate, hyaline, the enlarged portion covered with short (occasionally branched) rodlike to filamentous projections, 26–30 × 7–12 μ; no pleurocystidia; gill trama vinaceous brown in iodine; pileus trama with a thick subgelatinous pellicle (25–40 μ thick in KOH) of very slender hyphae, the tramal body of somewhat enlarged hyphae, the cells of which become vinaceous brown in iodine.

Habit, habitat, and distribution.—In troops on moss and carpets of conifer needles under Douglas fir in Oregon and northern California. Not uncommon in the fall.

Material studied.—Smith, 7826, 7886, 7979, 8836.

Observations.—Small forms of many species of Mycena have been found, and many of these appear to be genetically constant; at least one sees them from year to year. These, because they resemble normal fruiting bodies of the species in all other respects, have not been given recognition in this work. It is very likely that certain species produce large numbers of fruiting bodies when they fruit, and that, given poor nutrient conditions, a large number of small carpophores would be produced rather than a few of normal size. *M. pusilla* does not appear to fall into this category because of the additional characters which distinguish it from other terrestrial Mycenae with roughened cystidia.

FIG. 32. *M. pusilla:* 1, cheilocystidia; 3, spores. *M. alcaliniformis:* 2 cheilocystidia; 4, spores. *M. urania:* 5, spores; 6, cheilocystidia *M. psammicola:* 7, cheilocystidia; 8, spores. *M. atroalboides:* 9, cheilocystidia, 10, spores. *M. mutata.* 11, cheilocystidia and pleurocystidia, 12, spores

127. Mycena alcaliniformis Murrill

Mycologia, 8: 220. 1918

Prunulus alcaliniformis Murrill, North Am. Flora, 9: 331. 1916.

Illustrations:

Plate 35 E; Text fig. 32, nos. 2, 4.
Murrill, Mycologia, 4, pl. 68, fig. 9 (as *M. vexans*).

Pileus 1–2 cm. broad, obtusely conic to convex, becoming broadly convex, obtusely campanulate or with the disc flattened somewhat, margin appressed against the stipe at first, often flaring in age and often becoming crenate, surface moist, even, glabrous, very hygrophanous, conspicuously striate to the disc when moist, colors "fuscous" to "hair brown" over the disc and striae, grayish white elsewhere, fading quickly to pale cinereous; flesh very fragile, thin and membranous, grayish to whitish, taste not distinctive, odor faintly fragrant at first, soon fading; lamellae ascending-adnate with a slight tooth, subdistant to distant, 9–13 reach the stipe, one or two tiers of lamellulae, moderately broad, grayish, with whitish even margins; stipe 2–4 cm. long, 1–2 mm. thick, equal or tapered slightly at the base, hollow, very fragile, terete or compressed, base sparsely mycelioid, glabrous and polished above, translucent, concolorous with the pileus or paler, often whitish above.

Spores ellipsoid, 8–10 $\times$ 4–5 μ, amyloid (reaction strong); basidia four-spored, 26–29 $\times$ 7–8 μ; cheilocystidia clavate, hyaline, with obtuse fingerlike projections over the enlarged portion, often with wavy outlines or variously contorted; pleurocystidia not differentiated; gill trama vinaceous brown in iodine; pileus trama with a thin pellicle, beneath it a region of vesiculose cells (about half the thickness of the cap), the remainder of moderately broad cells, all but the pellicle vinaceous brown in iodine.

Habit, habitat, and distribution.—Densely gregarious on needle beds, especially those of white pine, in spring, late summer, and fall; Alabama, Massachusetts, New York, Michigan, and Oregon in the United States and Ontario in Canada.

Material studied.—Smith, 703, 711, 742, 3475, 4814, 5042, 8927. Overholts, September 6, 1935, Ontario. Stuntz, F 485.

Observations.—Murrill described the species as occurring on debris of spruce. In the Adirondacks I have found it exceptionally abundant under white pine. In Oregon I collected it sparingly under *Pinus contorta.* The very fragile consistency, conspicuously striate pileus, cheilocystidia, and fragrance readily separate it from other gray fragile Mycenae. The short-stemmed form of *M. atroalboides*, *M. pusilla*, and *M. constans* are all characterized by approximately the same stature and, apparently, by more or less the same color. From all these *M. alcaliniformis* is readily separated by the somewhat different cystidia, fragrance, and more fragile consistency.

128. Mycena urania (Fr.) Gillet

Champ. Fr., 1: 279. 1878

Agaricus uranius Fries, Syst. Myc., 1: 144. 1821.

Illustrations:

Plate 49 A; Text fig. 32, nos. 5–6 (p. 268).
Smith, Mycologia, 27: 601, fig. 2 c.

Pileus 4–10 mm. broad, conic to hemispheric at first, the margin appressed, becoming broadly umbonate to convex in age, surface hoary-pruinose but soon naked and dull, appearing rather dry, margin somewhat sulcate-striate or at times becoming crenate, color "dark plumbeous" or "deep Varley's gray" when fresh, fading through "lilac gray" and finally becoming pale drab, the margin usually pallid when fresh (dark grayish blue when young, becoming lighter and more bluish, but finally becoming pale drab); flesh thin, fragile, bluish gray to almost pallid, not changing when bruised, odor and taste not distinctive; lamellae broadly adnate to distant, white to whitish when young, becoming bluish or pallid gray, edges even and concolorous with the faces; stipe 1–3 cm. long, 0.5–1 mm. thick, equal, rigid, and fragile, tubular, pruinose and with a hoary unpolished appearance, at first "deep Varley's gray" to "pale violet-gray," but becoming paler (more or less concolorous with the pileus).

Spores 7–9 × 4–5 μ, ellipsoid, amyloid; basidia four-spored; pleurocystidia sporadic, scattered to abundant or apparently absent on some caps, similar to the cheilocystidia; numerous, 26–34 × 8–12 μ, clavate to subcapitate, the apex covered with short rodlike projections; gill trama homogeneous, the cells becoming distinctly inflated, vinaceous brown in iodine; pileus trama with a thin, poorly differentiated pellicle, the hyphae of which bear numerous short rodlike projections, hypoderm well differentiated, the remainder filamentous, all but the pellicle vinaceous red in iodine.

Habit, habitat, and distribution.—Gregarious on wet or swampy ground among mosses and decaying leaves in coniferous and hardwood forests, usually found during the late summer and fall. It is a rare species, known in North America from Michigan, North Carolina, and Tennessee.

Material studied.—Smith, 33–100, 33–548, 6374, 9824, 10776, 10829, Michigan, 1931.

Observations.—Fries described the stipe as flaccid and the gills as white. In my collections the stems were usually rigid enough to hold the small caps upright, although fruiting bodies from mossy places are often somewhat decumbent. In addition, the gills were always dark at maturity. Such differences do not seem sufficient to justify separating the American specimens as a distinct unit, particularly when it is remembered that the species is very poorly known in Europe. As interpreted here, it is characterized by its predominantly blue color, small stature, and echinulate cystidia. It is not closely related to *M. amicta* and *M. subcaerulea.* Its colors as well as other characters sharply distinguish it. Its relationships appear to be here with the small fragile gray to brown species having echinulate cheilocystidia.

129. Mycena psammicola (Berk. & Br.) Saccardo

Syll. Fung., 5: 275. 1887

Agaricus psammicola Berkeley & Broome, Ann. and Mag. Nat. Hist., Ser. 4, 17: 130. 1876.

Illustrations:

Plate 49 C; Text fig. 32, nos. 7–8 (p. 268).
Smith, Mycologia, 27, fig. 2 b.

Pileus (5) 10–15 mm. broad, obtusely conic but soon narrowly campanulate, becoming broadly campanulate, broadly conic or expanded and umbonate, the umbo usually abrupt and obtuse, margin appressed against the stipe when young, sometimes flaring slightly at maturity, moist, opaque at first but becoming translucent-striate, glabrous and smooth, hygrophanous and fading rapidly, sulcate when faded, "russet" to "tawny" over all when moist and fresh, the margin soon "ochraceous tawny" and the disc "Mars brown," sometimes the umbo nearly "fuscous," fading to "tawny olive" on the disc and "wood brown" on the margin, in age paler and "ochraceous tawny" to "cinnamon buff"; flesh concolorous with the surface, fragile, thin, taste mild or very faintly of radish, odor developing after the specimens have been collected, rather sharp and fragrant, somewhat like that of iodoform; lamellae ascending but bluntly adnate, slightly toothed in age, close, 17–20 reach the stipe, usually only two tiers of lamellulae, narrow and equal (1.5–2 mm.), "snuff brown" when young, becoming "cinnamon buff" in age, edges even and pallid; stipe 2–5 cm. long, 1–1.5 mm. thick, equal, tubular, cartilaginous and only moderately brittle, base very slightly white-strigose, glabrous above or the apex faintly pruinose, soon polished, concolorous with the cap or the apex slightly paler.

Spores narrowly ellipsoid, 7–8 × 3.5–4 μ, amyloid; basidia 26–28 × 6–7 μ, four-spored; cheilocystidia 20–32 (35) × 9–12 μ, covered with small rodlike projections, hyaline; pleurocystidia not differentiated; gill trama very pale vinaceous brown in iodine; pileus trama composed of a very thin, nongelatinous pellicle, beneath it a region of inflated cells which occupies half the tramal body and the remainder of filamentous tissue, all but the pellicle very pale vinaceous brown in iodine.

Habit, habitat, and distribution.—Scattered to gregarious on pine needles, late summer and fall; Ontario and Michigan. Rare.

Material studied.—Smith, 33–1017, 986, 1620, 4848, 6995, 14937. Kauffman, September 23, 1911, Michigan. Mains, 34–147.

Observations.—This species has a decided resemblance to species of the genus Galerina. Its hyaline spores, of course, prevent one from making the mistake of trying to key it out there. The russet to tawny or ochraceous colors in all parts are very peculiar for a Mycena. As Kühner has pointed out, *M. vitrea* var. *galericolor* Favre, in Kühner, very closely approaches my concept of *M. psammicola*, but it is described as having a nitrous odor, a pileus 3–4 cm. broad, and cheilocystidia 30–55 × 15–25 μ. These characters distinguish it rather sharply from the American fungus. My determination of the European species is based on Cooke's illustration and Rea's description. The rooting stipe was present in some carpophores of my collections but seemed to be merely a secondary development caused by the loose mat of needles and debris and did not appear to be a true pseudorhiza.

130. MYCENA METATA (Fr.) Quélet

Champ. Jura et Vosges, p. 106. 1872

Agaricus metatus Fries, Syst. Myc., 1: 144. 1821.

Illustrations:

Plate 50 A; Text fig. 32, nos. 11–12 (p. 268).

Lange, Flora Agar. Dan., 2, pl. 56 F (good).

Ricken, Die Blätterpilze, 2, pl. 110, fig. 9.
Smith, Am. Journ. Bot., 22, pl. 1, fig. 1.

Pileus (5) 10–25 mm. broad, equally high, obtusely conic to convex, sometimes campanulate, the margin often flaring and split radially, surface at first covered with a faint bloom, slowly becoming naked, glabrous, somewhat translucent-striate but usually with a rather dry appearance even before fading, usually "army brown" to "fawn color" or "avellaneous" over the disc, margin "warm buff" to ashy gray or fading and becoming whitish, disc sometimes "olive brown," sometimes the entire pileus pale cinereous with only a tinge of vinaceous; flesh thin, pallid to avellaneous, very fragile, odor sharp, rather faint (hardly alkaline), taste slight, acidulous; lamellae ascending-adnate, usually toothed, close, narrow to moderately broad, white at first, soon tinged creamy to "vinaceous buff" (tinged dull incarnate), edges even and pallid; stipe 3–9 cm. long, 1–2.5 mm. thick, very fragile, equal, tubular, strict or flexuous, sometimes bluish black when young but, if so, quickly fading to "wood brown" or paler avellaneous, apex pallid, faintly pruinose above, translucent when moist, glabrous except for the strigose base.

Spores 7–9 × 4–5 μ or 8–10 (12) × 4–6 μ, smooth, narrowly ellipsoid to somewhat pear-shaped, amyloid; basidia usually two- or three-spored, occasionally four-spored; pleurocystidia and cheilocystidia similar, rare to scattered on sides, abundant on edges of the lamellae, 27–38 × 9–17 μ, clavate to capitate, pedicel usually slender, head echinulate; gill trama homogeneous, dark vinaceous brown in iodine; pileus trama with a thin but sharply differentiated pellicle, a distinct hypoderm and the remainder filamentous, all but the pellicle deep vinaceous brown in iodine.

Habit, habitat, and distribution.—Scattered to gregarious under conifers, common in the fall; Pennsylvania, New York, Michigan, Idaho, Washington, Oregon, and California in the United States and from Nova Scotia to Ontario in Canada.

Material studied.—Smith, 32–601, 32–636, 32–645, 32–651, 64, 90, 744, 1253, 3328, 3390, 3444, 3715, 4381, 6155, 7784, 7945, 8029, 8059, 8106, 8624, 8741, 8813, 8836, 8919, 8926, 9043, 9076, 14156, 14679, 18119. Kauffman, Idaho and Michigan. Mains, 31–666, 5187, November 16, 1931; August 9, 1932; September, 1935. Overholts, 19398. Slipp, UIFP: 3231, 3675. Wehmeyer, 518, 518a, 518b. Weir, October 9, 1920.

Observations.—This species is characterized by its vinaceous-brown color, faint but sharp odor, echinulate cystidia, and habitat. The colors may be quite gray at times, but the flesh tints in such specimens are nearly always evident after the caps have faded. Along the Pacific coast I have made occasional collections of a very small form (pilei 3–8 mm.), which apparently is constant. It differs from *M. tenella* particularly in its habitat and stature. Forms with relatively long stipes are likely to be found where the moss is deep, but, to judge from my field experience to date, there is little likelihood of such forms being confused with *M. filopes* or *M. iodiolens*. The pleurocystidia and vinaceous colors as well as the larger size distinguish it from *M. psammicola*.

Kühner has called this fungus *M. vitrea* var. *tenella*. The species he places in *M. metata* is the one here treated as *M. leptocephala*. It is quite obvious that the true Friesian concept of *M. metata* will never be known. More than likely Fries himself allowed more variation for the species than would have been justified had he known the microscopic characters of these agarics as we now know them. Hence the only sensible procedure, if the name is to be retained, is to adopt the concept of some later worker. Specimens from Romell in the Atkinson Herbarium are as described above. The cystidia are present on both the sides and the edges of gills. Kühner has followed the concept of Schroeter (1889). It will probably always be a matter of opinion which microscopic interpretation of a Fresian species should be followed, but in my estimation that of Romell, which is also that of Lange, Kauffman, and Ricken, is the one best established by usage and the closest to the Friesian concept.

131. Mycena piceicola A. H. Smith

Mycologia, 31: 273. 1939

Illustrations:

Plate 51; Text fig. 33, nos. 1–2.
Smith, Mycologia, 31, fig. 1 K, fig. 3.

Pileus 2–3.5 cm. broad, ovoid to obtusely conic at first, becoming broadly convex or broadly ovoid in age, margin appressed against the stipe in buttons and sometimes wavy in age, surface hoary-pruinose at first, soon naked and polished, subhygrophanous, translucent-striate when moist, lubricous, even, color dark livid gray to "fuscous" or "hair brown," the margin usually paler and sometimes whitish, fading to "drab" on the disc or sordid ashy gray with a pallid margin;

Fig. 33. *M. piceicola:* 1, cheilocystidia; 2, spores. *M. hudsoniana:* 3, cheilocystidia; 6, spores. *M. plicosa:* 4, cheilocystidia; 5, spores. *M. subplicosa:* 7, spores; 8, cheilocystidia. *M. iodiolens:* 9, cheilocystidia; 10, spores. *M. filopes:* 11, cheilocystidia; 12, spores

fading slowly; flesh watery gray, thin, fragile, taste mild, odor slight, subfarinaceous, and hardly distinctive; lamellae adnate, developing a slight tooth, close to subdistant, 15–18 reach the stipe, narrow, 2.5 mm. broad, whitish to pallid or grayish and with pallid even edges, intervenose; stipe (2) 4–6 (8) cm. long, 1.5–2 mm. thick, equal or the base slightly inflated, fragile, tubular, usually strict, at first covered with a hoary bloom but soon naked and polished and somewhat translucent, the base white-strigose, at first dark bluish gray but soon fading to "drab" or sordid grayish brown below, apex pallid.

Spores ellipsoid, pointed at one end, 6–8 × 3.5–5 μ or 7–9 × 4–5 μ, amyloid but the reaction very weak; basidia four-spored, 26–28 × 5–7 μ; cheilocystidia cylindric to clavate with obtuse short rodlike projections over the enlarged portion, at times the projections elongated and branched into contorted filamentous processes, (18) 21–27 (30) × 5–9 (11) μ; pleurocystidia not differentiated; gill trama sordid vinaceous brown in iodine, pileus trama with a thin surface pellicle, which becomes yellow in iodine, beneath it a region of enlarged cells appearing somewhat cellular in tangential section, the remainder of loosely interwoven hyphae, all tissue beneath the pellicle vinaceous brown in iodine.

Habit, habitat, and distribution.—Gregarious or in troops of hundreds of fruiting bodies on needle beds under spruce and more rarely under Douglas fir; Washington, Oregon, and California. It was particularly abundant around Lake Tahkenitch, Oregon, after a period of cold weather in November, 1935. It has been found in Michigan also.

Material studied.—Smith, 32–600, 3324, 3442, 3543, 3548, 8030, 8740, 8814, 9070, 17066. Mains, September 4, 1934, Michigan.

Observations.—After observing a large number of specimens, one realizes that the colors and color changes are quite distinctive, even though they are very difficult to match in a color chart or to describe clearly in common terms. When one pulls the fruiting bodies from their attachment among the needles, the base of the stipe does not snap off, as it does in *M. plicosa.* It can, of course, be broken off if carelessly picked; the distinctive point is that in *M. plicosa*, no matter how careful one is, it is almost impossible to pull the stipe from its substratum without breaking it. I have never seen reddish stains on specimens of *M. piceicola.* The rather small cheilocystidia, with their frequently contorted projections, also distinguish it from such of its relatives as *M. metata*, *M. hudsoniana*, and *M. subplicosa.* It is most likely to be confused with *M. subplicosa*, but it lacks pleurocystidia.

132. Mycena hudsoniana Smith

Contrib. Univ. Mich. Herb., 5: 24. 1941

Illustrations:

Text fig. 33, nos. 3, 6 (p. 275).
Smith, Contrib. Univ. Mich. Herb., 5, pl. 6; pl. 25, figs. 4–5.

Pileus 2–5 cm. broad, obtusely conic, becoming somewhat expanded, usually broadly umbonate, margin appressed against the stipe when young, glabrous and polished, moist, "fuscous" on the disc, "drab" toward the whitish margin (blackish on the disc at first, dark gray toward the margin), hygrophanous and fading to "pale smoke gray," closely striate when moist, somewhat sulcate in age or when faded; flesh thin, firm but fragile, concolorous with the surface, odor mild or faintly fragrant, taste not distinctive or only slightly subnauseous; lamellae ascending-adnate or with a decurrent tooth (hooked), close to crowded, 25–30 reach the stipe, narrow (scarcely 2 mm. broad), color "pale smoke gray," edges even and pallid; stipe 3–5 cm. long, 1.5–3 mm. thick, equal, hollow, very fragile, with a faint hoary bloom at first, soon becoming polished and watery, concolorous with the pileus or slightly paler above.

Spores 8–11 (12) $\times$ 5–6 μ, amyloid, narrowly ellipsoid or pip-shaped; basidia four-spored or occasionally two-spored; cheilocystidia and pleurocystidia similar and fairly abundant, capitate to abruptly clavate, the enlarged portion very finely echinulate, hyaline, 32–50 $\times$ 15–40 μ; gill trama sordid vinaceous brown in iodine; pileus trama with a thin pellicle, a strongly differentiated hypoderm, and a moderately broad region of floccose tissue, vinaceous brown in iodine.

Habit, habitat, and distribution.—Gregarious on needle beds under conifers in the Hudsonian life zone, Olympic Mountains, Washington. The type was collected near Deer Lake, June 24, 1939 (14613). Additional material was found near or below Deer Lake on May 11 (13334) and June 13 (14301). On May 28 the species was gathered at an elevation of 4500 feet on the Boulder Lake Trail (13827).

Material studied.—Smith, 13334, 13827, 14301, 14473, 14613.

Observations.—The spore size of *M. hudsoniana* seems to be somewhat variable. In deposits the spores usually measure 9–12 $\times$ 5–6 μ, although in one pileus from the type collection the measurements were 7–10 × 5–7 μ. Occasional two-spored basidia were found in nearly all pilei examined, and they probably account for the variability in spore size. The sterigmata of the two-spored basidia were of the same size and shape as those of four-spored individuals, and as a result it was difficult to be certain of the two-spored condition or to be sure that spores from the bispored basidia were larger than those produced on the other basidia.

Mycena hudsoniana is very closely related to *M. subplicosa* Karsten, but can easily be distinguished by its more fragile stipe, which readily splits longitudinally into several segments when broken or even while still intact and growing. It is also distinguished by its much larger size, larger cystidia, lack of ochraceous tints on the margin of the pileus in age, and, apparently, by its habitat. In its fragile stipe it is similar to *M. plicosa* but differs in its larger spores and in not staining reddish when bruised or in age. As previously pointed out, however, the latter character should not be emphasized. When revived in KOH, specimens of *M. plicosa* have a thicker pellicle, but neither species is viscid when fresh.

133. Mycena plicosa (Fr.) Gillet

Les Hymén., p. 270 1874

Agaricus metatus β plicosus Fries, Syst. Myc., 1: 145. 1821.
Agaricus plicosus Fr., Epicr. Syst. Myc., p. 110. 1838.

Illustrations:
Plates 50 B, 52; Text fig. 33, nos. 4–5 (p. 275).
Fries, Icon. Sel. Hymen., 1, pl. 81, fig. 4.
Smith, Mycologia, 27, fig. 1 c.

Pileus 1–2(3) cm. broad, convex to broadly subconic at first, the margin appressed or slightly incurved, becoming hemispheric to expanded and broadly umbonate, surface dry and densely pruinose at first, slowly becoming naked and fairly moist, glabrous, when young blackish or very dark grayish brown, becoming "drab" to "avellaneous" or dull lead color and unicolorous, hardly hygrophanous, sulcate-striate to scalloped on the margin or halfway to the disc; flesh grayish to pallid, very rigid but brittle, thin, odor and taste not distinctive; lamellae close to subdistant, 20–32 reach the stipe, one or two tiers of short individuals, narrow to broad (2.5–4 mm.), adnate and with a slight tooth at times, drab to grayish or creamy or with a faint flesh tint discernible in age, often staining reddish brown where bruised, margin concolorous with the faces; stipe 2–5 cm. long, 1–2 (3.5) mm. thick, strict and rigid, equal or slightly enlarged below, hollow, very brittle, terete but often compressed in large individuals and then longitudinally twisted-striate, densely pruinose over all at first, base practically naked or slightly strigose, soon becoming polished above, concolorous with the pileus or paler.

Spores 7–9 × 3.5–4 μ, ellipsoid to subpyriform, smooth, weakly amyloid; basidia four-spored; pleurocystidia scattered to rare, embedded, clavate, 26–34 × (5) 7–13 μ, apices covered with short rod-like projections, hyaline; cheilocystidia similar to pleurocystidia, abundant; gill trama homogeneous, sordid brown (weakly amyloid) in iodine; pileus trama covered with a well-developed gelatinous pellicle (when revived in KOH), hypoderm fairly well differentiated, all but the pellicle sordid brown in iodine; stipe having roughened caulocystidia near the apex.

Habit, habitat, and distribution.—Gregarious to scattered under spruce and pine late in October or November in the vicinity of Ann Arbor, Michigan. It usually fruits during periods of warm wet weather between the heavy frosts of late fall. Hesler (5171) collected what appears to be this species under hemlock at Highlands, North Carolina, July 27, 1934.

Material studied.—Smith, 32–653, 77, 1167, 6157, November 16, 1931, Michigan. Hesler, 5171.

Observations.—As previously pointed out (Smith, 1927), the close, thin lamellae of my collections are not in agreement with the Friesian concept of the species. *M. plicosa* is rather distinct among the fragile gray Mycenae having echinulate cystidia. Its stipe is so fragile that it nearly always breaks at its point of attachment to the debris, even when one is very careful in collecting it. The collector must dig carpophores of this species out of the debris rather than pull them out, as he does in gathering most gray Mycenae. The reddish-brown stains on the gills appear to be an additional distinguishing character, although it would not be surprising to find collections in which these were not evident. When stains are present, one is likely to key the fungus out along with the short-stiped form of *M. atroalboides*, a species with which it has relatively little in common. The more numerous pleurocystidia, as well as the rather thick pellicle, allow dried specimens of *M. plicosa* to be readily identified. In fresh specimens the fragile stipe and particularly the late occurrence of the fruiting bodies should distinguish it from variations of *M. atroalboides*.

134. Mycena subplicosa Karsten

Acta Soc. pro Fauna et Flora Fenn., 9: 2. 1893

Illustrations: Plate 47 B; Text fig. 33, nos. 7–8 (p. 275).

Pileus 1–2 cm. broad, obtusely conic, becoming campanulate or expanded-umbonate, margin appressed against the stipe when young, often flaring somewhat in age, striate to the disc when moist, somewhat sulcate when faded, surface densely hoary-pruinose when young but soon naked, scarcely hygrophanous, when moist "fuscous" on the disc, pale grayish toward the whitish margin, fading to "hair brown" on the center and the margin cinereous, in age the margin often sordid ochraceous; flesh thin, scarcely fragile, grayish to pallid, odor and taste mild; lamellae ascending-adnate, close, 23–26 reach the stipe, narrow to moderately broad (2–2.5 mm.), whitish, becoming gray, strongly intervenose, edges even and pallid; stipe 3–4 cm. long, 1.5 mm. thick, equal, tubular, rigid-cartilaginous but moderately fragile (not splitting readily when collected), base white-strigose, apex frosted at first, soon naked and smooth, concolorous with the pileus or paler.

Spores narrowly ellipsoid, pointed at one end, 6–8 (9) $\times$ 3.5–4 μ, amyloid; basidia four-spored, 20–22 $\times$ 4–5 μ; cheilocystidia saccate to pedicellate with a globose head, the head more or less covered with short rodlike projections, 25–30 $\times$ 8–15 μ, abundant and usually forming a sterile band, hyaline; pleurocystidia similar to cheilocystidia, scattered; gill trama vinaceous brown in iodine; pileus trama with a thin adnate pellicle, a well-defined hypoderm, and the remainder of filamentous hyphae, all but the pellicle vinaceous brown in iodine; stipe tissue vinaceous brown in iodine.

Habit, habitat, and distribution.—Gregarious under spruce on needle beds during October and November; New York, Michigan, Washington, Oregon, and California.

Material studied.—Smith, 32-662, 3329, 3341, 3438, 7943, 7944, 7980, 8053, 8494, 8833, 16496, 17013, 18122, 18126, 18128, 18150, 18177. Kauffman (as *M. atroalba*), New York.

Observations.—Karsten did not describe the cystidia for his species, and as a result it cannot be stated definitely that the material I have referred to it is properly identified. My determination is based on the tendency of the pileus to become sulcate-striate, the small spores, and the stature, all characters which distinguish Karsten's species. As is indicated by its arrangement here as well as by its name, it appears to be most closely related to *M. plicosa*. It is distinguished from *M. plicosa* by its less fragile stipe, a tendency to develop yellow tints along the margin of the pileus, the thinner pellicle, and the less pronounced amyloid reaction of the trama of the gills and pileus.

135. MYCENA ATROALBOIDES (Pk.) Saccardo

Syll. Fung., 5: 276. 1887

Agaricus atroalboides Peck, Ann. Rep. New York State Mus., 27: 93. 1875.
Prunulus atroalboides Murrill, North Am. Flora, 9: 329. 1916.

Illustrations:

Plates 47 A, 53 C; Text fig. 32, nos. 9–10 (p. 268).
Lange, Flora Agar. Dan., 2, pl. 54 I (short-stiped form).
Smith, Am. Journ. Bot., 22, pl. 2, fig. 2.

Pileus 1–2 cm. broad, ovoid to convex or obtusely conic when young, becoming broadly conic or with a distinctly flattened disc, expanding and finally plane or with a low flattened umbo, margin connivent or curved in slightly when very young, flaring or recurved in age and frequently wavy or crenate, surface moist, at first conspicuously hoary, the bloom often persisting until late maturity, smooth or somewhat rugulose, translucent-striate toward the margin when moist, subhygrophanous and fading slowly, color "Chaetura black," "bone brown," or "fuscous" when young, the margin at first "cinnamon drab" to "drab," at maturity paler but retaining the faint cinnamon tinge, fading slowly to "pale smoke gray" over all, sometimes spotted reddish brown—particularly in wet weather; flesh thin but cartilaginous, concolorous with the surface (blackish to pallid), and occasionally stained reddish brown, taste and odor mild or very slightly of radish; lamellae ascending-adnate and with a distinct decurrent tooth, becoming horizontal and shallowly adnexed, close to subdistant, 19–25 reach the stipe, three tiers of lamellulae, narrow to subventricose in age (1.5–3 mm.), at times strongly intervenose, pallid grayish to dark gray, sometimes white when young, sometimes staining reddish brown in age or entirely rubescent, often strongly intervenose, edges pallid and even; stipe 2–4 cm. (when growing on needle carpets) or 5–12 cm. long (when growing in sphagnum), 1–2 mm. thick, equal, often flexuous, tubular, cartilaginous-brittle, with a whitish bloom at first, soon naked and polished, base densely white-strigose, concolorous with the pileus but soon paler, especially above, often with an abundant clear watery juice.

Spores ovoid to ellipsoid, 7–9 (10) × 3–4 (4.5) μ, amyloid (reaction weak); basidia two- or four-spored, 22–24 × 5–6 μ; cheilocystidia abundant, 19–25 (30) × 6–12 μ, clavate to subcapitate with the inflated portion covered with short or long prolongations, pleurocystidia absent or very rare, if present similar to the cheilocystidia; gill trama deep vinaceous brown in iodine; pileus trama with a well-differentiated pellicle, beneath it a distinct hypoderm (occupying about half the trama) the cells of which have brown contents, the remaining tissue filamentous and hyaline; stipe tissue vinaceous brown in iodine.

Habit, habitat, and distribution.—Densely gregarious on needle beds under conifers or in sphagnum bogs in the late summer or fall. It is common in the bogs of the northeastern United States and is frequently found on needle beds under spruce. On the Pacific Coast it is sporadic and often very abundant under Douglas fir and spruce. I have found it in the United States in Michigan, Washington, and California, and have examined material from Tennessee, New York, and Idaho. Specimens from British Columbia in Canada have also been studied.

Material studied.—Smith, 32–509, 32–510, 32–511, 32–512, 33–944, 33–945, 708, 750, 755, 863, 909, 923, 1250, 2510, 2598, 4882, 7776, 7777, 7825, 7869, 8004, 8202, 8242, 8276, 14196, 14943. Atkinson, 14665. Beebe, 85, British Columbia, Canada. Hesler, 14086. Gruber, Oregon. Slipp, UIFP, 2551, 2599.

Observations.—The colors of the pileus in this species nearly always possess a tinge of cinnamon, which gives a very characteristic appearance to the dominant dark gray. In dry weather one often finds large numbers of individuals in which there are no reddish spots. At other times the gills may become entirely reddish brown, and the stipe also may assume these colors. I have studied this point carefully, for I have followed the fruiting of this species regularly in one bog near Ann Arbor over a period of ten years. The stipe is typically short, as is shown in Plate 53 C. This is the most common form along the Pacific coast. Peck's type consists of long-stiped specimens from sphagnum, but, to judge from my own experience, this should not be considered the "typical" form of the species. In many of our sphagnum bogs the moss is not present under the dense clumps of black spruce. *M. atroalboides* frequently fruits along the line separating the moss from the needle carpet under these trees or just at the edge of the circle or arc formed by their outermost branches. Those specimens which develop in the dense shade of the tree, but on a flat needle carpet, have short stipes. Those growing beside them in the moss develop long stipes, as do most other agarics, including *Laccaria laccata*, which frequently grow in sphagnum.

Occasionally I have observed the gills separating from the stipe and forming a collar around it, but this is apparently not a constant enough character to be of any aid in recognizing the species. The bloom, which often remains throughout the development of the pileus during a relatively dry season, is more pronounced than in most Mycenae.

136. Mycena filopes (Fr.) Quélet

Champ. Jura et Vosges, p. 106. 1872

Agaricus filopes Fries, Syst. Myc., 1: 142. 1821.
Mycena albogrisea Peck, Bull. New York State Mus., 116: 27. 1907.
Prunulus albogriseus Murrill, North Am. Flora, 9: 327. 1916.

Illustrations:

Plate 54 A; Text fig. 33, nos. 11–12 (p. 275).
Beardslee and Coker, Journ. Elisha Mitchell Sci. Soc., 40, pl. 23 (lower figs.); pl. 24.
House, Bull. New York State Mus., 219–220, fig. 1, p. 234.
Konrad et Maublanc, Icon. Sel. Fung., 3, pl. 233, II.
Lange, Flora Agar. Dan., 2, pl. 57 G (small).
Ricken, Die Blätterpilze, 2, pl. 110, fig. 6 (as *M. vitilis*).

Pileus (5) 10–20 (25) mm. broad, cylindric to ovoid when young, becoming obtusely conic, campanulate, or with a flaring margin and a long obtuse umbo, surface at first hoary from a faint bloom, soon polished and moist, glabrous, the umbo "fuscous" (dark grayish brown) at first, fading to "benzo brown" (drab with a tinge of reddish), margin watery gray or paler, gradually fading over all to pallid gray with a pale brownish-gray disc, translucent-striate when moist, sometimes becoming sulcate in age, margin entire; flesh thin, rather fragile, no odor, taste mild, no color changes noted; lamellae close to scarcely subdistant, narrow, adnate, pallid brownish; stipe (5) 8–12 (15) cm. long, 1–2 mm. thick, equal, tubular, very brittle, covered with a hoary bloom at first, polished and translucent in age, base white-strigose and sometimes slightly rooting, the apex often bluish black at first, the remainder fuscous, fading to pallid gray or pallid watery white above, in age dark brownish drab toward the base.

Spores 7–9 × 4–5 μ, smooth, narrowly ellipsoid to ovoid, amyloid;

basidia four-spored; pleurocystidia scattered, 35–44 × 12–22 μ, capitate, with an obese or a slender pedicel, the head covered with short rodlike projections; cheilocystidia forming a sterile band, hyaline, similar in size and shape to pleurocystidia; gill trama homogeneous and vinaceous red in iodine; pileus trama with a well-differentiated pellicle and hypoderm, the remainder floccose but hyphae quite enlarged, in iodine turning vinaceous red in all parts except the pellicle.

Habit, habitat, and distribution.—Gregarious on conifer needles and humus, also on mossy areas in hardwood forests, abundant in some localities but generally rare in North America. It is known from Alabama, North Carolina, Tennessee, New York, Ohio, Michigan, Washington, and Oregon. It fruits during both the summer and fall months.

Material studied.—Smith, 32–530, 33–854, 33–915, 2742, 3276, 3473, 3499, 3517, 3563, 3614, 7823, 16233, 16436, 16951. Atkinson, six packets, New York. Burke, 1942, Alabama. Hesler, 12944, 11792, 14205. Kauffman, two collections. Slipp, UIFP: 2673.

Observations.—This is a slender gray to somewhat avellaneous species which develops a long stipe even when growing in open places. The lack of any distinctive odor readily separates it from *M. iodiolens*. In addition, the margin of the pileus in *M. iodiolens* usually consists of a very narrow sterile membrane, that is, the gills do not extend to the edge of the cap. *M. atroalboides* has a short stipe when growing in similar habitats, and frequently develops reddish stains in age. These are all secondary characters, it is true, and in certain groups of species would not be given much emphasis. However, ten years of field experience have convinced me that all three species are quite readily recognizable in North America, and that intergradations are not particularly numerous.

Mycena albogrisea Peck appears to be a small form of *M. filopes*. I have examined the type and found the spores to measure 8–11 × 4–5 μ. A few 11–13 × 6–7 μ were also observed. The remaining microscopic characters are as given in the description. No bispored basidia were seen. There is a resemblance between the type specimens of *M. albogrisea* and large dried carpophores of *M. capillaris*, but I have never observed the latter either with a stipe 1 mm. thick or occurring on carpets of conifer needles. Some may prefer Peck's name for this species as a convenient way of avoiding the confusion which exists in the use of the names *M. vitilis* and *M. filopes* in Europe. This expedient has much to recommend it. Peck thought the fungus belonged in the section Basipedes of Fries, but his view does not appear to be correct. In the specimen studied the stipe tissue seemed to be continuous with that of the pileus, and the strigose base of the stipe was also not typical for Basipedes.

137. Mycena iodiolens Lundell

Uppsal. Hymen.-Flora, 22: 7. 1932

Mycena graveolens Kauffman and Smith, Pap. Mich. Acad. Sci., Arts, and Letters, 17: 181. 1933.

Illustrations: Plates 54 B, 55; Text fig. 33, nos. 9–10 (p. 275).

Pileus 5–15 (25) mm. broad and high, obtusely conic to narrowly campanulate, becoming broadly campanulate or expanded-umbonate, the umbo sharp or obtuse and occasionally papillate, the margin appressed against the stipe when young and often flaring somewhat at maturity, sometimes crenate, a distinct narrow sterile margin present in nearly all specimens, surface conspicuously hoary when young, soon naked, hygrophanous, when moist "fuscous" to "light drab" around the umbo, near "tilleul buff" on the margin, striate, fading to pale drab gray or "vinaceous fawn," the umbo persistently darker on the disc, pallid toward the margin and fading to pale gray or gray with a vinaceous tinge; flesh membranous, thin, moderately fragile, pallid, taste not distinctive, odor lacking at first but shortly after specimens are gathered becoming strong and similar to that of iodoform; lamellae ascending and narrowly adnate, close, 22–26 or up to 32 reaching the stipe, usually very narrow but subventricose in large caps (1.5–3 mm.), whitish or tinged gray, edges even and pallid; stipe 5–12 cm. long, 1–2.5 (3) mm. thick, equal, cartilaginous and tough but becoming flaccid, sometimes moderately fragile, tubular, hoary at first, soon polished except for the white-strigose base, "avellaneous" or "olive brown," often pallid toward the apex, sometimes "ecru drab" toward the apex and "cinnamon drab" below.

Spores ovoid to broadly ovoid in two-spored forms, ellipsoid in four-spored forms, 8–11 × 5–7 μ or 7–10 × 4–6 μ, amyloid; basidia 18–22 × 5–6 μ; cheilocystidia forming a conspicuous sterile band along the gill edge (edge heteromorphous), saccate, and without a pedicel, 18–22 × 12–15 μ, hyaline, apices studded with short rodlike projections; pleurocystidia absent; gill trama dark vinaceous brown in iodine; pileus trama with a thin pellicle, a region of inflated cells (the hypoderm) beneath it and the remainder floccose, all parts beneath the pellicle reddish wine color in iodine.

Habit, habitat, and distribution.—Scattered to subcespitose on sticks and debris; throughout northeastern and central North America and along the Pacific coast from Washington to California. This is a common species in Michigan and along the Pacific coast during late fall after many agarics have ceased to fruit.

Material studied.—Smith, 32–90, 32–348, 32–362, 32–400, 32–457, 32–482, 32–487, 32–537, 32–570, 33–1074, 716, 781, 910, 1058, 2514, 2542, 2571, 2835, 3313, 3449, 3532, 3581, 3666, 4663, 5040, 5050, 5086, 7145, 7739, 7889, 7928, 8750, 8868, 8811, 9060, 9112, 14466, September 27 and November 17, 1931, October 29, 1932, Michigan. Atkinson, 13711, 13717, 13810 (all as *M. ammoniaca*). Brown, 1925, Lake Quinault, Washington. Kauffman, September 14, 1907 (as *M. leptocephala*). Gruber, Idaho. Mains, 32–263, 32–359, 32–437, 32–715, 32–798, 32–813, 34–67, 34–156, 5050. Wehmeyer, 816.

Observations.—The manner in which this species fades and its appearance when faded are almost as distinctive as the strong odor of iodoform which develops after the fruiting bodies have been collected. The faded portions of the cap appear pale gray and have a curious sheen, as if covered with a faint bloom. This appearance is caused by the way the moisture escapes from between the cells of the pellicle and is not due to hairs or cystidia, as is usual when a fungus appears pruinose. The two-spored form is more common than the four-spored form in the regions where I have collected.

For a lengthy account of this species and its related forms the reader is referred to Kühner (1938). He regards it as the typical form of *M. vitilis.* The names "*M. vitilis*" and "*M. filopes*" have been used in various concepts so much that at times one is tempted to discard both. Since in his *Monographia* (1857, p. 63) Fries described *M. vitilis* as "inodorus," it is reasonable to assume that he placed an odorless Mycena under that name. Apparently this is also the opinion of Lundell, who has been trained in the Friesian concepts. Rather than become entangled in the complex nomenclature of the fungi in this series it seems better to use a recent name which definitely applies to the fungus in question and has the added advantage of being admirably descriptive.

138. Mycena peltata (Fr.) Gillet

Les Hymén., p. 270. 1874

Agaricus peltatus Fries, Epicr. Syst. Myc., p. 110. 1838.

Illustrations: Text fig. 34, nos. 5–6 (p. 289).

Pileus 8–20 mm. broad, convex to obtusely campanulate when young, usually convex to plane at maturity, occasionally with an obtuse umbo, sometimes with a shallow depression over the disc, densely pruinose when young, soon naked, glabrous, surface lubricous to subviscid from a very thin adnate but somewhat gelatinous pellicle, nearly opaque or but faintly striatulate when moist, striae more conspicuous as the colors become lighter, disc usually "fuscous" at first, margin nearly "hair brown," in age evenly "drab gray" (dark gray to grayish brown, becoming paler and finally cinereous), rather evenly colored; flesh thin, pliant, grayish or pallid, no odor or taste; lamellae broadly adnate with a decurrent tooth, close, moderately broad, white to pale grayish at the base, edges pallid; stipe 3–5 or 8–15 cm. long, 1–2 mm. thick, equal or with a small bulb at the base, tubular, rather tough and cartilaginous, densely white-fibrillose below, glabrous above, lubricous (but never viscid) when wet, concolorous with the pileus or paler.

Spores 8–10 (11) × 3.5–4 μ, narrowly ellipsoid or curved slightly toward the apiculus, amyloid; basidia four-spored; pleurocystidia not differentiated; cheilocystidia basidium-like but soon developing contorted fingerlike prolongations, sometimes with wavy outlines, 28–33 × 6–9 μ; gill trama homogeneous, vinaceous brown in iodine; pileus trama with a thin, somewhat gelatinous pellicle, the hyphae of which have numerous short rodlike projections, hypoderm well differentiated, the cells filled with a dark-brown pigment, the remaining tissue filamentous, all but the pellicle vinaceous brown in iodine.

Habit, habitat, and distribution.—Gregarious on grassy hummocks and on sphagnum, late fall; Alabama and Michigan.

Material studied.—Smith, 1148, 1157. Burke, Alabama.

Observations.—*Mycena peltata* shows the variation in length of stipe when growing in exposed areas and in deep moss that is characteristic of *M. atroalboides*. It appears to be very closely related to *M. latifolia* both by its gill attachment and by its roughened cheilocystidia. It is more cartilaginous, however, and lacks the conspicuous pleurocystidia.

The pellicle of *M. peltata* is not the typically gelatinous type, and as a result the pileus is not truly viscid when fresh. In late fall, the time this species fruits, cold nights are rather frequent and more than likely many of the fruiting bodies are frosted. The pellicle of the frosted carpophores may possibly show more of a tendency to gelatinize when mounted in KOH than that of normal material.

Stipe Long; Cystidia Typically Smooth

139. Mycena vitilis (Fr.) Quélet

Champ. Jura et Vosges, p. 106. 1872

Agaricus vitilis Fries, Epicr. Syst. Myc., p. 113. 1838.

Illustrations:

Plate 56; Text fig. 34, nos. 1–2.
Konrad et Maublanc, Icon. Sel. Fung., 3, pl. 233, I.
Lange, Flora Agar. Dan., 2, pl. 52, fig. D (very good).

Pileus 5–15 (20) mm. broad, obtusely conic with an appressed margin when young, becoming campanulate or somewhat umbonate, the margin plane or becoming recurved, surface hoary at first but soon polished and lubricous, sometimes subviscid but the pellicle adnate or separable only in shreds, glistening when dry, margin even, slightly striate, color "drab" when fresh, fading to pale gray or nearly white in age, sometimes with a strong brownish tint when fresh and fading to "clay color," at least around the disc; flesh thin but pliant, grayish or pallid, cartilaginous, odor and taste not distinctive; lamellae attached by a tooth or narrowly adnate, close to subdistant, narrow, equal, white or grayish, edges concolorous and often slightly eroded; stipe 6–12 cm. long, (1) 1.5–2 mm. thick, equal, cartilaginous and tough, flexuous or straight but usually curved toward the base at least, tubular, rooting in the debris or attached to sticks, buried portion white-strigose, surrounded with a thin subgelatinous layer, which causes it to be lubricous to the touch, bluish black at first, soon gray, nearly concolorous with the pileus, apex somewhat fibrous-striate.

Spores 9–11 × 5–6 μ, ellipsoid, amyloid; basidia four-spored; pleurocystidia not differentiated or occasionally present near the edge and similar to cheilocystidia; cheilocystidia 32–46 × 8–14 μ, fusoid-ventricose or with two to several obtuse fingerlike projections arising from the apex; gill trama vinaceous brown in iodine, the subhymenium

FIG. 34. *M. vitilis:* 1, cheilocystidia; 2, spores. *M. pullata:* 3, spores; 4, cheilocystidia. *M. peltata:* 5, cheilocystidia; 6, spores. *M. magna:* 7, spores. *M. polygramma:* 8, spores; 9, cheilocystidia

of narrow, interwoven hyphae, the central portion of long, cylindric, moderately broad cells; pileus trama with a fairly thick subgelatinous pellicle, a well-differentiated hypoderm, and a filamentous tramal body, all except the pellicle vinaceous brown in iodine.

Habit, habitat, and distribution.—Scattered to gregarious on debris in hardwood or mixed conifer and hardwood forests. I have examined material from New York, Michigan, Washington, Oregon, and California. Along the Pacific coast it is occasionally rather abundant in red-alder slashes. In eastern North America it is quite common late in the season along with *M. semivestipes* and *M. pullata.*

Material studied.—Smith, 32–572, 32–620, 33–1139, 1257, 2547, 3200, 3268, 8327, 13767, 15529, 17070; two collections, 1931, Michigan. Kauffman, New York, Michigan, and Oregon. Mains, 32–502.

Observations.—Kühner uses the name "*M. filopes*" for this fungus. Since Lange's concept has been generally accepted by mycologists, it appears better, in the interests of a stable nomenclature, to use it here. The species is one of the easiest Mycenae to recognize, but it lacks a single fixed character clearly separating it from all others. Its grayish to brownish (but never vinaceous) colors, long and usually flexuous stipe, pliant consistency, lubricous feel, fairly large spores, and more or less fusoid-ventricose cystidia amply distinguish it. It appears to be closely related to *M. polygramma* and, in fact, may at times be mistaken for that agaric if the stipe is more striate than usual. *M. vitilis* lacks the pseudorhiza of *M. polygramma* and has more conspicuous cheilocystidia and, generally, a more slender stature. *M. pullata* is practically identical with *M. vitilis* in all characters except color.

140. Mycena pullata (Berk. & Cke.) Saccardo

Syll. Fung., 5: 277. 1887

Agaricus (*Mycena*) *pullatus* Berkeley & Cooke, Grevillea, 11: 69. 1882.

Illustrations:

Plate 57; Text fig. 34, nos. 3–4 (p. 289).

Cooke, Ill. Brit. Fungi, pl. 232 (no. 237).

Pileus 5–20 mm. broad, obtusely conic or somewhat ovoid when in button stages, the margin appressed, becoming broadly campanulate to expanded-umbonate, the margin often flaring or recurved somewhat, surface hoary but soon becoming naked and polished, usually appearing comparatively dry, fresh specimens even but soon becoming sulcate along the margin, colors dark vinaceous brown or tinged blackish purple, becoming paler but remaining vinaceous to purplish brown, in age sordid grayish brown with only a tinge of vinaceous ("army brown," "bone brown," "vinaceous brown," or "purple drab," fading to "vinaceous drab" or "Natal brown"); flesh moderately thick over the disc, tapered evenly to the margin, sordid vinaceous brown, rather pliant and cartilaginous, odor and taste not distinctive; lamellae ascending and narrowly adnate, narrow to moderately broad (somewhat ventricose at times), close to subdistant, whitish to cinereous or drab, usually concolorous with the pileus in age; stipe 5–12 (15) cm. long, 1–2 mm. thick, terete, strict, or flexuous, almost always curved near the base, equal, tubular, very rigid and cartilaginous, more or less concolorous with the pileus ("purple drab" to "vinaceous drab" or "vinaceous fawn"), often grayish below, faintly fibrillose-pruinose above, remainder sparsely longitudinally fibrillose-striate or at times glabrous and polished, white-mycelioid at the base.

Spores 8–10 (11) × 5–6.5 (7) μ, ellipsoid, smooth, amyloid; basidia four-spored; pleurocystidia not differentiated; cheilocystidia smooth at first and clavate to fusoid-ventricose, developing fingerlike prolongations, which often become branched; gill trama homogeneous, pale vinaceous in iodine; pileus trama with a well-developed pellicle appearing subgelatinous in KOH, hypoderm well differentiated, remainder of rather compactly interwoven hyphae, all except the pellicle pale vinaceous in iodine.

Habit, habitat, and distribution.—Solitary to scattered among fallen leaves in hardwood forests, often in company with *M vitilis*, late fall; Tennessee, New York, and Michigan.

Material studied.—Smith, 32–533, 32–569, 32–592, 32–635, 33–1077, 33–1138, 78, 158, 1171, 1256, 6139, 6158, 14966, 15528; five collections between 1929 and 1931, Michigan. Atkinson, 19515, 23992 (as *M. polygramma*). Hesler, 13157.

Observations.—The rather pliant cartilaginous consistency, small pileus, long stipe, habit, and vinaceous-brown to purplish-brown color are outstanding macroscopic characters. The microscopic characters do not readily separate it from either of the two most closely related species. *M. pullata* is intermediate between *M. polygramma* and *M. vitilis* and is distinct from both by its colors. Color is here used as a specific difference because ample collections of both *M. pullata* and *M. vitilis* made over a period of years have shown it to be constant.

Rea (1922) gives the spores of *M. pullata* as 6 × 3 μ. These measurements need to be verified. I have never detected a truly nitrous odor in any of my collections. If the English account of this species is accurate, then the American collections will have to be regarded as a separate species. They have been placed here because of the stature and the tinge of purple or vinaceous in the stipe and pileus.

141. Mycena polygramma (Fr.) S. F. Gray

Nat. Arr. Brit. Plants, 1: 619. 1821

Agaricus polygrammus Fries, Syst. Myc., 1: 146. 1821.

Illustrations:

Plates 58–59; Text fig. 34, nos. 8–9 (p. 289).
Atkinson, Stud. Am. Fungi (1911), fig. 96 (very slender form).
Lange, Flora Agar. Dan., 2, pl. 53, fig. F (very good).
Smith, Mycologia, 29, fig. 2 a.

Pileus (1.5) 2–4 cm. broad, obtusely conic to ovoid at first, becoming conic to campanulate or nearly convex with an abrupt small conic umbo, at times plane with a conic umbo, margin slightly incurved, frequently crenate, in age flaring or recurved and undulate, surface white-canescent at first, the bloom often persisting until near maturity, glabrous, lubricous in age, color "fuscous black" beneath the bloom or merely "fuscous," fading slowly to "drab" or paler grayish, nearly "pinkish buff" at times, margin opaque and frequently sulcate, the surface often more or less uneven and appearing streaked with glistening lines, not hygrophanous; flesh very hard and cartilaginous, watery grayish to white, rather thin, no odor, taste mild; lamellae narrowly adnate or with a short-decurrent tooth, close, 30–38 reach the stipe, broad anteriorly (4–7 mm.), white or whitish, in age flushed with pink, often with sordid-brownish stains, edges pallid and even; stipe 6–15 cm. long, 2–5 mm. thick, very brittle and cartilaginous, equal, tubular, sometimes with a well-developed pseudorhiza, base white-strigose and often staining reddish brown, closely silvery longitudinally striate, sometimes twisted-striate, "fuscous" or paler grayish brown beneath the silvery covering, at times nearly glabrous and smooth or glabrous and longitudinally grooved, apex pallid and faintly powdered.

Spores broadly ellipsoid, 7.5–10 × 5–6 μ, amyloid; basidia four-spored, 26–30 × 7–8 μ; cheilocystidia scattered to abundant, aciculate or the midportion somewhat enlarged and the apex forked or branched, giving rise to two or several contorted fingerlike projections; pleurocystidia not differentiated; gill trama dark vinaceous brown in iodine; pileus trama with a thick adnate pellicle, beneath it a region of inflated cells irregularly arranged (the hypoderm), their contents dark brown, the remaining tissue filamentous, all except the pellicle vinaceous brown in iodine.

Habit, habitat, and distribution.—Gregarious to subcespitose under hardwoods, particularly maple and basswood, from June to October; North Carolina, Massachusetts, New York, and Michigan. Rare in North America.

Material studied.—Smith, 5036, 6259. Atkinson 10225, Blowing Rock, North Carolina, and various collections from New York.

Observations.—The cheilocystidia, colors of the pileus, cartilaginous consistency, and striate stipe distinguish *M. polygramma*. *M. vitilis*, *M. pullata*, and *M. praelonga* have probably all been misidentified as *M. polygramma* in North America because of the striations occasionally present on their stipes. *M. vitilis* is quickly distinguished by its cystidia; *M. pullata*, by its color; and *M. praelonga*, by its habitat on sphagnum. *M. megaspora* is also very close to *M. polygramma*, but is readily distinguished by its much larger spores and generally smooth stipe or by having fibrils only toward the base. It is also distinguished by its cystidia. These are the usual clavate-roughened type with numerous short rodlike projections over the apices. The cheilocystidia of *M. polygramma* are somewhat intermediate between the smooth and the roughened types. Both species have the same stature, colors, consistency, sordid-reddish stains in age, and well-developed pseudorhiza. The stipes of *M. polygramma* may be glabrous and polished in off-season forms, and these are indistinguishable macroscopically from carpophores of *M. megaspora*.

Stipe Long; Cheilocystidia Roughened

142. Mycena magna Murrill

Mycologia, 8: 220. 1916

Prunulus magnus Murrill, North Am. Flora, 9: 338. 1916

Illustration: Text fig. 34, no. 7 (p. 289).

"Pileus very large for the genus, broadly convex, not fully expanding, gregarious to subcespitose, reaching 5 cm. broad; surface dry, glabrous, striate, avellaneous, slightly darker on the disk when dry:

lamellae adnate, broad, rather distant, whitish: spores subglobose, smooth, hyaline, 6–7.5 μ: stipe compressed, equal, smooth, glabrous, avellaneous, 5–8 cm. long, 2–4 mm. thick.

"Type collected on the ground in woods near Seattle, Washington, October 20–November 1, 1911, *W. A. Murrill 463* (herb. N. Y. Bot. Gard.).

"Distribution: Washington and Oregon."

I have not seen fresh material. The spores of the type are strongly amyloid, smooth, broadly ellipsoid to subglobose, and are borne on two-spored basidia. They measure 7–9 × 6–7.5 μ. Pleurocystidia and cheilocystidia are present but embedded and difficult to locate. They measure 28–34 × (6) 8–12 μ and have short projections over the clavate apex. The structure of the pileus trama is similar to that of *M. longipes*, except that the lactifers are less numerous.

Mycena longipes and *M. magna* impress me as being the same species. Mounts of spores from the types of both were compared, and enough intergradation of both size and shape was noted to make one question a separation on spore characters alone. The pleurocystidia of *M. magna* are so difficult to locate and so scattered that I hesitate to use their presence as distinctive. It seems best, however, to recognize both species—at least until a critical study can be made from fresh specimens. *M. rugulosiceps* should be readily distinguished by its larger spores and lignicolous habit.

143. Mycena longipes Murrill

Mycologia, 8: 220. 1916

Prunulus longipes Murrill, North Am. Flora, 9: 338. 1916.

Illustrations: Text fig. 35, nos. 1–2.

"Pileus convex, umbonate, rather tough, solitary, 2.5 cm. broad; surface smooth, dry, glabrous, very pale avellaneous, isabelline on the umbo, margin entire, pallid, finely striate: lamellae deeply sinuate with a decurrent tooth, subcrowded, arcuate, white: spores broadly ellipsoid, smooth, hyaline, granular, 8–10 × 6–7 μ: stipe smooth, glabrous, polished, equal, hollow, snow-white above, very pale avellaneous below, 15 cm. long, 2–3 mm. thick.

"Type collected in leaf-mold under redwoods in Muir Woods, California, November 22, 1911, *W. A. Murrill 1132* (herb. N. Y. Bot. Gard.).

"Distribution: Known only from the type locality."

FIG. 35. *M. longipes:* 1, cheilocystidia; 2, spores. *M. megaspora:* 3, spores; 4, cheilocystidia. *M. borealis:* 5, pleurocystidia; 6, spores. *M. fagetorum:* 7, cheilocystidia; 8, spores

Material studied.—Smith, 32–672, 33–515, 526, 9391, 16578, May 28, 1940, Michigan.

Observations.—I have found single specimens in Michigan, Washington, and California that apparently belong here, but have not secured enough material to justify making a critical study. The spores of the type measured 7–9 (10) × 5–6.5 μ in one fruiting body and 8.5–10.5 × 6–7 μ in another. The basidia of the latter were two-spored; those of the former, mostly four-spored. The spores are strongly amyloid, smooth, hyaline, and broadly ellipsoid. The cheilocystidia are 27–34 × 8–12 μ and clavate and are covered with short, often contorted, projections. The smooth cystidia previously mentioned (Smith, 1937) were very likely merely large immature basidia. The pileus trama is characterized by a well-developed pellicle, the threads of which bear numerous short projections. A well-differentiated hypoderm is also present, and the remainder of the tissue is floccose. Both the gill and the cap tissue were characterized by the presence of contorted lactiferous hyphae. The iodine reactions were inconclusive except on the spores.

This species bears the same relationship to *M. rugulosiceps* that *M. hemisphaerica* bears to *M. galericulata*. *M. longipes* superficially resembles pale forms of *M. megaspora*, particularly in the long cartilaginous stipe, but again the small spores clearly distinguish it. *M. longipes* apparently occurs in Michigan. Occasional specimens of a long-stiped dark-gray Mycena have been found with the characteristic small spores. These had radicating stems. Murrill did not get the base of the stipe in his specimen and failed to mention its character in the description. Since there is a strong tendency among the species of this group to develop a pseudorhiza it would not be surprising to find such a structure typical of *M. longipes*.

144. Mycena megaspora Kauffman

Pap. Mich. Acad. Sci., Arts, and Letters, 17: 182. 1933

Illustrations: Plates 60, 61; Text fig. 35, nos. 3–4 (p. 295).

Pileus 1–4 (5) cm. broad, convex to obtusely campanulate when young, becoming obtuse, broadly convex, campanulate or expanded and plane to umbonate, umbo varying from nearly papillate in some to very broad and low in others, margin connivent to the stipe when young or incurved slightly, often remaining somewhat decurved in age and occasionally becoming wavy, surface often hoary at first, soon naked and polished, very finely radially wrinkled-striate and sometimes slightly translucent-striate, more or less sulcate in age, lubricous to subviscid when wet, color blackish over all when young, the margin soon becoming pallid, fading slowly, and finally "drab" or "avellaneous," the sordid brownish colors more conspicuous in age and sometimes becoming sordid yellowish brown near the margin, subhygrophanous; flesh white, thick on the disc but tapered abruptly and equal (1.5 mm.) over the remainder, cartilaginous and tough, no odor, taste mild or slightly astringent; lamellae adnate, adnexed, or subsinuate, subdistant to close, 22–30 reach the stipe, narrow to moderately broad, ventricose and up to 5–8 mm., in age whitish to pallid or stained reddish, edges even, stipe 5–12 cm. long, 1–4 (5) mm. thick, equal above a long (3–10 cm.) tapered pseudorhiza, fistulose, very cartilaginous and tough, sometimes with a faint bloom when young, soon naked and polished, at times translucent when moist, concolorous with or paler than the pileus below, the upper portion pallid to whitish, occasionally twisted and faintly longitudinally striate.

Spores irregular in shape, broadly ellipsoid to ovoid, 12–17.5 × 6–8 (9) μ, amyloid; basidia two-spored or occasionally three-spored, 40–54 × 7–9 μ; cheilocystidia embedded but abundant and often forming a sterile band, 38–46 × 14–22 μ, clavate, enlarged portion covered with short rodlike projections or the projections elongated and branched or contorted, hyaline; pleurocystidia absent or present only near the gill edge; gill trama vinaceous brown in iodine; pileus trama consisting of a thin pellicle, an area of enlarged hyphal cells beneath it (the hypoderm), and the remainder filamentous, all but the pellicle vinaceous brown or bright reddish vinaceous in iodine, the inflated cells under the pellicle filled with a dark-brown sap; stipe dark vinaceous red in iodine.

Habit, habitat, and distribution.—Scattered to gregarious or in small clusters from a common pseudorhiza during the spring and fall; Michigan, New York, and Washington in the United States and Ontario in Canada. It has been found regularly on swampy muck soil under *Vaccinium corymbosum* and *Rhus vernix* and also on sphagnum in open bogs. Bisby has collected it in Ontario.

Material studied.—Smith, 31-47, 31-56, 31-63, 31-84, 31-117, 31-122, 32-240, 32-241, 32-280, 32-375, 32-486, 32-827, 33-47, 33-827, 33-1058, 1318, 3981, 4257, 6345, 7695, 11068, 13688, 14637, Atkinson, 24094. Bisby, Kenora, Ontario. Mains, 31-119, 34-114. Kauffman, two collections.

Observations.—The colors of this species are like those of the other members of the group. A very pale form, however, has been found in which the pileus is "buffy brown" at first and fades to "pale olive buff." Usually the pileus is some shade of sordid grayish brown at the same stage in typical fruiting bodies. The reddish spots on the gills are found most frequently on old specimens that have endured much wet weather. The consistency is very cartilaginous, resembling that of *M. polygramma*, and in this respect the fungus differs from most bog-dwelling Mycenae.

It appears that *M. megaspora* is related to *M. polygramma* on the one hand and to *M. galericulata* on the other. I have never been able to determine with certainty that the pseudorhiza of *M. megaspora* originates from decaying roots of *Vaccinium corymbosum*, but such a connection is very likely. *M. galericulata* is typically a lignicolous species which frequently develops a long pseudorhiza, particularly if the fruiting bodies occur on a very rotten log. It has similar cystidia and only slightly smaller spores. Its gills, however, are often evenly flushed with pale pink, a character not observed for *M. megaspora*. Collection 14637 from the Olympic Mountains appears to be a four-spored form of *M. megaspora*. Its spores are 10–12 × 5–7 μ.

Stipe and Cap Intermediate in Size

145. Mycena griseiconica Kauffman

Pap. Mich. Acad. Sci., Arts, and Letters, 11: 199. 1930

Illustrations: Text fig. 27, nos. 5–6 (p. 240).

Pileus 1–3.5 (5) cm. broad, up to 2.5 cm. high, at first narrowly elliptic, then broadly conic to campanulate, at maturity broadly convex or with an obtuse conic umbo, margin appressed against the stipe at first, surface smooth and with a faint bloom in young specimens, soon polished, moist, translucent-striate, subhygrophanous, sulcate-striate in age, bluish gray when young, becoming "hair brown" on the disc and "tilleul buff" on the margin, fading to silvery gray or the disc, remaining more brownish (near "avellaneous"); flesh thin, pliant or scarcely fragile, grayish, no odor, taste mild; lamellae ascending-adnate at first, horizontally adnate in convex pilei, subdistant, about twenty reach the stipe, narrow at first (2–2.5 mm.) but broader (4 mm.) in age, two tiers of lamellulae, pale glaucous whitish, slightly intervenose, edges even and pallid; stipe 5–10 cm. long, 1–2.5 mm. thick, equal, strict, hollow, cartilaginous but fragile, glabrous except for the white-strigose base, apex naked, even, "hair brown" toward the base, concolorous with the gills near the apex.

Spores ellipsoid, 9–13 (14) × 4–5.5 (6) μ, amyloid; basidia four-spored, 33–38 × 7–8 μ; cheilocystidia (36) 40–54 × 9–14 μ, very abundant, obovoid and tapering to a pedicel, the apex provided with one or several (up to five) long, cylindric prolongations (8–12 × 2–3 μ), hyaline, ventricose portion smooth; pleurocystidia absent; gill trama vinaceous brown in iodine; pileus trama with a thin pellicle, beneath it a well-differentiated hypoderm and the remainder filamentous, all but the pellicle vinaceous brown in iodine.

Habit, habitat, and distribution.—Gregarious to subcespitose on debris in Douglas fir and mixed conifer and alder forests during November and June; Washington and Oregon. Kühner (1938) has found it in Europe. One collection from J. E. Bier, November 24, 1939, apparently belongs here.

Material studied.—Smith, 14252, 14609. Stuntz, F 419. Bier, November 24, 1939, British Columbia.

Observations.—Although I have not seen abundant material and only the type collection has been located in Kauffman's specimens, I am sure that *M. griseiconica* is one of the larger Mycenae. It is slightly less cartilaginous than *M. galericulata* but by no means as fragile as *M. Abramsii*. *M. Abramsii* and *M. griseiconica* are remarkably similar in many respects, but for the present at least they are kept separate. Some pleurocystidia have always been found in collections of *M. Abramsii*. In the type they measure 60–80 × 12–16 (20) μ and are very abundant. This difference is striking when the types of both species are compared, since Kauffman's specimens lack pleurocystidia. On the basis of their scattered smaller pleurocystidia the eastern collections identified as Murrill's fungus are somewhat intermediate between the two. *M. Abramsii* has always impressed me as being related to *M. stannea*, whereas Kauffman's specimen appears to be related to *M. megaspora*. Kühner (1938), however, has arranged *M. griseiconica* beside *M. aetites* (*M. stannea* of this work), the position in which I have placed *M. Abramsii*.

146. Mycena fagetorum (Fr.) Gillet

Les Hymén., p. 274. 1874

Agaricus fagetorum Fries, Epicr. Syst. Myc., p. 106. 1838.

Illustrations:

Plate 62; Text fig. 35, nos. 7–8 (p. 295).

Kauffman, Agar. Mich., 2, pl. 162 (as *Collybia alcalinolens*).
Lange, Flora Agar. Dan., 2, pl. 56 E.
Ricken, Die Blätterpilze, pl. 109, fig. 9.

Pileus 10–30 (35) mm. broad, convex to obtusely campanulate when young, the margin usually incurved, soon broadly convex, with a broad low umbo or nearly plane, the margin often reflexed in age, surface smooth or slightly uneven, glabrous, lubricous when moist, subhygrophanous, color "drab" or darker on the disc at first, the margin whitish, becoming "wood brown" to "buffy brown" or "avellaneous" on the disc, and finally "mouse gray" to "light grayish olive" or pale cinereous over all, sometimes becoming nearly white (in general sordid grayish brown to dark gray at first and fading to avellaneous, pale cinereous, or nearly white in age), translucent-striate when fresh, sometimes becoming sulcate-striate; flesh thick on the disc, moderately thin on the margin, watery gray, odor and taste merely slightly farinaceous or not distinctive; lamellae adnate with a tooth or adnexed in age, often separating from the stipe but adhering to each other and thus forming a collar, close to subdistant, 25–34 reach the stipe, three or four tiers of lamellulae, narrow but becoming ventricose and broad (3–4 mm.), pale ashy gray, occasionally with an incarnate tinge, edges concolorous with the faces, often stained with sordid-brownish spots; stipe 4–7 cm. long, 2–4 mm. thick, equal or sometimes enlarged above, terete or compressed, fragile to fairly cartilaginous, base usually curved and prolonged beneath the leaves into a white-strigose pseudorhiza, surface pruinose at first and hoary, soon polished and moist, translucent, "Quaker drab" (bluish gray) when fresh, soon more or less concolorous with the pileus, sometimes pallid in age.

Spores (7) 8–11 × 3–4.5 μ, subcylindric to narrowly pear-shaped, smooth, faintly amyloid; basidia four-spored; pleurocystidia not differentiated; cheilocystidia clavate, smooth, or with obtuse contorted projections scattered over the apex, 30–42 × 7–11 μ; gill trama homogeneous, vinaceous red in iodine; pileus trama with a thin nongelatinous pellicle, a slightly differentiated hypoderm, and the remainder filamentous, vinaceous red in iodine.

Habit, habitat, and distribution.—Scattered to subcespitose among fallen leaves in open oak woods; Michigan, Nova Scotia. I have found this species in three localities near Ann Arbor. In one it appears quite regularly in late October and November.

Material studied.—Smith, 33–1141, 33–1141a, November 16, 18, 1931, Michigan. Kauffman, November 8, 1911, November 13, 1926, Michigan. Mains, two collections. Wehmeyer, 128.

Observations.—One would not be likely to confuse this species with other gray fragile or cartilaginous Mycenae, but it does lack an outstanding character. The moderately fleshy consistency, almost homogeneous flesh of the pileus, and Collybia-like stature serve to distinguish it from other fragile gray species, and its shorter stipe, narrow spores, and habitat separate it from the long-stiped cartilaginous species.

European material is described as having certain characters which are not typical of American collections. Kühner mentions in particular the noncespitose habit, farinaceous odor, gelatinous subhymenium, and slightly different cheilocystidia. Considering these characters in the order named, I have found the fungus fruiting prolifically in just the manner Kühner described, and I do not believe that the additional variations observed here are in the least significant, particularly since they are well within the expected range of variation. The farinaceous odor is a character of some value and should not be ignored. In my notes for the season of 1931 I recorded it as slightly farinaceous.

In 1933 one vigorously developing specimen had a distinctly alkaline odor, but in my other collections it was merely marked "not distinctive." In view of this situation it does not appear desirable to place much emphasis on either the odor or the taste. When my specimens are revived in KOH, the subhymenium shows a tendency to gelatinize, although the character does not appear distinct enough to be useful taxonomically. It can be demonstrated best near the gill edge. Both types of cystidia figured by Kühner can be found on a single gill in many of the American collections. Those illustrated are typical of mature pilei.

Kühner admits the American material into his species concept, but the differences between our descriptions required some explanation.

Typically Lignicolous Species with Elongated Cystidia

147 Mycena borealis A. H. Smith

Mycologia, 27: 586. 1935

Illustrations:

Plate 63; Text fig. 35, nos. 5–6 (p. 295).

Smith, Pap. Mich. Acad. Sci., Arts, and Letters, 21, pl. 26, fig. 1.

Pileus 1–4 cm. broad, button stages obtusely conic or the midportion ventricose, becoming broadly conic to obtusely campanulate, with a prominent obtuse umbo in age, margin appressed against the stipe at first and becoming plane or reflexed in age, sometimes crenate, frequently splitting radially, surface glabrous, lubricous to subviscid, very finely radially wrinkled, translucent-striate when moist, somewhat sulcate in age, pellicle separable only in shreds, colors gray with a tint of cinnamon or the margin nearly white ("cinnamon drab" to "fuscous" on the umbo), subhygrophanous, fading to pale gray in age; flesh thick on the disc, whitish to grayish, fragile, no odor, taste slightly farinaceous; lamellae adnate with a slight tooth, subdistant to distant, 18–22 reach the stipe, broad (3.5–5 mm.), intervenose, white, pruinose from cystidia, not staining reddish, edges even and pallid but pruinose under a lens; stipe 3–5 cm. long, 2–4 mm. thick, cartilaginous, equal, hollow, often compressed, beautifully white-fibrous when young, glabrescent and rather transparent in age, base white-strigose, concolorous with or paler than the pileus.

Spores ellipsoid to subovoid (8) 9–10 × (4) 5–6 μ, amyloid; basidia four-spored, 32–34 × 7–8 μ; cheilocystidia and pleurocystidia similar, or the former merely clavate (50) 60–120 × 8–14 μ, often projecting 50–75 μ, smooth and fusoid-ventricose at first, nearly cylindric in age and the elongated neck becoming studded with numerous small rodlike projections over the apical region (the projections about 2 μ long), hyaline; gill trama very faintly vinaceous brown in iodine; pileus trama with a well-developed subgelatinous pellicle of narrow hyphae, below it a region of vesiculose cells (the hypoderm), the remainder homogeneous and of narrower hyphae, very pale vinaceous brown in iodine except for the pellicle; tissue of the stipe also very pale vinaceous brown in iodine.

Habit, habitat, and distribution.—Gregarious to subcespitose on wood of conifers during the summer and fall; Alabama, Tennessee, New York, and Michigan in the United States and Nova Scotia, New Brunswick, and Ontario in Canada. Very abundant at times. Hesler has found that it is particularly common near the summit of Mt. Le Conte in the Great Smoky Mountains National Park.

Material studied.—Smith, 456, 481, 520, 1120, 4474, 7353, 9731, 10856. Atkinson, 13772, 24327. Atkinson, Herb. G. Hay, 123. Burke, 1942, Alabama. Hesler, 6513, 9059, 10694, 13922, 14255, June 12, 1938, Indian Gap, Great Smoky Mountains National Park. Kauffman, September 14, 1944, Lake Placid, New York (as *M. galericulata*). Mains, 32–123, 32–474, 5051. Wehmeyer, 759, 759 a–g.

Observations.—During seasons when it fruits sparingly small fragile specimens are sometimes found. Although these have the appearance of an ordinary gray Mycena, they can readily be identified by the peculiar cystidia. No other Mycena is known to have greatly elongated pleurocystidia with echinulate apices. The fungus is apparently limited to conifer wood or debris. It appears to be most closely related to the species in this group. In stature, consistency, and habit it is almost identical with *M. niveipes*.

148. Mycena excisa (Lasch) Gillet

Les Hymén., p. 275. 1874

Agaricus excisus Lasch, Linnaea, 4: 534. 1829.

Illustrations:

Text fig. 36, nos. 1–4 (p. 304.)

Beardslee and Coker, Journ. Elisha Mitchell Sci. Soc., 40, pl. 20 (lower figures).

Fries, Icon. Sel. Hymen., 1, pl. 81, fig. 1.

Pileus 15–35 mm. broad, broadly conic to obtusely campanulate, the margin sometimes incurved at first, becoming broadly umbonate to nearly flat, surface densely pruinose at first and appearing dry, soon polished and moister, "fuscous" (blackish) on the disc, margin paler and tinged with brown, fading slowly to dark gray or sordid grayish brown in age, margin entire, opaque when moist, sometimes very faintly translucent, becoming radially rugulose at maturity; flesh thin, tough and cartilaginous, grayish brown, no odor or taste; lamellae broadly and deeply adnexed, in age ventricose and broad, subdistant to close, white, becoming faintly fuscous or grayish, edges white or grayish; stipe 3–5 cm. long, 2–3 mm. thick, short, cartilaginous and tough, equal, hollow, pale fuscous below, pallid above, evenly white-pruinose at first, polished in age, often compressed and twisted.

Spores 8–10 (12) $\times$ 6–7 μ, broadly ellipsoid, smooth, amyloid; basidia four-spored; pleurocystidia scattered to abundant, 50–70 $\times$ 10–15 μ, fusoid-ventricose with obtuse apices or subcylindric, smooth, or with an occasional hump or obtuse projection on the elongated portion, rarely clavate and with several fingerlike prolongations, cheilocystidia similar to pleurocystidia or clavate and with obtuse projections or fingerlike prolongations, 35–46 $\times$ 10–14 μ; gill trama homogeneous, vinaceous brown in iodine; pileus trama with a thin adnate pellicle (when revived in KOH the pellicle distinctly gelati-

FIG. 90. *M. excelsa:* 1–2, cheilocystidia; 3, pleurocystidia; 4, spores. *M. tenuiceps:* 5, spores; 6, cheilocystidia; 7, pleurocystidia

nous), a well-differentiated hypoderm (the cells with brown contents), and the remainder filamentous, all parts except the pellicle vinaceous brown in iodine.

Habit, habitat, and distribution.—Gregarious on humus in mixed woods of fir and alder and scattered to subcespitose on decaying fir logs during October and November; Michigan, Washington, and California in the United States and Ontario and Manitoba in Canada. Kauffman collected it in oak woods in Michigan. Apparently it is quite rare in North America.

Material studied.—Smith, 2544, 2574, 3113, 8474, 14632, 16712, 17425. Kauffman, three collections, Michigan.

Observations.—*Mycena excisa* is very similar to *M. galericulata* in all its characters except the cystidia; in fact, Kauffman confused the two, as is evidenced by his Michigan collections. *M. excisa* has both the smooth and the roughened types of cheilocystidia, a feature which Kauffman overlooked and which I did not fully appreciate in my first study. The cystidia illustrated were all drawn from a single section of one pileus and show the extreme variation. The colors of the pileus and stipe are not distinctive, and the consistency is more cartilaginous than that of either *M. niveipes* or *M. algeriensis.*

When revived in KOH, the pellicle gelatinizes very distinctly, so that in a study of herbarium specimens one might be misled into believing he has a different and distinctly viscid fungus. When sections of fresh material are mounted in water, this character is not evident. However, fruiting bodies which have been exposed to continued wet weather often feel somewhat lubricous because of the continued soaking, and in them the pellicle is usually slightly gelatinized. One must learn to make allowances for this condition.

149. **Mycena tenuiceps,** sp. nov.

Illustrations: Text fig. 36, nos. 5–7.

Pileus 1–3 cm. latus, obtusus demum planus vel subdepressus, glaber, fuligineus, ad marginem pallidus, striatus; lamellae adnatae vel subdecurrentes, angustae (3 mm. ±), subdistantes, pruinosae, pallide cinereae, stipes 1.5–6 cm. longus, 1.5–2.5 mm. crassus, aequalis, fragilis, fistulosus, fuligineus, fibrillose punctatus, glabrescens; sporae 6–8 × 3.5–5 μ; pleurocystidia et cheilocystidia 50–95 × 10–16 μ, subventricosa. Specimen typicum in Herb. Univ. Mich. conservatum. Legit A. H. Smith, n. 10759, prope Grassy Patch, Great Smoky Mountains National Park, Tenn., Sept. 1, 1938.

Pileus 1–3 cm. broad, obtuse when young, the margin incurved, becoming plane to slightly depressed in age, the margin either remaining decurved or becoming variously lobed, wavy, or recurved, surface moist and glabrous, dark watery gray on the disc, pale watery gray toward the margin, translucent-striate, moist, opaque and more or less wrinkled when faded; flesh thin and lax, grayish, no odor, taste not recorded, no latex present, and no color changes noted; lamellae adnate to hooked and subdecurrent, narrow (3 mm. ±), subdistant, two or three tiers of short individuals, pruinose from the cystidia, pallid gray, almost concolorous with the pileus; stipe 1.5–3 cm. long, 1.5–2.5 mm. thick, equal, fragile, hollow, pale watery gray or concolorous with the pileus, densely innately fibrillose-punctate to somewhat striate, pruinose to white-scurfy above, more or less glabrescent in age, base sparsely white-mycelioid at point of attachment, not rooting.

Spores 6–8 × 3.5–5 μ, subellipsoid (usually narrowed toward the apiculus), smooth, hyaline, amyloid; basidia four-spored; pleurocystidia and cheilocystidia abundant, 50–90 × 10–16 μ, subclavate to subfusoid, finally nearly cylindric, smooth, hyaline; gill trama of interwoven hyphae, faintly amyloid; pileus trama with a poorly differentiated pellicle (the cells radially arranged and 4–8 μ thick), no pilocystidia seen, hypoderm distinct, of one to two rows of enlarged cells, which are filled with a smoky-brown content, the filamentous tramal body below the hypoderm well developed, contorted metallic-appearing lactifers scattered through it, all parts weakly amyloid; stipe when young more or less covered with long mucronate caulocystidia measuring 80–120 × 10–20 μ.

Habit, habitat, and distribution.—Single to gregarious on beech logs, during late August and early September; Great Smoky Mountains National Park, Tennessee. The type (10759) was found at Grassy Patch, September 1, 1938, and an additional collection was made at Indian Camp Creek, August 30, 1938.

Observations.—The fungus resembles *Collybia lacerata* in its thin flabby pileus, but the cystidia readily distinguish it. It appears to be closely related to *M. scabripes* and *M. trichoderma* but lacks the pilocystidia of the latter and has a different habitat as well as smaller stature. The narrow gills, lignicolous habitat, and squatty stature also separate it from ***M. scabripes.*** From ***M. algeriensis*** it is readily distinguished by its squatty appearance, plane to slightly depressed pileus, and dark gills. Both, however, are quite similar in their microscopic characters.

150. MYCENA ALGERIENSIS R. Maire in Kühner

Encyc. Myc., 10: 490. 1938

Illustrations: Plates 64–66; Text fig. 37, nos. 1–4 (p. 308).

Pileus 1–3 (4) cm. broad, up to 2.5 or 3 cm. high, sharply or obtusely conic when young, becoming narrowly campanulate or conic and in age broadly campanulate, occasionally convex with a low umbo, margin appressed against the stipe at first, often flaring in age and splitting, surface glabrous, radially rugulose, translucent-striate when moist and mature, glistening and lubricous, color "fuscous black" over all at first, becoming "fuscous," gradually becoming paler to near "drab" (sordid grayish brown) and at that time translucent-striate, hygrophanous, fading to sordid brownish gray and becoming slightly sulcate; flesh thin, soft, and fragile, dark grayish or when faded whitish, not changing color when bruised. odor and taste not distinctive; lamellae ascending-adnate and sometimes attached only by a tooth, close to crowded (or subdistant in small caps), 18–25 reach the stipe, three tiers of lamellulae (one of which extends nearly to the stipe), narrow (2 mm. or up to 4 mm. in large caps), whitish to pale grayish, much paler than the pileus, sometimes crenulate, under a lens pruinose from cystidia, edges even or somewhat eroded; stipe 4–7 (12) cm. long, 2–3.5 mm. thick, equal or slightly and evenly enlarged downward, hollow, very fragile and readily splitting lengthwise when snapped off, base usually densely white-strigose, densely or sparsely coated with white-fibrillose flecks over at least the lower two thirds, glabrescent in age, concolorous with the pileus (blackish to pale gray but never white), becoming translucent in age and sometimes compressed, occasionally slightly longitudinally striate.

Spores ellipsoid 7–8 (10) × 5–6 μ (or 7–10 × 6–7 μ) in the two-spored form, amyloid; basidia two- or four-spored, both may be found on the same pileus, 24–28 × 5–6 μ; cheilocystidia broadly fusoid-ventricose to ovoid, 38–52 × 9–18 μ; pleurocystidia very abundant, hyaline, broadly ventricose with obtuse apices, becoming subcylin-

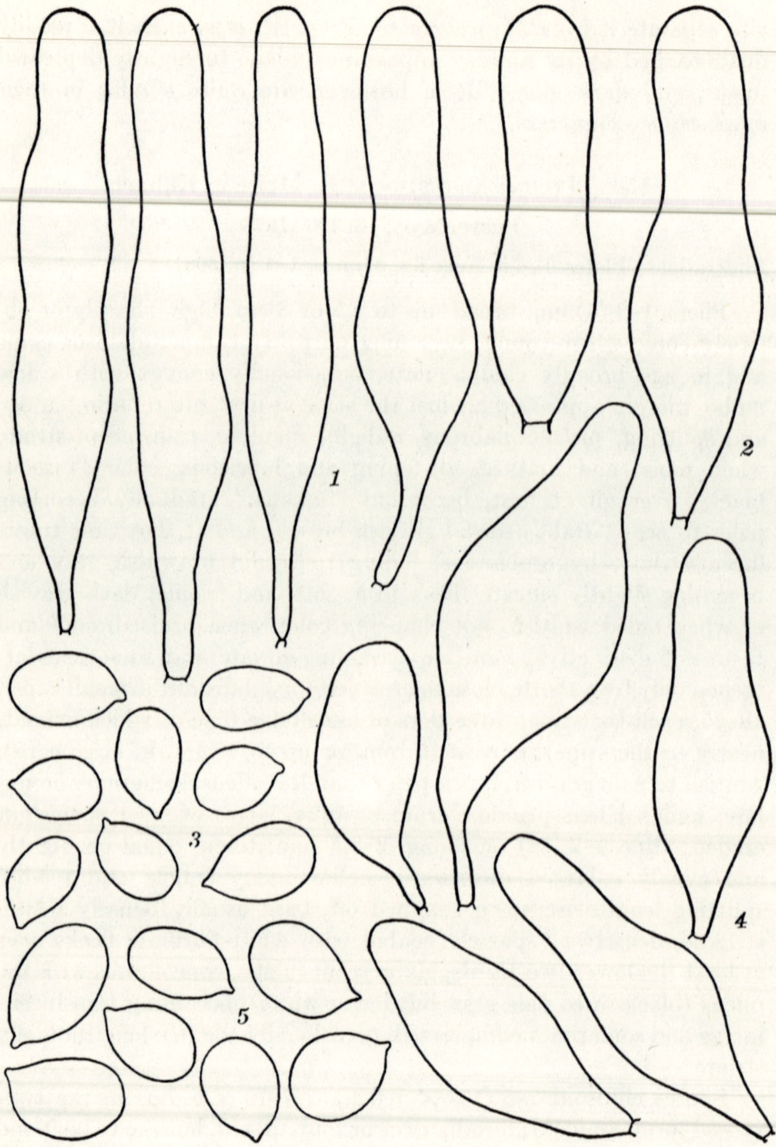

FIG. 37. *M. algeriensis:* 1, pleurocystidia; 2 and 4, cheilocystidia; 3, spores.
M. niveipes: 5, spores

dric, the apices remaining very obtuse, only rarely branched at the apex, 60–94 (100) $\times$ 9–12 μ; gill trama vinaceous brown in iodine; pileus trama with a poorly formed pellicle, below it a region of strongly inflated cells (the hypoderm) filled with dark-brown fluid, the remainder of the tissue filamentous.

Habit, habitat, and distribution.—Gregarious to densely cespitose on logs, particularly of elm and alder, during the spring and fall months; New York, Michigan, and Washington in the United States and Ontario in Canada.

Material studied.—Smith, 32–358, 32–359, 32–360, 32–399, 32–503, 32–523, 32–586, 32–641, 33–679, 30, 82, 82a, 420, 444, 933, 1049, 3970, 4328, 6432, 7080, 7418, 9538, 9590, 10614, 10675, 10977, 13592, 13686, 14126, 15055, 15126, 16304, September 14, October 30, 31, November 18, 19, 1931; June 9, 1941. Atkinson, 18538, New York. Bisby, 3373. Jackson, August 23, 1934. Mains, 5066.

Observations.—The outstanding characters of the species are its cystidia, its blackish colors, which become sordid brownish gray in age, the soft fragile consistency, and the white-fibrillose flecks, which at first cover the stipe. The American collections differ from the description of the species as given in Kühner (1938) in certain characters: The stipe hardly becomes whitish at any stage, and he made no mention of the softness of the gills. The gills in American collections showed no signs of becoming ventricose. They appear whitish because of the contrast with the dark pileus, but in reality they are very pale gray, especially toward the base. No nitrous odor was ever noted in any of my collections. These differences do not seem sufficient to justify establishing another species but, if constant, they may necessitate classifying the American material as a variety.

The question of the status of the name *M. dissiliens* should be considered here. Kauffman and Smith (1933) reported under this name a Mycena without pleurocystidia. This fungus has not been collected since, and no further study of it has been possible. Without additional information it should not be given further consideration. My identification of the collections reported in 1935, which was based on Kauffman's notes, applies to *M. algeriensis*, as Kühner suggested. Since the concept of *M. dissiliens* is not clearly established in Europe, and since in North America several different species have been found with the peculiar character of the stipe (see also *M. hudsoniana*), it appears best to drop the name *dissiliens*.

151. Mycena niveipes Murrill

Mycologia, 8: 221. 1916

Mycena polygramma var. *albida* Kauffman, Rep. Mich. Acad. Sci., 13: 219. 1916.

Prunulus niveipes Murrill, North Am. Flora, 9: 332. 1916.

Illustrations:

Plates 67–68; Text figs. 37, no. 5 (p. 308); 38, nos. 1–2.
Lange, Flora Agar. Dan., 2, pl. 51 E (short-stemmed form).
Smith, Mycologia, 29, fig. 3 a.

Pileus (1.5) 2–7 cm. broad, ellipsoid when young, convex to obtusely conic or becoming nearly plane at maturity, the margin appressed against the stipe at first by a narrow sterile band, which soon becomes lacerated, in age the margin either spreading or slightly recurved, surface smooth, translucent-striate, moist, color "clove brown" to "olive brown" or sordid drab when young and moist, hygrophanous, fading to whitish or various shades of sordid grayish brown on the disc, margin white and sometimes sulcate in age, often splitting; flesh thin and very fragile, grayish to whitish, taste acidulous to subfarinaceous, odor either nitrous or entirely lacking; lamellae ascending-adnate or toothed, close to subdistant (22–35 reach the stipe), broad (5–8 mm.), white or faintly bluish gray at first, soon fading to white, in age occasionally flushed with pink, edges even or slightly fimbriate; stipe 4–10 cm. long, 2.5–7 mm. thick, at times somewhat radicating, very fragile, equal, hollow, base covered with white mycelium or somewhat strigose, midportion covered with a dense white appressed fibrillose coating, soon longitudinally fibrous-striate or in age becoming glabrous, apex somewhat scabrous at first, pale bluish cinereous when young, becoming either sordid or shining white.

Spores subglobose to ellipsoid, 8–10 (11) × 5–6 (7) μ, amyloid; basidia two-, three-, or four-spored, 28–30 × 6–7 μ; cheilocystidia and pleurocystidia similar and abundant, 50–90 × 8–15 μ, smooth, broad at the base, with subacute to acute apices, those on the gill edge often shorter (40–60 μ) and more nearly fusoid-ventricose; gill trama yellowish or very faintly vinaceous brown in iodine; pileus trama with a well-developed pellicle, a distinct hypoderm, and the remainder filamentous, yellowish in iodine.

Habit, habitat, and distribution.—Single to gregarious or sub-

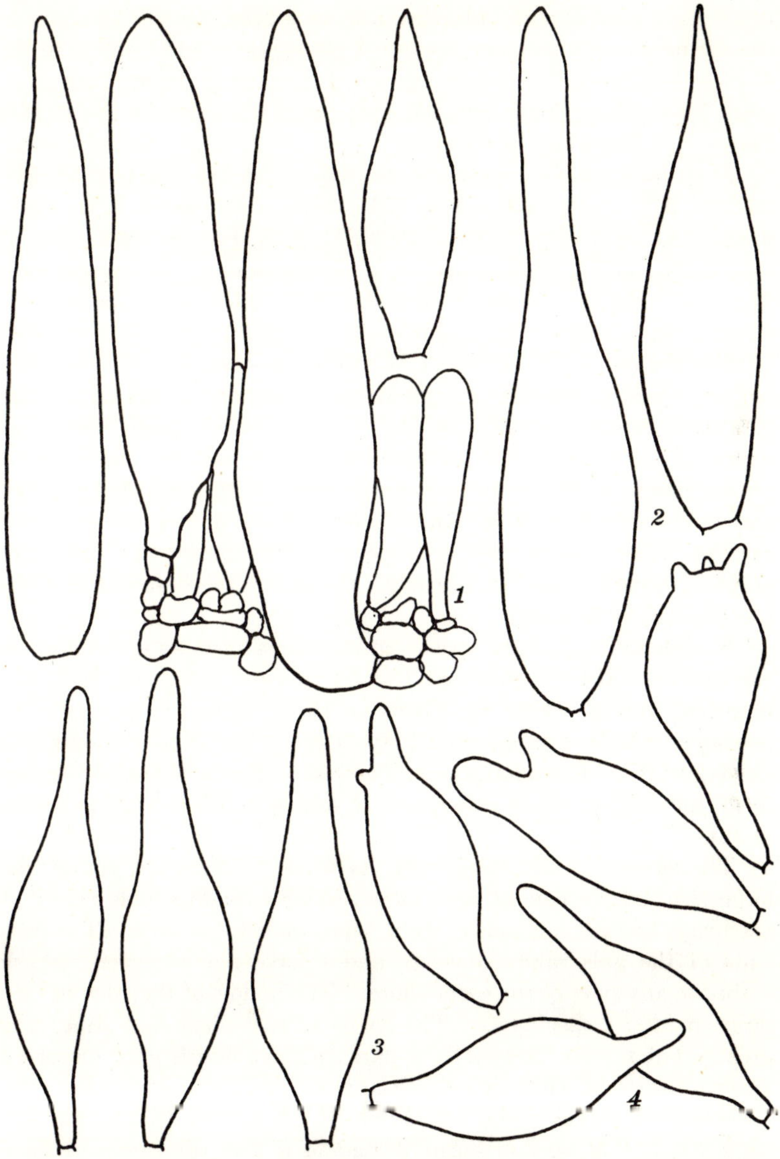

FIG. 38. *M. niveipes:* 1, pleurocystidia; 2, cheilocystidia. *M. alcalina:* 3, pleurocystidia; 4, cheilocystidia

cespitose on old logs of oak, elm, ash, or maple, usually common in the spring or early summer; eastern United States and Canada to the Rocky Mountains. I have examined specimens from Tennessee, New York, Ohio, Michigan, Missouri, and Colorado in the United States and Manitoba in Canada.

Material studied.—Smith, 32–42, 32–62, 32–110, 32–112, 32–365, 33–126, 971, 1324, 1450, 1474, 3967, 3976, 4206, 6246, 6271, 6291, 6321, 7243, 8617, 9537, 15071, 16303, May 28, June 2, 1932. Atkinson, 15195, 23923, 23927 (filed as *M. leptocephala*). Bisby, June 21, 27, 1926. Hesler, 11388. Overholts, 1537, 1655 (small spores, 7.5–8.5 × 5–6 μ), 1989. Sharp, 3825.

Observations.—The spores of Murrill's type measure 8–10 × 5–7 μ and are broadly ellipsoid, the cystidia are abundant on the sides of the gills and measure 60–90 × 8–12 μ. They are not truly fusoid (as previously described) but, at least in revived material, have broad obtuse bases and taper to long narrow necks with somewhat pointed tips. The cheilocystidia are 38–50 × 10–14 μ, ventricose, and smooth. The basidia are four-spored. The spores are distinctly amyloid, but the reactions of the gill trama and the flesh of the pileus are practically negative.

Kauffman studied the species in its robust form as it develops in the spring. Murrill collected the late-season form, which is not luxuriant and in which the fibrils on the stipe are either not conspicuous or soon disappear entirely, leaving the stipe glabrous and shining white. The odor is more likely to be absent in the late-season form, but I have found collections of robust odorless fruiting bodies in June.

The silvery striate stipe, with its white fibrils when young, the large size, fragile consistency, nitrous odor (usually), pale colors at maturity, and conspicuous cystidia serve to distinguish it. The flesh tints of the gills rarely develop, and I have not observed reddish stains on any part of the carpophore. The hollow of the stem is very large in robust specimens. The colors of the pileus and stipe vary greatly but mostly merely from dark to light bluish gray or sordid grayish brown to white or whitish in age.

Kühner uses the name "*M. Jacobi* R. Maire" for this fungus and suggests that *M. rugosa* sensu Bresadola is also the same. I have one collection, 8617, made on November 11, 1937, from the base of a madroña tree in the Siskiyou National Forest of northern California, that appears to be very close to Bresadola's fungus. I have tentatively filed it under *M. niveipes.* The pilei were up to 5 cm. broad and very dark gray over all even at maturity, the stipe was smooth, no distinctive odor was present, and the gills were drab gray at maturity. In its stature and consistency it is practically identical with *M. niveipes.* If the characters mentioned above are found to be constant, it may be desirable to give it at least the rank of a variety. Bisby's report of *M. Jacobi* was based on odorless specimens of *M. niveipes.* Overholts (1655) collected another variant of the species on May 8, 1914, near Creve Coeur, Missouri. The following description is taken from his notes: Pileus 2.5–5 cm. broad, campanulate, at first watery brown, becoming grayish and slightly tinted rose color, moist, glabrous, widely wrinkled-striate on the margin; flesh white, taste and odor strongly alkaline; lamellae adnexed or adnate with a decurrent tooth, broad, rather distant, white; stipe central, equal, hollow, splitting into two or three segments, shining, white-villose at base, 4–9 cm. long, 3–5 mm. thick; spores 7.5–8.5 × 5–6 μ.

Sections of the material verified Overholts's spore measurements and revealed the presence of the numerous typical pleurocystidia and cheilocystidia. The collection appears to be distinguished by its smaller spores and, possibly, by a more homogeneous pileus, although the latter character should be restudied from fresh material.

152. Mycena macrocystidiata Singer

Ann. Myc., 34: 430. 1936

Illustrations: Plate 69; Text fig. 39, nos. 3–5 (p. 316).

Pileus 1–5 cm. broad, conic when young, becoming obtuse or broadly campanulate in age, sometimes umbonate and the margin plane, surface hoary but soon becoming naked and moist, color "plumbeous black" at first, fading through "bone brown" to paler grayish brown, only subhygrophanous, slightly translucent-striate when moist, frequently very finely radially wrinkled-striate nearly to the disc, margin sometimes whitish at first; flesh thin, fairly pliant, pallid, taste acidulous, odor nitrous; lamellae adnate or with a decurrent tooth, close but appearing subdistant in broadly expanded caps, moderately broad and becoming ventricose in age, white at first, in age sometimes assuming a grayish-incarnate tinge; stipe (1.5) 4–11 cm. long, 1–3 (5) mm. thick, equal or compressed, tubular, fragile, surface with a hoary bloom at least toward the apex, sometimes longitudinally striate, white-strigose at the base, color deep "plumbeous black" when young (especially toward the apex), fading to "army brown," "bone brown," or pale grayish, often white in age.

Spores 8–10 × 5–6 μ, ellipsoid, amyloid; basidia four-spored; pleurocystidia abundant, 60–90 (110) × 8–16 μ, smooth, hyaline, subcylindric to fusoid-ventricose; cheilocystidia similar but shorter (up to 70 μ); gill trama homogeneous, vinaceous brown in iodine; pileus trama with a well-differentiated pellicle and hypoderm, the remainder filamentous, all but the pellicle pale vinaceous brown in iodine.

Habit, habitat, and distribution.—Gregarious to cespitose on conifer logs and debris; most abundant during late May and early June in the vicinity of Ann Arbor. I have examined material from Tennessee, Pennsylvania, and Michigan.

Material studied.—Smith, three collections, 32–12, May 25, 1932, Michigan. Hesler, 3816, 11415. Overholts, 22749.

Observations.—In his notes Overholts described the pileus as pale watery brown to "Saccardo's umber" or "snuff brown," and the spores as measuring 8–10 × 5–7 μ. Thus it appears that the same range in color noted for *M. alcalina* is present in this species. The collections just listed have been identified as Singer's species chiefly on the basis of the very large cystidia and the odor. Kühner reduced *M. macrocystidiata* to synonymy with *M. metata* (*M. leptocephala* in this work). The cystidia of the latter fungus, which has been collected abundantly in North America, have never approached the measurements given by Singer. Consequently, although the species described here is a robust Mycena, it does not seem wise to separate it from *M. macrocystidiata.* The variations in stature observed for *M. alcalina* locally are no greater than those indicated by Singer's original description and my notes. *M. macrocystidiata* and *M. niveipes* have many characters in common, but their appearance readily distinguishes them. Compare Plates 67 and 68 with Plate 69.

Typically Lignicolous Species; Cystidia Not Greatly Elongated

153. Mycena alcalina (Fr.) Quélet

Champ. Jura et Vosges, p. 105. 1872

Agaricus alcalinus Fries, Syst. Myc., 1: 142. 1821.
Mycena vexans Peck, Ann. Rep. New York State Mus., 38: 85. 1885.

Prunulus vexans Murrill, North Am. Flora, 9: 329. 1916.

Illustrations:

Plates 70–71; Text figs. 38, nos. 3–4 (p. 311), 39, nos. 1–2 (p. 316).
Bresadola, Icon. Mycol., 5, pl. 242.
Fries, Icon. Sel. Hymen., 1, pl. 81, fig. 3.
Konrad et Maublanc, Icon. Sel. Fung., 3, pl. 232.
Lange, Flora Agar. Dan., 2, pl. 51, fig. H (very good).
Ricken, Die Blätterpilze, 2, pl. 110, fig. 4.

Pileus (0.5) 1–3 (4) cm. broad, (0.5) 1–3 cm. high, evenly and obtusely conic, the margin usually flaring somewhat at maturity, sometimes convex or with a broad obtuse umbo, small forms often conic-campanulate or papillate, color "bone brown" to "plumbeous black" when young, "dusky brown" to "pale drab gray" at maturity or dull brownish, "Saccardo's umber" to "snuff brown" (blackish at first, becoming very sordid yellowish brown or drab gray in age), the margin frequently whitish or pallid, surface at first evenly covered with a conspicuous bloom, giving it a bluish-gray cast or a glaucous sheen, soon polished and then lubricous, subhygrophanous, obscurely striate when young, striate to the disc at maturity, the margin sometimes slightly sulcate; flesh thin or moderately thick under the disc, fragile but cartilaginous, white or grayish, odor strongly alkaline, sometimes lacking, taste acidulous but hardly distinctive; lamellae moderately close to subdistant (20–27 reach the stipe), adnate or with a slight-decurrent tooth, narrow to moderately broad (2–3.5 mm.), pure white or grayish at first, remaining white or becoming cinereous, sometimes spotted with sordid reddish-brown stains in age, intervenose at times, edges concolorous with the surfaces or remaining white; stipe (1.5) 4–9 (11) cm. long, (1) 1.5–2.5 (4) mm. thick, concolorous with the pileus or paler, especially at the apex, and often sordid yellowish brown in age, covered with a bloom and appearing bluish at first, soon polished and lubricous, not viscid, tubular to hollow, often compressed in large specimens, cartilaginous and brittle, base sparsely white-mycelioid to strigose.

Spores (7.5) 8–10 (11) × 4.5–6 (7) μ, ellipsoid to broadly ellipsoid, smooth, distinctly amyloid; basidia usually four-spored, two-spored forms occasionally abundant; pleurocystidia numerous, scattered or rare, sometimes apparently absent, fusoid-ventricose to subcylindric, the apices sometimes forked, hyaline, smooth, (35) 40–60 × 8–15 (20) μ; cheilocystidia abundant, hyaline, subclavate with an

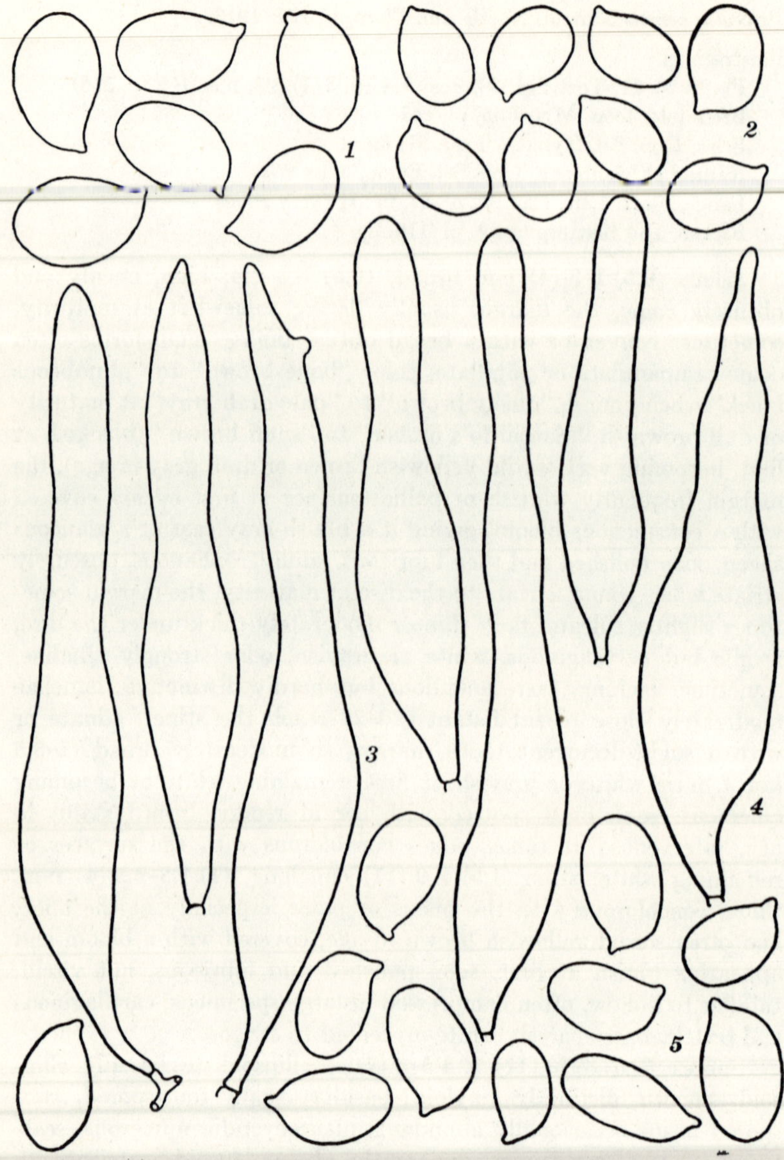

FIG. 39. *M. alcalina:* 1, spores of two-spored form; 2, spores of four-spored form. *M. macrocystidia:* 3, pleurocystidia; 4, cheilocystidia; 5, spores

abruptly narrowed apex, broadly fusoid-ventricose or sometimes the apex obtuse and furnished with two or more obtuse fingerlike projections, sometimes the apex forked or branched, 28–40 × 9–20 μ; gill trama pale yellow or sometimes vinaceous red in iodine; pileus trama with a rather thick pellicle and a well-differentiated hypoderm, the remainder filamentous, yellowish or slightly vinaceous brown in iodine.

Habit, habitat, and distribution.—Gregarious to subcespitose on decaying wood of conifers and densely gregarious on humus under conifers, particularly larch. I have found it most abundant in the spring in northeastern North America, but it does occur during the fall months as well. I have examined specimens from Alabama, Tennessee, Pennsylvania, New York, Michigan, Colorado, Montana, Washington, Oregon, and California in the United States and from Nova Scotia, Ontario, and Manitoba in Canada.

Material studied.—Smith, 31–100, 32–2 (bispored), 32–3, 32–3a, 32–3b, 32–5, 32–5a–h, 32–6 (bispored), 32–13 (bispored), 32–18, 32–28, 32–57, 32–68, 32–73, 32–89, 32–100, 32–106, 33–50, 33–181, 33–529, 33–668, 33–1102, 657, 1301, 3609, 4673, 6216, 6247, 6756, 8102, 13087, 13335, 13593, 13826, 13956, 14079, 14201, 14230, 16181. Atkinson, 13442, 14161, 24113. Burke, 1942, Alabama. Flett, Washington. Hesler, 12945. Kanouse, Michigan. Kauffman, collections from Pennsylvania, New York, Michigan, Montana, Colorado. Ruth Macrae, Ontario. Mains, 32–120, 33–122, 33–180, 5079, June 9, 1933. Stuntz, F 413. Wehmeyer, 657.

Odorless form.—Smith, 3708, 3709, 3758, 3811, 3827. Atkinson, 22172, 22672, 24063, 24070. Overholts, 1791.

Observations.—This is perhaps the commonest species in the genus in North America, but at the same time it is one of the most puzzling because of the variability of its distinctive characters. Ordinarily, the presence or the absence of a distinctive odor or of pleurocystidia in Mycena is regarded as taxonomically significant, but in *M. alcalina* as I have found it neither character appears to be sufficiently constant to serve as an infallible mark of distinction. The odor is often weak and detectable only if the flesh is crushed. Old specimens, or those which have developed slowly, may seem to lack it entirely. I have found occasional collections of perfectly fresh, rapidly developing specimens with no odor and only a slight acidulous taste which were identical in every other respect with typical material. By "typical specimens" I refer to those in which pleurocystidia are lacking or are very rare. In California such odorless forms are sometimes abundant on redwood logs during November and December. Their gills have been observed to become stained with sordid reddish brown in age. Such a change has been noticed only rarely in Michigan material; it appears to be correlated with wet weather.

I have been unable to demonstrate that the presence or the absence of pleurocystidia is taxonomically significant in this species. All gradations have been found in material collected in Michigan. This was a distinct surprise because in other species of the genus, as well as in most other gill fungi, the presence or the absence of pleurocystidia has proved to be a valuable taxonomic character.

The cheilocystidia vary considerably in shape. In some collections they are obtuse and with one or more fingerlike prolongations; in others this type is found along with typically fusoid-ventricose cystidia with pointed apices; and in still others the fusoid-ventricose pointed type is the only one present. In general, the pleurocystidia and cheilocystidia on any one cap are similar in size and shape. In the form which Peck described as *M. vexans*, however, the pleurocystidia are narrowly fusoid and the cheilocystidia obtuse and more or less branched.

Although the tissue of both gills and pileus gives the typical amyloid reaction, I have found it to be an unreliable character. In selecting specimens for iodine tests one should be careful to take only those which have been properly dried.

After studying this species locally for a period of years, I believe it is made up of a large number of races which are only beginning to be differentiated morphologically. In their present state of incomplete differentiation they are not recognizable as distinct units of the agaric flora and should all be grouped under one name. If one attempts to distinguish forms on the basis of the presence or the absence of pleurocystidia, he can generally recognize four conditions: cystidia absent, rare, scattered, or abundant. With these one might attempt to correlate the odor and make a series of forms based on the strength of the odor and whether pleurocystidia are present or absent, until all the possible combinations have been made. Such a procedure would be arbitrary and would only cause confusion.

154. **Mycena incarnatifolia,** sp. nov.

Illustrations: Plate 72; Text fig. 40, nos. 1–2 (p. 320)

Pileus 1.5–3 cm. latus, obtuse conicus demum campanulatus vel umbonatus, glaber, striatus, hygrophanus, sordide albidus demum pallide vinaceus, margine albidus, expallens subcandidus, crenatus; lamellae confertae, angustae demum latae, adnatae, candidae dein pallide incarnatae; stipes 3–5 cm. longus, 2–3 mm. crassus, cavus, deorsum albostrigosus, sursum pruinosus, subcandidus; sporae 8–10 × 5–6 μ; basidia tetraspora; cheilocystidia 40–60 × 9–12 (15) μ, fusoide ventricosa. Specimen typicum in Herb. Univ. Mich. conservatum. Legit A. H. Smith, n. 15267, prope Ann Arbor, Mich., Sept. 9, 1940.

Pileus 1.5–3 cm. broad, obtusely conic, becoming campanulate or broadly expanded with an obtuse umbo, surface moist, faintly pruinose when young, glabrous, translucent-striate, hygrophanous, whitish at first, soon pale grayish vinaceous over the disc and with a white margin ("tilleul buff" in buttons but soon "pale vinaceous fawn" or tinged brighter pink over the disc), nearly snow white when faded and then opaque, occasionally with rusty stains in age, the margin appressed against the stipe by a rather broad sterile band, the sterile portion becoming crenate in age (as in *M. haematopus*), flesh thin but firm, watery white, becoming snow white, with a scant hyaline watery juice, odor and taste mild, lamellae ascending-adnate, narrow to moderately broad (2.5–3.5 mm.), broadest near the stipe and tapered evenly to the margin, close, 22–27 reach the stipe, one or two tiers of lamellulae, pure white at first but soon flushed evenly with clear pale pink, edges minutely fimbriate but even and whitish, not spotted; stipe 3–5 cm. long, 2–3 mm. thick, terete or compressed, hollow, with a scanty clear watery juice, rather firm but brittle, base white-strigose, upper portion hoary-pruinose, hyaline white over all, becoming opaque and somewhat glabrescent, occasionally with scattered rusty stains but base remaining white in age.

Spores 8–10 × 5–6 μ, ellipsoid, smooth, amyloid; basidia four-spored; pleurocystidia not differentiated; cheilocystidia abundant, hyaline or with a granular-appearing highly refractive content, 40–60 × 9–12 (15) μ, fusoid-ventricose with pointed evenly tapered apices or the apices forked or sparsely branched; gill trama homogeneous, not colored in iodine; pileus trama not amyloid, with a thin pellicle, the cells of which give off scattered short branches, hypoderm hardly differentiated from the tramal body.

Habit, habitat, and distribution.—Cespitose on a black-cherry

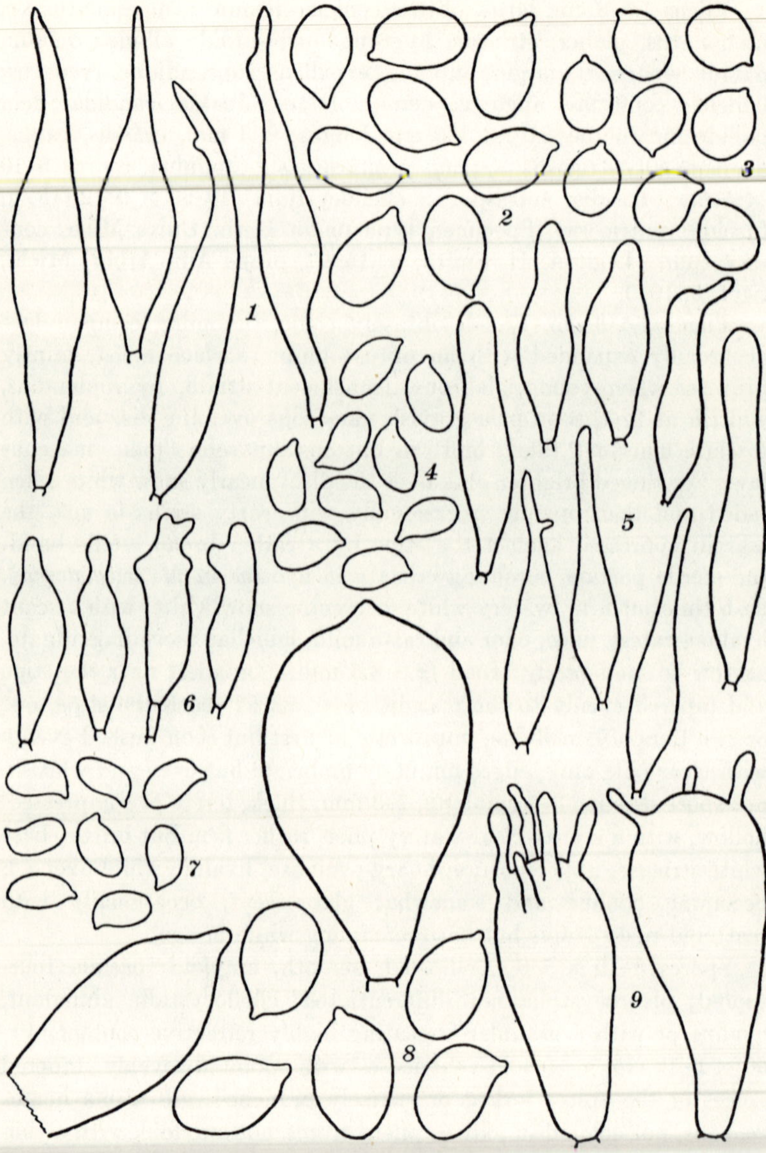

FIG. 40. *M. incarnatifolia:* 1, cheilocystidia; 2, spores. *M. collybiiformis:* 3, spores. *M. laevigata:* 4, spores; 5, cheilocystidia. *M. semivestipes:* 6, cheilocystidia; 7, spores. *M. parabolica:* 8, spores; 9, cheilocystidia

(*Prunus serotina*) log; September 9, 1940, Ann Arbor, Michigan. Known only from the type locality.

Observations.—Although the species has been found but once, I have no hesitation in describing it. Its characters separate it sharply from any other known Mycena. The white to pale-pink color, hyaline watery juice, crenate margin of the pileus, cespitose habit, fusoid cheilocystidia, and pink gills at maturity are very distinctive. In general bearing it is very similar to *M. haematopus*, but can hardly be considered a white form of it. Many of the carpophores, which were perfectly fresh when collected, had developed under very favorable weather conditions. The old pilei were infected with a species of Spinellus, a very common occurrence in Mycenae. The presence of a hyaline watery juice in the stipe excludes the possibility that *M. incarnatifolia* is an aberrant form of *M. haematopus*. Both species were found on the same cherry log.

Mycena incarnatifolia appears to be more closely related to *M. rubrotincta*. It is readily distinguished, however, by its more aciculate cystidia, the paler color throughout its development, the pink gills, and the white stipe, which does not become darker at the base in age. It is much more brittle than *M. laevigata*, and its gills are not so broadly adnate. In fact, carpophores of *M. incarnatifolia* are almost waxy in consistency. The white form which Ricken mentions under his *M. polygramma* may be the same as *M. incarnatifolia*. Ricken's figure (Pl. 111, fig. 7) of *M. polygramma* appears to me to represent some other species; which one, I am not prepared to say.

155. Mycena collybiiformis Murrill

Mycologia, 8: 220. 1916

Prunulus collybiiformis Murrill, North Am. Flora, 9: 335. 1916.

Illustration: Text fig. 40, no. 3.

"Pileus thin, rather tough, convex to expanded and at length umbilicate or depressed, cespitose, 2–4 cm. broad; surface dry, glabrous, cinereous, darker and rugose on the disk; margin entire, pallid, becoming slightly striate with age or on drying: lamellae adnate, plane, white, distant, rather narrow, interveined: spores ellipsoid, smooth, hyaline, 6–7 × 4–5 μ: stipe cylindric, somewhat enlarged at the base, hollow, smooth, pallid, subglabrous, whitish-mycelioid at the base, 3–5 cm. long, 2–5 mm. thick.

"Type collected on an oak stump at Port Jefferson, Long Island,

New York, August 5, 1902, *C. H. Peck & F. S. Earle 839* (herb. N. Y. Bot. Gard.).

"Distribution: Long Island, New York, Louisiana."

This is the original account of the fungus. I have examined the type and found the spores to measure 5–6 (7) × 4–5 μ. They are broadly ellipsoid to subglobose, smooth, and weakly amyloid. Occasionally small oil droplets have accumulated around the periphery of the spores and caused them to appear very minutely echinulate. The gill trama and that of the pileus did not revive very well, but both turned yellow in iodine. The pileus trama appeared to be practically homogeneous beneath a thin pellicle, and no cystidia were observed on the gills or the pileus. The species appears to be well characterized by its small, very broadly ellipsoid, weakly amyloid spores and its lack of cystidia.

156. Mycena laevigata (Lasch) Quélet

Champ. Jura et Vosges, Suppl. 4: 326. 1876

Agaricus laevigatus Lasch, Linnaea, 3: 388. 1828.
Mycena vulgaris var. *caespitosa* Kauffman and Smith, Pap. Mich. Acad. Sci., Arts, and Letters, 17: 186. 1933.
Mycena pseudo-vulgaris Kühner, Encyc. Myc., 10: 393. 1938.

Illustrations:
Plate 73; Text fig. 40, nos. 4–5 (p. 320).
Bresadola, Icon. Mycol., 5, pl. 239 (very good).

Pileus 1–2 (4.5) cm. broad, conic to convex or with a low subconic umbo, remaining broadly conic or convex, often with a small papilla, rarely slightly depressed, margin connivent to the stipe when young, opaque, becoming closely striate to near the disc at maturity, surface lubricous or subviscid in age or when wet, at times distinctly viscid, glabrous, when young pale fuscous gray to watery gray on the disc and whitish toward the margin, at times the disc pale bluish gray, soon fading and whitish over all, with a tendency to become cream-colored or stained tawny in age; flesh thin, flaccid, and cartilaginous, white, odor and taste not distinctive; lamellae broadly adnate-subdecurrent, close, 22–25 reach the stipe, broad (3–4 mm.), white, or occasionally flushed pale incarnate, edges even; stipe 2–5 (10) cm. long, 1–2 mm. thick, equal, cartilaginous and brittle, tubular, glabrous except for the white-strigose, somewhat rooting base, when young bluish gray toward the apex and whitish below, soon fading to watery grayish white over all, lubricous to somewhat viscid.

Spores broadly ellipsoid, 6–8 × 3–4 μ, amyloid (reaction very strong); basidia four-spored, 23–25 × 5–6 μ; cheilocystidia abundant, short and clavate or becoming subfusoid to cylindric, sometimes greatly elongated into a narrow lanceolate apex, tips occasionally forked and occasionally one or two short protuberances may develop either on the inflated portion or on the neck, 30–40 × 9–12 μ (clavate type) or 35–55 × 5–8 μ (elongated type), both seen on the gills of a single pileus; pleurocystidia absent, gill trama faintly vinaceous brown in iodine; pileus trama with a thin gelatinous surface pellicle, a well-differentiated hypoderm made up of hyphae having walls that sometimes gelatinize slightly, the remaining tissue filamentous, vinaceous brown in iodine.

Habit, habitat, and distribution.—Cespitose to subcespitose on wood of conifers; Tennessee, New York, Michigan, Colorado, Washington, Oregon, and California in the United States and Ontario in Canada. Overholts found it in Colorado in June at an elevation of 11,000 feet; it was coming up through the snow around old stumps.

Material studied.—Smith, 3919, 4733, 4896, 7024, 8157, 8818, 17621, 18120, 18164, 18187, 18188, 18189, 18190, 18198. Hesler, 11905, 14112, 14775. Kauffman, two collections (as *M. collariata*), New York; Takilma, Oregon. Overholts, 1823, 1838.

Observations.—The color changes of this species are curious and quite distinctive. Fuscous-gray to bluish-gray shades are present over the disc of the pileus and over the upper part of the stipe in young carpophores. These fade rather rapidly, and the fungus soon becomes whitish. In age yellowish tints or discolorations develop. When dried the pilei are dull whitish over all or some show sordid discolorations. If very young specimens are properly dried, some of the gray may be retained. The viscidity of the cap is a character of doubtful value in this species. The pellicle may be thin and hardly gelatinous or relatively thick and rather gelatinous. Both these conditions are well shown in my specimens from Oregon and California. I have never found the stipe to be truly glutinous, but a thin peripheral zone of subgelatinous hyphae readily explains the lubricous to viscid feel. The cheilocystidia are apparently rather variable also. In old specimens they are usually much elongated. Considering the variation observed on a single pileus, I am not inclined to attach much weight to slight differences between collections.

A comparison of my collections of 1935, 1936, 1937, and 1941 with the one made at Rock River, Michigan, and described as *M. vulgaris* var. *cespitosa* by Kauffman and myself, shows that all are identical

Kauffman studied the Rock River collection when it was fresh; in his notes he described both the cap and the stipe as glutinous and stated that the fungus was a segregate of *M. vulgaris*. When revived in KOH, the fruiting bodies of *M. laevigata* from Oregon have more conspicuously gelatinous layers on their pilei and stipes than do those on the Michigan specimens. They were not more than subviscid to the touch when fresh.

Kühner, recognizing that the Michigan specimens were very distinct from *M. vulgaris*, gave them a new name, *M. pseudovulgaris*. However, he had not seen any fresh material. Therefore, since I had seen both and did not place them together when I published on *M. laevigata* (Smith, 1937), it was logical for him to assume that the two were different. From my observations on various collections in North America I would say that the color differences between American and European specimens do not appear to be of taxonomic importance.

My collections of *M. laevigata* from Ontario strongly resembled an Omphalia in the manner in which the gills were attached. Additional collections from the West Coast in 1937 clearly indicated, however, that such a resemblance is not a constant character. *M. semivestipes* frequently shows the same type of variation. In addition, I have compared my specimens with material from the Union of Soviet Socialist Republics sent to me by Dr. Rolf Singer. The cystidia, spores, and pellicle of the pileus and the outer layer of the stipe tissue are identical with those of my collection 8157 from Oregon. The dried specimens are also strikingly similar.

157. **Mycena semivestipes** (Pk.), comb. nov.

Omphalia semivestipes Peck, Bull. Torr. Bot. Club, 22: 200. 1895.
Prunulus semivestipes Murrill, North Am. Flora, 9: 334. 1916.
Omphalia curvipes Peck, Bull. Torr. Bot. Club, 34: 345. 1907.
Prunulus curvipes Murrill, North Am. Flora, 9: 329. 1916.
Mycena subalcalina Atkinson, in Douglas, Am. Journ. Bot., 5: 37. 1918.

Illustrations:

Plate 74; Text fig. 40, nos. 6–7 (p. 320).
Hard, The Mushroom, Edible and Otherwise, fig. 91 (as *M. alcalina*).
Smith, Mycologia, 29: 351, fig. 2 b (as *M. tintinnabulum*).

Pileus 1–3.5 cm. broad, very obtuse to convex when young, the margin usually incurved, becoming expanded and slightly umbonate, convex or plane at maturity, occasionally the disc slightly depressed, surface smooth and glabrous, subviscid to lubricous when wet, pellicle hardly separable, opaque at first but becoming striate nearly to the apex, colors black or blackish at first, subhygrophanous, fading slowly to various shades of sordid grayish brown, sometimes becoming sordid whitish or grayish in age or the margin tinged ochraceous, in wet weather often staining sordid reddish brown or flushed sordid brownish vinaceous over large patches; flesh thin but cartilaginous and rather tough, concolorous with the pileus or paler, often stained reddish brown, odor very strongly nitrous at first but soon disappearing and at times apparently absent even in button stages, taste farinaceous to bitter; lamellae narrow to moderately broad (2–3 mm.), close, subdistant on broadly expanded pilei, 20–28 reach the stipe, three tiers of lamellulae, horizontally adnate or with a decurrent tooth, white to pallid or at times pallid fuscous in age, often flushed with sordid pink in age and frequently with reddish-brown spots, edges even and whitish to grayish; stipe 2–6 cm. long, 1–3 mm. thick, tubular, terete or sometimes compressed above, equal or nearly so, tough and cartilaginous, pruinose but soon polished, base white-strigose or strigose over the lower one third, concolorous with the pileus or paler, becoming reddish or blackish brown at the base in age, sometimes more or less sordid yellowish toward the apex.

Spores 4–5 × 2.5–3 μ (5–6 × 3 or 3–4 × 2 μ), ellipsoid, amyloid; basidia four-spored; pleurocystidia rare to scattered, similar to the cheilocystidia; cheilocystidia embedded in the gill edge and often inconspicuous, 23–32 × 7–11 μ, saccate to fusoid-ventricose, occasionally with knoblike processes at the apex or smooth and often with wavy to rather irregular outlines, sometimes difficult to distinguish from immature basidia; gill trama with a subgelatinous subhymenium, the central portion vinaceous red in iodine; pileus trama with a thin gelatinizing pellicle, a slightly differentiated hypoderm and the remainder floccose and of filamentous hyphae, all but the pellicle becoming reddish violet in iodine.

Habit, habitat, and distribution.—Densely cespitose to densely gregarious on old hardwood logs and stumps; rather common during late fall and early winter. The type was collected in Newfoundland. I have examined material from Tennessee, New York, Ohio, Michigan, and Missouri in the United States and from Ontario in Canada.

Material studied.—Smith, 32–650, 33–918, 33–1135, 908, 1255, 1264, 6059, 10561, 11108, 15053, 15511, November 21, 1931, Michigan. Atkinson, 7842, 13776, 19436 (from M. E. Hard), 23505. Hesler,

2653, 5370, 7784, 8665, 10155, 10156, 10987. Kauffman, two collections (as *O. curvipes*). Overholts, 1452, 1526, 2315.

Observations.—The microscopic characters of *M. subalcalina*, *Omphalia curvipes*, and *O. semivestipes* are identical. All are characterized by a thin gelatinous pellicle, very small spores (4–6 × 2–3 μ), and cystidia which are smooth or somewhat contorted and branched but which do not develop numerous fingerlike processes. Macroscopically they all have a rather tough cartilaginous consistency, broadly adnate to subdecurrent, usually rather close gills, and the same habit. Because of the more or less incurved margin of the pileus one is likely at first to search for the species in either Collybia or Omphalia. The structure of the pileus and the obvious relationship to *M. tintinnabulum* place it in Mycena.

This fungus was previously identified as *M. tintinnabulum* (Smith, 1937). Kühner questioned this determination because of a difference in the cystidia of the European and American material. In European specimens the cystidia bear numerous short fingerlike projections, which may become branched and quite contorted. In American material the cystidia are essentially of the fusoid-ventricose type but do fork or branch occasionally or become irregular in outline. This difference between material from the two continents is no greater than that which has been observed between different collections of *M. citrinomarginata*, and so I was at first inclined to disregard it. It must be admitted, however, that in this case the difference is constant. Material from Sweden and France as well as numerous collections from North America have been examined.

158. Mycena ochraceicinerea Murrill

Mycologia, 8: 221. 1916

Prunulus ochraceicinereus Murrill, North Am. Flora, 9: 333. 1916.

"Pileus thin, broadly convex, not umbonate, gregarious, 2 cm. broad; surface smooth, glabrous, shining, pale-ochraceous-cinereous, margin whitish, slightly striate: lamellae adnate with a decurrent tooth, very few inserted, broad near the margin, subdistant, pallid: spores oblong-ellipsoid, smooth, hyaline, about 7 × 4–5 μ: stipe cylindric, equal, smooth, glabrous, concolorous, nearly white at the apex, pruinose at the base, 3–4 cm. long, 2 mm. thick.

"Type collected on decayed wood at Edgewater, New Jersey,

September 20, 1902, *F. S. Earle & W. A. Murrill 1419* (herb. N. Y. Bot. Gard.).

"Distribution: Known only from the type locality."

Observations.—The spores of the type measure 7–9 × 6–7 μ, are broadly ellipsoid, smooth, hyaline, and amyloid. The basidia are four-spored. Cystidia are not differentiated. The cells on the gill edge are basidium-like, but in one or two instances the tip, which was prolonged into a short, contorted protuberance, indicates that the species belongs near *M. tintinnabulum*, from which the spores readily distinguish it. Nothing distinctive was observed in the structure of the gill and pileus trama. Neither revived very well, and both remained yellowish in iodine. I have never seen fresh material. Hesler's 14200, collected near Mt. Le Conte, Tennessee, May 24, 1942, belongs here and extends the known range.

159. **Mycena pseudoinclinata,** sp. nov.

Illustrations: Text figs. 41, no. 1 (p. 328); 42, no. 4 (p. 330).

Pileus 15–25 mm. latus, conico-campanulatus, demum late campanulatus, saepe umbonatus, griseus vel griseo-brunneus, sericeus, subhygrophanus, striatus demum sulcatus; sapor et odor subfarinaceus; lamellae griseae vel subincarnatae, subdistantes, latae; stipes 2.5–6 cm. longus, 1.5–3 cm. crassus, sursum pallidus, deorsum subbrunneus, fibrillosus, cavus; sporae 8–11 × 5–6 μ; pleurocystidia et cheilocystidia 23–30 × 5–7 μ. Specimina typica in Herb. Univ. Mich. et Herb. Univ. Tenn. conservata. Legit L. R. Hesler, n. 12940, prope Cades Cove (Great Smoky Mountains National Park), Tenn., Oct. 19, 1940.

Pileus 1–3 (5) cm. broad, obtusely conic to campanulate, the margin either straight or curved in against the gills, becoming broadly conic to broadly campanulate, at times nearly plane, often splitting radially part way to the disc or disc variously cracked, surface moist and glabrous, striate, becoming sulcate, subhygrophanous, fuscous to dark watery gray at first, sometimes appearing silky, the margin pallid or whitish, gradually becoming whitish over all; flesh thick on the disc, thin elsewhere, grayish to whitish, fragile, odor and taste subfarinaceous; lamellae ascending-adnate, usually with a slight tooth but readily seceding, close to subdistant, white or pale grayish, sometimes evenly flushed with an incarnate tinge, edges even; stipe 2.5–6

FIG. 41. *M. pseudoinclinata:* 1, cheilocystidia. *M. inclinata:* 2 and 5, cheilocystidia; 4, spores. *M. maculata:* 3, spores; 6, cheilocystidia. *M. occidentalis:* 7, spores; 8, cheilocystidia. *M. hemisphaerica:* 9, cheilocystidia; 10, spores. *M. galericulata:* 11, spores

cm. long, 1.5–3 mm. wide, pallid above, pale brown or tawny below in age, hollow.

Spores 8–11 × 5–6 μ, broadly ellipsoid, smooth, strongly amyloid; basidia four-spored; pleurocystidia and cheilocystidia present, 23–30 × 5–7 μ, embedded, clavate and smooth, or ventricose and subcapitate, at times with fingerlike projections or irregular humps at the narrowed apex.

Habit, habitat, and distribution.—Densely cespitose; known from Tennessee, Pennsylvania, and Michigan.

Material studied.—Smith, 18801. Hesler, 10996, 12940, 14780. Kauffman, 1929, Michigan; two collections, Sept. 12, 1924, Mt. Gretna, Pennsylvania.

Observations.—This species appears to be identical with *M. inclinata* in all characters except its cystidia and the lack or smaller number of fibrillose flecks on the stipe. Both pleurocystidia and cheilocystidia are present. They are alike in size and shape, but are rather difficult to locate because they do not project beyond the basidia. They are clavate and smooth or ventricose and subcapitate. A few were found to have obtuse fingerlike projections or irregular humps scattered over the upper portion. Hesler reported that the fruiting bodies were densely cespitose—about 150 in a cluster. The difference between the cystidia of this species and those of typical specimens of *M. inclinata* is similar to that between the cystidia of *M. tintinnabulum* and *M. semivestipes*. Since the latter are kept separate, the collections listed above have been described as a new species.

160. Mycena alnicola Smith

Contrib. Univ. Mich. Herb., 5: 23. 1941

Illustrations:

Text fig. 42, nos. 1–2 (p. 330).

Smith, Contrib. Univ. Mich. Herb., 5, pl. 5, fig. 2, pl. 25, figs. 1–3.

Pileus 1–2.5 cm. broad, ellipsoid at first, soon becoming obtusely conic, finally broadly campanulate or convex, the margin appressed at first, surface glaucous from a dense bloom giving the cap a pale-bluish sheen, soon polished and moist, color "benzo brown" to "fuscous" on the disc, the remainder "light drab" or the extreme margin white, becoming "drab gray" or paler in age, hygrophanous but scarcely changing color when moisture escapes, translucent-striate

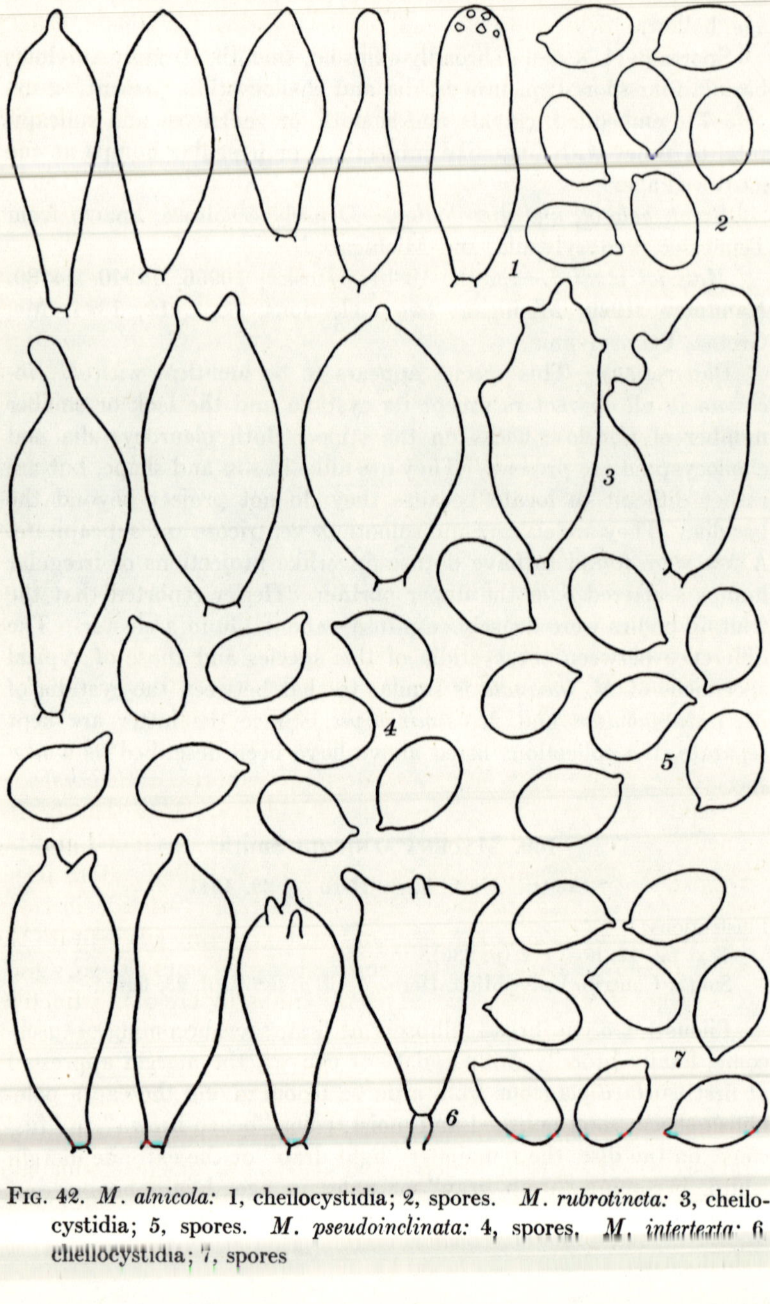

FIG. 42. *M. alnicola:* 1, cheilocystidia; 2, spores. *M. rubrotincta:* 3, cheilocystidia; 5, spores. *M. pseudoinclinata:* 4, spores. *M. intertexta:* 6 cheilocystidia; 7, spores

when mature and moist, becoming sulcate-striate; flesh thin and fragile, gray, odor and taste mild; lamellae ascending, adnate, usually not toothed, subdistant, 20–30 reach the stipe, two tiers of lamellulae, narrow to moderately broad (2 mm. ±), equal, color "pale gull gray," intervenose, edges even; stipe 4–6 cm. long, 1.5–2 mm. thick, fragile, equal, hollow, base somewhat strigose, the upper part at first covered with a dense white bloom, soon polished and moist, "dark Quaker drab" to "dark mouse gray" (with a distinctly bluish cast), at first concolorous with pileus at maturity.

Spores (6) 7–9 (10) × 4–5 μ, amyloid, ellipsoid, smooth; basidia four-spored; cheilocystidia 26–40 × 9–15 μ, clavate to broadly fusoid, becoming elongated, subcylindric, fusoid-ventricose or sometimes with one or two fingerlike protuberances, the apices smooth or with a slight resinous incrustation when revived in KOH; pleurocystidia similar to cheilocystidia, rare to scattered; gill trama of inflated cells, homogeneous, pileus trama with a thin pellicle, a strongly differentiated hypoderm, and a narrow region of filamentous tissue, all but the pellicle vinaceous brown in iodine; stipe and gill trama also pale vinaceous brown in iodine.

Habit, habitat, and distribution.—Cespitose to scattered on Alnus logs; Olympic Mountains, Washington. The type (14559) was collected near the Elwha River Ranger Station, June 23, 1939. Additional material was collected along the Clearwater River.

Material studied.—Smith, 13237, 13691, 14203, 14442, 14562, 16581, 16582, 17235. C. A. Brown, November 4, 1925, Washington.

Observations.—This species is intermediate between *M. parabolica* sensu Kauffman and *M. griseiconica* Kauffman. It is sharply distinct from both because of its smaller spores. It was found growing along with *M. griseiconica* in the Elwha River Valley. Both pileus and stipe of young specimens show considerable variation in color, but nearly always a strong shade of bluish gray prevails beneath the hoary bloom. In age the entire fruiting body becomes glaucous gray. The colors of the fragile gray Mycenae ordinarily are not distinctive, but in this species the bluish-gray cast is more pronounced than usual.

161. Mycena alnicola var. odora, var. nov.

Pileus 1–3 cm. latus, conicus demum campanulatus vel subplanus, canescens, fuscus dein cinereus, striatus; caro cinerea; odor subraphanoideus; lamellae subdistantes, angustae; stipes 4–6 cm. longus, 1.5–2.5 mm. crassus, glaber, cinereus; sporae 7–9 (10) × 4.5–5.5 μ; cheilocystidia 36–48 (53) × 9–14 μ, mucronata.

Pileus 1–3 cm. broad, ellipsoid to obtusely conic, becoming campanulate or in age nearly plane, surface hoary and dark fuscous at first, becoming paler in age and finally cinereous, faintly translucent-striate when fading; flesh thin, fragile, grayish, odor and taste subraphanoid; lamellae subdistant, narrow, ascending, pallid; stipe 4–6 cm. long, 1.5–2.5 mm. thick, equal, glabrous, fragile, base white-strigose, cinereous over all or apex pallid.

Spores 7–9 (10) $\times$ 4.5–5.5 μ, smooth, amyloid; basidia four-spored; cheilocystidia abundant, 36–48 (53) $\times$ 9–14 μ, fusoid-ventricose, mucronate or more or less clavate and with several to many short protuberances (all types observed on a single gill edge); pleurocystidia similar to cheilocystidia but rare, pileus and gill trama strongly amyloid.

Habit, habitat, and distribution.—On a dead alder tree, August 19, 1941; Anderson Creek, Mt. Baker National Forest, Washington, A. H. Smith, 16284. Known only from the type locality.

Observations.—In addition to the distinctive odor, the variety is characterized by the more variable cheilocystidia. The degree of variability appears about equal to that found in *M. citrinomarginata*.

162. Mycena parabolica (Fr.) Quélet

Champ. Jura et Vosges, p. 242. 1872. Sensu Kauffman,
Agar. Mich., p. 800. 1918

Illustrations:

Plate 75; Text fig. 40, nos. 8–9 (p. 320).
Fries, Icon. Sel. Hymen., 1, pl. 80, fig. 3.
Smith, Am. Journ. Bot., 22, pl. 3, fig. 1 (as *M. rugosa*).

Pileus 2–4 (6) cm. broad, obtusely campanulate to broadly conic when young, becoming broadly conic or campanulate in age, usually umbonate, margin connivent with the stipe at first, spreading or flaring slightly in age, surface densely pruinose at first, soon naked, deeply and regularly sulcate-striate to the disc, somewhat translucent-striate when moist, hygrophanous, "sooty black" at first, soon becoming watery gray, fading to pale ashy gray over all, margin sometimes paler in both fresh and faded fruiting bodies; flesh thin and fragile, pale gray, odor and taste not distinctive; lamellae ascending-adnate or with a tooth, subdistant, narrow to moderately broad, equal, dark gray with a whitish border at first, in age pale gray over all or whitish, intervenose; stipe 4–10 cm. long, 2–5 mm. thick, terete or compressed, equal, hollow, fragile to cartilaginous, pruinose above when young and the lower portion covered with small white fibrillose flecks, soon naked and glabrous, polished and translucent in age, base white-strigose and somewhat rooting, concolorous with the pileus over all or paler gray.

Spores (8) 9–11 × 5–7 μ, broadly ellipsoid, smooth, amyloid; basidia four-spored, 30–35 (40) × 6–8 μ; pleurocystidia scattered to rare and soon collapsing, similar to the cheilocystidia but usually remaining unbranched; cheilocystidia clavate to mucronate, in age developing obtuse fingerlike projections over the apex (1–6), sometimes merely remaining saccate, 35–52 × 9–18 μ, hyaline; gill trama dark vinaceous brown in iodine, made up of short and moderately broad regularly arranged cells; pileus trama with a thin pellicle, a well-differentiated hypoderm, and a thick tramal body of filamentous hyphae, all but the pellicle vinaceous brown in iodine.

Habit, habitat, and distribution.—Gregarious to subcespitose on wood of either hardwood or conifer trees; New York, Pennsylvania, Michigan, Colorado, Wyoming, Idaho, Washington, and Oregon in the United States and Ontario in Canada. It fruits from July to November, but is rather sporadic. During some seasons it is common.

Material studied.—Smith, 32–326, 32–544, 33–1124, 33–1125, 1174, 4671, 5065, 5188, 5260, 6164, 8101, 8216, 9283, 11109, 13590, 15502, 15504, 15532, 18144, 18217, October 26, 1937, Oregon; November 18, 1931, Michigan; June 15, 1933; October 8, 1940. Flett, Washington. Gruber, Idaho. Kauffman, four collections, Colorado; three collections, Michigan; one collection, Wyoming. Mains 5182, 5259. Overholts, 1361, 17596. Stuntz, F 465.

Observations.—The pleurocystidia are often difficult to find in mature or old specimens but can usually be demonstrated in young pilei. They collapse very readily and are hard to revive in KOH. As seen in young pilei, they are usually devoid of the fingerlike prolongations characteristic of the cheilocystidia, but there is a tendency to produce them. If one happens to get a pileus in just the right stage of development, they can readily be observed.

The colors of the entire fungus are blackish gray at first and become watery gray before fading to pale ashy gray. The generally even coloration, along with the fragile consistency, is characteristic among the lignicolous cespitose Mycenae and distinguishes this species from all except *M. alnicola*, from which it is separated by the larger spores.

This is the species reported under the name "*M. rugosa*" (Smith, 1935). At the present time it seems that both *M. rugosa* and *M. parabolica* are not clearly understood in Europe. It might, indeed, be best to drop both names, since the accounts of various workers disagree. Kühner (1938) does not recognize either one. My earlier identification of *M. rugosa* was based largely on Rea (1922) and on Cooke's figure, which he cites. The tough consistency emphasized by Fries was disregarded because the specimens in hand appeared to be of about the same consistency as *M. inclinata*—and *M. inclinata* and *M. galericulata* have frequently been confused in both Europe and America. After continued study of *M. parabolica* it became apparent that its consistency usually approaches that of *M. stannea* more closely than that of *M. galericulata*. The study of additional collections also sheds more light on the characteristics of the cheilocystidia. In the form first identified as *M. rugosa* the cystidia ordinarily were provided with fingerlike prolongations, whereas in Kauffman's specimens my first examinations indicated that in his *M. parabolica* the tendency was more toward the smooth type. Continued study of fresh material, as it became available, showed this difference to be more apparent than real, and I now regard it as a variation within the species and am placing all my collections under *M. parabolica* sensu Kauffman. In my estimation this concept is very closely in line with that of Fries. At the time Kauffman published his account he did not have a clear idea of what constituted a cystidium and his remark stating that the cystidia were sometimes found on the edge of the gills but that sterile cells were lacking must be disregarded. Cheilocystidia as given in the description are present in his specimens.

163. **Mycena rubrotincta,** nom. nov.

Mycena tenuicula Murrill, Mycologia, 8: 221. 1916 (not *M. tenuicula* Karsten in Saccardo, 14: 84. 1899).

Prunulus tenuicula Murrill, North Am. Flora, 9: 334. 1916.

Illustrations:

Plate 76; Text fig. 42, nos. 3, 5 (p. 330).

Smith, Am. Journ. Bot., 22, pl. 1, fig. 3.

Pileus 5–20 (30) mm. broad, very obtusely conic, convex, or with the apex flattened and the sides nearly parallel, becoming broadly convex to plane in age, at times with a broad umbo, margin appressed against the stipe at first, spreading in age, surface pruinose but soon naked, moist, opaque, or faintly translucent-striate, becoming sulcate in age, hygrophanous, color variable, "dark vinaceous drab," "benzo brown," "hair brown," pale "vinaceous brown" on the disc, margin pale gray to whitish, fading to pale gray or whitish but vinaceous tints usually persisting on the disc and seldom present near the margin, becoming dingy rufescent-spotted in age or where bruised, no odor, taste slightly farinaceous; lamellae broadly adnate to arcuate, subdistant to distant, 18–21 reach the stipe, two tiers of lamellulae, moderately broad (2.5–3.5 mm.), whitish, becoming cinereous, and finally reddish-spotted in age, intervenose at times, edges even and pallid; stipe 3–6 cm. long, 1–3 mm. thick, equal or slightly enlarged above, the bases sometimes connate, apex pruinose, midportion with scattered delicate fibrils at first, base sparsely or densely white-strigose, glabrous and naked above in age, dark gray or blackish over all at first, frequently developing dull-vinaceous tints, pale gray in age except for the base, which becomes sordid purplish or reddish brown.

Spores broadly ellipsoid, (7) 8–10 × 5–6 μ, amyloid (reaction very strong); basidia four-spored, 22–24 × 7–8 μ; cheilocystidia and pleurocystidia similar, abundant on the edges, rare on the sides, soon collapsing, 40–60 × 10–18 μ, clavate to obovoid, smooth or with one to several short obtuse often contorted projections, frequently mucronate; gill trama very faintly vinaceous brown in iodine or remaining yellowish; pileus trama with a moderately thick pellicle, the region beneath it of somewhat enlarged cells, the remainder filamentous, pale vinaceous brown in iodine.

Habit, habitat, and distribution.—Cespitose to densely gregarious on stumps and debris of hardwoods, sporadic, common during some seasons in the late summer and early fall if the rainfall is heavy; North Carolina, Tennessee, Connecticut, New York, and Michigan in the United States and Ontario in Canada.

Material studied.—Smith, 32–301, 32–507, 32–566, 32–621, 32–630, 32–669, 33–1051, 33–1052, 935, 979, 6402, 10928, 11085, 15203, 15246, 15260, 15355. Atkinson, 23759. Hesler, 6510, 10699, 11442. Kauffman, New York. Kelly, 1382.

Observations.—The colors vary from sordid blackish brown to grayish brown, but nearly always have a reddish tinge over the disc. The development of the white hairs over the basal part of the stipe is directly dependent on the position from which the fruiting bodies arise. If they come from the upper surface of a log or stick, the hairs remain inconspicuous. If they come from the side or underside of a log or develop beneath the bark and come out through a crack, the white hairs are well developed. In most of the specimens I have seen the reddish tinge has been a conspicuous character on some part of the carpophore, but it should not be considered an infallible character. During the season of 1940 the species occurred in quantity on old canes of a species of Rubus and on the smaller branches at the top of a fallen oak tree. Apparently because of the shade and the very moist condition of the habitat the stipes were so elongated (8–10 cm.) that the fruiting bodies had the stature of *M. iodiolens*. In nearly every carpophore the base of the stipe was characterized by a dense cottony mat of mycelium at the point of attachment. The stains on the pilei and lamellae of these specimens were not reddish but, instead, a very dirty brown. Typical material was found on logs near by, and an intergrading series was established which clearly proved that these characters of the carpophores were the result of unusual conditions.

A collection which appears to be an even more aberrant form of *M. rubrotincta* was found on an oak stump during September, 1940 (Smith, 15355). When the cap was moist, the disc had a tinge of pale gray; when faded, it was white with no reddish tinge. The gills were close and rather broad. The cheilocystidia were more variable and narrower than usual for the species, measuring 9–11 μ wide and 28–42 μ long. The spores measure $7\text{–}9 \times 5\text{–}6\ \mu$. The stipes were white and distinctly fibrillose-scurfy at first; in age they were striate, much as in *M. niveipes*. Pleurocystidia were not present, and the cap was covered with a thin somewhat gelatinous (in KOH) pellicle. I have referred the collection to *M. rubrotincta* in spite of the lack of reddish tints, the pale color, and the slightly different cystidia. It may represent a distinct species related to *M. incarnatifolia* but more lubricous to subviscid (like *M. laevigata*); further information is needed. The poorly known *M. sudora* should be a fungus of this general type, with a somewhat viscid pileus.

Mycena rubrotincta appears to be closest to if not identical with *M. intertexta*. The habitat and the tendency of *M. rubrotincta* to develop a strong reddish cast apparently separate it from the following species. When one considers that other cespitose Mycenae can adapt themselves to either a conifer or a hardwood substratum and that the tendency to stain reddish may not be a taxonomically valuable character, it is evident that the two are distinguished by rather poor characters at best. I have not seen fresh specimens of *M. intertexta.* Kauffman made one collection, on spruce wood (North Elba, New York), which he identified as a variety of *M. haematopus.* It may be *M. intertexta.*

164. Mycena intertexta (Berk. and Curt.) Saccardo

Syll. Fung., 14: 83. 1899

Agaricus (*Mycena*) *intertextus* Berkeley and Curtis, Ann. and Mag. Nat. Hist., 3 (4): 286. 1859.
Prunulus intertextus Murrill, North Am. Flora, 9: 328. 1916.
Mycena avellanea Murrill, Mycologia, 8: 220. 1916.
Prunulus avellaneus Murrill, North Am. Flora, 9: 329. 1916.

Illustrations:
Text fig. 42, nos. 6–7 (p. 330).
Beardslee and Coker, Journ. Elisha Mitchell Sci. Soc., 40, pl. 27.

"Densissime caespitosus; pileo hemispherico, umbonato, demum expanso striato; stipitibus villo connatis, subfragilibus; lamellis distantibus, crassiusculis, adnatis. Curt. nos. 1741, 2557, 2558. On pine, South Carolina, M. A. Curtis.

"Pileus 3–8 lines across, carnoso-membranaceous, convex, umbonate, then expanded, pellucid, fusco-cinereous, margin paler, striate; stems 2 inches high, ½ a line thick, fistulose, smooth above and fuscous, pale below and joined intimately together by matted down; gills white, thick, adnate.

"A very beautiful and singular species."

This is the original description. The following account is quoted from Coker, who has studied the species critically:

"Densely cespitose on rotting pine; cap 0.5–2.1 cm. broad, hemispheric or subexpanded, often somewhat umbonate, glabrous, not viscid, pallid tan or deeper cinereous brown, the center often darker; margin striate. Flesh thin, not very brittle; odor and taste slightly musty.

"Gills subdistant, broadly adnate and mostly arcuate-decurrent by a tooth, pallid tan, in age yellowish brown.

"Stem 2–8 cm. long, 1–1.5 mm. thick, glabrous, hollow, color of the cap and gills, the bases joined by dense, tawny floccose fibers which are very conspicuous where the stems run under the bark.

"Spores (of No. 2955) smooth, elliptic, 4–5.5 × 6.8–8.6 μ. Basidia (of No. 3601) 4-spored, hymenium 22–30 μ thick; cystidia abundant on the gill margins and not rare on the sides, hyaline, pyriform, 11–15 × 22–30 μ, often with 1 or 2 spines on the end. Threads of the gill-trama 5–15 μ thick, very irregular.

"Not rare on rotting pine logs and stumps and appearing through worm-holes or from between the cracked bark, also on trash containing twigs and leaves of pine."

Coker examined cotype material. An examination of the type of *M. avellanea* Murrill, which was collected on a balsam-fir log near Lake Placid, New York, revealed that it had spores measuring 7–9 (10) × 5–6 (7) μ. Pleurocystidia were present and identical in shape with those illustrated by Coker. They measure 10–15 × (26) 34–43 μ. The cheilocystidia were similar or had more projections from the apex. Since there are no characters by which to distinguish these two, *M. avellanea* is placed in synonymy with *M. intertexta*.

165. Mycena inclinata (Fr.) Quélet

Champ. Jura et Vosges, p. 105. 1872

Agaricus inclinatus Fries, Epicr. Syst. Myc., p. 107. 1838.
Mycena galericulata var. *calopoda* Saccardo, Syll. Fung., 5: 268. 1887.
Mycena calopus Ricken, Die Blätterpilze, p. 437. 1915.
Mycena Rigelliae Velenovský, České Houby, p. 318. 1920–22.

Illustrations:

Plates 77–79; Text fig. 41, nos. 2, 4–5 (p. 328).

Beardslee and Coker, Journ. Elisha Mitchell Sci. Soc., 40, pls. 25, 26 (lower figs.), 28 (as *M. galericulata*).

Bresadola, Icon. Mycol., 5, pls. 235 (as *M. calopoda*), 236.

Charles, Vera K., Some Common Mushrooms, fig. 25 (as *M. galericulata*).

Clements, Minn. Plant Studies, IV. Minn. Mushrooms, fig. 17 (as *M. galericulata*).

Cole, Emma L. Taylor, Guide to the Mushrooms, plate opposite p. 77 (as *M. galericulata*).

Fries, Icon. Sel. Hymen., 1, pl. 80 (as *Agaricus galericulatus* var. *calopus*).

Güssow and Odell, Mushrooms and Toadstools, pl. 55, fig. 2 (as *M. galericulata*).

Hard, The Mushroom, Edible and Otherwise, fig. 89 (as *M. galericulata*), fig. 92 (as *M. stannea*).

Konrad et Maublanc, Icon. Sel. Fung., 3, pl. 231.

Lange, Flora Agar. Dan., 2, pl. 55 E and E′.

McIlvaine and Macadam, One Thousand American Fungi, pl. 10, fig. 5 (as *M. galericulata*).

Marshall, The Mushroom Book, plate opposite p. 55 (as *M. galericulata*).
Moffatt, Bull. Chicago Acad. Sci., 7, pl. 4 (as *M. galericulata*).
Murrill, Mycologia, 10, pl. 8, fig. 5 (as *M. galericulata*).
Ricken, Die Blätterpilze, 2, pl. 111, fig. 3 (as *M. calopus*).
Thomas, Field Book of Common Gilled Mushrooms, pl. 13, no. 90 (as *M. galericulata*).

Pileus 1–3 (5) cm. broad, obtusely conic to campanulate, the margin either straight or curved in against the gills, expanding to broadly conic or broadly campanulate, sometimes nearly plane, rarely gibbous, frequently splitting radially part way to the disc, or the disc variously cracked, the margin usually becoming crenate, surface moist and glabrous, hygrophanous, fuscous to dark watery gray at first, the margin pallid or whitish, gradually changing to whitish over all (the disc usually a bit grayish), translucent-striate, hygrophanous, and fading to a dead whitish gray or pale cinereous; flesh thick on the disc, thin elsewhere, grayish to whitish, fragile, odor and taste slightly farinaceous; lamellae ascending-adnate, usually with a slight tooth but readily seceding, close to subdistant, 26–35 reach the stipe, about three tiers of lamellulae, broad (3–6 mm.), white or pale grayish, sometimes evenly flushed with an incarnate tinge, edges even; stipe 5–10 cm. long, 1.5–3 (4) mm. thick, equal, fragile, tubular or hollow, base densely strigose, lower portion at first covered with an appressed white fibrillose coating, which breaks up into fibrillose flecks and eventually disappears, sometimes remaining appressed and causing the stipe to appear fibrillose-striate, apex pruinose but soon polished, pallid above, pale to dark gray toward the base at first, yellowish to tawny brown below in age, basal fibrils also becoming tawny.

Spores 7–9 (10) $\times$ 5–6.5 (7) μ, broadly ellipsoid, smooth, strongly amyloid; basidia four-spored; pleurocystidia not differentiated; cheilocystidia embedded in the gill edge and very inconspicuous, clavate to subcapitate, 24–36 $\times$ 5–10 μ, apices covered with contorted slender or obese projections; gill trama homogeneous, pale yellowish to sordid brown in iodine; pileus trama with a distinct pellicle, a well-differentiated hypoderm, and a filamentous tramal body, pale yellowish to sordid brownish in iodine.

Habit, habitat, and distribution.—Densely gregarious to cespitose on decaying hardwood logs and stumps during the spring and fall. In eastern North America it is abundant in the area bounded by Nova Scotia, Ontario, Manitoba, Missouri, North Carolina, and New York. In Oregon it has been found on oak, but it appears to be generally rare along the Pacific coast.

Material studied.—Smith, 32–34, 32–538, 32–541, 32–607, 32–609, 32–628, 33–1084, 1361, 4962, 5030, 6220, 6279, 6367, 7699, 9528, 15064, 15101, 15127, 15567, October 8, 10, 1931, June 19, Michigan, 1935. Atkinson, collections of *M. galericulata*, abundant material from Ithaca and vicinity. Hesler, 3643, 14088. Kauffman, Ithaca, New York. Routien, 960.

Observations.—This species can best be recognized by its crenate pileus and the fibrillose flecks usually adhering to the lower portion of the stipe. The tendency of the base to become tawny in age is an additional fairly reliable character, as is also the habitat on hardwood. *M. intertexta*, a closely related species found on conifer wood, also becomes tawny over the basal portion of the stipe. The cheilocystidia of *M. inclinata*, which are very small and difficult to demonstrate, are very irregular in shape. As a rule, they are some variation of the clavate-roughened type, but one can frequently find individuals which are not much more than a contorted filament and no longer than a basidium. The apices of the broader individuals are usually furnished with a variable number of contorted fingerlike processes or irregular humps. The spores are intermediate in size between those of *M. galericulata* and *M. hemisphaerica*. These two species can nearly always be distinguished by their glabrous polished stems.

During cold wet weather in late fall a form, very abundant locally, occurs in which the cap, gills, and stipe become stained sordid reddish to purplish brown. Sometimes the entire fruiting body is dark purplish brown. The habit, habitat, spores, cheilocystidia, fibrillose flecks on the stipe, and colors of the young fruiting bodies all indicate identity with *M. inclinata*.

166. Mycena subinclinata Murrill

Mycologia, 30: 371. 1938

Prunulus subinclinatus Murrill, *ibid.*, p. 367.

"Pileus convex to expanded, scattered or gregarious, not densely cespitose, about 3–5 cm. broad; surface glabrous, radiate-sulcate and rugose, fawn-colored to dull watery-brown, darker on the umbo; context white, taste farinaceous, lamellae almost free, white, broad, entire; spores not examined; stipe slender, equal, hollow, smooth, glabrous, white, whitish-myceliod below, 6–10 cm. long.

"Type collected by Erdman West and W. A. Murrill on a dead oak log in woods at Gainesville, Fla., November 9, 1932 (No. F 15708)

Also collected in similar situations in November and December, 1932 (F 15692, F 15693, F 15709 and F 15695). A fine species, abundant at times on oak logs, suggesting *M. galericulata* and *M. inclinata.*"

Through the kindness of Dr. Murrill I have been able to examine material collected by him at Gainesville, November 7, 1938. The spores were found to be strongly amyloid and broadly ellipsoid, and to measure 5–7 × 4.5–5.5 μ. The basidia are four-spored. Cheilocystidia of the clavate-echinulate type are present, but are not conspicuous and measure 28–34 × 6–9 μ. The pileus and gill tramae are strongly amyloid and similar in organization to those of *M. inclinata*. The white stipe and small spores distinguish it from *M. inclinata* and the cheilocystidia from *M. pseudoinclinata.* It is also close to *M. hemisphaerica*, but the color of the stipe should distinguish it. Since I have not seen fresh material I hesitate to refer *M. subinclinata* to synonymy with *M. hemisphaerica.* Spores from deposits should be compared to see if the apparent difference in size is constant.

167. Mycena maculata Karsten

Symb. ad Myc. Fenn., 29: 87. 1889

Mycena rugosoides Peck, Bull. New York State Mus., 67: 22. 1903.
Prunulus rugosoides Murrill, North Am. Flora, 9: 334. 1916.

Illustrations:
- Plates 53 A, 80; Text fig. 41, nos. 3, 6 (p. 328).
- Bresadola, Icon. Mycol., 5, pl. 238 (as *M. parabolica*).
- Peck, Bull. New York State Mus., 67, pl. M, figs. 17–34.

Pileus (1) 2–4 (5) cm. broad, varying in shape from obtusely conic to convex at first, soon expanding to campanulate or broadly convex, at maturity more or less expanded, usually with a distinct umbo, umbo abruptly conic in some and very broad and low in others, margin connivent to the stipe when young, sometimes flaring or recurved in age, surface glabrous, lubricous, often opaque when young but becoming translucent-striate nearly to the disc before fading, often somewhat wrinkled or sulcate striate in age, colors sordid blackish brown or nearly black at first, becoming paler sordid brown to brownish gray in age and nearly always with sordid reddish-brown spots ("light seal brown" to "snuff brown" when young, becoming "hair brown" to "avellaneous" or pale "vinaceous buff," the margin frequently "smoke gray" or paler watery gray in age, some collections pale watery gray over all from the beginning); flesh thickish under the umbo but abruptly thinner over the marginal area (0.15 mm.), cartilaginous and firm, dark or pale watery gray, changing slowly to sordid reddish brown when cut or bruised, no odor, taste mild to faintly farinaceous; lamellae bluntly adnate, becoming toothed and somewhat sinuate, occasionally rather broadly adnexed, sometimes attached to each other and forming a collar, close to subdistant, 17–24 reach the stipe, in about three tiers, narrow, becoming moderately broad (4–5 mm.), whitish to pale gray, soon stained with reddish spots or sometimes entirely rubescent in age, edges even and pallid; stipe usually 4–8 cm. long and 2–5 mm. thick, occasionally much longer, often with a long pseudorhiza (1–5 cm.), densely white-strigose over the lower portion, glabrous above, sometimes twisted, nearly equal, hollow, cartilaginous, apex pallid, the remainder concolorous with or paler than the pileus, the base becoming stained reddish brown to purplish or the entire lower portion sordid wine color.

Spores ellipsoid, 7–9 (10) × 4–5 (6) μ, amyloid, basidia 30–35 × 7–8 μ, four-spored; cheilocystidia embedded and inconspicuous, 20–28 × 6–12 μ, very irregular in shape and markings, clavate to contorted, some with short rodlike projections on the upper part, some with irregular branched fingerlike protuberances, and some with wavy walls and an elongated contorted apex; no pleurocystidia; gill trama hyaline or very faintly vinaceous brown in iodine; pileus trama with a thin pellicle, the region beneath it composed of hyphae with only slightly enlarged cells, the remainder filamentous, yellowish to slightly vinaceous brown in iodine.

Habit, habitat, and distribution.—Gregarious to cespitose on wood and debris of both coniferous and deciduous trees. I have examined material from North Carolina, Tennessee, New York, Michigan, Colorado, Washington, Oregon, and California in the United States and from Ontario in Canada. In northeastern North America it is usually gregarious to subcespitose, whereas along the Pacific coast it is generally cespitose.

Material studied.—Smith, 33–891, 33–946, 326, 766, 777, 916, 938, 1066, 3089, 3120, 3137, 3366, 3502, 3681, 3717, 3754, 4477, 4791, 7388, 9303, 15087, 15510, 15520, 17540. Atkinson, 23758. Henry, 1262. Hesler, 9607, 10693, 14207, 14254. Kauffman, three collections (as *M. rugosoides*). Mains, 34–148, September, 1932, Munising, Michigan. Overholts, 1357, 1934. Parlin, 15199, 15501, 15376, Maine.

Observations.—The color of the pileus is variable much as in *M. megaspora.* Hesler (11537) has found a very pale to whitish form at Newfound Gap in the Great Smoky Mountains National Park. The striae of the pale forms are very conspicuous, whereas those of the darkly colored pilei are not. Certain of these color forms may be distinct genetically and fairly constant, but the differences among them are not sufficient to justify taxonomic recognition. The pseudorhiza in this species, as in others, varies according to the substratum. On a hard surface it may be lacking, but on a soft one, such as a very rotten log, it is usually quite long.

The spores are slightly smaller than those of *M. galericulata*, and the reddish stains are usually much more pronounced, but otherwise the two species are almost indistinguishable. *M. maculata* as I have found it is generally more inclined to a cespitose habit. The short-stiped specimens Kauffman placed in *M. tinctura* apparently belong here, as does the variety he described in the *Agaricaceae of Michigan* under "*M. haematopoda.*" Although a watery juice is present in the stipe of this species, the fungus does not belong in the section Hydropus. Such a watery juice is present in varying amounts in all the larger Mycenae I have tested. The relationships of *M. maculata* are obviously with the other small-spored species that resemble *M. galericulata* in color and consistency.

168. Mycena occidentalis Murrill

Mycologia, 8: 221. 1916

Prunulus occidentalis Murrill, North Am. Flora, 9: 337. 1916.

Illustrations: Plates 53 D, 81; Text fig. 41, nos. 7–8 (p. 328).

Pileus 1–2 cm. broad, broadly conic to convex, often with a flattened disc, becoming broadly campanulate to broadly convex, in age sometimes nearly plane, margin appressed against the stipe at first, surface hoary but soon polished and naked, when wet lubricous to subviscid, conspicuously translucent-striate to disc at maturity when moist, at times slightly rugulose around the center, color occasionally "fuscous" when young but more often dark watery gray, becoming somewhat drab gray on the disc and pale watery gray along the margin, fading (often on the disc first) to ashy brown or brownish gray; flesh thin except under the disc, brittle or moderately pliant, dark gray, becoming pallid, no odor, taste mild; lamellae adnate at first, developing a slight tooth or becoming adnexed, subdistant to distant,

18–20 reach the stipe, two tiers of lamellulae, narrow to moderately broad, whitish to grayish, sometimes grayish at the base and whitish at the margin, edges even; stipe 2–4 (6) cm. long, 1.5–2.5 mm. thick, equal or either the base or apex enlarged slightly, tubular, cartilaginous but fragile, base white-strigose to nearly glabrous, glabrous above, the apical region hoary at first but soon polished and translucent, slightly transversely undulate, concolorous with the pileus or a paler gray.

Spores ellipsoid, 7–8 (9.5) × 4–5 (5.5) μ, amyloid (reaction strong); basidia four-spored, 32–36 × 6–7 μ; cheilocystidia embedded and inconspicuous, 26–30 × 5–10 μ, very irregular in shape, smooth at first and clavate to fusoid, soon with obtuse protuberances developing and in age often variously elongated and branched, the branches sometimes developing very short blunt projections; gill trama very faintly vinaceous brown in iodine or remaining yellowish; pileus trama with a distinct thin pellicle, a slightly differentiated hypoderm, and the remainder of the tramal body of filamentous hyphae, hardly turning brownish in iodine.

Habit, habitat, and distribution.—Subcespitose to cespitose on conifer logs and debris; Colorado, Washington, Oregon, and California. It fruits in the high mountains during the summer and along the coast in the fall. It is very common—at least during certain seasons. Hardy has found it in British Columbia, Canada.

Material studied.—Smith, 3004, 3409, 3595, 3710, 8010, 9186, 9240, 9306, 17321 (pale form), 17590, 18116. Flett, Washington. Gruber, Idaho. Hardy, 82. Overholts, 1350 A.

Observations.—The center of the pileus in the typical form sometimes becomes depressed in age or in drying, but the character is of no value taxonomically. Plate 53 D shows the way the stems are attached to the substratum in collections found on hard surfaces such as decorticated sticks. Those illustrated in Plate 81 were from a rotten log. In these the bases are more or less held together by white radiating hairs, but no typical pseudorhiza seems to be present. A whitish form of what appears to be this species was found on Mt. Angeles in the Olympic Mountains, Washington (17321). It seems to have no other distinctive character.

The lack of reddish stains on the gills and cap and the absence of a pseudorhiza under conditions which would normally cause the development of such a structure distinguish *M. occidentalis* from *M. maculata*. In addition, the cheilocystidia of *M. occidentalis* manifest a tendency to remain smooth, as do those of *M. pseudoinclinata*. It is not so pronounced as in that species, however. From *M. hemisphaerica*, *M. occidentalis* is separated with more difficulty. The cheilocystidia seem to bc fairly distinctive, however, as shown in the text figures. In *M. occidentalis* they are much more difficult to find, and may at times be mostly smooth and clavate to fusoid-ventricose. In *M. hemisphaerica* they are regularly more echinulate. There are also differences in stature, colors, and manner of growth. At first I was inclined to regard both *M. maculata* and *M. occidentalis* as varieties of *M. hemisphaerica*, but, in view of the emphasis given to the markings of the cheilocystidia in this work, they have been retained as distinct species.

169. Mycena hemisphaerica Peck

Ann. Rep. New York State Mus., 46: 104 (24). 1893

Mycena atroumbonata Peck, Bull. New York State Mus., 157: 29. 1912.
Prunulus hemisphaericus Murrill, North Am. Flora, 9: 329. 1916.
Prunulus subtenuipes Murrill, *ibid.*, p. 334.
Prunulus fuliginosus Murrill, *ibid.*, p. 335.
Prunulus atroumbonatus Murrill, *ibid.*
Mycena fuliginosa Murrill, Mycologia, 8: 220. 1916.
Mycena subtenuipes Murrill, *ibid.*, p. 220.
Mycena tenacipes Kauffman and Smith, Pap. Mich. Acad. Sci., Arts, and Letters, 17: 185. 1933.

Illustrations:
Plate 82; Text fig. 41, nos. 9–10 (p. 328).
Lange, Flora Agar. Dan., 2, pl. 56 D (as *M. parabolica*).
McDougal, Mushrooms, plate opposite p. 104 (as *M. galericulata*).
Smith, Am. Journ. Bot., 22, pl. 4, fig. 5

Pileus 1–5 cm. broad, conic, obtuse or acorn-shaped at first, sometimes convex, becoming irregularly campanulate; conic with a flaring margin, broadly convex or plane, and more or less umbonate, shape variable, margin appressed against the stipe when young and often flaring or recurved in age, surface often hoary at first, soon polished, very finely radially wrinkled-striate, moist but opaque when young, becoming translucent-striate before fading, subhygrophanous, nearly black or "fuscous" to "bister" when young, becoming "wood brown" to "avellaneous," finally fading to sordid grayish or grayish brown toward the margin, the disc frequently remaining darker (often blackish); flesh thin except under the disc, pallid to dark grayish, no odor, taste slightly rancid-farinaceous; lamellae adnate, often with a tooth, thin, close at first, subdistant in age, narrow to moderately broad (if ventricose, then appearing widely sinuate and very broad), dark gray to whitish at first, pallid in age, edges even; stipe (2) 5–9 (12) cm. long, 1–3 mm. thick, hollow, very cartilaginous and tenacious, equal, terete or compressed, base white-strigose, rooting somewhat, upper portion pruinose, glabrous and shining in age, "fuscous" to "storm gray" when young, paler in age—especially at the apex.

Spores ellipsoid to broadly ellipsoid, 7–9 (10) × 4–5 μ, amyloid (reaction very strong); basidia two- or four-spored, (25) 32–38 (40) × 7–8 μ; cheilocystidia embedded and inconspicuous, cylindric to clavate, the upper portion covered with rodlike projections, which often elongate to narrow contorted fingerlike processes, 20–28 × 5–11 μ (variable in size and shape); gill trama vinaceous brown in iodine; pileus trama with a thin subgelatinous pellicle, beneath it a fairly well differentiated hypoderm, the remainder filamentous, all except the pellicle vinaceous brown in iodine.

Habit, habitat, and distribution.—Gregarious to cespitose on wood of either conifers or deciduous trees; Alabama, Tennessee, Maine, New York, Ohio, Michigan, and Missouri. It is very abundant in northern Michigan during wet seasons.

Material studied.—Smith, 32–36, 32–551, 32–557, 32–611, 32–612, 32–629, 33–327, 1259, 1673, 6276, 6366, 11039, 14941, 14968, October 4, 1929, Ann Arbor, Michigan (fide Kauffman as *M. dissiliens*). Burke, Alabama, 1942. Hesler, 12955, 14087; Kauffman, November 15, 18, 19, 1929. Kauffman and Smith, Rock River, Michigan (as *M. tenacipes*). Mains, 5181. Overholts, 906, 1530.

Observations.—The outstanding characters of this species are the blackish colors when young, the cartilaginous consistency of all parts, the rather small spores, and the roughened cystidia. The lamellae do not stain reddish when bruised or in age—at least I have not seen them do so. *M. hemisphaerica* and forms of *M. inclinata*, which occur locally late in the fall, are very similar in nearly all their characters, but can be separated by the fibrils on the stipe of immature fruiting bodies of *M. inclinata* and by its sordid stains, which become most conspicuous in wet weather. The gills in *M. hemisphaerica* sometimes appear thickish, but this is not an important character. Rather, it appears to be an abnormality that is not uncommon in Mycena and Collybia. I have frequently observed that such common species as

C. palustris have very thick gills. The spore size varies somewhat, and when one has in hand a pileus with both two- and four-spored basidia, the variation appears to be great. In some collections they measure 6–8 μ long, but usually they are 7–9 μ and, on two-spored forms, 8–10 μ long.

Mycena fuliginosa, *M. tenacipes*, *M. subtenuipes*, and *M. atroumbonata* all appear to me to belong in one species—*M. hemisphaerica.* The spores in the type of *M. atroumbonata* measure 8–10 × 4–5 μ, but two-spored basidia were quite numerous and thus nullified the difference in spore size. The spores of *M. fuliginosa* measure 6–8 × 3–4 μ, those of *M. tenacipes*, 7–9 × 4–5 μ; and those of *M. subtenuipes*, 7–9 (10) × 5–6 μ. There seems to be a significant difference in the width of the spores of *M. fuliginosa* and *M. subtenuipes*, but it must be remembered that these measurements were not made from spore deposits or even from fresh material. In the fresh specimens I have studied the measurements are usually 7–9 × 4–5 μ, exactly intermediate between the two. I have observed a greater variation in spore size in *M. haematopus.*

Mycena maculata has been separated from *M. hemisphaerica* by the dark reddish-brown to purplish-brown stains that develop on all parts of the fruiting body of the former. The distinction appears to be a poor one in the light of my observations on the genus as a whole, but it does serve a useful purpose here. Peck and Kauffman apparently gave it considerable weight. I have collected both species under identical conditions in the field and cannot attribute the difference to environmental factors. Even in dry seasons the stains are evident in *M. maculata.* Both species are frequently very abundant in Michigan.

170. Mycena galericulata (Fr.) S. F. Gray

Nat. Arr. Brit. Plants, 1: 619. 1821

Agaricus galericulatus Fries, Syst. Myc., 1: 143. 1821.
Prunulus galericulatus Murrill, North Am. Flora, 9: 336. 1916.
Collybia ligniaria Peck, Ann. Rep. New York State Mus., 54: 145. 1901.
Prunulus ligniarius Murrill, North Am. Flora, 9: 333. 1916.
Gymnopus dentatus Murrill, *ibid.*, p. 366.
Collybia dentata Murrill, Mycologia, 8: 218. 1916.
Mycena Atkinsoni House, Bull. New York State Mus., 219–220: 233. 1920.
Mycena atridisca Murrill, Mycologia, 8: 220. 1916.
Prunulus atridiscus Murrill, North Am. Flora, 9: 334. 1916.

Illustrations:

Plate 83; Text figs. 41, no. 11 (p. 328); 43, no. 1.
Bresadola, Icon. Mycol., 5, pl. 234.
House, Bull. New York State Mus., 219–220, fig. 2 (as *M. Atkinsoni*).
Kauffman, Agar. Mich., 2, pl. 170 (as *M. inclinata*).
Lange, Flora Agar. Dan., 2, pl. 56 C.
Ricken, Die Blätterpilze, 2, pl. 111, fig. 1.

Pileus 2–4 (5) cm. broad, obtusely conic when young, becoming broadly campanulate or with an obtuse conic umbo and spreading margin, umbo sometimes disappearing, margin more or less incurved at first, soon spreading, surface more or less radially rugulose, scarcely translucent-striate on the margin, lubricous but not viscid, with a cartilaginous feel, frequently splitting radially to the disc, somewhat "buffy brown" on the margin, fading gradually (not hygrophanous) to pale sordid tan or dirty cinnamon brown (near "Sayal brown"); flesh thick on the disc, tapering evenly to the margin, watery gray becoming pallid, very cartilaginous, odor and taste somewhat to strongly farinaceous; lamellae adnexed (often broadly adnexed or sinuate), close to subdistant, 26–36 reach the stipe, three or four tiers of lamellulae, strongly intervenose, moderately broad (5–7 mm.), white or grayish white, soon flushed evenly with pale pink, not reddish-spotted, edges even and pallid; stipe 5–9 (12) cm. long, 2–4 (6) mm. thick, equal, very cartilaginous, hollow, glabrous, naked, smooth or twisted-striate, often with a long pseudorhiza, pale grayish white and naked above, pale fuliginous below, base becoming only slightly sordid brownish in age but not with reddish stains.

Spores ellipsoid, 8–10 (11) × 5.5–7 μ, amyloid; basidia four-spored, 34–40 × 7–9 μ, sterigmata stout; cheilocystidia numerous, 32–40 × 8–12 μ, clavate to capitate, the apices or the entire enlarged portion furnished with rodlike projections, which become elongated in age and somewhat branched; pleurocystidia not differentiated; gill trama of subparallel hyphae, vinaceous brown in iodine; pileus trama with very thin pellicle, the hypoderm narrow and of only slightly enlarged hyphae, the remainder of the flesh floccose, deep vinaceous brown in iodine except for the pellicle.

Habit, habitat, and distribution.—Scattered to gregarious or subcespitose, usually on decaying hardwood logs, stumps, or debris, frequent during the spring and fall. It occurs throughout eastern North America and also along our Pacific coast. I have examined material from Alabama, Tennessee, Pennsylvania, Maine, Massachusetts,

FIG. 43. *M. galericulata:* 1, cheilocystidia. *M. rugulosiceps:* 2, spores; 4, cheilocystidia. *M. radicatella:* 3, spores; 5, cheilocystidia. *M. Brownii:* 6, spores; 9, cheilocystidia. *M. misera:* 7, cheilocystidia; 8, spores

New York, Illinois, Michigan, Missouri, and Washington in the United States and from Ontario and Manitoba in Canada.

Material studied.—Smith, 32–357, 32–373, 32–541, 32–560, 32–562, 32–587, 32–613, 32–634, 33–1072, 7675, 8627, 9310, 9316, 10981, 11082, 13451, 13595, 13689, 14076, 14472, 14967, 14977, 15348, 15509, 15524, 15753, 16584, 18853, September 3, 1929, Michigan. Burke, 1942, Alabama. Flett, Washington. Hesler, 3818, 7825, 15075. Kauffman, Mt. Hood, Oregon (as *M. inclinata*); 1926, Mt. Gretna, Pennsylvania (not typical); October 17, 1914, Michigan. Overholts, 876, 1531, 11082. Also types of species listed in synonymy.

Observations.—The pellicle of the pileus is very thin, and at times is difficult to demonstrate under the microscope. Fresh specimens sometimes appear to have a rather thick separable though nongelatinous pellicle, but this is caused by the removal of the hypoderm along with the tissue above it. The odor is usually not very pronounced, and the taste usually is only mildly farinaceous. Kühner (1938) describes the odor and taste as somewhat radishlike or farinaceous. I have one collection (14976) from Olympic Hot Springs, Olympic National Park, Washington, in which the odor was strongly raphanoid, but which otherwise appears to belong here.

No other species has been so frequently misidentified in North America as *M. galericulata*. Most of the early workers had no clear concept of it. Among Peck's collections I found nearly all the cartilaginous lignicolous Mycenae classified under this name, including one which he considered a variety with rough spores. His variety *longipes* and variety *expansus* very likely belong to some other species, but the specimens could not be found. Most investigators, including Kauffman (1918), placed *M. inclinata* under *M. galericulata*. The stipe of *M. galericulata* is smooth and polished at all times, whereas that of *M. inclinata* is more or less covered with white flecks when young. This is the classical distinction between the two, and it is also a reliable character—at least for young material. The stipe of *M. inclinata* soon becomes glabrous as well as rusty brown over the basal portion.

Peck recognized *M. galericulata* as a distinct species, but, because of its somewhat incurved cap margin, placed it in Collybia, naming it *C. ligniaria*. The spores of Peck's type measure 7–10 × 6–7 μ, and the cheilocystidia are clavate and roughened, as in *M. galericulata*. No pleurocystidia were found. Murrill (1916) placed Peck's species in Mycena (Prunulus). Apparently his interpretation of *C. ligniaria* was based on material sent to him by Peck. I examined these specimens at the New York Botanical Garden and found that they possessed abundant pleurocystidia similar to those of *M. niveipes*. Although I reëxamined Peck's type later, I failed to find in it any fruiting bodies with pleurocystidia. Thus Murrill actually did not have correctly determined specimens of *C. ligniaria* at hand when he made his study. However, the material Murrill received from Peck and the fruiting bodies in Peck's type are indistinguishable macroscopically, and it appears that Peck's mistake really did not mislead Murrill.

Among Murrill's new species of Collybia I found that *C. dentata* had the same microscopic characters as *M. galericulata* and the same appearance when dried. The only detail in the original description that does not check equally well with *M. galericulata* is the habitat on lawns. Because of the manner in which the base of the stipe in the type specimen is broken off one can obtain no information about its mode of attachment, but since in all other respects the carpophore is obviously a Mycena of the *M. galericulata* group, it is safe to assume that his specimen came from buried wood. I have occasionally collected such fruiting bodies of *M. galericulata* in waste places where there was no evidence of lignicolous debris above ground, and they were always attached to buried sticks or roots. See Buller (1934) for additional information on this point.

Mycena Atkinsoni also appears to be identical with *M. galericulata*. Its microscopic characters are the same as those given in the foregoing description of the species. The specimen reported from Michigan under this name was found on very rich humus, but differs in no other respect from a typical *M. galericulata*. In the original description the stipe is described as becoming tawny or chestnut below. This is a darker color than I have observed, but, since the species is generally described as having a stipe with a brownish base, the difference is one of only a slight degree of coloration, and can hardly be significant.

The type of *M. atridisca* has spores 7–10 × 5–6.5 μ, instead of 6–7 × 4–5 μ, as given by Murrill. It also has the typical cheilocystidia of *M. galericulata*. The iodine reaction of the spores and tissues of the pileus and gills was not characteristic for *M. galericulata*, but the specimens were obviously very poorly dried, and hence it does not appear advisable to place much emphasis on their iodine reactions. Hesler's 14200 appears to be a form in which the cheilocystidia are very rare to absent. It was collected May 24, 1942, near Mt. Le Conte in Tennessee.

171. Mycena rugulosiceps (Kauff.) Smith

Mycologia, 29: 342. 1937

Collybia rugulosiceps Kauffman, Pap. Mich. Acad. Sci., Arts, and Letters, 5: 126. 1926.

Illustrations:

Plate 84, Text fig. 49, nos. 2, 4 (p. 349).

Kauffman, Pap. Mich. Acad. Sci., Arts, and Letters, 5, pl. 6.

Pileus (2.5) 3.5–6 (7) cm. broad, obtuse when young, becoming broadly campanulate to convex or more or less plane, the umbo often disappearing, the margin frequently more or less incurved and in age often becoming elevated, surface covered with a hoary bloom at first, soon becoming shining and polished, lubricous when wet, margin opaque but becoming faintly translucent-striate when nearly mature, becoming radially wrinkled to rugulose or sulcate, color "blackish mouse gray" when young, becoming "drab" on the disc and watery gray along the margin, sometimes entirely "pale smoke gray," subhygrophanous, fading slowly to "avellaneous" with a sordid-brownish darker disc; flesh thick on the disc, tapering gradually and thin over the marginal area, whitish to cinereous, very tough and cartilaginous, odor and taste not distinctive; lamellae adnate to somewhat adnexed, at times sinuate in age, usually with a decurrent tooth, close but becoming subdistant, 28–36 reach the stipe, broad (5–7 mm.), nearly equal or tapered at the margin, in age sometimes subventricose, strongly intervenose, rigid and firm, white or pale grayish, becoming sordid gray or flushed with pinkish and appearing grayish pink, sometimes stained sordid reddish brown in spots, edges even; stipe variable in length, depending on the habitat, 5–8 cm. long in open places and 8–12 cm. long when growing from buried wood or on the debris on the forest floor in wet places, 4–8 mm. thick, sometimes with a short pseudorhiza, hollow, very tough and cartilaginous, glabrous, smooth or more or less twisted-striate, blackish to pale or dark sordid gray or grayish brown below, whitish or merely pallid above.

Spores ellipsoid, 8–10 × 5–6 (7) μ, amyloid; basidia four-spored, 38–40 × 8–9 μ when the sterigmata begin to form; cheilocystidia embedded, 26–34 × 7–12 μ, clavate to subcapitate, the enlarged portion furnished with numerous wavy filamentous projections 5–10 × 1.5 μ, hyaline; pleurocystidia not differentiated; gill trama vinaceous brown in iodine; pileus trama composed of a thin pellicle of very narrow hyphae which do not gelatinize in KOH, beneath this a poorly formed hypoderm not sharply separated from the remainder of the filamentous trama, all but the pellicle staining brownish in iodine.

Habit, habitat, and distribution.—Gregarious to cespitose on stumps, logs, and debris of alder, Douglas fir, and rhododendron; Washington, Oregon, and California. It was common during the fall of 1935 but very rare during the season of 1937. In the spring of 1939 it was found sparingly in Washington. Kauffman (1926) reported it as common around Mt. Hood in October.

Material studied.—Smith, 2575, 2744, 3003, 3124, 3361, 3474, 3607, 3684, 3838, 8021, 8621, 13690, 16312, 16384, 16523, 16794. Gruber, Oregon. Kauffman, Oregon.

Observations.—Kauffman described the pileus as hygrophanous and "cinnamon drab" to "avellaneous," with the umbo "Sayal brown." These were the colors of fading pilei in my collections. Even in continuous wet weather the colors usually fade as the fruiting body matures. The pileus and the lower part of the stipe are frequently about the same color—as Kauffman described them. In young specimens these parts are usually blackish. The pinkish tints of the gills are most frequently present in the forms with avellaneous pilei, and the habitat is not necessarily limited to wood and debris of conifers.

Kauffman commented on the trama of the pileus as follows: "The surface layer of the pileus is corticate, composed of one or two series of globose pyriform, brown cells." I have examined the type and found the cap trama to be made up of a thin pellicle of very narrow hyphae beneath which is a region of inflated hyphae (the hypoderm), the cells of which may appear pseudoparenchymatous if a good tangential section is obtained. This is the layer Kauffman referred to, although, from his comment, one might easily suppose that he was describing the type of cuticle found in *M. rorida* or in species of Conocybe. In iodine the trama of the cap and the gills of the type become vinaceous brown and the spores bluish. Kauffman illustrated the long-stiped form. That on Plate 84 is of the form collected around stumps in open places.

The lack of farinaceous taste and the presence of sordid-reddish stains on the gills appear to distinguish the species from robust forms of *M. galericulata*. The two are very closely related, however, and may not be distinct. Kühner (1938, p. 329) has discussed the relationship of *M. rugosa* to *M. galericulata* and, with Lange, is inclined to regard the former as merely a large *M. galericulata* with ashy rather than incarnate gills at maturity. Both comment on not finding *M. galericulata* distinctly cespitose. It is possible that Fries confused *M. galericulata* and *M. hemisphaerica*, a situation which would explain his accounts of the habit of the former.

172. Mycena radicatella (Pk.) Saccardo

Syll. Fung., 5: 275. 1887

Agaricus radicatellus Peck, Ann. Rep. New York State Mus., 31: 32. 1879.
Prunulus radicatellus Murrill, North Am. Flora, 9: 323. 1916.
Prunulus adirondackensis Murrill, *ibid.*, p. 332.
Mycena adirondackensis Murrill, Mycologia, 8: 220. 1916.
Mycena subviscida Kauffman and Smith, Pap. Mich. Acad. Sci., Arts, and Letters, 17: 184. 1933.

Illustrations: Plate 85; Text fig. 43, nos. 3, 5 (p. 349).

Pileus 1–2 (3) cm. broad, obtusely conic when young, becoming campanulate to subexpanded, in age often with a broad umbo, margin usually incurved slightly at first, soon straight and finally somewhat recurved, striate, glabrous and shining, when wet subviscid from the thin separable pellicle, appearing whitish but actually "deep olive buff," disc distinctly grayish in young specimens but soon fading; flesh thin but flexible and tough, white, no odor, taste mild to subnauseous; lamellae adnate or joined together next to the stipe to form a collar, close, becoming subdistant when the pileus is broadly expanded, intervenose, whitish or flushed very pale pinkish (not stained) in age, edges pallid and entire; stipe 2–4 cm. long, (1) 2–3 mm. thick, equal or slightly thickened at the base, which is white-strigose and usually furnished with a pseudorhiza, cartilaginous and tough, terete, hollow, glabrous and polished, subviscid when wet (not furnished with a gelatinous outer layer), whitish or pallid above, pale olive buff below.

Spores broadly ellipsoid to subglobose, 6–8 × 5.5–7 μ (8–9 × 5–7 μ in the type), amyloid (reaction strong); basidia 30–34 × 7–8 μ, two-, three-, or four-spored; cheilocystidia clavate to subcapitate, the enlarged portion covered with slender rodlike projections, tapered below to a slender pedicel, 35–40 × 9–14 μ; gill trama purplish red in iodine (amyloid reactions very strong); pileus trama with a well-differentiated subgelatinous pellicle, a distinct hypoderm, and the remainder filamentous, all but the pellicle purplish red in iodine.

Habit, habitat, and distribution.—Scattered to gregarious or subcespitose, particularly on logs and debris of maple or aspen but on wood of other deciduous trees as well, during the summer and fall; Tennessee, New York, and Michigan in the United States and Nova Scotia and Ontario in Canada. Hesler and Meyer (13781) found it on a balsam log at Clingman's Dome in the Great Smoky Mountains National Park.

Material studied.—Smith, 32–525, 33–689, 226, 478, 485, 499, 747, 786, 869, 952, 995, 1051, 1362, 4587, 6275, 6314, 6383, 9542, 10698, 15097, 15433. Hesler, 14196, 14201, 14204, 14252, and Hesler and Meyer, 13781. Wehmeyer, 869.

Observations.—The pale colors, tough consistency, subviscid pileus, glabrous stipe at all stages of development, and usually the presence of a pseudorhiza, along with the roughened cheilocystidia and broad, short spores, distinguish it. The pseudorhiza is best developed when the fruiting bodies arise from a soft substratum. The colors are always pale grayish to whitish and rather sordid.

An examination of the type has shown that the spores are smooth and otherwise as described above. The apparent roughness Peck saw was no doubt caused by numerous very small oil drops arranged around the periphery just inside the spore wall, a not uncommon occurrence in hyaline-spored agarics. The type specimen is small, but it is obviously a carpophore of a species in the series containing *M. galericulata*. *M. subviscida* was described as a new species from Michigan. Up to that time neither Kauffman nor I had examined the type of *M. radicatella*, and we did not suspect its identity with *M. subviscida*. Murrill apparently did not examine the spores of the type, since he copied Peck's error in his account in the *North American Flora*. *M. adirondackensis* Murrill is a luxuriant form of *M. radicatella*. The pellicle of the specimens in the type of Peck's species is well developed. Consequently it is very likely that the species was subviscid when fresh. Murrill described the pileus of *M. adirondackensis* as sulcate, but this character probably developed as a result of aging or of being exposed to the weather for some time before the fruiting bodies were collected. The iodine reactions of all parts are similar in the types of all three species and are very pronounced. In the field I have been able to recognize only one species with these characters. In the absence of significant differences it is necessary to reduce *M. subviscida* and *M. adirondackensis* to synonymy with *M. radicatella*. Smith and Wehmeyer (1936) reported this species under the name *M. atroumbonata*.

SECTION OMPHALIARIAE

The species of this section are distinguished from the gray species of the Deminutivae and the Typicae by their horizontal, broadly adnate, or more or less decurrent gills. In a certain measure the section also bridges the gap between the Typicae of Eumycena and the gray species of Glutinipes. It contains more than a single phylogenetic series. The range in the microscopic characters noted for the Typicae is also present here.

There follows a list of species which might be sought for in this section because of the manner in which the gills are attached, but which have been placed elsewhere in order to keep closely related species together: *M. semivestipes*, *M. rubrotincta*, *M. laevigata*, *M. ochraceicinerea*, *M. latifolia*, *M. peltata*, *M. pusilla*, and *M. tenuiceps*. All these are in the Typicae. For small species on the bark of living trees see the section Corticolae (pp. 66–76). *M. insignis* and *M. clavicularis* of Glutinipes may be sought for here if their stipes have dried somewhat or have been severely rain-washed.

KEY TO SPECIES

1. Cheilocystidia smooth, fusoid to filamentous or basidium-like 2
1. Cheilocystidia clavate, with roughened apex or apex elongated and ventricose portion verrucose 18

2. Pleurocystidia scattered to abundant 3
2. Pleurocystidia, if present rare, and usually near gill edge 8

3. Spores small, broadly ellipsoid to globose (4–6 × 3–5 μ) 4
3. Spores larger or more narrowly ellipsoid 5

4. Spores subglobose, 5–6 × 4–5 μ (spores not amyloid ?) *Omphalia clavata*
4. Spores 4–5 × 3–3.5 μ, amyloid................ 178. *M. praedecurrens*

5. Spores pseudo-echinulate (with fine pores through spore wall—stain spores and observe with oil-immersion lens); pleurocystidia subcylindric, prominent .... 232. *M. cineraria*
5. Spores smooth, pleurocystidia ventricose, at least when young 6

6. Pileus viscid.......................... 185. *M. pseudoclavicularis*
6. Pileus moist, usually hygrophanous 7

7. Pileus turbinate; pleurocystidia 40–55 × 7–11 μ .... 189. *M. turbinata*
7. Pileus convex-depressed; pleurocystidia 25–30 × 7–12 μ 188. *M. pseudogrisea*

8. Stipe about 1 mm. or more thick 10
8. Stipe usually distinctly less than 1 mm. thick 9

9. Pileus usually dark gray or grayish brown; cheilocystidia elongating and becoming filamentous .... 174. *M. speirea*
9. Pileus very pale gray and soon fading to whitish; cheilocystidia remaining more or less ventricose .... 173. *M. hiemalis*

10. Lignicolous species; on old logs and stumps 11
10. Species terrestrial or merely attached to bits of debris 15

11. Surface of pileus covered by tangled mass of pilocystidia, which are clavate to somewhat elongated (15–30 × 7–11 μ) 12
11. Surface of pileus with thin pellicle or made up of radially arranged enlarged cells (corresponding to hypoderm) 13

12. Spores 6–7.5 × 3.5–4 μ; on coniferous wood
198. *M. marginella* var. *rugosodisca*
12. Spores 4–5 × 3–3.5 μ; on wood of Magnolia .... 201. *M. seriflua*

13. Spores amyloid, 8–10 × 4–5 μ .... 184. *M. fuliginella*
13. Spores not amyloid, smaller 14

14. Spores 3.5–5 μ, globose .... 191. *M. ulmicola*
14. Spores 5–7 × 5–6 μ .... 176. *M. thujina*

15. Cheilocystidia absent 16
15. Cheilocystidia present 17

16. Pileus with thin pellicle of radially arranged hyphae 186. *M. serotina*
16. Pileus covered by tangled pilocystidia having dark contents
202. *M. umbrina*

17. Cheilocystidia ventricose to subcylindric; spores amyloid
187. *M. arenaria*
17. Cheilocystidia saccate to elongated; spores not amyloid 175. *M. pallida*

18. Pleurocystidia abundant; cheilocystidia with ventricose portion roughened .... 124. *M. latifolia*
18. Pleurocystidia absent 19

19. Pileus trama with distinctly gelatinous layer beneath hypoderm
179. *M. subconcolor*
19. Pileus trama without gelatinous layers 20

20. Taste strongly rancid-farinaceous .... 180. *M. cinerella*
20. Taste mild or only slightly farinaceous 21

21. Taste slightly farinaceous; stipe lubricous
181. *M. cinerella* var. *subviscida*
21. Taste mild; stipe moist or dry 22

22. Pileus 2–5 mm. broad; cheilocystidia 3–7 μ broad; scattered on old canes of Rubus .... 177. *M. Brownii*
22. Pileus larger; strictly terrestrial 23
23. Spores 6–8 × 3–4 μ (4-spored) .... 183. *M. concolor*
23. Spores (7) 8.5–10 × 4–5.5 μ .... 182. *M. misera*

173. Mycena hiemalis (Fr.) Quélet

Champ. Jura et Vosges, p. 110. 1872

Agaricus hiemalis Fries, Epicr. Syst. Myc., p. 119. 1838.
Omphalia corticola Peck, Ann. Rep. New York State Mus., 44: 18. 1891.
Omphalopsis corticola Murrill, North Am. Flora, 9: 312. 1916.

Illustrations:

Plate 53 B; Text fig. 14, nos. 5–6 (p. 154).
Beardslee and Coker, Journ. Elisha Mitchell Sci. Soc., 40, pl. 23 (upper figures).
Fries, Icon. Sel. Hymen., 1, pl. 85, fig. 1.
Lange, Flora Agar. Dan., 2, pl. 54 A.
Peck, Ann. Rep. New York State Mus., 44, pl. 2, figs. 8–12.

Pileus (2) 5–15 mm. broad, conic to convex, often papillate, the disc usually depressed around the papilla, surface smooth, glabrous, translucent-striate when moist, coarsely sulcate at least along the margin in age, the margin often irregular or wavy, color variable, disc grayish to avellaneous, the margin whitish, sometimes quite dark over the central portion and the margin pale grayish in buttons but soon fading to whitish; flesh membranous but fragile, pallid avellaneous, odor and taste not distinctive; lamellae subdistant to distant, (6) 9–15 reach the stipe, one or two tiers of lamellulae, broadly adnate or with a decurrent tooth, narrow to broad, edges even; stipe 5–7 mm. long, less than 1 mm. thick, equal, inserted on the substratum, fragile, whitish, very finely pruinose-pubescent over all under a lens but soon glabrous, base scarcely strigose.

Spores 8–9 (10) × 4.5–6 μ, ellipsoid to somewhat ovoid, not amyloid; basidia two-spored; pleurocystidia not differentiated; cheilocystidia rare, clavate to fusoid-ventricose, hyaline, 26–33 × 7–9 μ; gill trama homogeneous, yellowish in iodine; pileus trama yellowish in iodine, pellicle thin, hypoderm not distinct, tramal body filamentous.

Habit, habitat, and distribution.—Single or scattered on tree trunks of soft maple, willow, or elm; Tennessee, New York, and Michigan. Rare.

Material studied.—Smith 15530. Hesler, 5225, 6298, 13824. Mains, November 19, 20, 1931.

Observations.—Characteristically small species such as this often exhibit great variation in many of the macroscopic characters usually emphasized in classifying members of this genus, and hence it is not surprising to find what appear to be rather great differences in the descriptions of this species by different authors. The width of the gills is a good example. It varies from narrow to very broad, and the attachment varies accordingly. As interpreted here, this fungus is merely a small grayish or whitish Mycena with fusoid cheilocystidia, a relatively short stipe, and a tendency to grow on tree trunks or pieces of bark. The stipe is likely to be more pubescent downward and a bit longer than that of *M. corticola.* Kühner (1938) has given rather complete accounts of both typical material and related forms. Except for the two-spored basidia, my collection is closest to his typical form.

A study of Peck's type of *Omphalia corticola* revealed that in all probability it belongs here. Cystidia were not found on the gill edges, but the material was very scanty, and since cheilocystidia are sometimes rare in *M. hiemalis*, their apparent absence in *O. corticola* cannot be given a great deal of emphasis. The base of the stipe in the latter species was described as brownish, but this could have been a discoloration caused by age.

174. Mycena speirea (Fr.) Gillet

Les Hymén., p. 280. 1874

Agaricus speireus Fries, Syst. Myc., 1: 159. 1921.
Omphalia speirea Quélet, Champ. Jura et Vosges, p. 240. 1872.
Omphalia camptophylla Berkeley, English Flora, 5: 62. 1836.
Omphalia tenuistipes Lange, Dansk Bot. Arkiv, 6 (5): 16. 1930.

Illustrations:
Plate 86 A; Text fig. 44, nos. 2–3 (p. 371)
Bresadola, Icon. Mycol., 5, pl. 224, fig. 4.
Fries, Icon. Sel. Hymen., 1, pl. 78, fig. 2.
Lange, Flora Agar. Dan., 2, pl. 61 E (as var. *tenuistipes*).
Smith, Pap. Mich. Acad. Sci., Arts, and Letters, 19, pl. 43, fig. 2.

Pileus 5–10 mm. broad, convex or obtuse when young, the margin straight, becoming plane, slightly umbilicate or papillate and slightly depressed around the papilla, surface very faintly pruinose and soon naked, moist or appearing dry, lubricous when wet, "fuscous" on the disc, "drab" toward the whitish margin, fading to pale grayish brown or yellowish brown at times, the margin always paler, usually conspicuously striate, sometimes widely plicate, often radially wrinkled; flesh very thin, pliant, odor and taste not distinctive or merely slightly farinaceous; lamellae horizontally adnate or arcuate-decurrent, narrow to moderately broad, subdistant, 9–12 reach the stipe, two tiers of lamellulae, pallid to grayish, intervenose, edges even and pallid; stipe 1–3 (6) cm. long, filiform or up to 1 mm. thick, equal, flexuous, cartilaginous and firm, tubular, the apex pallid or at times yellowish to greenish yellow, pallid grayish below or in age yellowish over all, base usually crooked and conspicuously white-strigose.

Spores 7.5–9 (10) × 4–5.5 μ, subovoid, often with a prominent curved apiculus, nonamyloid; basidia two- or four-spored; pleurocystidia not differentiated; cheilocystidia abundant, resembling sterile basidia when young but becoming elongated, (23) 26–38 × 5–8 μ, subcylindric in age and somewhat irregular in outline, hyaline; gill trama of interwoven hyphae, sordid yellowish in iodine; pileus trama with a thin, often poorly differentiated pellicle, a distinct hypoderm, and the remainder of interwoven filamentous hyphae, yellowish in iodine.

Habit, habitat, and distribution.—Single to scattered on pieces of bark or on sticks usually partly buried in the mud or soil in wet places; not common, usually found during late spring and early summer but also fruiting in the fall; Michigan and Washington.

Material studied.—Smith, 32–21, 32–79, 32–453, 33–132, 1354, 3284, 4459, 8322, 13446, 13876, 14078, 14128, 15227.

Observations.—Previously (Smith, 1934) this fungus has been referred to Omphalia. In view of the broader scope given to the genus Mycena by Kühner, with which I am in agreement, this species is properly classified as a Mycena. Like many of these small, inconspicuous agarics, *M. speirea* is readily recognizable once one has become familiar with it, but is not always easy to identify the first time it is collected. The yellow color in the apex of the stipe is a fairly reliable character, but young specimens must be observed. All stages from pallid to grayish yellow, bright yellow, and greenish yellow have been found near Ann Arbor. In age the stipe may become sordid yellow over all, but that is a discoloration. The cheilocystidia often appear to be undifferentiated on maturing pilei, but are readily distinguishable on caps just past maturity. *M. speirea* is nearly always found along with *M. acicula*, and this should serve as a clue to its identity.

175. **Mycena pallida** (Murr.), comb. nov.

Omphalopsis pallida Murrill, Mycologia, 9: 41. 1917.
Mycena phaeophylla Kühner, Encyc. Myc., 10: 590. 1938.

"Pileus conic to convex, becoming umbilicate, gregarious, 1.5 cm. broad; surface dry, smooth, minutely silky, not striate, pallid, pale-avellaneous on the disk; lamellae decurrent, especially when young, arcuate, distant, broad, white; spores ellipsoid, smooth, hyaline, 7–8.5 × 4–5 μ; stipe very slender, cartilaginous, equal, smooth, hyaline, white, 2–3 cm. long, 1 mm. thick.

"Type collected in soil on a roadside bank at Lake Placid, Adirondack Mountains, New York, October 3–14, 1912, *W. A. & Edna L. Murrill 1093* (herb. N. Y. Bot. Gard.)."

Observations.—The spores of the type are as Murrill described them. They are nonamyloid. Cheilocystidia are rare, 26–38 × 7–10 μ, and more or less cylindric to saccate. They have thin walls and a hyaline content. The basidia are four-spored, and no pleurocystidia were seen. The gill trama is compact and of narrow interwoven hyphae 5–8 μ in diameter. The pileus trama is homogeneous beneath a slightly differentiated pellicle, from which numerous filamentous hyphae project to produce the minutely silky appearance mentioned by Murrill. The trama of neither the pileus nor the gills is amyloid.

The apparent discrepancy in the color of the gills in the descriptions of Murrill and Kühner is doubtless insignificant. In other similarly colored agarics the gill color frequently varies from whitish to dark gray or brownish gray. The nonamyloid spores, stature, and hairs over the surface of the pileus, along with the smooth cheilocystidia, are the important diagnostic characters. *Mycena serotina* is distinguished by the well-developed pellicle, the hyphae of which lack the projections found in *M. pallida.* At the time Kühner was completing his work I had not examined the type of *O. pallida.*

176. **Mycena thujina,** sp. nov.

Illustrations: Plate 86 D; Text fig. 45, nos. 9–10 (p. 377).

Pileus 5–15 mm. latus, convexus vel late convexus, pruinosus dein glaber, subhygrophanus, umbrino-brunneus demum cinnamomeus; lamellae subdistantes, arcuatae vel dente decurrentes, angustae; stipes (2) 3–7 cm. longus, 1–1.5 mm. crassus, cartilagineus, glaber, sordide brunneus; sporae 5–7 × 5–6 μ, nonamyloideae; cheilocystidia 28–40

$\times$ 8–12 μ. Specimen typicum in Herb. Univ. Mich. conservatum. Legit A. H. Smith, n. 4444, prope Lake Timagami, Ontario, Sept. 2, 1936.

Pileus 5–15 mm. broad, convex to broadly convex, the margin straight, in age the disc flattened or very shallowly depressed and the margin somewhat crenate at times, surface moist, hoary pruinose when young but soon glabrous and polished, subhygrophanous, faintly striatulate before fading, color deep dull grayish brown fading to pale cinnamon brown or buff ("bister" on disc, "snuff brown" toward the margin or evenly colored "bister" at first but fading to "clay color" or "cinnamon buff," sometimes becoming "cinnamon" to "pinkish cinnamon" before losing moisture); flesh thin, firm, no odor or taste, concolorous with pileus; lamellae close to subdistant, 14–19 reach the stipe, one or two tiers of lamellulae, arcuate, becoming decurrent by a long tooth, narrow to moderately broad, whitish or tinged cinnamon buff, edges even or slightly fimbriate under a lens; stipe (2) 3–7 cm. long, 1–1.5 mm. thick, tubular, cartilaginous-fragile, pruinose but soon glabrous and polished, concolorous with the pileus, base echinulate-strigose with white fibrils.

Spores 5–7 $\times$ 5–6 μ, subglobose to broadly ellipsoid, smooth, white in mass, nonamyloid; basidia 24–26 $\times$ 6–7 μ; pleurocystidia rare to absent, usually near the gill edge; cheilocystidia abundant to scattered, 28–46 $\times$ 8–12 μ, fusoid-ventricose to narrowly fusoid, apex occasionally branched; gill trama homogeneous, reddish brown in iodine (not vinaceous); pileus trama with a thin adnate pellicle, the hyphae of which give off numerous short filamentous projections (which cause the pruinose appearance of the cap), beneath this a distinct hypoderm of enlarged cells which intergrade gradually with the rest of the trama, reddish to reddish brown in iodine.

Habit, habitat, and distribution.—Single or scattered on logs of cedar (*Thuja occidentalis*) or other conifers. Early summer and fall. Apparently not common. The type was collected in Ontario, Canada. It has also been found in Michigan and California. The California collection (3672) was from a redwood log.

Material studied.—Smith, 33–438, 33–531, 3672, 4444. Mains, 5064.

Observations.—The structure of the pileus indicates that this fungus belongs in Mycena. The nonamyloid spores, however, distinguish it from most other members of the Omphaliariae, the gray species in particular. In its consistency it bears some resemblance to

Omphalia campanella, but the nonamyloid spores again distinguish it. In iodine the spores of *M. thujina* assume a strong yellow to yellowish-brown tint. There appear to be two color forms if one examines my specimens, but this difference is more than likely due to the length of time the fruiting bodies stood before being collected. In No. 4444 (the type) the colors were "cinnamon" at maturity, and the specimens dried a dull buffy brown. No. 33–531, from Whitefish, Michigan, is similar in color to the type. No. 33–438 from Au Train, and Mains, No. 5964, from Leland, Michigan, dried more or less bister, the color of the fresh young carpophores.

177. **Mycena Brownii,** sp. nov.

Illustrations: Text fig. 43, nos. 6, 9 (p. 349).

Pileus 2–5 mm. latus, 3–6 mm. altus, conicus, glaber, griseus vel fuscus, valde striatus; lamellae subdistantes, latae, late adnatae vel subdecurrentes, pallide griseae; stipes 6–15 mm. altus, 0.5 mm. crassus, pallidus, aequalis, bulbosus, glaber; sporae 9–11 × 4–6 μ; cheilocystidia 21–32 × 3–7 μ, irregularia. Specimen typicum in Herb. Univ. Mich. conservatum. Legit C. A. Brown, prope Lake Quinault, Washington, Nov. 1, 1925.

Pileus 2–5 mm. wide and 3–6 mm. high, conic with an appressed margin, remaining conic or the margin flaring somewhat, occasionally convex, surface moist, glaucous gray to fuscous over the disc, paler toward the margin, conspicuously striate when moist, opaque when faded; flesh thin and membranous, moderately pliant, odor and taste not recorded; lamellae subdistant, broad, arcuate or arcuate-decurrent, pale grayish, edges pallid; stipe up to 1.5 cm. long, 0.5 mm. thick, pallid, equal above a distinctly enlarged abruptly bulbous base, glabrous above, bulb only faintly mycelioid and grayish to pallid.

Spores 9–11 × 4–6 μ, narrowly ovoid, smooth, amyloid; basidia four-spored; pleurocystidia not differentiated; cheilocystidia embedded in the gill edge, 21–32 × 3–7 μ, often contorted or irregularly branched; gill and cap tramae weakly amyloid (pale sordid brown in iodine); pileus trama with a thin adnate pellicle made up of slender cells with numerous short rodlike projections arising from their walls, hypoderm well differentiated, the cells 18–32 μ in diameter and having dark smoky-brown contents, filamentous portion of trama greatly reduced in thickness (20–30 μ); stipe with scattered contorted hairs (similar to cheilocystidia); no gelatinizing layer present.

Habit, habitat, and distribution.—Scattered on decaying stems of Rubus; November 1, 1925, Lake Quinault, Washington. Known only from the type locality.

Observations.—This fungus was among Kauffman's unidentified collections. The habit is that of *M. brevipes*, for it is scattered on dead sticks, but the two have little else in common. Its relationships are obviously with *M. concolor*, from which the bulbous stipe, spores, and cheilocystidia distinguish it. It is named in honor of its collector, Dr. C. A. Brown, who accompanied Kauffman on his expedition to Lake Quinault in 1925.

178. Mycena praedecurrens Murrill

Mycologia, 4: 165. 1912

Omphalopsis praedecurrens **Murrill, North Am. Flora, 9: 314. 1916.**

Illustration: Murrill, Mycologia, 4, pl. 68, fig. 4.

"Pileus conic to subturbinate when young, then umbonate, and at length nearly plane, densely gregarious to subcespitose, reaching 1.5 cm. broad and nearly 1 cm. high; surface glabrous, very slightly viscid when wet, avellaneous, with darker avellaneous umbo; margin straight, appressed, usually striate, often yellowish-white: lamellae long decurrent, distant, nearly plane, inserted, entire, white with an ashy tint, acute at each end; spores ovoid, smooth, hyaline, 5×3–$3.5\ \mu$; stipe enlarged at the apex, subglabrous, gelatinous-white, avellaneous at the base, slightly viscid when wet, stuffed, about 4 cm. long and 2 mm. thick.

"Type locality: The Bronx, New York City.

"Habitat: on a mossy bank filled with slender roots in moist deciduous woods.

"Distribution: Known only from the type locality."

The type consists of a good collection. The spores measure 4–5 $\times$ 3–3.5 μ and are ellipsoid, smooth, hyaline, and amyloid. The basidia are four-spored. The pleurocystidia are narrowly fusoid-ventricose, with obtuse apices and rather thick, slightly tapered necks. They are smooth and hyaline, and frequently the walls are irregular in outline (the material was revived in KOH). The irregularity may be caused by their not reviving completely. Cheilocystidia are numerous and similar to the pleurocystidia. The tramae of the gills and the pileus are characterized by the presence of numerous lactiferous hyphae, which had a very "metallic" appearance when mounted in KOH. The pileus is practically homogeneous beneath a thin non-gelatinous pellicle. No truly gelatinous layers were present on either the stipe or the pileus.

This is a peculiar species which appears to be well characterized by the cystidia, the small amyloid spores, and the metallic lactifers of its flesh. Murrill described both the cap and the stipe as slightly viscid. This may have been a false viscidity produced by damp weather, or it may actually have been caused by a very thin gelatinous layer which was torn off in sectioning, as usually happens in *M. quinaultensis*. It is very doubtful, however, whether such a gelatinous layer ever covered the pileus. It would be very unusual for an agaric to have a nongelatinous pellicle beneath one that was gelatinous. The account given by Murrill in the *North American Flora* has been quoted.

179. **Mycena subconcolor,** sp. nov.

Illustrations: Text fig. 45, nos. 2, 4 (p. 377).

Pileus 1–2.5 cm. latus, convexus demum late convexus vel subumbilicatus, atro-fuscus dein brunneo-griseus, striatus; lamellae late adnatae demum decurrentes, subdistantes, griseae; stipes 3–4 cm. longus, 1–2 mm. crassus, glaber, brunneo-griseus, sursum pallidus; sporae 8–10 (12) × 5–6 (7) μ, ellipsoideae; basidia tetraspora; cheilocystidia subclavata, subechinulata. Specimen typicum in Herb. Univ. Mich. conservatum. Legit A. H. Smith, n. 8465, prope Darlingtonia, Smith River Canyon, Siskiyou National Forest, Calif., Nov. 6, 1936.

Pileus 1–2 cm. broad, convex with a connivent margin at first, remaining broadly convex or the disc becoming slightly depressed, hoary at first, soon naked and glabrous, opaque at first, translucent-striate at maturity, hygrophanous, olivaceous black over all when young, becoming sordid brownish gray and then opaque; flesh grayish, thin, moderately fragile, odor and taste mild; lamellae bluntly adnate, becoming slightly decurrent, subdistant, 18–23 reach the stipe, usually one tier of lamellulae, dull gray (almost concolorous with the pileus), sometimes strongly intervenose, edges concolorous with the sides and even; stipe 3–4 cm. long, 1–2 mm. thick, glabrous and polished, equal, cartilaginous, dull brownish gray below at first, becoming stained reddish brown at the base and finally bister, upper portion pallid grayish.

Spores 8–10 (12) × 5–6 (7) μ, ellipsoid, smooth, hyaline, amyloid; basidia four-spored; pleurocystidia not differentiated; cheilocystidia basidium-like at first; in age some developing short projections or the walls becoming quite irregular in outline, hyaline, embedded; gill trama homogeneous, vinaceous brown in iodine, subhymenium not gelatinous; pileus trama with a thin nongelatinous pellicle, a well-differentiated hypoderm, the cells of which have a sordid-brown content, a wide band (70–150 μ) of moderately broad hyphae, which are distinctly gelatinous, occurring beneath the hypoderm, and the remainder of the flesh (100–200 μ thick) made up of floccose filamentous hyphae, subhymenium not gelatinous, all but the pellicle and gelatinous portion vinaceous brown in iodine.

Habit, habitat, and distribution.—Scattered on a lawn; November 6, 1937, Darlingtonia, Smith River Canyon, Siskiyou National Forest, California, A. H. Smith, 8465. Known only from the type locality.

Observations.—This is a most curious species. The large cap and short stipe make it appear very top-heavy. In color it is indistinguishable from *M. concolor* or *M. misera*, but is readily separated from both by the gelatinous layer beneath the hypoderm of the pileus. There is also a distinct difference in the cheilocystidia. Those of *M. subconcolor* are scarcely differentiated from the basidia, or often have only four contorted projections, which one might possibly mistake for sterigmata. The mild taste of *M. subconcolor* serves to distinguish it from *M. cinerella*, as do the smaller cheilocystidia and the gelatinous layer in the pileus.

180. Mycena cinerella Karsten

Rysslands Finlands och den Skandinaviska Halföns Hattsvampar, p. 113. 1879

Agaricus (*Mycena*) *cinerellus* Karsten, Hedwigia, 18: 22. 1879.

Illustrations:

Plate 86 C; Text fig. 45, nos. 7–8 (p. 377).
Lange, Flora Agar. Dan., 2, pl. 61 H.
Smith, Mycologia, 28, fig. 1, no. 1 (small carpophores).

Pileus (3) 8–15 (25) mm. broad, obtusely conic to convex, in age often broadly umbonate, the umbo usually somewhat flattened, surface glabrous and moist, lubricous margin appressed against the stipe at first, entire or becoming notched in age and sometimes crenate, translucent-striate, color dark watery gray to pale watery gray, somewhat hygrophanous and fading to ashy whitish or brownish ashy; flesh thin, watery gray, cartilaginous and tough, odor and taste strongly farinaceous or rancid-farinaceous when crushed or chewed; lamellae close to subdistant, 18–26 reach the stipe, two or three tiers of lamellulae, moderately broad (2–3 mm.), horizontally adnate or arcuate, soon developing a pronounced decurrent tooth, sometimes ascending but then with a distinct decurrent tooth, occasionally separating from the stipe and forming a collar around it, whitish to grayish, edges even and pallid; stipe 2–5 cm. long, (0.7) 1–2.5 mm. thick, equal, hollow, terete or compressed, strict or flexuous, cartilaginous and brittle, glabrous or polished, apex at first faintly pruinose, base sparsely strigose, concolorous with the pileus or paler.

Spores 7–9 × 4–5 μ, ellipsoid, smooth, amyloid (reaction very weak in some collections); basidia four-spored, or occasionally two- and four-spored; pleurocystidia not differentiated; cheilocystidia embedded and inconspicuous, 22–36 × 5–11 μ, nearly filiform, with numerous contorted branches or protuberances, clavate with fingerlike prolongations scattered over the apex and occasionally on the pedicel, or capitate and the head furnished with rodlike projections or crooked fingerlike prolongations; gill trama homogeneous, vinaceous brown in iodine; pileus trama with a well-differentiated pellicle, hypoderm differentiated but not very well developed, hyphae 10–20 μ broad, all but the pellicle vinaceous brown in iodine.

Habit, habitat, and distribution.—Gregarious on needles under pine and Douglas fir; Michigan, Washington, Oregon, and California. It usually fruits during the fall months, but was collected in the Olympic Mountains of Washington in June, 1939. To judge from my own experience, it is not common. It would not be at all surprising, however, to find it very abundant occasionally.

Material studied.—Smith, 3918, 6161, 7933, 8018, 8920, 8924, 9091, 14624, 18145, 18224. Mains, 34–179, September 14, 1935.

Observations.—This species is quite easily recognizable if one finds typical specimens and does not allow the rather decurrent gills of the older pilei to mislead him. The exceptionally strong and rancid-farinaceous odor and taste allow it to be readily distinguished from other fragile gray Mycenae. The consistency is a bit more cartilaginous than usual for fungi of this type, and in this character it approaches *M. pusilla.* The stature is rather variable. One frequently finds what I consider to be off-season forms, which are small and quite unlike the usual form in stature. In Hygrophorus such forms have sometimes been given names. This has not been done in Mycena because most of them appear to represent seasonal variations.

In a previous account (Smith, 1936) the spores were described as yellowish in iodine and hence nonamyloid. Tests on various collections since that time have demonstrated that the spores actually are weakly amyloid, as Kühner described them. The separation of the gills from the stipe and their adherence to one another to form a collar around the stipe is apparently unusual and without particular taxonomic significance. The one species supposedly distinguished by this character, *M. collariata*, is very poorly known in Europe.

In collection 9091 from Kerby, Oregon, November 26, 1937, the spores were found to measure 8–12 × 4–5 μ, and a mixture of two- and four-spored basidia was noted on many pilei.

181. Mycena cinerella var. subviscida Kauffman and Smith

Pap. Mich. Acad. Sci., Arts, and Letters, 17: 178. 1933

Illustrations: Text fig. 45, nos. 5–6 (p. 377).

Pileus 5–10 mm. broad and high, conico-campanulate or obtuse, broadly convex or obtusely umbonate at maturity, glabrous, striatulate, subviscid, color sordid brownish cinereous; flesh membranous, grayish, taste slightly farinaceous, no odor; lamellae ascending, arcuate with a decurrent tooth, close, moderately broad, becoming ventricose, whitish then tinged ashy, edges pallid and even; stipe 3–4 cm. long, 1 mm. thick, equal, rather cartilaginous, apex pruinose, base slightly strigose, remainder glabrous, grayish or brownish gray over all, subviscid to the touch when wet.

Spores ellipsoid, 7–8 (8–10) × 4–5 μ, amyloid, basidia two- or four-spored, 20–22 × 7–8 μ; cheilocystidia embedded or projecting slightly, clavate to subcapitate, with or without a pedicel, 26–38 × 9–15 μ, enlarged portion smooth at first but soon becoming roughened with rodlike projections; no pleurocystidia; pileus trama with a well-differentiated subgelatinous pellicle, a distinct hypoderm, and the remaining flesh of filamentous hyphae, all but the pellicle dark vinaceous brown in iodine; gill trama vinaceous brown in iodine.

Habit, habitat, and distribution.—Gregarious on sphagnum, September; Michigan. Known only from the type locality.

Observations.—The outer cortex of the stipe gelatinized slightly, producing the viscidity noted. The gelatinization is not sufficient, however, to justify placing the species in Glutinipes. This variety differs from typical material chiefly in the more lubricous pileus and stipe and the lack of the strong odor and taste. It differs from *M. pusilla* in having a thinner pellicle of the cap and a distinct hypoderm. Its more arcuate gills and smaller spores are additional distinguishing characters. Because of the difference in the structure of the cap I doubt whether var. *subviscida* is truly related to *M. pusilla*, and so have retained it here.

182. **Mycena misera** (Fr.), comb. nov.

Agaricus (*Collybia*) *misera* Fries, Monogr. Hymen., p. 290. 1863.
Collybia misera Gillet, Les Hymén., p. 309. 1874.
? *Mycena pseudo-picta* Lange, Dansk Bot. Arkiv, 6 (5): 15. 1930.

Illustrations:
Plate 87 C; Text fig. 43, nos. 7–8 (p. 349).
Bresadola, Icon. Mycol., 5, pl. 207, fig. 2.
Fries, Icon. Sel. Hymen., 1, pl. 70, fig. 4.
Smith, Pap. Mich. Acad. Sci., Arts, and Letters, 19, pl. 38, fig. 1.

Pileus 4–10 (15) mm. broad, very obtuse to convex at first and with an incurved margin, usually with a broad umbo at maturity, sometimes nearly plane, surface conspicuously pruinose at first, becoming naked and lubricous, evenly fuscous over all (beneath the pruina), the disc remaining fuscous, the margin becoming drab or with a more brownish cast, subhygrophanous, faintly striate when moist, fading to sordid cinereous and becoming opaque; flesh thin, sordid fuscous, fairly cartilaginous, taste slightly farinaceous, no odor; lamellae distant to subdistant, broad, adnate or arcuate-adnate, sometimes short-decurrent, drab but with a pallid sheen, edges minutely eroded and concolorous with the faces; stipe 1–3 cm. long, 1–1.5 mm. thick, equal, tubular, cartilaginous, concolorous with the pileus, pruinose over all at first, apex somewhat furfuraceous, base hardly strigose and tending to stain brownish in age.

Spores (7) 8.5–10 × 4–5.5 μ, ellipsoid, amyloid, basidia four-spored; pleurocystidia not differentiated; cheilocystidia embedded and inconspicuous, clavate to subcapitate or irregular in outline, 26–34 × 7–12 μ, the upper portion bearing numerous rodlike projections, which become elongated, contorted, and sometimes branched in age; gill trama homogeneous, very dark sordid brown in iodine; pileus trama with a thin nongelatinous pellicle, the hyphae of which bear numerous short projections, hypoderm well differentiated, the remainder of filamentous hyphae.

Habit, habitat, and distribution.—Gregarious on humus; Michigan, Idaho, Washington, and California.

Material studied.—Smith, 32–637, 32–638, 32–639, 32–640, 32–641, 3706, 3881, 15506. Slipp, UIFP: 2600.

Observations.—*Mycena misera* is very similar to *M. concolor* in many respects, but differs in its obtuse pileus which has an incurved margin, and in its slightly larger spores. The illustrations of Fries (1868) and Bresadola (1928) picture the Michigan material very well. Murrill described *M. pallida*, another agaric which apparently has the same stature and colors as *Collybia misera* sensu Bresadola but which, in contrast to my collections, has smooth cheilocystidia.

In both Michigan and northern California a form has been collected in which the pileus is almost milk white. A watery-gray tinge, however, is usually present over the disc. The pilei are hygrophanous and fade to a dead white. The microscopic characters are as given in the description of the species. The spores of one collection measured 8–9 × 3–4 μ and in another were 7–8 × 3.5–4 μ. In the dark form they are 7–9 or 8–10 × 4–5 μ. Pale forms such as this have been found in other normally dark-gray to blackish Mycenae and do not appear to be sufficiently distinct to justify taxonomic recognition

183. Mycena concolor (Lange) Kühner

Encyc. Myc., 10: 371. 1938

Omphalia picta var. *concolor* Lange, Dansk Bot. Arkiv, 6 (5): 22. 1930.

Illustrations:

Plate 86 B; Text fig. 44, nos. 1, 4.
Lange, Flora Agar. Dan., 2, pl. 61 I.
Smith, Pap. Mich. Acad. Sci., Arts, and Letters, 21, pl. 25, fig. 3.

Pileus 5–12 mm. broad, conic or obtuse when young, usually conic or campanulate when expanded but at times varying toward convex, the margin appressed when young, surface covered with a grayish bloom at first, soon polished and lubricous, striate nearly to the apex, "fuscous" to "bone brown" on the disc, the margin grayish to brownish with fuliginous striae, hygrophanous, pale cinereous when faded; flesh thin, firm, fuliginous to pallid, no odor, taste mild or bitterish; lamellae distant to subdistant or appearing close at first, moderately broad, arcuate, and with a long-decurrent tooth, glaucous gray to whitish, intervenose, edges even; stipe (1) 2–4 cm. long, (0.5) 1–2 mm. thick, equal, fragile, tubular, hoary at first, soon polished, dark fuscous at first, becoming pale brownish gray to nearly white

FIG. 44. *M. concolor:* 1, cheilocystidia; 4, spores. *M. speirea:* 2, cheilocystidia; 3, spores. *M. fuliginella:* 5–6, cheilocystidia; 8, spores. *M. pseudoclavicularis:* 7, pleurocystidia; 9, spores. *M. arenaria:* 10, cells from surface of pileus

Spores 6–8 × 3–4 μ, narrowly ellipsoid to subcylindric, weakly amyloid; basidia four-spored; pleurocystidia not differentiated, cheilocystidia 26–32 × 7–10 μ, embedded, clavate, hyaline, apex covered with short rodlike projections which in age become elongated and contorted or branched; gill trama homogeneous, sordid brownish in iodine; pileus trama with a thin, somewhat gelatinous pellicle (in KOH), a poorly differentiated hypoderm, the enlarged cells of which are filled with a dark-brown fluid, and the remaining flesh of moderately broad hyaline hyphae, subhymenium very thin, all but the pellicle yellowish to sordid brown in iodine.

Habit, habitat, and distribution.—Gregarious or scattered in moist places, late fall; Tennessee, Michigan, and California. Apparently it is a rather rare species; it fruits on old fern hummocks in wet places.

Material studied.—Smith, 33–605, 3707, 3843. Hesler, 12829.

Observations.—The long-decurrent tooth of the gills appears to be a reliable character, as are also the straight margin of the pileus and its generally conic shape. The species is distinct enough, but because of the lack of an outstanding character to separate it from its relatives, many collectors may have difficulty in identifying it. When it was first reported from Michigan I had not seen typical specimens of *Omphalia picta* and thought it best to use Lange's name. Since that time fresh material of *O. picta* has been collected and studied. It is entirely different from *M. concolor*, as Kühner suspected. Kühner's disposition of Lange's variety appears to me to be the best solution for the situation and is the one followed here. *M. concolor* might be confused with *M. cinerella*, but the taste and odor of the latter species should readily distinguish it. *M. concolor* is very close to *M. misera* sensu Smith (see p. 369).

184. **Mycena fuliginella**, sp. nov.

Illustrations: Text fig. 44, nos. 5–6, 8 (p. 371).

Pileus 10–15 (20) mm. latus, obtuse conicus demum convexus vel umbonatus, glaber, valde striatus, fuligineus, hygrophanus, demum sordide cinereus, ad marginem laceratus; lamellae subdistantes, adnatae vel decurrentes, angustae, pallidae; stipes 2–3 (5) cm. longus, 1–1.5 mm. crassus, aequalis, glaber, fragilis, solidus vel fistulosus, sursum pruinosus, deorsum albo-strigosus; sporae 8–10 × 4–4.5 μ, subcylindricae; basidia tetraspora; cheilocystidia clavata vel fusoideo-ventricosa, 33–62 × (5) 7–12 μ. Specimen typicum in Herb. Univ.

Mich. conservatum. Legit A. H. Smith, n. 14072, prope Olympic Hot Springs, Olympic Mountains, Washington, June 5, 1939.

Pileus 10–15 (20) mm. broad, obtusely conic with an appressed margin at first, becoming more or less convex at times, in age the margin flaring or recurved somewhat and the disc with a low broad umbo, surface smooth and glabrous, moist, translucent-striate with broad dark striations, hygrophanous, "fuscous" on the disc and striae, the remainder drab or the margin pallid in some, becoming "Saccardo's umber" over the margin before fading (dark fuliginous to fuscous and developing a brownish cast upon maturing), fading to sordid brownish gray, the margin frequently lacerated in age; flesh concolorous with the surface, thin, fragile, odor and taste not distinctive; lamellae subdistant, 9–12 reach the stipe, adnate and with a pronounced decurrent tooth but readily seceding, broadest near the stipe (2 mm.), tapered evenly toward the margin of the cap, one or two tiers of lamellulae, whitish but grayish toward the bases, edges even; stipe 2–3 (5) cm. long, 1–1.5 mm. thick, equal, fragile, slender, solid, sometimes with a very slight tubule, with a watery unchanging juice, hyaline white, glabrous, or faintly frosted above, base densely white-strigose.

Spores 8–10 × 4–4.5 μ, subcylindric or slightly curved, very faintly amyloid; basidia four-spored; pleurocystidia not differentiated; cheilocystidia abundant, 33–62 × (5) 7–12 μ, clavate to narrowly fusoid-ventricose, the apex in many of the clavate individuals drawn out into a long neck, hyaline, smooth; gill trama of short interwoven greatly inflated cells intermixed with very slender threads, pale sordid brown in iodine; pileus trama without a distinct pellicle, surface layer several hyphae deep and formed by inflated radially arranged cells, which are filled with a dark-brownish fluid, the remainder of the trama filamentous, the hyphae rather broad, and dull to sordid brown in iodine.

Habit, habitat, and distribution.—Gregarious on logs and stumps of Douglas fir; Humid Transition and Canadian Zones, Olympic Mountains, Washington. The type was found on the Boulder Lake trail above Olympic Hot Springs in the Olympic National Park, June 5, 1939.

Material studied.—Smith, 12060, 14072, 14324.

Observations.—At first sight one would unhesitatingly refer this fungus to *M. concolor*, since the stipe of the latter is quite pale at times. I have not observed *M. concolor* having the conspicuously lacerated margin of the pileus of *M. fuliginella*, but this character will probably be found to be variable. *M. fuliginella* is easily distinguished from *M. concolor*, *M. misera*, and *M. subconcolor* by its cystidia. From all but *M. concolor* its spores also readily distinguish it. It is very close to *M. pallida* Murrill, but the lack of a pellicle and the presence of a very distinct intracellular pigment in the enlarged cells forming the surface of the pileus distinguish it. It also differs from *M. pallida* in its longer and narrower spores on four-spored basidia and in its habitat. *M. fuliginella* is a typically lignicolous species.

The margin of the pileus splits very readily because of the lack of binding tissue to hold the large cells together. Hence weather conditions can very easily cause the character to be pronounced or very obscure, depending on whether they are constant during the development of the carpophores or change rapidly.

185. **Mycena pseudoclavicularis,** sp. nov.

Illustrations: Plate 88; Text fig. 44, nos. 7, 9 (p. 371).

Pileus 1–2 cm. latus, obtuse conicus vel convexus, glaber, viscidus, fuscus demum cinereus; lamellae confertae dein subdistantes, latae, pallidae, late adnatae; stipes 3–4 cm. longus, 1–2 mm. crassus, aequalis, pallide cinereus vel fuscus, non viscidus; sporae 6–8 (9) × 3.5–4.5 μ; pleurocystidia et cheilocystidia fusoide ventricosa, 22–48 (60) × 8–11 μ. Specimen typicum in Herb. Univ. Mich. conservatum. Legit A. H. Smith, n. 8925, prope Takilma, Ore., Nov. 20, 1937.

Pileus 1–2 cm. broad, very obtusely conic to convex, becoming broadly convex or the disc flattened, margin appressed or connivent when young, surface viscid (as in *M. vulgaris*), glabrous, polished in age, dull or very faintly pruinose at first, color evenly drab to pale fuscous, fading slowly to dull sordid gray, more or less unicolorous at all times, faintly striate to opaque when moist, margin slightly sulcate in age; flesh thin but pliant-cartilaginous, pallid to gray, no odor, taste mild; lamellae close to subdistant, broad, horizontally adnate to arcuate, in age with a decurrent tooth, broad, white to pale gray, edges even and pallid, no reddish stains developing; stipe 3–5 cm. long, 1–2 mm. thick, equal, cartilaginous-pliant and strict, tubular, glabrous, and polished, concolorous with the pileus or paler above, apex faintly pruinose, base hardly strigose or merely woolly-strigose, lubricous when wet but without a gelatinous sheath.

Spores smooth, ovoid, 6–8 (9) × 3.5–4.5 μ, strongly amyloid;

basidia four-spored; pleurocystidia scattered to abundant but embedded in the hymenium and easily overlooked, 32–46 × 8–11 μ, smooth, subfusoid, originating deep in the subhymenium, cheilocystidia similar to pleurocystidia or up to 60 μ in length because of the long pedicel, projecting 5–15 μ; gill edges not gelatinizing; gill and pileus trama strongly amyloid; pellicle of the pileus gelatinous, made up of very narrow (2–3 μ) smooth hyphae.

Habit, habitat, and distribution.—Gregarious under ponderosa pine, October and November; Oregon.

Material studied.—Smith, 8925 type, 18113, 18117, 18131.

Observations.—At first sight I mistook this species for a form of *M. clavicularis.* The specimens were collected after a heavy rain, and I thought that the gluten had been washed off the stipe and that the pellicle of the pileus had gelatinized because of the prolonged wet weather. A careful study showed, however, that there were no grounds for such assumptions. The cystidia separate this species from *M. clavicularis.* From *M. laevigata* it differs in both habitat and cystidia—even if one overlooks the stature and colors. Actually, the species appears to be most closely related to *M. latifolia* by its habit, habitat, consistency, and broadly adnate gills. It differs in having a thick gelatinous pellicle over the pileus and in its smooth cheilocystidia. The most satisfactory method of studying the cystidia is first to section and then to crush the sections under a cover glass just enough to spread the basidia well without destroying the outline of the gills.

186. **Mycena serotina** (Peck), comb. nov.

Omphalia serotina Peck, Bull. Torr. Bot. Club, 34: 98. 1907.
Omphalopsis serotina Murrill, North Am. Flora, 9: 314. 1916.

"Pileus submembranous, convex, sometimes slightly depressed in the center or subumbilicate, widely striate on the margin when fresh and moist, slightly striate when dry, grayish-brown, grayish-white or subcinereous; lamellae rather broad, subdistant, adnate or slightly decurrent, white; stem slender, hollow, glabrous, slightly villose-tomentose at the base, pallid, spores narrowly elliptic, 8–10 μ long, 4–5 μ broad.

"Pileus 1–2 cm. broad; stem 1.5–2.5 cm. long, 1 mm. thick."

Among fallen leaves in woods; the type locality is near Boston, Massachusetts. I have examined the type and found the spores to measure 6–8 × 2.5–3 μ. Peck gave them as 8–10 × 4–5 μ. It is quite likely that my measurements are based on immature spores and that Peck's are correct for those that are fully matured. The pileus trama is homogeneous beneath a thin pellicle, and no pleurocystidia or cheilocystidia were discovered. Unfortunately, iodine reactions were not taken. To judge by the appearance of the dried carpophores, the species has much the appearance of *M. pseudoclavicularis* and finds its logical position here in the Omphaliariae.

187. **Mycena arenaria,** sp. nov.

Illustrations: Text figs. 44, no. 10 (p. 371); 45, nos. 1, 3.

Pileus 8–15 mm. latus, convexus demum subumbilicatus vel papillatus, striatus, griseus; lamellae subdistantes, arcuatae demum decurrentes, angustae; stipes 1–3 cm. longus, 1–2 mm. crassus, fuligineus, pruinosus, aequalis; sporae 7–9 × 4–5 μ, amyloideae; cheilocystidia obtusa, subcylindrica. Specimen typicum in Herb. Univ. Mich. conservatum. Legit A. H. Smith, n. 1507, prope Chelsea, Mich., July 19, 1935.

Pileus 8–15 mm. broad, convex at first, soon plane or the disc slightly depressed, the margin bent in slightly at first, sometimes becoming elevated in age, occasionally the disc furnished with a papilla or a small umbo, surface glabrous and moist, "fuscous" to pale "drab" at first (disc darker than margin), faintly translucent-striate but becoming conspicuously striate in age, subhygrophanous and becoming more or less cinereous when faded; flesh watery-fragile, grayish, no odor, taste mild or very slightly raphanoid; lamellae subdistant, 18–25 reach the stipe, one or two tiers of lamellulae, arcuate-decurrent, becoming deeply decurrent, narrow to moderately broad, pallid gray to white, edges even and pallid; stipe 1–3 cm. long, 1–2 mm. thick, fuliginous but usually paler than the cap, white-pruinose over all when young but soon glabrous, hollow, watery-fragile, with a narrow tubule.

Spores 7–9 × 4–5 μ, white in mass, subellipsoid, smooth, distinctly amyloid; basidia four-spored; pleurocystidia absent or occasional near the gill edge and similar to cheilocystidia; cheilocystidia scattered to abundant, 30–48 × 7–12 μ, smooth, hyaline, cylindric to subventricose, the apices obtuse; gill trama filled with lactifers, yellowish in iodine; pileus trama homogeneous, numerous pilocystidia scattered over the surface (in form and size more or less like the cheilocystidia), yellowish in iodine, numerous lactifers present, caulo

FIG. 45. *M. arenaria:* 1, cheilocystidia; 3, spores. *M. subconcolor:* 2, cheilocystidia; 4, spores. *M. cinerella* var. *subviscida:* 5, cheilocystidia; 6, spores. *M. cinerella:* 7, cheilocystidia; 8 spores. *M. thujina:* 9, cheilocystidia; 10, spores.

cystidia abundant, somewhat similar to those on the gill edge, and frequently clustered.

Habit, habitat, and distribution.—Scattered on sandy soil along a road through an oak woods; July 10, 1935, Chelsea, Michigan. A. H. Smith, 1507, same locality, July 22, 1935, 1590. Chelsea, 18354.

Observations.—The younger pilei remind one of *M. misera*, but the smooth subcylindric cystidia easily distinguish this fungus from that species. *M. arenaria* closely resembles *M. pallida*, but is readily distinguished by its amyloid spores. A few fruiting bodies of a species similar in appearance though differing in having globose weakly amyloid spores 5–7 μ in diameter was also found with the material cited above. As yet my information on it is too incomplete to enable me to describe it.

188. **Mycena pseudogrisea** (Murr.), comb. nov.

Omphalopsis pseudogrisea Murrill, North Am. Flora, 9: 314. 1916.
Omphalia pseudogrisea Murrill, Mycologia, 8: 220. 1916.

"Pileus convex-depressed, gregarious, 1–2 cm. broad; surface glabrous, smooth, subhygrophanous, fuliginous, paler when dry, margin concolorous, faintly striate, appressed when young: lamellae inserted, not forking, decurrent, distant, broad, arcuate to plane, pallid: spores ovoid, pointed at one end, smooth, hyaline, uniguttulate, 6–8 × 4 μ: stipe cylindric, smooth, glabrous, subconcolorous, apparently solid, 3–4 cm. long, 1–2 mm. thick."

Type collected on a rotten stump at Englewood, New Jersey, by F. S. Earle (1911). Known from New Jersey and southern New York. I have examined the type and found the tramae of the pileus and gills to be homogeneous. Both pleurocystidia and cheilocystidia are present but scattered. They are the fusoid-ventricose, smooth type and measure 25–30 × 7–12 μ. The basidia are four-spored, and the spores measure 6–7 × 3.5–4 μ. They are smooth, hyaline in KOH, and bluish in iodine (hence amyloid). In age or when dried the pilei are infundibuliform owing to the manner in which the margin becomes uplifted. This is a rather tall cartilaginous, lignicolous species which should be quite easily recognized by its habitat and strongly decurrent gills.

189. **Mycena turbinata** (Murr.), comb. nov.

Omphalopsis turbinata Murrill, North Am. Flora, 9: 313. 1916.
Omphalia turbinata Murrill, Mycologia, 8: 220. 1916.

"Pileus convex to turbinate, usually umbonate, gregarious, 8–14 mm. broad; surface glabrous, hygrophanous, pale-grayish, becoming lighter when dry, margin thin, pallid, even or faintly subplicate, appressed when young: lamellae long-decurrent, subdistant, rather broad, concolorous: spores ellipsoid, pointed at one end, smooth, hyaline, 7–8 × 4 μ: stipe subcylindric, smooth, slightly pruinose, concolorous above, darker below, firm, nearly solid, 3 cm. long, 1–1.5 mm. thick."

The type was collected on mosses on the ground at the New York Botanical Garden. I have quoted Murrill's description. The type is well preserved and consists of a good collection. The conic cap, long-decurrent gills, and pale brown ("clay color") colors are distinctive of the dried carpophores. The spores measure 7–8 × 3.5–4 μ, are ellipsoid, smooth, hyaline in KOH, and bluish in iodine (amyloid). The basidia are four-spored. Pleurocystidia and cheilocystidia are present; they are smooth and fusoid-ventricose but vary to near cylindric in age. They project only slightly from the hymenium, and measure 40–55 × 7–11 μ. The tramae of the pileus and gills are each apparently homogeneous. They did not revive well, however, and the presence of a pellicle and a hypoderm could not be accurately determined.

SECTION FLOCCIPES

The three species grouped here are distinguished by their non-amyloid spores and floccose-scabrous stipes. They constitute a related series of fungi which appear to have been derived from the members of the Typicae having floccose stipes, such as *M. scabripes* and *M. trichoderma*. The latter have amyloid spores.

KEY TO SPECIES

1. Gills with dark-brown edges; spores 6–8 × 4–5 μ .. 192. *M. Kauffmanii*
1. Gills not marginate; spores globose 2

2. Pleurocystidia 60–90 μ long; spores 5–7 μ .... 190. *M. floccipes*
2. Pleurocystidia rare (only near edge); spores 3.5–4 μ .. 191. *M. ulmicola*

190. Mycena floccipes (Fr.) Kühner

Encyc. Myc., 10: 540. 1938

Agaricus floccipes Fries, Epicr. Syst. Myc., p. 87. 1836.
Collybia floccipes Gillet, Les Hymén., p. 318. 1874.
Mycena atribrunnea Murrill, Mycologia, 8: 220. 1916.
Prunulus atribrunneus Murrill, North Am. Flora, 9: 329. 1916.
Mycena maura R. Maire, Bull. Soc. Myc. Fr., 44: 41. 1928.

Illustrations:
Plate 87 A; Text fig. 46, nos. 1–3, 8.
Kauffman, Agar. Mich., 2, pl. 169 (as Collybia).
Maire, Bull. Soc. Myc. Fr., 44, pl. 2, figs. 12–15, pl. 3, figs. 20–24.

Pileus 3–15 (22) mm. broad, sharply to obtusely conic or somewhat ovoid when young, becoming campanulate or expanded-umbonate, margin appressed against the stipe at first or somewhat incurved (often curved in against the gills in partly expanded pilei), frequently splitting radially in age, surface glabrous, moist or lubricous, color "bister" or "snuff brown" when fresh (blackish brown), fading slowly to nearly "buffy brown" (dull olive brown), or "fawn color" (with a tinge of vinaceous), opaque at first, striate when fading; flesh thin, grayish, moderately cartilaginous, odor and taste not distinctive; lamellae narrowly adnate-uncinate to nearly free, close to subdistant, 23–30 reach the stipe, moderately broad (3 mm.), but distinctly ventricose at times, thickish, white, pruinose under a lens from the large cystidia, edges whitish; stipe 1–7 (10) cm. long, 1–2 (3) mm. thick, equal, hollow, fragile, somewhat rooting and strigose over the basal portion, when young covered with a dense minute brown scurfiness, in age polished over all or the apex pruinose, whitish or becoming pure white but in age often sordid yellowish, at least near the base.

Spores typically subglobose to globose, apiculus usually prominent, hyaline, smooth, 5–7 μ, nonamyloid; basidia four-spored; pleurocystidia and cheilocystidia similar, both abundant, subventricose to subcylindric, apices obtuse, hyaline, 60–90 $\times$ 8–15 μ, smooth; caulocystidia numerous, at first clavate to ventricose, the apices obtuse, thin-walled, contents hyaline or sordid brownish, 18–50 $\times$ 10–20 μ; gill trama homogeneous, nonamyloid (merely pale brown in iodine); pileus trama with a surface layer which in tangential section appears to be made up of vesiculose cells having a dark-brown content, the remainder of the flesh homogeneous, of narrower hyphae, a poorly

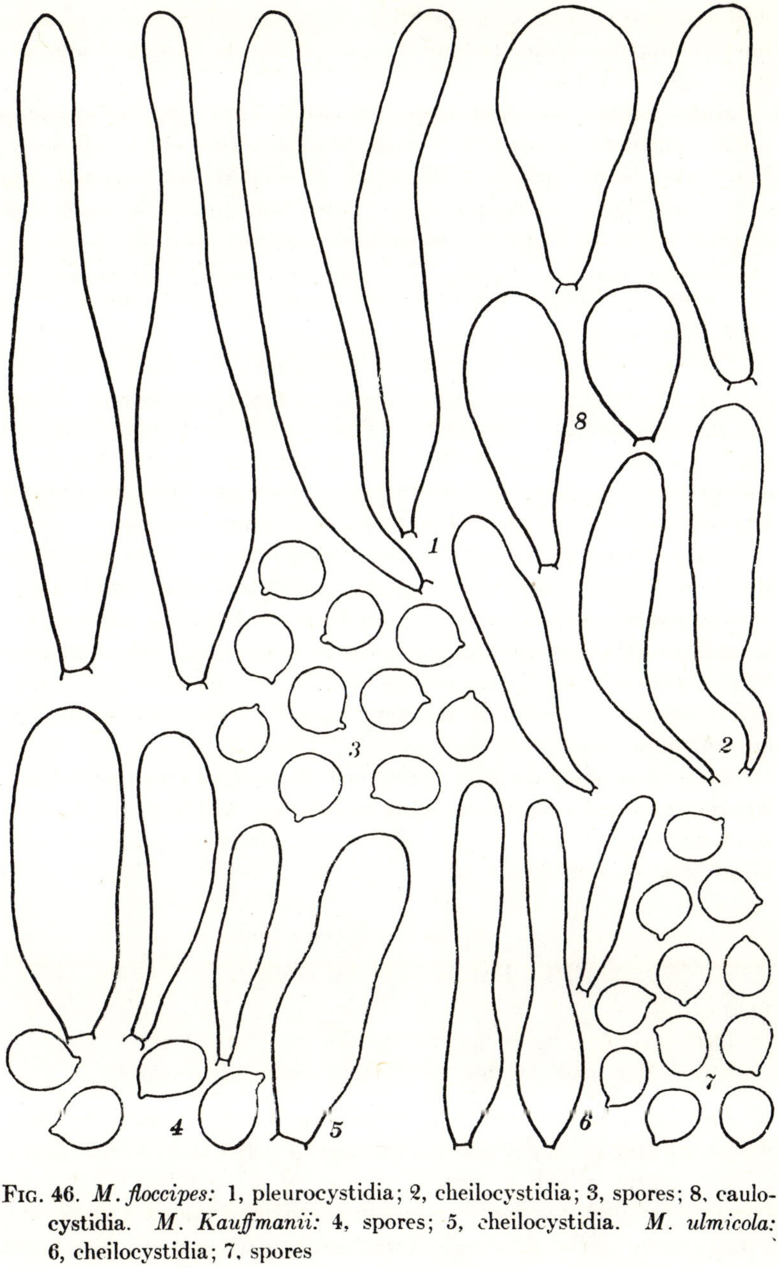

FIG. 46. *M. floccipes:* 1, pleurocystidia; 2, cheilocystidia; 3, spores; 8, caulocystidia. *M. Kauffmanii:* 4, spores; 5, cheilocystidia. *M. ulmicola:* 6, cheilocystidia; 7, spores

differentiated pellicle of narrow (3–4 μ) hyphae demonstrable on some pilei (its hyphae serving to bind the vesiculose cells together), all parts pale yellow in iodine.

Habit, habitat, and distribution.—Scattered to gregarious on damp soil but particularly on partly buried pieces of bark of elm and hickory; Tennessee, New York, and Michigan, also in Africa. Around Ann Arbor, Michigan, the species is very abundant during the latter half of June, when the weather is warm and showery.

Material studied.—Smith, 32–34, 32–248, 32–364, 32–458, 32–530, 717, 1313, 1377, 1394, 1414, 1481, 6372, 10233, 15096, September 5, 6, 1932, Michigan.

Observations.—According to my experience, the floccose appearance of the stipe is most noticeable on young carpophores. Many hyaline caulocystidia are intermingled with those having brownish content. In age nearly all the caulocystidia are hyaline, and many collapse. When Murrill described *M. atribrunnea*, he undoubtedly had old specimens in which the stipes had become glabrous.

The tendency of the pilei to split radially is correlated with their structure. The pellicle is very poorly developed and is made up of narrow hyphae which serve to bind together the large cells of the hypoderm (these may appear to form a cuticle). The enlarged hyphae forming the hypoderm are radially arranged and, of course, separate readily when the slender threads of the binding hyphae are not numerous.

In my tests on well-dried specimens no positive amyloid reaction was observed on stipe, pileus, gills, or spores. Although *M. floccipes* is a very distinct species, carpophores with glabrous stipes may be a bit difficult to identify.

191. Mycena ulmicola Smith

Mycologia, 31: 281. 1939

Illustrations:

Plate 89 A; Text fig. 46, nos. 6–7 (p. 381).

Smith, Mycologia, 31, figs. 1 L; 2 D, E (cystidia and spores).

Pileus 1–2.5 cm. broad, obtusely conic, becoming umbonate with a flaring or recurved margin or sometimes nearly plane, with a narrow sterile margin, which is curved in slightly at first but which soon becomes straight and more or less lacerated and finally deeply split, surface even to slightly rugose, translucent, striatulate on the margin when moist, color "fuscous" on the disc at first and "buffy brown" toward the margin, becoming paler and often watery gray before fading, hygrophanous, fading to "olive buff" or whitish; flesh thin, pliant, watery gray, odor and taste mild; lamellae adnate or subdecurrent, at times readily seceding, crowded, 18–25 reach the stipe, lamellulae in three or four tiers, narrow, white, edges even and pallid, sometimes crisped or wavy; stipe 2–3 cm. long, 1–1.5 mm. thick, equal, solid, brittle, white-strigose around the base, densely pruinose to subpubescent over all, whitish above, pallid to grayish or sordid yellowish near the base in age.

Spores globose to subglobose, 3.5–4 μ, smooth, hyaline, nonamyloid; basidia four-spored, 20–22 $\times$ 5–6 μ; cheilocystidia hyaline, subcylindric to subventricose with obtuse apices, 40–50 (60) $\times$ 9–12 μ, numerous; pleurocystidia present only near the gill edge and similar to the cheilocystidia; caulocystidia very numerous, 40–50 $\times$ 8–10 μ, filamentous with obtuse apices; gill trama yellowish in iodine; pileus trama without a differentiated pellicle, the surface region composed of a compact mass of radially arranged hyphae which are one or two times the diameter of the hyphae forming the remainder of the tramal body, in tangential section the surface region appearing cellular owing to the cut ends of the hyphae which form it, both stipe and pileus trama yellowish to yellowish brown in iodine.

Habit, habitat, and distribution.—Gregarious on elm logs which have not lost their bark and are still solid. It usually fruits in June and, apparently, does not favor logs which are more or less covered with moss. Hesler found it in August at Highlands, North Carolina.

Material studied.—Smith, 6284, 6303, 9645. Hesler, 14688.

Observations.—This species reminds one to a certain extent of *M. marginella* var. *rugosodisca*, but it lacks a copious watery juice, grows on a different kind of wood, and has different spores. Because of the absence of a pellicle over the pileus one might consider the cap to be corticated, and hence try to place the species in the following section. Its natural affinities, however, are with *M. floccipes*. It has a superficial resemblance to *M. fuliginella*, but is never so dark; its spores are nonamyloid, and they differ decidedly in size and shape. Kühner groups these species in the subgenus Paramycena and includes, along with *M. floccipes* and *M. Kauffmanii*, *M. subalpina* von Höhnel, which has spores somewhat similar to those of *M. fuliginella* but nonamyloid. He considers *M. pseudoradicata* Lange and Møller to be synonymous with von Höhnel's species.

192. Mycena kauffmanii Smith

Mycologia, 27: 588. 1936

Illustrations: Plate 87 B; Text fig. 46, nos. 4–5 (p. 381).

Pileus 1–3.5 cm. broad, obtusely conic to campanulate when young, frequently with a small, almost papillate umbo, becoming broadly conic to plane with an obtuse umbo, the margin slightly incurved at first, splitting readily and in age frequently presenting a ragged appearance, disc slightly rugose in large specimens, surface striate, subvelvety, and dry but relatively transparent at maturity, color "fuscous black" to "fuscous" on the umbo when young, fading to "mummy brown" or "drab," margin paler and "avellaneous" to "drab," not hygrophanous; flesh thickish on the disc and tapered abruptly, whitish, rather fragile, odor and taste not distinctive; lamellae adnate, becoming broadly uncinate to adnexed, close to subdistant, 25–28 reach the stipe, two to four tiers of lamellulae, broad and becoming ventricose (4–5 mm.), white, the edges dull brown and denticulate; stipe 1–5 cm. long, 1–3 mm. thick, equal, with a long (2–4 cm.), tapering pseudorhiza, stuffed but soon hollow, tough and elastic, covered over all with a brown pruinose scurfiness, somewhat glabrescent or with adhering patches of scurfiness in age, pallid.

Spores bluntly ovate to broadly ellipsoid, $6–8 \times 4–5\ \mu$, nonamyloid; basidia four-spored, $28–30 \times 6–7\ \mu$; cheilocystidia subcylindric, clavate or somewhat fusoid, apices obtuse, $34–48 \times 8–12\ \mu$, filled with a brown substance; gill trama yellowish in iodine; pileus almost lacking a pellicle, the hypoderm several cells deep, and its cells with brown contents, subcylindric cystidium-like cells project from this layer and cause the velvety appearance of the cap, tramal body of narrower hyphae, all yellowish in iodine; caulocystidia abundant to scattered, with brown contents, $30–44 \times 5–10\ \mu$, variously shaped; stipe tissue yellow in iodine.

Habit, habitat, and distribution.—Scattered to gregarious around elm and ash stumps or on humus in elm and ash swamps, usually in June; Michigan and Ontario. Not common.

Material studied.—Smith, 32–233, 32–540, 1316, 1348, 1408, 4694, 6292, 6381, 9524, 15103.

Observations.—The long pseudorhiza tapers to a thread and is easily broken, which causes great difficulty in ascertaining its attachment. It probably arises from buried wood; at least I have never been able to locate a sclerotium. The brown-edged denticulate gills, blackish to dirty-brown color, scurfy stipe, and pseudorhiza distinguish the species macroscopically. This is the *M. denticulata* of Kauffman in the *Agaricaceae of Michigan.*

SECTION CORTICATAE

The species of this section have the cuticle formed by a palisade of clavate or enlarged cells instead of the usual pellicle of narrow radially arranged hyphae. In one species the spores are amyloid. The fungi of the previous section approach those included here to some extent, but there is a fundamental difference in the organization of the cuticle. In the species of the previous section the pellicle is poorly developed, and the hypoderm may thus appear to be the cuticle and to be made of vesiculose hyaline cells. In radial sections these cells are seen to be considerably elongated and to occur as segments of filaments, whereas in the Corticatae the cells which actually form the cuticle present the same appearance regardless of whether the sections of the cap are radial or tangential.

KEY TO SPECIES

1. Spores amyloid .... 193. *M. hymenocephala*
1. Spores nonamyloid 2

2. Spores 6–7.5 × 3.5–4 μ .... 195. *M. wyomingensis*
2. Spores 5–6 × 4.5–5.5 μ .... 194. *M. ludoviciana*

193. **Mycena hymenocephala** (Smith), comb. nov.

Collybia hymenocephala Smith, Pap. Mich. Acad. Sci., Arts, and Letters, 26: 61. 1941.

"Pileus 2–3 cm. broad, convex or obtuse when young, the margin at first incurved, broadly convex to plane in age, glabrous, 'olive-brown' to 'buffy brown' when fresh, hygrophanous, fading to 'tilleul buff' (whitish), or more olivaceous gray, atomate and glistening when faded, not striate; flesh concolorous with the surface in moist or faded condition, thin, very soft and fragile, odor not distinctive, taste slightly farinaceous; lamellae depressed-adnate but toothed, subdistant (18–20 reach the stipe), 2–3 tiers of short individuals, moderately broad (2.5–3 mm.), 'tilleul buff' or becoming darker and grayish in age, edges slightly uneven; stipe 4–5 cm. long,

2–3 mm. thick, equal, solid, cartilaginous, fragile, longitudinally appressed-fibrillose striate (not with superficial fibrils and not pruinose near the apex), color pallid to ashy brown (almost concolorous with the faded pileus); spores 6–7 × 5–5.5 μ, broadly ellipsoid, white in mass, dark blue in iodine, smooth; basidia four-spored, 26–28 × 7–8 μ; pleurocystidia and cheilocystidia not differentiated; gill trama homogeneous, pale sordid yellowish brown in iodine; pileus trama corticated by a palisade of inflated pedicellate cells 25–36 × 20–25 μ, the remainder homogeneous.

"Scattered on wet ground near the edge of a swampy area, Dexter, Michigan, September 23, 1938 (11050, 18650, 18651).

"... The lax flesh, the pale sordid olive-brown color of the pileus when moist, the atomate glistening appearance of faded specimens, the palisade of inflated cells over the pileus, and the strong bluish reaction of the spores in iodine are all outstanding. In stature it resembles *Collybia ludoviciana* Murrill, which, however, is much more cartilaginous (dried material of both compared) and its spores are yellowish in iodine. *C. fissilis* Maire is also readily distinguished by the iodine reaction of its spores."

In view of the broad concept of Mycena adopted in this work, it was necessary to transfer *C. hymenocephala* to Mycena, where it falls in the Corticatae. Because of its very strongly amyloid spores it does not appear to be related closely to the other members of the section.

194. MYCENA LUDOVICIANA Murrill
Mycologia, 8: 220. 1916

Prunulus ludovicianus Murrill, North Am. Flora, 9: 330. 1916.

Illustrations: Text figs. 47, no. 1; 48, nos. 5–6 (p. 389).

"Pileus fleshy, firm, convex to expanded, gibbous, solitary, 3 cm. broad; surface moist, glabrous, scarcely striate, dark-tan or nearly fuscous: lamellae free or nearly so, crowded, of medium breadth, whitish: spores subglobose, smooth, hyaline, 6–7 μ: stipe cylindric, equal, glabrous, pallid, shining, solid, 5 cm. long, 5 mm. thick.

"Type collected on the ground in a wet thicket at New Orleans, Louisiana, September 8, 1908. *F. S. Earle 132* (Herb. N. Y. Bot. Gard.).

"*Distribution:* Known only from the type locality."

FIG. 47. *M. ludoviciana:* 1, cells from surface of pileus. *M. brevipes:* 2, spores; 3, pleurocystidia. *M. marginella:* 4, pleurocystidia and cheilocystidia; 5, spores

The type has been examined and the following details observed. The spores are subglobose, 5–6 × 4.5–5 μ, smooth, hyaline and nonamyloid, and have slightly thickened walls. The pleurocystidia are fairly abundant but collapse readily and are difficult to revive. They measure 50–78 × 10–16 μ, are subcylindric, and have obtuse apices. Some are slightly ventricose toward the base. Their walls are smooth and thin, and the contents of the cell are hyaline in KOH. The cheilocystidia are very abundant, measure 28–35 × 12–20 μ, and are saccate. They also collapse readily. The gill trama is homogeneous and yellowish in iodine, and is made up of rather narrow interwoven hyphae. The pileus trama is homogeneous beneath a well-differentiated cuticle of inflated cells 18–30 μ thick. These are not arranged in a palisade, although some are pear-shaped and stand upright. Their contents are yellowish brown when revived in KOH. The remainder of the trama is yellowish in iodine.

The type specimen bears no resemblance to *M. floccipes*. The spores of the two are similar in shape but the cystidia appear to be distinctive. The slightly thickened wall of the spores of *M. ludoviciana* also seems to be characteristic. I have not seen fresh material.

195. Mycena wyomingensis Smith & Arenberg

Mycologia, 33: 50. 1941

Illustrations: Text fig. 48, nos. 1–4.

Pileus 1.5–2.5 cm. broad, convex to somewhat umbonate and more or less expanded, margin slightly incurved at first, surface hoary when young, appearing rather dry, dark blackish brown over all, not fading; flesh rather tough and cartilaginous (but not reviving as in Marasmius), odor slight and not distinctive, taste mild; lamellae close to subdistant, rather broad, sinuate, in three tiers, whitish, edges even or nearly so, no color change when bruised; stipe 3–5 cm. long, 2–2.5 mm. thick, with a pseudorhiza 2–4 cm. long, cespitose, surface pallid and finely pruinose-pubescent near the apex from numerous caulocystidia, more or less white-mycelioid toward the base, pseudorhiza mycelioid and with debris clinging to it, very cartilaginous.

Spores ellipsoid, 6–7.5 × 3.5 μ, nonamyloid, smooth; basidia four-spored, 22–24 × 5–6 μ; cheilocystidia and pleurocystidia scattered, fusoid ventricose to narrowly ovoid or somewhat clavate, (30) 45–60 × 12–20 μ, hyaline, thin-walled, smooth, the cheilocystidia usually

FIG. 48. *M. wyomingensis:* **1**, pleurocystidia; **2**, spores; **3**, cells from surface of pileus; **4**, caulocystidia. *M. ludoviciana:* **5**, spores; **6**, pleurocystidia

lacking an elongated pedicel; gill trama of narrow compactly arranged hyphae, yellowish in iodine; pileus trama with a cuticle of a palisade of clavate cells 22–35 × 8–10 μ, their contents dark smoky brown, hyaline thin-walled pilocystidia project at intervals from the palisade and measure 30–50 × 7–10 μ, tramal body homogeneous, composed entirely of compactly arranged narrow hyphae, yellowish in iodine; caulocystidia 40–70 × 7–10 μ, subcylindric, hyaline, thin-walled; stipe tissue yellowish in iodine.

Habit, habitat, and distribution.—Cespitose on rotten wood under spruce and fir, June, 1938; University of Wyoming Science Camp, 10,000 feet elevation. Known only from the type locality.

Observations.—This species is very closely related to *Mycena trichoderma* Josserand in Kühner, but differs in the more compact palisade layer over the pileus, the cespitose habit of growth, rooting stipe, and smaller spores. The spores do not turn distinctly yellowish in iodine, as happens in many species with nonamyloid spores. *M. lenta* Maire is also rather closely related but is said to be rimose, as in certain species of Inocybe, and to lack pleurocystidia. In addition, its colors are apparently more brownish. In Kühner's classification *M. wyomingensis* would fall in the group *Spuriae* if the spores were amyloid.

SECTION HYDROPUS

Mycena brevipes is placed here because of the numerous lactifers in the trama of the pileus and gills. It is not known whether the species has a colored or a hyaline latex or, for that matter, any latex at all. Kühner has suggested that this group ought to be considered a distinct genus, and Singer has recognized it as such. I have seen fresh specimens of only *M. marginella* and *M. umbrina* and, on the basis of these species and var. *rugosodisca* of *M. marginella*, am not inclined to make such a distinction. *M. atroalboides*, which has a rather copious juice at times, is also keyed out here, but is related to species in the Typicae.

KEY TO SPECIES

1. Gill edges sordid brownish .... 198. *M. marginella*
1. Gill edges concolorous with faces 2

2. Cheilocystidia regularly echinulate or apices covered with short rod-like protuberances .... 135. *M. atroalboides*
2. Cheilocystidia smooth or intermediate between rough and smooth types 3

3. Spores up to 6 μ long 4
3. Spores regularly more than 6 μ long 5

4. Spores globose, 3–4 μ; carpophores usually on hemlock
196. *M. fuliginaria*
4. Spores broadly ellipsoid, 4–5 × 3–3.5 μ; on Magnolia; southern United States .... 201. *M. seriflua*

5. Pleurocystidia abundant; fusoid-ventricose .... 200. *M. brevipes*
5. Pleurocystidia absent or, if present, rare and only near gill edges 6

6. Pileus covered with tangled pilocystidia 7
6. Pileus without pilocystidia over surface; spores 9–11 × 5–6.5 μ
197. *M. taxodii*

7. Spores 6–7.5 × 3.5–4 μ; cheilocystidia abundant
198. *M. marginella* var. *rugosodisca*
7. Spores 7–9 × 4.5–6 μ; gills distant; cheilocystidia apparently absent
202. *M. umbrina*

196. Mycena fuliginaria (Fr.) sensu Kühner

Encyc. Myc. 10: 531. 1938

Collybia succosa Peck, Bull. Buffalo Soc. Nat. Sci., 1: 46. 1873.
Collybia atramentosa Kalchbrenner, Icon. Sel. Hymen. Hung., p. 15. 1873.
Mycena succosa Saccardo, Syll. Fung., 5: 293. 1887.
Collybia nigrescens Quélet, Bull. Soc. Bot. Fr., 23: 325. 1876.
Mycena atramentosa von Höhnel, Fragm., III, Mitt. n. 97. 1907.

Illustrations:

Bresadola, Icon. Myc., pl. 216, fig. 1.
Kalchbrenner, Icon. Sel. Hymen. Hung., pl. 6, fig. 2 (as *Collybia atramentosa*).

"Pileus 1–3 cm. broad, subcartilaginous, campanulate to convex, *cinereous-brown to fuliginous, minutely pubescent*, margin incurved and surpassing the gills. Flesh thickish, white at first *becoming purplish-black where wounded.* Gills adnate with a slight decurrent tooth, becoming emarginate, moderately broad, tapering in front, close, whitish, *turning blackish where bruised.* Stem 2–5 cm. long, 2 mm. thick, equal, cartilaginous, compact except the stuffed axis, often curved, *clothed with a fine, fuliginous pubescence*, becoming blackish. Spores minute, globose-ovoid, 3–4 micr. in diameter, smooth, white. Cystidia none; *sterile cells* on edge of gills, abundant, slender, subfiliform. Odor and taste not marked.

"Scattered or caespitose. On decayed wood, logs, etc., mostly on hemlock, in coniferous regions. Marquette, Munising, South Haven, New Richmond. July–September.

"Easily distinguished by the change of color when bruised. This change is due to lactiferous tubes containing a juice which turns blackish on exposure to the air. These tubes are specialized hyphae interspersed throughout the trama of the pileus, gills and stem. Under the microscope it may be seen that the pubescence is composed of elongated hyaline cells. The presence of a juice which exudes on wounding the plant is unusual in this genus, and reminds one of a section of the genus Mycena; but the incurved margin of the young pileus indicates its relationship with Collybia."

This is Kauffman's description of *Collybia succosa* in his *Agaricaceae of Michigan.* I have examined his collections. As Kühner has reported, the spores of the species are distinctly amyloid, but the tissues are not. The name Kühner decided upon has been used here, and his synonymy is included. I have never seen fresh material of this fungus and so can add nothing from my own experience. Hesler, who has found specimens on several occasions, sent the following notes on his collection 6521.

"Pileus 1–3 cm. in diameter, convex-campanulate, expanding somewhat (in a few), umbo at times prominent and obtuse, pale brown to buff-brown, becoming stained bluish-black when handled or wounded, fibrillose, not viscid, not hygrophanous, margin pellucid-striate; flesh dark, soon black when cut; odor and taste none; gills white at first, then with a flesh tint, bluish-black when wounded, emarginate or emarginate-adnexed, edges ciliate, close, broadest behind (about 2 mm.); stipe 2–4 cm. $\times$ 1.5–3 mm. rigid, hollow, white-frosted (pruinose) throughout, concolor, watery (the juice staining the stipe on wounding, as in cap and gills), equal or tapering upward slightly, base expanded into a slight disc, disc white-strigose, the hairs turning black by the juice of stem, dry. Spores subglobose to short-elliptical, 3–4 $\times$ 3 μ; cystidia on edges of gills, 47–54 $\times$ 9.5–11 μ; trama parallel, 'vessels' conspicuous, blackish, in cap and gills. Remarkable for the presence in stipe gills and pileus of a copious watery juice which stains flesh of sporophore bluish-black when wounded; stains the fingers dark blue; stains white paper as follows: quickly purple, then becoming bluish, finally bluish black."

Murrill cited the illustration in *Mycologia*, 6, pl. 137, fig. 9, as this species, but an examination of the figure leads me to believe it represents *M. haematopus* instead.

Material studied.—Atkinson, 15370, 24187 (as *M. atramentosa*); six packets (as *M. succosa*), 2418, 2896, 4594, 5334, 13392, 13499. Hesler, 6521. Kauffman, Michigan (as *M. succosa*).

197. Mycena taxodii Murrill

Bull. Torr. Bot. Club, 67: 235. 1940

Prunulus taxodii Murrill, *ibid.*, p. 234.

"Pileus conic to broadly convex, often slightly umbilicate or truncate, cespitose or closely gregarious, 1–1.5 cm. broad; surface glabrous, black when young, umbrinous or fumosous in the older stages, distinctly sulcate-striate to the small central disk, which is very rugose; margin straight, entire, paler; context very thin, blackish, mild, odorless; lamellae adnate, broad, ventricose, inserted, distant, interveined, entire, cinereous, blackish near the context, blackening when bruised but not bleeding; spores irregular, subglobose to ellipsoid, smooth, hyaline, granular, 4–6 × 3–4 μ; stipe equal, smooth, glabrous, whitish-shaggy at the base, blackish to avellaneous, not bleeding, about 3 × 0.1 cm.

"Type collected by W. A. Murrill on the base of dead standing trunks of pond cypress in a cypress bog near Orange Heights, Alachua Co., Fla., Nov. 9, 1938 (*F 18376*). The trees were killed by a grass fire during a drought. So far as I now remember, these are the first specimens of *Mycena* to be reported on *Taxodium* in America. Their color is quite peculiar; while the fresh spores suggest dried English peas."

Dr. Murrill sent me a portion of the type for microscopic study. The spores are amyloid, 9–11 × 5–6 μ, ellipsoid, smooth, and hyaline. The content is frequently granular and organized into an irregular body. When stained, the thin outer wall is clearly visible; in unstained mounts in KOH it usually stands out very clearly. Cheilocystidia are abundant and are intermediate between the clavate-roughened and the fusoid types. Ordinarily they have two to four obtuse contorted projections arising from the apex, but typical clavate-roughened individuals as well as of some that are fusoid-ventricose were noted. No pleurocystidia were seen. The tramae of the gills and pileus were nonamyloid and not otherwise distinctive.

The pileus is homogeneous beneath a hypoderm of more or less enlarged cells having dark-brown contents. The pellicle, which is very poorly developed, is represented by narrow hyphae, which bind the cells of the hypoderm together. The upper surface of the enlarged cells (where exposed) and the narrow hyphae are both covered with short rodlike projections. Scattered lactifers are present in the stipe. The fungus appears to be most closely related to *M. fuliginaria*.

198. Mycena marginella (Fr.) Quélet

Champ. Jura et Vosges, p. 343. 1873

Agaricus marginellus Fries, Syst. Myc., 1: 113. 1821.

Omphalia marginella Josserand and Maire, Bull. Soc. Linn. Lyon, 10: 115. 1931.

Illustrations: Text fig. 47, nos. 4–5 (p. 387).

Pileus 1–2 (3) cm. broad, obtuse to broadly convex when young and with a somewhat incurved margin, sometimes with a rather pronounced umbo, sometimes with a somewhat depressed disc, surface glabrous, even or slightly wrinkled, appearing rather dry and velvety, hardly striate but the margin frequently cracks or splits radially in age, color "fuscous" over the disc, "drab" or paler toward the margin, not fading appreciably; flesh thin, watery (if the cap surface is cut with a sharp instrument, drops of a hyaline liquid ooze out), brittle, odor and taste mild; lamellae close to crowded, 26–35 reach the stipe, two or three tiers of lamellulae, narrow, equal, broadly adnate to arcuate-subdecurrent, sometimes interveined, edge pruinose under a lens from projecting cheilocystidia and dull sordid brown to fuliginous, faces pallid; stipe short, 1.5–2.5 (3) cm. long, 1–2 mm. thick, equal or the base slightly enlarged and with scattered mycelial hairs, rather brittle-cartilaginous, dark gray to blackish brown at first, becoming grayish brown to almost hyaline gray, dull and pruinose at first, becoming more or less glabrous and polished in age.

Spores 6–7.5 × 3.5–4 μ, ellipsoid, smooth, the wall very delicate, very weakly amyloid; basidia four-spored; pleurocystidia present only near the gill edge and similar to the cheilocystidia; cheilocystidia abundant, of two types, saccate and measuring 35–46 × 15–20 μ, or fusoid-ventricose with obtuse apices and 40–60 μ, contents of both kinds sordid brownish; gill trama regular, with numerous lactifers, hardly amyloid; pileus trama with a surface covering of irregularly arranged saccate cells 30–40 × 9–20 μ, which have dull-brown contents, their walls occasionally with a thin incrustation when revived in KOH, beneath the surface layer the trama proper composed of very broad hyphae (15–28 μ in diameter), as well as interwoven narrow connective hyphae and numerous lactifers 5–12 μ in diameter, nonamyloid; surface of stipe furnished with short or long caulocystidia (26–57 × 8–15 μ), which are subcylindric with obtuse apices or subclavate, and are filled with a dark-brown pigment.

Habit, habitat, and distribution.—Scattered to gregarious on conifer logs, spring and fall; North Carolina and Washington.

Material studied.—Smith, 3176, 13920, 14127, 14284, 14607, 14736, 16183, 16299. Hesler, 10751.

Observations.—This species is very easily recognized by its marginate gills and the drops of liquid which ooze from the cap when it is cut with a sharp instrument. In my specimens the amyloid reactions have been very unsatisfactory. The spores usually remain hyaline, and very rarely have I found groups of immature individuals which gave a pale-grayish reaction. They are therefore classed here as weakly amyloid. I have never observed them to be yellowish in iodine. The reactions of the gill trama and the flesh of the pileus were also inconclusive.

Fries, in *Systema Mycologicum*, 1: 113, obviously took his description and concept of the species from Persoon, but in his later works apparently applied the name to a different fungus. I have cited the authority of the species in accordance with the rules which require that Fries be used as the starting point, but have retained the concept of Persoon and that of Fries in *Systema Mycologicum*, which is also the concept established by Josserand and Maire (1931).

199. Mycena marginella var. **rugosodisca** (Pk.), comb. nov.

Agaricus (*Omphalia*) *rugosodiscus* Peck, Bull. Buffalo Soc. Nat. Sci., 1: 48. 1873.

Mycena rugosodisca Saccardo, Syll. Fung., 5: 293. 1887.

Omphalia rugosodisca Peck, Ann. Rep. New York State Mus., 45: 34. 1893.

Omphalia marginella var. *rugosodisca* Josserand and Smith, Mycologia, 29: 721. 1937.

The variety differs only in its nonmarginate gills.

Material studied.—Smith, 399, 4489, 9763, 10636. Hesler, 6514, 9128, 9061. Kauffman, three collections, 1905, Michigan. Overholts, 5952. Sharp, 6520. Wehmeyer, 686.

Observations.—This variety, which is quite abundant on coniferous stumps and logs in northeastern North America, extends south into the Great Smoky Mountains of North Carolina and Tennessee. An interesting parallel situation is found in *M. haematopus.* In that species a variety with marginate gills exists in addition to the typical variety, but their taxonomic ranking has been reversed, the non-marginate fruiting bodies being classed as typical and those that have marginate gills as the variety. Peck described a variety *laevidisca* under *O. rugosodisca*, but it does not deserve taxonomic recognition. I have found it on fir stumps, and the cap may be either smooth, slightly rugose, or distinctly rugose on carpophores in a single group.

200. Mycena brevipes Murrill

Mycologia, 8: 220. 1916

Prunulus brevipes Murrill, North Am. Flora, 9: 328. 1916.

Illustrations: Text fig. 47, nos. 2–3 (p. 387).

"Pileus conic to convex, not umbonate, solitary, 8 mm. broad; surface dry, glabrous, pale-gray, striate, margin pallid, entire, appressed when young: lamellae adnexed or nearly free, crowded, narrow, broader near the margin, white: stipe very short for the genus, slightly tapering downward, smooth, dry, glabrous, white, attached to the substratum by a broad, circular mat of white mycelium, scarcely 1 cm. long, 1–2 mm. thick."

Habit, habitat, and distribution.—Single on hardwood sticks; Tennessee.

Observations.—Upon reëxamining the type I obtained the following data: The spores measure 7–9 × 5–6 μ, are broadly ellipsoid, and definitely amyloid. Pleurocystidia and cheilocystidia were found to be fairly abundant, but nearly all were so badly collapsed that they could very easily be overlooked in mounts of revived material. KOH revived them better than any lactic-acid solution, and a careful study of the sections mounted in KOH and stained with phloxine showed the cystidia to be fusoid-ventricose, smooth, and about 35–50 × 8–13 μ. The cystidia illustrated in text figure 47, no. 3, are camera-lucida drawings of the more nearly normal individuals. The basidia revive very well; they measure 24–28 × 7–8 μ and are four-spored. The pileus trama consists of a thin pellicle and a tramal body apparently made up almost entirely of metallic-appearing lactiferous hyphae 4–9 μ thick. Sections of the pellicle were not good, and no details could be made out. The tramal body revived well, as one would expect. The gill trama was also filled with "metallic" lactiferous hyphae, although some floccose tissue was visible. The trama of both the pileus and the gills remained yellow in iodine.

When I first examined the type, the only stained preparations made were in lactic acid. Because of the clearing action of the acid, in addition to its failure to revive the pleurocystidia, I overlooked the pleurocystidia in that examination, and the significance of the structure of the pileus escaped me. As a result, I misdetermined a group of collections from the western United States and published (1937) an account of them under Murrill's name. For further data on these collections see *M. subcana* (p. 75). So far as is now known, *M. brevipes* should be recognizable by its cap trama, cystidia, and spores. It remains to be ascertained whether or not a colored latex is present in fresh specimens. Since none was noted, the specimens have been placed here provisionally.

201. Mycena seriflua Murrill

Bull. Torr. Bot. Club, 66: 37. 1939

Galactopus serifluus Murrill, *ibid.*, p. 32.

"Pileus convex to subexpanded, umbilicate, gregarious, 1.5 cm. broad; surface smooth, glabrous, opaque, melleous, blackish at the center; margin incurved, even, entire; context rather thick, opaque, pallid, odorless, somewhat mawkish, bleeding watery drops when cut; lamellae adnate with a small decurrent tooth, medium distant, broad behind, inserted, entire, white, unchanging; spores ovoid, smooth, hyaline, l-guttulate, about $3 \times 2\,\mu$; cystidia none; stipe equal, smooth, pallid and pruinose above, subconcolorous and subglabrous below, whitish-tomentose at the base, 2×0.2 cm.

"Type collected by W. A. Murrill on a rotten magnolia log in a hammock at Gainesville, Fla., Sept. 14, 1938 (*F 18104*). An aberrant species with incurved margin like *Gymnopus* and bleeding from the context more than from other parts of the hymenophore. Dried specimens would be referred to *Gymnopus* without hesitation. *Galactopus rugosodiscus* (Peck) Murr. is probably its nearest relative."

The original account is here quoted. I have examined a portion of the type and found the spores to be distinctly amyloid and to measure 4–5 (6) $\times$ 3–3.5 μ. They are smooth and broadly ellipsoid.

The pileus trama is homogeneous, nonamyloid, and somewhat "glassy" or subgelatinous when revived in KOH. The surface is covered with a loose or tangled mass of hyaline clavate pilocystidia measuring 7–11 × 15–26 µ. The gill trama is nonamyloid and appears to be slightly gelatinous in KOH. The basidia are four-spored. Pleurocystidia are absent, but cheilocystidia are abundant, they measure 26–52 × 9–14 µ and are narrowly clavate to subcylindric. Their walls, which are thin, collapse readily. The relationships of the species are as pointed out above by Murrill.

202. **Mycena umbrina,** sp. nov.

Illustrations: Text fig. 49, nos. 1–4.

Pileus aquosus, 3–5 mm. latus, convexus, umbrinus, impolitus; lamellae distantes, decurrentes, angustae, pallide griseae; stipes 18–20 mm. longus, 1 mm. crassus, aequalis, aquosus, subumbrinus; sporae 7–9 × 4.5–6 µ, amyloideae. Specimen typicum in Herb. Univ. Mich. conservatum. Legit A. H. Smith, n. 33–538, prope Rock River, Mich. June 17, 1933.

Pileus 3–5 mm. broad, convex, in age broadly convex but hardly depressed, dark blackish brown over all, with blackish striations, margin becoming slightly paler in age, surface unpolished to subtomentose, moist, subhygrophanous; flesh dark brown, thin, very watery, no odor, taste not recorded; lamellae distant, arcuate-decurrent, narrow, pale gray, edges even; stipe 18–20 mm. long, 1 mm. thick, equal, watery, and very fragile, smoky gray over all or pallid near the apex, pruinose.

Spores 7–9 × 4.5–6 µ, subovoid, amyloid (reaction strong), smooth; basidia four-spored; pleurocystidia and cheilocystidia not differentiated, gill trama homogeneous, some lactifers present, yellowish in iodine; pileus trama corticated, with a very compact layer of clavate to subfusoid pilocystidia (as in *M. marginella*), irregularly arranged, their contents dark brown, beneath this a region of broad hyphae with elongated cells also with dark-brown contents, remainder floccose, not amyloid in any part; clavate caulocystidia scattered over the apex of stipe, their contents also dark-colored (cystidia on cap and stipe 26–58 × 8–18 µ).

Habit, habitat, and distribution.—Single in a bed of moss; Rock River, A. H. Smith, 33–538, June 17, 1933, Michigan. Known only from the type collection.

FIG. 49. *M. umbrina:* 1–2, caulocystidia; 3, cells from covering of pileus; 4, spores. *M. rorida:* 5, cells from hymeniform layer of pileus; 6, cheilocystidia; 7, spores. *M. clavicularis:* 8, pleurocystidia; 9, spores. *M. insignis:* 10, spores; 11, cheilocystidia

Observations.—Because of its watery juice and the organization of the pileus, *M. umbrina* appears to be very closely related to *M. marginella.* It can be distinguished at once by its distant gills, larger spores, and lack of cheilocystidia. In view of the striking microscopic characters I have no hesitation in describing it, although the material is very scanty. The two carpophores collected were in perfect condition.

SUBGENUS GLUTINIPES

GREAT emphasis has been placed here on the outer gelatinous layer of the stipe. This layer not only affords a very convenient character by which to recognize the group, but brings together three distinct phylogenetic lines which are clearly derived from various groups within the sections Typicae and Omphaliariae. The most discordant elements are *M. insignis* and *M. rorida.* The relationships of these species are obscure, and so a practical grouping is preferred here to any questionable natural arrangement. Hence they are placed together in the section Diversiformes. The species in the sections Viscosae and Caespitosae form two homogeneous natural series. It is interesting to note that the latter section is made up entirely of American species. *M. trojana* and *M. fumosiavellanea,* which have been described by Murrill from the American tropics and which have been found to belong in this subgenus, are grouped in the special section (pp. 453–470), along with his other tropical species.

KEY TO SECTIONS

1. Pileus dry or very soon becoming so; no gelatinous layers in or on pileus .... Section Diversiformes 4
1. Pileus viscid, often hoary at first; gelatinous layers present on cap as pellicle or just under thin nongelatinous pellicle 2

2. Conspicuously cespitose species with lignicolous habitats, rarely gregarious (certain varieties in Viscosae grow gregariously or scattered on mossy logs) .... Section Caespitosae 6
2. Humus-dwelling forms, usually gregarious or scattered........... 3

3. Stipe with yellow or greenish to greenish-gray tints . . Section Viscosae 13
3. Stipe pallid or grayish to fuscous .... Section Fuliginellae 18

KEY TO SPECIES

DIVERSIFORMES

4. Pileus with palisade layer of pear-shaped cells forming surface layer 203. *M. rorida*
4. Pileus not as above 5

5. Pileus dark gray to blackish on disc, pale gray in age; cheilocystidia clavate-echinulate .... 220. *M. clavicularis*
5. Pileus sordid milk white or cream-colored, disc darker than margin; cheilocystidia with thornlike projections (not echinulate) 204. *M. insignis*

CAESPITOSAE

6. Spores small, not over 6 μ long 7
6. Spores over 6 μ long (6–7 μ in *M. lilacifolia*) 10
7. Pileus tinged with orange or yellow and gills bordered pale yellow or orange .... 209. *M. texensis*
7. Pileus not with orange or yellow 8
8. Pileus plicate-striate; cheilocystidia without pedicels 212. *M. subepipterygia*
8. Pileus not plicate; cheilocystidia with long contorted pedicels...... 9
9. Pileus gray; gills somewhat decurrent and broadest at stipe; pleurocystidia abundant .... 208. *M. euspeirea*
9. Pileus gray; gills adnate, broadest in middle; pleurocystidia rare 207. *M. hondurensis*
10. Fruiting body brilliant yellow to orange 11
10. Fruiting body white to grayish 12
11. Gills yellow, margined with orange .... 210. *M. Leaiana*
11. Gills lilac or pallid .... 211. *M. lilacifolia*
12. Pileus gray, fading to white .... 156. *M. laevigata*
12. Pileus white, gregarious to subcespitose; northern United States 206. *M. Austini*
12. Pileus white, cespitose; southern United States .... 205. *M. glutinosa*

VISCOSAE

13. Cheilocystidia with one or more prominent thornlike projections; often quite irregular in shape .... 213. *M. griseoviridis*
13. Cheilocystidia mostly aciculate .. 214. *M. griseoviridis* var. *cascadensis*
13. Cheilocystidia merely roughened with fingerlike prolongations or rodlike projections 14
14. Taste mild or farinaceous 16
14. Taste strongly rancid-farinaceous, very disagreeable 15
15. Odor subfarinaceous, disagreeable .... 215. *M. viscosa*
15. Odor of iodoform .... 216. *M. viscosa* var. *iodiolens*
16. Usually growing on mossy conifer logs or stumps or on debris of conifers .... 219. *M. epipterygia* var. *lignicola*
16. On humus and carpets of conifer needles 17

17. Pileus dark olive gray, not fading to white .... 217. *M. epipterygioides*
17. Pileus decidedly yellowish and fading to white or pale pearly gray 218. *M. epipterygia*

FULIGINELLAE

18. Pleurocystidia large and conspicuous, projecting 20–30 μ or more beyond hymenium 19
18. Pleurocystidia clavate-echinulate; pileus dry .... 220. *M. clavicularis*
18. Pleurocystidia absent or, if present, almost embedded in hymenium .. 20

19. Taste very strong and disagreeable, resembling raw cucumber 226. *M. tenax*
19. Taste not distinctive .... 225. *M. quinaultensis*

20. Odor fruity .... 224. *M. odorifera*
20. Odor not fruity 21

21. Pleurocystidia present (inconspicuous) 22
21. Pleurocystidia absent 23

22. Gill edge furnished with fusoid cheilocystidia .... 156. *M. laevigata*
22. Gill edge furnished with clavate-roughened cheilocystidia 223. *M. pelliculosa*

23. Gills more or less decurrent or arcuate .... 222. *M. vulgaris*
23. Gills ascending-adnate and more or less toothed .. 218. *M. epipterygia*
23. Gill attachment variable; pileus conic; stipe bulbous.. 221. *M. militaris*

SECTION DIVERSIFORMES

203. Mycena rorida (Fr.) Quélet

Champ. Jura et Vosges, p. 108. 1872

Agaricus roridus Fries, Syst. Myc., 1: 156. 1821.

Illustrations:

Plate 89 B; Text fig. 49, nos. 5–7 (p. 399).

Beardslee and Coker, Journ. Elisha Mitchell Sci. Soc., 40, pl. 10 (lower figs.).

Konrad et Maublanc, Icon. Sel. Fung., 3, pl. 228, II (good).

Lange, Flora Agar. Dan., 2, pl. 54 D.

Pileus (2) 5–10 (15) mm. broad, globose or very broadly convex at first, becoming more or less plane and often with a depressed disc at maturity, margin straight when young, spreading in age and frequently crenate, surface dry and appearing very finely furfuraceous to subpruinose, striatulate on the margin at times, becoming sulcate-striate, colors variable, pale fuscous on the disc when young, brownish toward the whitish margin, fading through pale brown to tan and finally whitish or yellowish white, usually rather sordid in age; flesh thin and moderately fragile, pallid, odor and taste not recorded; lamellae arcuate-adnate, becoming decurrent in age, subdistant, 14–18 reach the stipe, narrow to moderately broad, white, edges even; stipe 2–3 (5) cm. long, 1 mm. ± thick, equal, elastic, base strigose, bluish black near the apex at the very first but quickly fading to whitish, base becoming sordid brownish in age, when young embedded in a thick glutinous sheath, the gluten gradually becoming aggregated toward the base in relatively large masses.

Spores narrowly ellipsoid, 8–10 × 4–5 μ or 9–12 × 4.5–6 μ (four- and two-spored basidia respectively), amyloid (reaction strong); basidia 26–30 × 5–7 μ; cheilocystidia abundant, smooth, fusoid-ventricose to nearly cylindric, often irregular in outline, 26–34 × 6–10 μ; pleurocystidia not differentiated; gill trama reddish to purplish brown in iodine, subhymenium and gill edges not gelatinous; pileus trama corticated by a palisade of inflated pedicellate cells with sordid-brown contents, the cells 25–40 × 15–30 μ above the pedicel, pedicel 12–30 × 3–5 μ, a narrow band of slender hyphae immediately beneath the palisade layer, the remainder of the trama of floccose tissue composed of more or less enlarged hyphae, all parts except the palisade layer reddish to purplish brown in iodine; stipe tissue dark vinaceous brown to purplish in iodine.

Habit, habitat, and distribution.—Gregarious on needle beds under conifers (often in large numbers) and scattered on small sticks or branches of conifer wood, not rare. Fruiting during spring, summer, and fall. Kauffman found it at Echo Lake, Montana, in 1928. I have collected it in North Carolina, New York, Michigan, Washington, Oregon, and California in the United States and in Nova Scotia and Ontario in Canada.

Material studied.—Smith, 327, 732, 767, 3278, 3515, 3845, 3922, 3975, 4778, 4863, 5027, 6305, 7507, 8044, 8350, 8769, 13735, 14124, 14454, 14475, 14639, 16501, 16657, 16702. Kauffman, Montana. Slipp, UIFP: 3037, 3111, Idaho. Wehmeyer, 516a, 642.

Observations.—Both the two-spored and the four-spored forms occur in North America; in fact, one often finds pilei with both two- and four-spored basidia. The four-spored form is more abundant along the Pacific coast, whereas the two spored form is more frequent in eastern North America. The manner of fruiting and the habitat are a bit peculiar at times. In the vicinity of Lake Tahkenitch, Oregon, it frequently occurs on dead twigs of cedar trees six to ten feet or more above ground instead of on the forest floor, where one would expect to find it. There was no evidence that this agaric was in any way parasitic on the trees.

Mycena rorida is one of the most distinctive Mycenae both macroscopically and microscopically. The pale colors of the cap and its dry, somewhat pruinose appearance, in addition to the voluminous gelatinous sheath over the stipe, are reliable field characters. The spores, cystidia, and corticated pileus at once distinguish it microscopically.

204. Mycena insignis Smith

Contrib. Univ. Mich. Herb., *5*, p. 25. 1941

Illustrations:

Text fig. 49, nos. 10–11 (p. 399).

Smith, Contrib. Univ. Mich. Herb., no. 5, pl. 5, fig. 1; pl. 25, figs. 6–7.

Pileus 5–10 mm. broad when young, convex or flattened somewhat on the disc, becoming broadly convex or plane in age, margin appressed against the stipe when young, flaring slightly at maturity and sometimes wavy in age, surface subviscid to viscid and shining when wet, soon dry and dull or very faintly pruinose under a lens, with a very thin but tenacious separable pellicle, faintly translucent-striate to the disc, color more or less dull milk white or the disc "cartridge buff" to "tilleul buff," sordid whitish over all in age; flesh thin, moderately fragile, pallid, odor and taste not recorded; lamellae bluntly adnate when young, soon with a short- or a long-decurrent tooth, close to subdistant, 12–14 reach the stipe, narrow but broadest at point of attachment, 1.5–2 mm. ±, equal, cartilaginous, pliant, shining and viscid when wet, soon dry and dull, base sparsely white-strigose, dull white above.

Spores narrowly ellipsoid, 6–7.5 × 3 μ, amyloid; basidia four-spored, 20–22 × 5–6 μ; cheilocystidia very abundant, 30–37 × 6–10 μ, clavate to fusoid, often forked, the apices evenly tapered to sharp points, often with occasional short projections arising anywhere from near the base to the apex and in an irregular manner, hyaline; no pleurocystidia; gill trama of floccose tissue, subhymenium not gelatinous, gill edges not gelatinous, pale vinaceous brown in iodine; pileus trama with a well-differentiated hypoderm beneath the thin nongelatinous pellicle, the remainder (one half of the trama) floccose, all but the pellicle pale vinaceous brown in iodine, the pellicle composed of very slender threads, which give off short branches having tips that gelatinize somewhat.

Habit, habitat, and distribution.—Gregarious in troops of fifty to several hundred fruiting bodies on needle beds of Douglas fir during June; Olympic Mountains of Washington. The type collection was found above Olympic Hot Springs along the trail to Mt. Appleton at about 2,700 feet elevation, June 5, 1939. Collection 14471 was made along the trail to Sol Duc Park, at an elevation of 3,500 to 4,000 feet, on June 20, 1939.

Material studied.—Smith, 13101, 14471.

Observations.—This is one of the most peculiar species of Mycena collected in the course of the present study. It is easily noticed in the field because of the whitish color assumed by the conifer needles over the area on which the fruiting bodies will develop. In addition, the fruiting bodies themselves have very unusual characters. The cheilocystidia are peculiar, and the pellicle of the pileus is most deceiving. One studying only the dried material would never suspect the species of having either a separable or a viscid pellicle, since the pellicle does not gelatinize in the usual manner when mounted in water or KOH. When plants are collected in dry weather the viscidity of both cap and stipe might not be noticeable. During wet weather, however, both are distinctly gelatinous to the touch and, in fact, remind one of wet specimens of *M. epipterygia*. The gelatinous character of the stipe is caused by a sparse coating of long, slender hyphae ($2 \mu -$) with subgelatinous walls. These hyphae project out from the stipe at first, but as they elongate they become appressed. Their tips become more or less gelatinous when wet.

In general stature this species resembles *M. citrinella*, which, however, apparently belongs in the Viscosae. There is considerable confusion in regard to its distinguishing characters.

SECTION CAESPITOSAE

205. Mycena glutinosa Beardslee

Mycologia, 26: 257. 1934

Illustrations:

Text fig. 50, nos. 8–9 (p. 411).

Beardslee, Mycologia, 26, p. 258.

"Pileo sub-membranaceo, campanulato-convexo, disco demum depresso, striatulato, glabro, viscoso, pellicula gelatinosa, secernibili tecto, albo; stipite albo, glutinoso, pellicula secernibili tecto, basi fibrilloso, lamellis albis, angustis, non confertis, adnatis demum decurrentibus.

"Pileus 1–3 cm. latus. Stipes 6–8 cm. longus. Sporae ellipsoideae 6–8 × 3.5–4 μ. Ad truncos, dense caespitosa, Oviedo Fla. Cystidia rara, parva, acuta.

"This seems to be a very distinct species. It was found at Oviedo Fla., growing in dense masses on old logs. The plants are pure white and both pileus and stipe are very viscid, and both have a tough, gelatinous cuticle, which can easily be stripped off entire. The pileus is convex, and striatulate to the depressed center. The lamellae appear adnate at first but soon seem more or less arcuate decurrent. It could easily be considered an *Omphalia*, but in-so-much as several species of *Mycena* which are closely related to this plant are abnormal for *Mycena* in the same way it has seemed best to place this species with them."

Habit, habitat, and distribution.—Growing in dense masses on old logs; Florida.

Material studied.—Beardslee, December, 1932, Florida. Murrill, September 17, 1938, Florida.

Observations.—The original account is quoted above. The following observations were made from a cluster of fruiting bodies sent to me by Mr. Beardslee: The pellicle on both the pileus and the stipe is easy to demonstrate in sections revived in KOH. The basidia are very small, 14–16 × 4–5 μ, and four-spored. Pleurocystidia are scattered, measure 32–48 × 6–12 μ, and are rather variable in shape. Some are ventricose, with a rounded cylindric neck ending in a rounded apex, and are furnished with a long contorted base; others are narrowly fusoid, with a sharp-pointed apex and the base not at all or only slightly elongated. In some of either type one or two obtuse rodlike projections occur near the apex. The cheilocystidia are of two types. They are either similar to the pleurocystidia or basidium-like, with the apex slightly roughened or of irregular outline. Those of the latter type are more difficult to locate, since they are embedded in the gill edge. The subhymenium is distinctly gelatinous. The floccose portion of both the pileus and the gill trama turns dark reddish brown in iodine and the spores dark blue. Numerous metallic lactiferous hyphae were observed in the tissue of the stipe.

206. Mycena Austini (Pk.) Kühner

Encyc. Myc., 10: 394. 1938

Agaricus (*Omphalia*) *Austini* Peck, Ann. Rep. New York State Mus., 28: 48. 1876.

Omphalia Austini Saccardo, Syll. Fung., 5: 336. 1887.

Omphalopsis Austini Murrill, North Am. Flora, 9: 312. 1916.

"White, rather tenacious; pileus convex or hemispherical, glabrous, striate, deeply umbilicate, sometimes perforate, viscid when moist; lamellae subdistant, decurrent; stem slender, equal, hollow, smooth, villose at the base; spores elliptical, .00025′ long.

"Plant gregarious, about 1′ high, pileus 3″–6″ broad.

"Prostrate dead trunk of a small spruce tree. Providence, Saratoga county. August.

"Dedicated to *Mr. C. F. Austin.*"

Material studied.—The type and Kauffman, September 8, 1914, North Elba, New York.

Observations.—The type specimens closely resemble those of *M. glutinosa* in their macroscopic characters. The spores of Peck's type were found to measure 7–8 × 3.5–4 μ and to be hyaline, smooth, ellipsoid, and strongly amyloid. The basidia were four-spored. Pleurocystidia were scattered, measured 26–28 × 5–7 μ, were subfusiform, and had acute apices that sometimes forked. For some reason or other no good sections of the gill edges were obtained. However, they were apparently gelatinous, and the cheilocystidia were similar to the pleurocystidia or more branched. The subhymenium was gelatinous, and thick gelatinous layers were found over both the pileus and the stipe, thus indicating that both were slimy when fresh. The tramae of the gills and pileus turned slightly yellowish brown in iodine. Lactiferous hyphae were not observed.

There is some question in my mind whether *M. glutinosa* is distinct from *M. Austini*, but I have never seen either one when fresh. The difference in substratum and distribution, the apparent lack of metallic hyphae in the flesh of *M. Austini*, and its slightly different reaction in iodine lead me to keep the two separate—at least until future studies clarify the situation.

207. Mycena hondurensis A. H. Smith

Contrib. Univ. Mich. Herb., 1: 22. 1939

Illustrations:

Text fig. 50, nos. 6–7 (p. 411).

Smith, Contrib. Univ. Mich. Herb., 1, pl. 1, figs. 1–2.

Pileus 1–2 cm. broad, convex to nearly plane in age, the disc broadly depressed, margin connivent with the stipe or slightly incurved at first, surface glabrous, viscid, striate on the margin or nearly to the disc, pale brownish gray over the center, the margin somewhat paler; flesh thin, membranous, pliant, rather tough; lamellae bluntly adnate, subdistant to close, 13–15 reach the stipe, narrow, pallid grayish brown to whitish, edge not differently colored; stipe 2–4 cm. long, 1.5–2.5 mm. thick, equal or slightly enlarged below, glabrous or the apex faintly pruinose, glutinous, tough, hollow, pale brownish gray below, pallid above.

Spores ovoid, 3.5–4.5 × 2.5–3 μ, smooth, amyloid; basidia four-spored, 13–17 × 5–6 μ; cheilocystidia and pleurocystidia similar, 35–50 × 8–14 μ, fusoid-ventricose with long contorted pedicels, smooth, hyaline in KOH; gill trama with a central strand of floccose tissue flanked on either side by the gelatinous subhymenium; pileus trama with a thick layer of gelatinous hyphae over the surface, the hyphae 1.5–2 μ in diameter and with numerous clamp connections, the central portion of floccose tissue with the cells of its hyphae 8–10 μ in diameter, the subhymenium made up of gelatinous narrow hyphae; the floccose tissue of both cap and gill trama faintly vinaceous brown in iodine.

Habit, habitat, and distribution.—Densely cespitose on rotten logs during July; El Cayo District, British Honduras. Known only from the type locality.

Observations.—*Mycena hondurensis* is apparently most closely related to *M. euspeirea* (Berk. & Curt.) Saccardo. I have examined specimens of the latter at the New York Botanical Garden which were collected by Earle (419) in Cuba. The points of similarity between the two are as follows: Both have a thick layer of gelatinous hyphae over the pileus and stipe; both are cespitose and have rather small spores, which become bluish in iodine; and they have similar cystidia. They differ in their gill characters: The gills of *M. euspeirea* are very broad and distinctly adnate-decurrent. Those of *M. hondurensis* are narrow, bluntly adnate, and broader in the middle than at the point of attachment. The appearance of the dried specimens is very different. *M. euspeirea* is pallid to whitish, whereas fruiting bodies of *M. hondurensis* are dark grayish. When revived for sectioning the pilei of *M. euspeirea* are very fragile and gelatinous, and sections soon disintegrate in mounts of KOH. Sections of *M. hondurensis*, however, revive well, are tenacious, and remain firm in KOH. Pleurocystidia are very abundant in the latter and rare in the former; there is a slight difference in spore size.

208. Mycena euspeirea (Berk. & Curt.) Saccardo

Syll. Fung., 5: 287. 1887

Agaricus (*Mycena*) *euspeireus* Berkeley and Curtis, Journ. Linn. Soc., 10: 285. 1868.
Agaricus (*Omphalia*) *subpellucidus* Berkeley and Curtis, *ibid.*, p. 286.
Agaricus (*Omphalia*) *anthiceps* Berkeley and Curtis, *ibid.*
Hiatula caespitosa Berkeley and Curtis, *ibid.*, p. 293.
Omphalia paraguariensis Spegazzini, Anal. Soc. Ci. Argent., 16: 245. 1883.
Omphalopsis euspeirea Murrill, North Am. Flora, 9: 316. 1916.

"Pileus thin, convex, more or less umbilicate, densely cespitose, 1–2.5 cm. broad; surface glabrous, white, subfuscous at the center, margin striate-sulcate, depressed, often elevated with age: lamellae adnate-decurrent, distant, broad, white: spores ellipsoid, smooth, hyaline, $5 \times 3\,\mu$: stipe slender, glabrous, white, pellucid hollow, dilated at the apex, strigose at the base, 2.5–3 cm. long.

"Type locality: Cuba.

"Habitat: On decayed logs in woods.

"Distribution: Cuba, Porto Rico, and Jamaica; also in Paraguay."

Observations.—I have studied Earle's specimens, as previously stated (p. 409), and have found the microscopic characters to be the same as those for *M. hondurensis*, except that the spores are a little larger and the pleurocystidia more abundant. Since my concept has been based on the material at the New York Botanical Garden, Murrill's synonymy and description are given above. I have not seen fresh specimens and have not had an opportunity to study the type of *M. euspeirea* or the types of the species Murrill placed in synonymy with it. For a comparison with *M. hondurensis* see page 409.

209. Mycena texensis Smith

Mycologia, 29: 341. 1937

Illustrations: Text fig. 50, nos. 1, 3.

Pileus 8–15 mm. broad, oval to convex becoming broadly umbonate, the disc sometimes slightly depressed in age, margin appressed against the stipe at first, surface glabrous, viscid, striate, margin somewhat sulcate in age, color nearly white when young, disc becoming bluish fuscous or dark grayish brown tinged with orange and the margin pale cinereous to whitish or tinged orange yellow;

FIG. 50. *M. texensis:* 1, cheilocystidia; 3, spores. *M. Leaiana:* 2, cheilocystidia; 4, pleurocystidia; 5, spores. *M. hondurensis:* 6, pleurocystidia × 1650; 7, spores. *M. glutinosa:* 8, spores; 9, cheilocystidia and pleurocystidia

flesh thin and pliant, odor and taste not known; lamellae arcuate-decurrent, subdistant to close, 20–24 reach the stipe, moderately broad (3 mm.), whitish or pale yellowish to orange, margin a deeper orange; stipe 4–8 cm. long, 1–1.5 mm. thick, equal, viscid, with a dense orange-pruinose covering toward the apex, the base densely white- or grayish-strigose and rooting somewhat in the substratum, yellowish to orange above, becoming whitish, sordid grayish, or sordid grayish brown toward the base.

Spores ovoid, 4.5–6 × 3.5 μ, amyloid; basidia two-spored or occasionally four-spored, 18–22 × 4–5.5 μ; cheilocystidia fusoid-ventricose at first, soon with the apex more or less lobed or divided, sometimes furnished with obtuse knoblike protuberances, variously contorted and irregular in outline, 22–32 × 5–9 μ; pleurocystidia absent; gill trama with a gelatinous subhymenium in revived material, the floccose central strand vinaceous brown in iodine; pileus trama with a thick gelatinous surface pellicle, the tissue beneath floccose and vinaceous brown in iodine, the subhymenium gelatinous.

Habit, habitat, and distribution.—Densely cespitose on oak logs and stumps during rainy weather in September; Florida and Texas.

Observations.—When sections are mounted in iodine the surface of the gelatinous pellicle appears bluish because of the large number of spores that have accumulated there. *M. texensis* is very closely related to *M. Leaiana*. The gelatinous pellicle over the pileus and stipe, the orange-pruinose stipe and yellowish to orange gills with brighter edges, and the densely cespitose habit all clearly relate it to that species. It differs in the very small spores, the short narrow basidia, and the grayish colors of the pileus as well as in the shape and distribution of the cystidia. I have not seen fresh material.

210. Mycena Leaiana (Berk.) Saccardo

Syll. Fung., 9: 38. 1891

Agaricus (*Mycena*) *Leaiana* Berkeley, Lond. Journ. Bot., 4: 300. 1845.
Collybia Leaiana Fairman, Proc. Rochester Acad. Sci., 2: 155. 1893.
Prunulus Leaianus Murrill, North Am. Flora, 9: 333. 1916.

Illustrations:

Plate 90; Text fig. 50, nos. 2, 4–5 (p. 411).
Beardslee and Coker, Journ. Elisha Mitchell Sci. Soc., 40, pl. 12 (lower)
Güssow and Odell, Mushrooms and Toadstools, pl. 56, fig. 1.

Hard, The Mushroom, Edible and Otherwise, fig. 94.
White, Conn. State Geol. and Nat. Hist. Surv., Bull. 15, pl. 11.

Pileus 1–4 cm. broad, obtusely campanulate when young, the margin sometimes bent inward slightly, soon broadly campanulate to convex, the disc sometimes finally slightly depressed, surface naked or scurfy-pruinose at times, viscid, pellicle separable, shining in age and glabrous, color brilliant reddish orange when young, becoming paler and more yellowish ("mikado orange" to "cadmium orange"), finally fading to nearly white, opaque or translucent-striate when moist; flesh rather thick, watery-whitish beneath the orange pellicle, pliant, odor subfarinaceous, taste slight and not distinctive; lamellae adnate, becoming sinuate, broad, finally ventricose (5–8 mm.), close to crowded, thickish at times, faces "light ochraceous salmon," readily staining orange yellow when bruised, edges brilliant reddish orange and even; stipe 3–7 cm. long, 2–4 mm. thick, equal or flaring slightly at the apex, even or transversely undulate, hollow, tough-cartilaginous, base densely strigose, covered with an orange scurfy pulverulence, glabrescent in age, viscid to lubricous, orange to yellow over all or pallid near the apex, with a scant watery orange juice.

Spores 7–9 (10) × 5–6 µ, ellipsoid, smooth, amyloid, basidia four-spored; pleurocystidia scattered to abundant, 32–46 × 8–13 µ, with a pale-orange content, ventricose and mucronate, smooth, sessile, or pedicellate; cheilocystidia very abundant, 26–38 × 6–14 µ, variable in shape, fusoid-ventricose, clavate, subcylindric, or with one or more knoblike protuberances on the upper portion; gill trama with a broad central strand of more or less parallel hyphae which become reddish brown in iodine, subhymenium broad, of narrow hyphae which gelatinize in KOH, nonamyloid; pileus trama covered with a thick gelatinous pellicle, the hyphae of which contain a brilliant orange substance, homogeneous below, and vinaceous brown in iodine; stipe with a thick gelatinous outer sheath, the cells with orange contents.

Habit, habitat, and distribution.—Cespitose on wood of various hardwoods, throughout eastern and central United States and Canada. In northern regions it occurs on the wood of the bog alder, *Alnus incana.* I have not found it along the Pacific coast. I have examined fresh material from North Carolina, Tennessee, Ohio, Indiana, Michigan, and Missouri in the United States and from Nova Scotia and Ontario in Canada.

Material studied.—Smith, 32–80, 32–110, 32–572, 32–641, 33–366,

33–403, 33–514, 33–771, 123, 414, 533, 1052, 4585, 5074, 6243, 6300, 6399, 7236, 7587, 7658, 9340, 9539, 10153, 10405, 15083, Michigan, October 1, 1931; September 6, 1932; June 5, 1941. Atkinson, eleven packets. Ellis and Everhardt, North American Fungi no. 3506. Hesler, 10410. Kauffman, Kentucky, New York, Michigan. Kelly, 1077, 1678, 1702. Mains, 5068. Oakes, July 20, 1922. Overholts, 869, 925, 1534, 1535, 4278, 4563, 9851, 11070. Povah, FP 14, FP 258, Fl. 150. Wehmeyer, 568, 568a.

Observations.—Beardslee and Coker pointed out that pleurocystidia as well as cheilocystidia are present. Both types have been found on all collections in the University of Michigan Herbarium. The gelatinous subhymenium is not always readily demonstrable in water mounts of fresh material but shows up well in sections revived in KOH. The mycelium is often orange, but tends to fade where exposed to light. The species is not closely related to Heliomyces, as has been suggested, at least no more so than *Hygrophorus laetus* or other Mycenae with gelatinous subhymenia.

211. **Mycena lilacifolia** (Pk.), comb. nov.

Agaricus lilacinus Peck, Ann. Rep. New York State Mus., 24: 63. 1872.
Agaricus lilacifolius Peck, *ibid.*, 29: 66. 1878.
Omphalia lilacifolia Peck, *ibid.*, 45: 94 (34). 1893.
Omphalina lilacifolia Murrill, North Am. Flora, 9: 346. 1916.
Clitocybe lilacifolia Singer, Lloydia, 5: 105. 1942.

Pileus 8–25 mm. broad, broadly convex with a flattened disc and an incurved margin when young, in age with a flattened or somewhat depressed disc, the margin usually remaining sloped abruptly downward (as in a truly conic species, thus causing the cap to be helmet-shaped), surface glabrous, viscid to glutinous, translucent-striate, "Naples yellow" on the disc, "cream color" toward the margin (bright to pale yellow), buttons often dull lavender but quickly changing to yellow, odor and taste not distinctive; lamellae "pale lilac," gradually becoming pallid or retaining a lilac tinge, subdistant to moderately close, 23–26 reach the stipe, two tiers of lamellulae, narrow (2 mm. ±), unequally decurrent, at first arcuate to subdecurrent, edges even; stipe 1–3 (4.5) cm. long, 1–2.5 mm. thick, equal or the base slightly enlarged, tubular, cartilaginous-fragile, glabrous, slimy-viscid, concolorous with gills when very young, lower portion soon becoming yellow but the basal mycelium remaining lilac, in age pallid above or with a persistent lilac tint.

Spores 6–7 × 3–3.5 μ, smooth, subellipsoid, nonamyloid; basidia four-spored; pleurocystidia and cheilocystidia not differentiated; gill edges not gelatinous; gill trama of a mixture of inflated thin-walled hyphal cells and numerous long, narrow contorted lactifers, subhymenium very thin, of narrow hyphae; pileus trama with a thick gelatinous pellicle, the hyphae of which have yellow contents, the remainder a mixture of floccose hyphae and lactifers, all parts, including the pellicle, reddish brown to more or less vinaceous brown in iodine.

Habit, habitat, and distribution.—Scattered, gregarious, or subcespitose on decaying wood of balsam, fir, or hemlock; Tennessee, New York, Michigan, Washington, and Oregon in the United States and Ontario in Canada. I have found it in both spring and fall, but it appears to be rather uncommon.

Material studied.—Smith, 33–376, 33–540, 16719, 16862. Gruber, Oregon. Hesler, 11536. Kauffman, six collections, New York and Michigan. Kelly, 1206.

Observations.—The carpophores have pallid gills immediately after being dried, but they often change slowly to yellowish brown. The pilei and stipes dry a dull yellow. This is a rather anomalous species by virtue of its gill color, nonamyloid spores, and deep reddish-brown flesh of the pileus in iodine. The iodine reactions of the flesh of the pileus are not vinaceous as in other members of this group. The relationships of the species appear to be here rather than in any other section of Mycena or in any other genus. The decurrent gills, which caused Peck to place it in Omphalia, cannot be regarded as an important generic character in this instance because many of the viscid-stiped Mycenae also have them. Singer (1942) has placed it in Clitocybe, but it does not appear to be related to the lilac-gilled species of that genus.

212. Mycena subepipterygia Murrill

Bull. Torr. Bot. Club, 67: 235. 1940

Prunulus subepipterygius Murrill, *ibid.*, 233. 1940.

Illustrations: Text fig. 51, nos. 1–2 (p. 410).

"Pileus convex to plane, depressed at the center, gregarious to cespitose, 1–2 cm. broad; surface slimy-viscid, smooth, glabrous, sulcate, white, umbrinous on the disk; context membranous, white; lamellae adnexed, narrow, tapering behind, medium distant, inserted,

FIG. 51. *M. subepipterygia:* 1, cheilocystidia; 2, spores. *M. viscosa:* 3, cheilocystidia; 9, spores. *M. epipterygioides:* 4, spores; 5, cheilocystidia. *M. epipterygia:* 6, cheilocystidia; 7, spores. *M. epipterygia* var. *lignicola:* 8, spores

entire, white, unchanging; spores ellipsoid, smooth, hyaline, 1-guttulate, 5–6 × 3–4 μ; cystidia subcylindric, smooth, hyaline, projecting 15–30 × 15 μ, occupying the entire edge of the gill; stipe smooth, white, glabrous, viscid, ridged at the apex, 1–3 × 0.1–0.2 cm.

"Type collected by West and Murrill on a hardwood log in Planera Hammock, eleven miles northwest of Gainesville, Fla., July 20, 1938 (*F 18363*). Also collected at the same place by West, Arnold and Murrill on an oak log, July 27, 1938 (*F 17910*). Evidently near *P. epipterygius* (Scop.) Murrill but having a shorter stem and smaller spores."

Through the kindness of Dr. Murrill I have been able to examine a part of the type. The spores, which in my mounts measured 5–7 × 3.5–4 μ, were smooth, ellipsoid, and not distinctly amyloid. This last character is very important in this group since in most species the spores are strongly amyloid. The basidia are short and broad, 16–18 × 7–9 μ, and four-spored. Pleurocystidia were not seen, but the cheilocystidia, as Murrill pointed out, form a sterile band along the edge and are very conspicuous. They measure 28–42 × 10–25 μ, and when young are saccate to subventricose or broadest at the base. The walls are delicate and collapse readily. In older specimens the apex frequently becomes drawn out to a point. The gill edge is only slightly gelatinous beneath the cystidia and does not fan out in mounts of sections, as it does in many other species. The subhymenium is gelatinous. The central strand of the gill trama is floccose and vinaceous red in iodine. The pileus trama is made up of a gelatinous pellicle, which in sections of revived material occupies three fourths of the thickness of the cap. The floccose portion, which becomes vinaceous red in iodine, is thus reduced to a narrow band. A gelatinous outer layer covers the stipe. In the youngest specimen examined large caulocystidia were observed over the surface. These measured 26–48 × 10–15 μ and were clavate to subcylindric. They collapsed readily and were difficult to find on the oldest fruiting body. Their presence no doubt caused the stipe to be pruinose at first, and may account for the ridges Murrill observed at the apex. The ridges were not distinct in the dried specimens. The striations of the pileus appear to be of the plicate type as found in some dark-spored agarics. They extend to the disc in dried material and form a very striking macroscopic character.

Although this species keys out near *M. euspeirea* and *M. hondurensis*, it cannot be considered closely related to them. The iodine reaction of the spores distinguishes it at once, as do its cheilocystidia and plicate-striate cap. For the same reasons it cannot be considered closely related to the members of the section Viscosae although, when dried, carpophores do remind one of a squatty *M. epipterygia*.

SECTION VISCOSAE

213. Mycena griseoviridis, sp. nov.

Illustrations: Plate 91; Text fig. 52, nos. 1–2.

Pileus 1–3.5 cm. latus, obtuse conicus vel convexus, crenatus, pruinosus, viscidus vel glutinosus, olivaceo-fuscus vel citrino-fuscus; sapor et odor farinaceo-rancidus; lamellae adnatae, confertae vel subdistantes, latae, griseo-albidae vel citrino-griseae, rufomaculatae; stipes (3) 5–9 cm. longus, 1–2.5 mm. crassus, aequalis, pruinosus, viscidus; luteo-citrinus demum pallide griseus; sporae 9–11 × 5–6.5 μ; basidia tetraspora; cheilocystidia 30–60 × 6–11 μ, distinctissima. Specimen typicum in Herb. Univ. Mich. conservatum. Legit A. H. Smith, n. 15498, prope Ann Arbor, Mich., Oct. 30, 1940.

Pileus 1–3.5 cm. broad, ovoid when young, soon becoming broadly conic to convex, in age broadly campanulate to convex, margin exceeding the gills and appressed against the stipe when young, soon lacerated or crenate and frequently flaring, conspicuously white-pruinose over all and only slowly becoming naked and polished, glutinous, pellicle separable, opaque to faintly translucent-striate, merely slightly sulcate along the margin in age, "deep olive" to "olive brown" to "clove brown" and finally blackish, sometimes "Chaetura black" with a hoary sheen when young, thin, tenacious, dark grayish citrine becoming greenish brown in age, odor and taste strong, reminding one of green cucumbers or much more disagreeable; lamellae ascending-adnate, with a slight tooth in age, close to subdistant, 16–22 reach the stipe, moderately broad, white with a greenish tinge, soon greenish gray and becoming spotted with reddish-brown stains, edges even and pallid; stipe (3) 5–9 cm. long, 1–2.5 (3.5) mm. thick, equal, terete or compressed, tenacious, tubular, base somewhat white-strigose, densely white pruinose over all at first, finally naked and shining, very viscid, "light dull green yellow" to "courge green," sometimes rather bright yellowish fading to pearly gray and usually sordid purplish brown toward the base.

Spores ellipsoid, 9–11 × 5–6.5 μ, amyloid (reaction weak); basidia

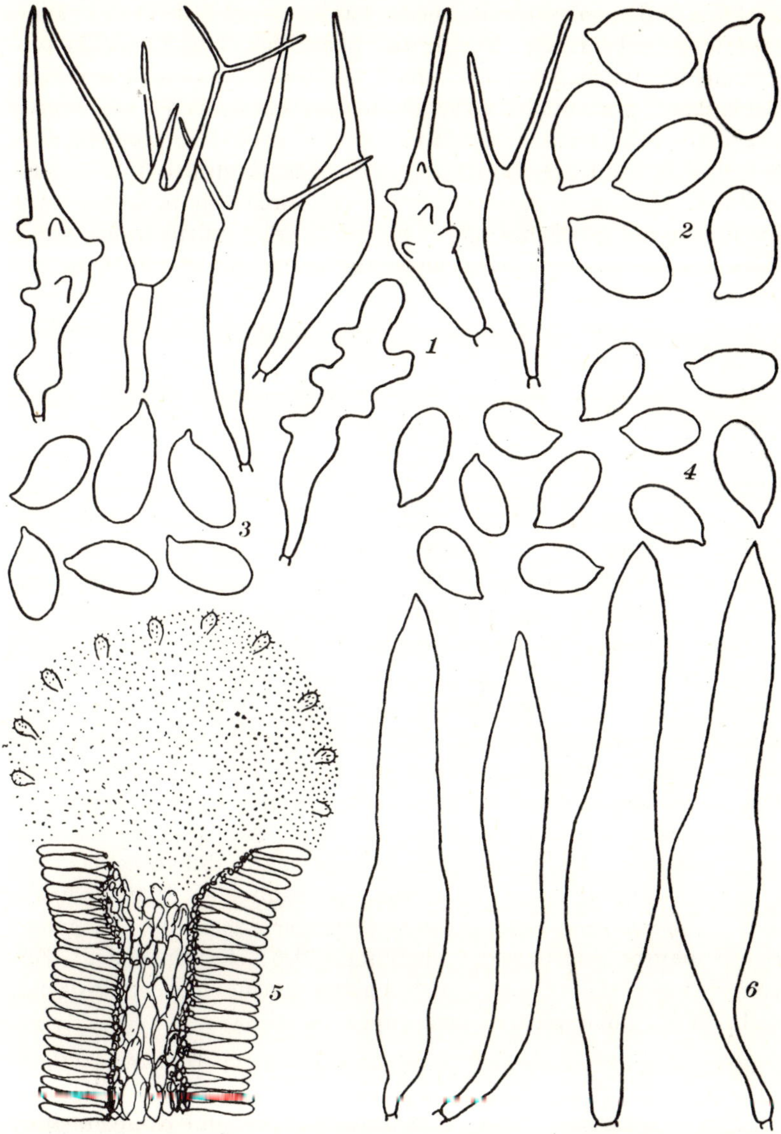

FIG. 52. *M. griseoviridis:* 1, cheilocystidia; 2, spores. *M. vulgaris:* 3, spores; 5, gill edge × 315. *M. tenax:* 4, spores; 6, pleurocystidia

32–36 × 7–8 μ, four-spored; cheilocystidia 30–60 × 6–11 μ, hyaline, clavate at first but soon fusoid, with one to four long needlelike projections 15–30 μ long and 2–3 μ thick, the projections simple or forked, the enlarged portion of the cystidium sometimes covered with obtuse irregular protuberances in addition to the needlelike projections (a few cheilocystidia are clavate and more or less contorted); no pleurocystidia; gill trama floccose and faintly brownish in iodine, subhymenium not gelatinous, gill edges gelatinizing; pileus trama with a thick gelatinous separable pellicle, the remainder of the tissue not appreciably differentiated, floccose and brownish in iodine; stipe with a gelatinous outer layer, the inner tissue vinaceous red in iodine.

Habit, habitat, and distribution.—Gregarious to scattered in oak or pine woods; Tennessee, Michigan, Oregon, and California in the United States and Quebec in Canada. It occurs either at low elevations in the fall of near melting snowbanks in the mountains in the summer.

Material studied.—Smith, 6159, 6165, 15498, 15516, November 18, 1931, Michigan. W. B. Cooke, 1557. O. E. Fischer, October 16, 1905, Detroit (determined as "*M. epipterygia* form A" by Kauffman). Grove, F 9719 (collected by Drayton), Quebec. Hesler, 13164.

Observations.—The cheilocystidia are the most distinctive character of the species, and they separate it at once from *M. epipterygioides* Pearson. There are also differences in the colors and the manner in which they change, but in view of the great variation of yellow, green, and gray in this group, I hesitate to emphasize them. One of two specimens of *M. viscosa* sent to me by Dr. Rolf Singer has cheilocystidia like those of *M. griseoviridis;* the other, like those of *M. epipterygia*. In the American collections this difference has been found constant, and serves to distinguish readily both *M. griseoviridis* and *M. epipterygia*. On the basis of Singer's specimen, *M. griseoviridis* may be said to occur in Europe also. Kauffman's *M. epipterygia* var. *A* also belongs here, but his specimens blackened in drying and now are not typical in their appearance.

214. Mycena griseoviridis var. **cascadensis,** var. nov.

Pileus 8–15 mm. latus, obtuse conicus demum campanulatus, striatus, viscidus, luteo-griseus vel luteo-brunneus, demum sordide luteus; lamellae pallide luteae, subdistantes, adnatae, angustae; stipes 1–3 cm. longus, 1–1.5 mm. crassus, laete luteus, viscidus; sporae 8–10 × 5–7 μ; cheilocystidia aciculata vel subventricosa,

32–48 × 5–9 μ. Specimen typicum in Herb. Univ. Mich. conservatum. Legit A. H. Smith, n. 16656, prope Noisy Creek, Baker Lake, Wash., Sept. 5, 1941.

Pileus 8–15 mm. broad, subovoid when young, becoming obtusely conic and finally obtusely campanulate, surface hoary at first but soon polished, translucent-striate with dark striations, sulcate in age, viscid when wet, pellicle thick and tenacious, completely separable, color variable, a mixture of yellow, brown, and gray, dark on disc when young, pale yellow on margin, pale yellowish over all except the disc and striae in age; flesh thin, pliant, odor and taste strongly farinaceous; lamellae pallid yellowish, subdistant, ascending, hooked, edges even; stipe short, 1–3 cm. long, 1–1.5 mm. thick, pale clear yellow, glutinous, glabrous, base faintly strigose.

Spores 8–10 × 5–7 μ, amyloid, ellipsoid to ovoid (a few subglobose); basidia four- or occasionally two-spored; gill edges heteromorphous, the cheilocystidia very abundant, awl-shaped to subventricose, smooth or with somewhat irregular walls, gelatinizing only in age (gelatinization confined mostly to the pedicel), 32–48 × 5–9 μ; pleurocystidia embedded and very difficult to locate, scattered, similar to cheilocystidia; gill and pileus trama weakly amyloid.

Habit, habitat, and distribution.—Scattered on conifer logs (*Abies sp.*); Noisy Creek, Baker Lake, Washington. Known only from the type locality.

Observations.—This fungus is readily distinguished from *M. epipterygia* var. *lignicola* by its awl-shaped cheilocystidia. In color, odor, and taste, as well as in general habit, the two are very similar and can easily be confused unless the gill edges are studied. From the typical variety of the species it is distinguished by its more intense colors, habitat, and simple or seldom-branched cystidia.

215. Mycena viscosa (Sec.) Maire

Bull. Soc. Myc. Fr., 26: 162. 1910

Agaricus alcalinus viscosus Secretan, Mycogr. Suisse, 2: 312. 1833.
Mycena splendidipes Peck, Bull. New York State Mus., 167: 28. 1913.
Prunulus splendidipes Murrill, North Am. Flora, 9: 330. 1916.
Mycena epipterygia var. *viscosa* Ricken, Die Blätterpilze, p. 419. 1915.

Illustrations:

Plate 92; Text fig. 51, nos. 3, 9 (p. 416).
Konrad et Maublanc, Icon. Sel. Fung., 3, pl. 227 (very good).
Peck, Bull. New York State Mus., 167, pl. 10.

Pileus 8–10 mm. broad, ovoid when young, becoming obtuse to convex, often rather broadly expanded or slightly umbonate in age, margin appressed against the stipe by a sterile narrow band which soon becomes lacerated or crenate as expansion takes place, margin finally flaring at times, surface covered with a completely separable tenacious pellicle, when young white-pruinose but finally naked and shining, very viscid, often sulcate-striate in age, color variable but usually yellowish gray, yellowish, or greenish gray, in age becoming sordid brownish; flesh thin, tenacious, concolorous with the surface, sordid reddish in age, odor strong—somewhat resembling that of fresh cucumbers—taste rancid-farinaceous and very strong; lamellae adnate to arcuate or with a distinct tooth, subdistant, 18–26 reach the stipe, two tiers of lamellulae, narrow to moderately broad, whitish, yellowish, or tinged greenish gray, edges even and pallid, often spotted reddish brown in age; stipe 3–7 (10) cm. long, 1–2 (3) mm. thick, equal, strict, or somewhat flexuous, tenacious, faintly pruinose over all when young but soon shining and glutinous, lemon yellow throughout or tinged greenish yellow, in age usually reddish at the base.

Spores drop-shaped to subovoid, 9–11 $\times$ 6.5–8 μ, on two-spored basidia, ellipsoid and 8–10 $\times$ 5–6.5 μ on four-spored basidia, amyloid; basidia 28–33 $\times$ 6–7 μ; cheilocystidia abundant, 40–60 $\times$ 6–9 μ, with long flexuous gelatinizing pedicels, apices clavate and covered with short contorted projections or obtuse irregular fingerlike processes; gill trama floccose, pale vinaceous brown in iodine, gill edge very gelatinous, subhymenium not gelatinous; pileus trama with a thick separable gelatinous pellicle, hypoderm differentiated (the cells moderately enlarged), the remainder of the trama filamentous, all but the pellicle vinaceous brown in iodine; stipe with an outer gelatinous layer, the portion within vinaceous red in iodine.

Habit, habitat, and distribution.—Gregarious under oak and pine; Alabama, Tennessee, Michigan, Oregon, and California. It usually occurs late in the fall after frosts.

Material studied.—Smith, 32–554, 32–617, 32–642, 32–657, 33–1143, 1125, 3712, 5035, 6158, 8109, 18118, November, 1932, Michigan. Burke, Alabama. Hesler, 10132.

Observations.—The strong odor and taste and changing flesh are the important characters which separate this species from *M. epipterygia*. Both occur in coniferous woods in North America. The colors of the American specimens are not whitish at first, but that seems to be the only difference between them and European material. *M. splendidipes* Peck is described as having yellow in the cap but is otherwise similar to *M. viscosa* as described by Kühner. In such a group as this, in which yellow, green, and gray colors occur in mixed and confusing proportions, I hesitate to use a single color variation as a specific character.

216. Mycena viscosa var. **iodiolens,** var. nov.

Aetate marcescens valde odorata, odore iodoformi similis; pileus 1–3 cm. latus, obtuse conicus vel campanulatus, glaber, viscidus, striatus, cinereus, rufo-maculosus; lamellae subdistantes, pallidae, angustae; stipes 5–9 cm. longus, 2–3 mm. crassus, flavidus; sporae 8–10 × 4.5–5.5 μ. Specimen typicum in Herb. Univ. Mich. conservatum. Legit A. H. Smith, n. 18205, prope Detroit, Ore., 1941.

Pileus 1–3 cm. broad, obtusely conic, becoming obtusely umbonate to nearly plane, glabrous, viscid, striate, pale gray to dark gray with hardly a tinge of yellow, in age very pale cinereous and spotted-reddish, striate, becoming sulcate-striate; flesh tenacious, thin, taste and odor disagreeable and subfarinaceous when first collected, within an hour the odor changing and becoming very strong and reminding one of iodoform; lamellae subdistant, ascending, hooked, pallid, narrow; stipe 5–9 cm. long, 2–3 mm. thick, glutinous, pliant, bright pale yellow, base strigose.

Spores 8–10 × 4.5–5.5 μ, narrowly ellipsoid, smooth, weakly amyloid; basidia four-spored; cheilocystidia subgelatinous and causing the gill edges to enlarge somewhat in KOH, filamentous, 30–60 × 4–6 μ, covered with short rodlike projections; no pleurocystidia; pileus trama with a very thick gelatinous pellicle; pileus and gill trama faintly vinaceous in iodine.

Habit, habitat, and distribution.—Gregarious under cedar and alder; Santiam River near Detroit, Oregon. Known only from the type locality.

Observations.—This variety is distinguished from other members of the group by the tardily developing odor of iodoform and the narrower spores, and apparently by the slender cheilocystidia. Since the differences in the last two characters are slight, the fungus has been given only varietal rank.

217. Mycena epipterygioides Pearson

Trans. Brit. Myc. Soc., 6: 135. 1918

***Mycena epipterygia* var. *epipterygioides* Kühner, Encyc. Myc., 10: 353. 1939.**

Illustrations:

Plate 93 B; Text fig. 51, nos. 4–5 (p. 416).

Lange, Flora Agar. Dan., 2, pl. 58 C (as *M. pelliculosa*).

Pileus 1–2.5 cm. broad, ovoid to obtusely conic, soon broadly conic to convex, sometimes nearly plane in age, often with a somewhat flattened disc or a low umbo, margin appressed against the stipe when young as well as exceeding the gills and often becoming somewhat crenate, often flaring in age, surface hoary-pruinose when young, finally naked, viscid to glutinous, pellicle completely separable, sulcate-striate at maturity, color "deep olive buff" to "citrine drab," fading to "dark olive buff" or "deep olive buff," never white; flesh thin, tenacious, dark olive brownish, odor faint, reminding one of fresh cucumbers, taste farinaceous but not strong; lamellae ascending but bluntly adnate, sometimes toothed; subdistant to close, 18–23 reach the stipe, two tiers of lamellulae, moderately broad, sometimes separating from the stipe but remaining attached to each other and thus forming a collar, white or gray-tinged, edges pallid and even, becoming reddish-spotted in age; stipe 4–6 (7) cm. long, 1.5–2.5 mm. thick, equal, straight or flexuous, glabrous, glutinous, base with scattered fibrils, color pale greenish yellow, whitish or grayish above in age and base frequently reddish brown.

Spores ellipsoid, 8–10 (12) $\times$ 5–6 μ, amyloid; basidia four-spored, 32–36 $\times$ 6–7 μ, sterigmata stout; cheilocystidia embedded in a gelatinous matrix, narrowly clavate, the upper portion with obtuse contorted projections which may or may not be branched, the head 5–8 μ thick, branches 2–4 μ thick, pedicels indistinct; no pleurocystidia; gill trama floccose, subhymenium not gelatinous, gill edge gelatinizing, floccose tissue vinaceous brown in iodine; pileus trama with a thick gelatinous pellicle, floccose below the pellicle but the hyphae moderately broad, all but the pellicle vinaceous brown in iodine; stipe with a thick outer gelatinous layer, the inner portion vinaceous red in iodine.

Habit, habitat, and distribution.—Gregarious under conifers or also under oak, usually appearing late in the fall after the first heavy frosts. I have collected it in Michigan, Washington, and Oregon. Hesler has found it in Tennessee and Burke in Alabama. In the vicinity of McKenzie Pass, Oregon, and also later during the season in 1937 at Takilma, Oregon, it literally covered the carpets of pine needles under open stands of ponderosa pine.

Material studied.—Smith, 3446, 6151, 6750, 7930, 8911, 18149, November 7, 1941. Burke, Alabama, 1942. Hesler, Tennessee.

Observations.—The color of the stipe when fresh varies from yellow to greenish, but in dried material it is always a pale yellow. My collections have been compared with material from England which Mr. Pearson very kindly sent me. The American specimens are similar to the English material in every respect except in the number of spores borne on a basidium and in spore size. I regard the slight difference in the width of the spores as correlated with the two-spored basidia and of no taxonomic significance.

There is some doubt in my mind whether or not Pearson's species is distinct enough from *M. viscosa* to justify the position given it here. Kühner has reduced both *M. viscosa* and *M. epipterygioides* to varieties under *M. epipterygia.* As I have frequently pointed out, the reddish stains that develop on the gills and stipe in old carpophores are an unreliable character in many groups of Mycenae. The colors of the pilei in this group are inconstant within certain limits, and one can reasonably allow considerable variation in the intensity of the taste of any species. These are all relative characters which the collector must learn to evaluate. The only justification I have for giving *M. epipterygia*, *M. viscosa*, and *M. epipterygioides* specific rank here is that they are recognizable as distinct entities in our flora and appear to be as constant in their characters as most other members of the genus.

218. Mycena epipterygia (Fr.) S. F. Gray

Nat. Arr. Brit. Plants, 1: 619. 1821

Agaricus epipterygius Fries, Syst. Myc., 1: 155. 1821.
Prunulus epipterygius Murrill, North Am. Flora, 9: 335. 1916.
Mycena paludicola Murrill, Mycologia, 8: 221. 1916.
Prunulus paludicola Murrill, North Am. Flora, 9: 336. 1916.

Illustrations:

Plate 95 A; Text fig. 51, nos. 6–7 (p. 416).
Atkinson, Mushrooms, Edible, Poisonous, etc., 1900, fig. 96.
Beardslee and Coker, Journ. Elisha Mitchell Sci. Soc., 40, pl. 11.
Lange, Flora Agar. Dan., 2, pl. 58 A.
Ricken, Die Blätterpilze, 2, pl. 109, fig. 12.

Pileus 8–20 mm. broad, ovoid at first, soon with a prominent umbo, margin appressed against the stipe at first, sometimes flared somewhat in unexpanded individuals, uneven at times or lacerated but not conspicuously crenate, surface glabrous, viscid, and with a tenacious separable pellicle, becoming plicate-striate at maturity, color at first "old gold" on the disc, "mustard yellow" toward the margin, fading slowly to whitish, and in age flushed or tinged with pink and gray (yellow dominant at first, a secondary tinge of greenish gray present but only enough to tone the yellow slightly, when dried the caps becoming pale gray or yellowish gray); flesh thin, pliant, yellowish, not staining when bruised or in age, odor faintly fragrant or lacking, taste mild; lamellae ascending-adnate, sometimes hooked, subdistant, 14–17 reach the stipe, two tiers of lamellulae, narrow, whitish or faintly yellowish, edges even and pallid; stipe 6–8 cm. long, 1–2 mm. thick, tubular, pliant, slimy-viscid, base faintly strigose, evenly "lemon chrome" when young, exposed portions fading to white (the portion protected by moss or debris remaining bright lemon yellow), becoming yellow again when dried, no reddish stains present even on very old specimens.

Spores ovoid, 8–10 (11) $\times$ 5–6 μ, amyloid; basidia four-spored, 22–24 $\times$ 7–8 μ; cheilocystidia clavate, the apices roughened with short rodlike projections, gelatinizing as in *M. vulgaris* (measurements were difficult to make because of the gelatinous pedicels), clavate, the upper portion 4–8 μ broad; no pleurocystidia; gill trama floccose, vinaceous brown in iodine, subhymenium not gelatinous in KOH, gill edge decidedly gelatinous; pileus with a thick gelatinous pellicle, a slightly differentiated hypoderm, and beneath it a region made up of rather broad filamentous hyphae, all but the pellicle vinaceous brown in iodine; stipe with a broad gelatinous sheath, reddish brown within when treated with iodine.

Habit, habitat, and distribution.—Scattered to gregarious under conifers during late August, September, October, and November; Tennessee, New York, Michigan, Washington, Oregon, and California in the United States and Manitoba in Canada.

Material studied.—Smith, 3154, 3208, 3271, 3331, 3378, 3450, 3451, 3478, 4172, 4214, 7940, 7972, 8448, 8628, 8664, 16768, 17641, 18060. Atkinson, numerous collections, vicinity of Ithaca. These were not studied in detail. Brown, 1925, Lake Quinault, Washington. Gruber, Oregon. Hesler, 3479, 3608, 8448. Kauffman, New York;

Lake Quinault, Washington; Mt. Hood, Oregon. Mains, 32–415, 34–193. Morse, January 7, 1934, California.

Observations.—The outstanding characters of this species are the faint odor (or lack of odor), mild taste, bright yellow stipe, greenish- to brownish-yellow cap, which becomes whitish to grayish or tinged incarnate, and lack of reddish stains on gills and cap. Lange (1936) described the cap as "livid brownish to livid white." "Livid white" characterizes the faded pilei very well. My material was anything but livid brownish when fresh. An examination of Lange's illustration at once makes it apparent, however, that no real discrepancy existed between the color of my specimens when they were fresh and the color of those from Denmark. Yellow is the dominant color in his illustration, which, when it is matched with Ridgway, is found to be remarkably close to "old gold."

An examination of the type of *M. paludicola* Murrill shows that it is unquestionably *M. epipterygia*. Its spores measure 9–11 × 5–6 μ and are smooth, hyaline, ellipsoid, and amyloid. Its basidia are four-spored, and the cheilocystidia gelatinize, producing the characteristic inflated gill edge. In Murrill's specimens the cheilocystidia have gelatinized to such an extent that their outlines are very indistinct. This frequently happens in old specimens of *M. epipterygia*. No pleurocystidia are differentiated. The pileus trama is covered with the remains of a viscid pellicle. This layer was overlooked in my first examination of the type but has since been clearly observed. Its hyphae have almost completely gelatinized, and its real thickness could be observed in only one or two sections. The stipe possessed similar remnants of gelatinous hyphae. The iodine reactions of the flesh of the pileus, gills, and stipe were a bit weaker than those of typical *M. epipterygia*, but still definitely amyloid.

After collecting and studying large quantities of this species from the pine forests along the ocean in Oregon and California in 1935 and 1937, I was able to recognize Murrill's "species" as merely old fruiting bodies of *M. epipterygia*. The telltale characters are the white apex and the yellow base of the stipe, the white margin of the pileus, its pale avellaneous tint elsewhere, and the viscidity of both the pileus and the stipe. I have observed hundreds of carpophores—all of them old—with these characters. Murrill's concept of *M. epipterygia* was apparently based on the form which is common on conifer logs in the eastern United States. In the West the almost continual rain during late fall frequently washes off the gelatinous layers from the pileus and stipe, so that old specimens are soft and rather fragile.

219. Mycena epipterygia var. **lignicola,** var. nov.

Illustration: Text fig. 51, no. 8 (p. 416).

Pileus 8–15 mm. latus, obtusus demum obtuse campanulatus, viscidus, striatus, olivaceus vel olivaceo-luteus, lamellae subdistantes, angustae, adnatae, pallidae vel pallide luteae; stipes 4–6 cm. longus, 1–1.5 mm. crassus, viscidus, clare luteus; sporae 9–11 (12) $\times$ 5.5–8 μ; cheilocystidia clavata. Specimen typicum in Herb. Univ. Mich. conservatum. Legit A. H. Smith, n. 17316, prope Mt. Angeles, Wash., Sept. 26, 1941.

Pileus 8–15 mm. broad, ovoid when young, soon provided with an abrupt obtusely conic umbo, pruinose when young, viscid, the pellicle tenacious and entirely separable, colors "citrine" to "dark citrine" on the disc or "picric yellow" on the disc and "naphthalene yellow" toward the whitish margin; flesh yellowish, not staining red when bruised, odor and taste slightly to rather strongly farinaceous; lamellae subdistant to distant, narrow, adnate by a tooth, white to pale yellow, edges even; stipe 4–6 cm. long, 1–1.5 mm. thick, equal, viscid, glabrous, "lemon chrome" to "citron yellow" at first and fading to "massicot yellow" (bright yellow fading to pale yellow), slightly strigose at the base.

Spores (9) 10–12 (13) $\times$ 5.5–8 μ, broadly ellipsoid to ovoid, amyloid; basidia two-, three-, and four-spored, usually two-spored in northeastern North America; cheilocystidia abundant, gelatinizing and forming a broad sterile band on the gill edge, clavate and covered with short rodlike projections; pleurocystidia not differentiated; subhymenium slightly gelatinous; gill trama weakly amyloid; pileus trama weakly amyloid beneath the pellicle.

Habit, habitat, and distribution.—Gregarious on conifer wood from Nova Scotia to Washington and southward where conifers are found.

Material studied.—Smith, 32–247, 33–372, 33–703, 5026, 9328, 16769, 16928, 17316. Brown, November 3, 1925, Lake Quinault, Washington. Hesler, 14253. Kauffman, several collections, Michigan; July 5, 1928, Montana. Mains, 32–122, 32–796, 34–51, 34–85. Wehmeyer, 615, 757.

Observations.—In one collection from Ermine Creek, Baker National Forest, Washington, Smith, 16928, a strong odor of iodoform developed after the specimens had stood for a few hours. The pilei of this collection were dark olive and dried an olive gray rather than the usual yellow or yellowish brown. However, the specimens came from a very well protected locality on a log in an old windfall. This may have had some effect on the colors and the manner in which the specimens dried, so that I hesitate to place much emphasis on the character, particularly in view of the great variability in color observed in other regions and in different seasons in this variety. *M. epipterygia* var. B of Kauffman's *Agaricaceae of Michigan* is apparently an additional form of var. *lignicola* with a nitrous odor. The favorite habitat of the variety and its forms is old hemlock logs well covered with Dicranum and other mosses, but in the western United States it is frequently found on the naked bark around knots on fallen trees of hemlock, balsam, or Douglas fir. The cheilocystidia readily distinguish var. *lignicola* from *M. griseoviridis* var. *cascadensis*. Kauffman and Smith noted considerable variation in both spore size and odor in the specimens collected at Rock River, Michigan, and reported under the name *M. epipterygia*. Mains found a collection of four-spored specimens south of Au Train, Michigan, August 27, 1932. They were on a mossy log; the spores measure 6–8 × 5–7 μ, and the cheilocystidia are covered with short projections. No odor was noted. In North America this form has usually been identified as *M. citrinella*.

SECTION FULIGINELLAE

220. Mycena clavicularis (Fr.) Gillet

Les Hymén., p. 257. 1874

Agaricus clavicularis Fries, Syst. Myc., 1: 158. 1821.
Prunulus clavicularis Murrill, North Am. Flora, 9: 330. 1916.

Illustrations:

Plates 93 C, 94 A; Text fig. 49, nos. 8–9 (p. 399).
Beardslee and Coker, Journ. Elisha Mitchell Sci. Soc., 40, pl. 9 (upper figs.).
Fries, Icon. Sel. Hymen., 1, pl. 84, fig. 1.

Pileus (5) 10–20 mm broad, convex to obtuse when young, rarely obtusely conic, becoming broadly convex or with a low obtuse umbo in age, sometimes becoming depressed on the disc, margin straight but seldom touching the stipe when young, spreading and uneven in age, surface pruinose and even when young, naked and wrinkled to sulcate in age, faintly translucent-striate when moist, glabrous, moist to lubricous but not viscid, color dark to light gray, often bluish gray at first ("fuscous" to "hair brown" or "drab," "Quaker drab," or "Chaetura black"), margin pale gray, usually near "pale smoke gray," not hygrophanous, fading slowly to sordid yellowish or whitish; flesh thin, pliant, and cartilaginous, grayish or pallid, odor and taste not distinctive; lamellae bluntly adnate to slightly arcuate, sometimes toothed, subdistant to distant, 15–20 reach the stipe, two or three tiers of lamellulae, narrow to moderately broad (2 mm. ±), white when young, soon becoming sordid pale gray, edges even and pallid; stipe 2–5 cm. long, 1–1.5 mm. thick, equal, the base narrowed slightly or a bit enlarged, elastic, cartilaginous, tubular, pruinose above when young, soon naked, viscid, glutinous in wet weather, bluish black to "fuscous" when young and fresh, usually concolorous with the pileus or paler in age, becoming slightly tinged with yellowish gray, base slightly mycelioid.

Spores ellipsoid, 7–8.5 (9) × 3–4 μ, amyloid; basidia four-spored or rarely two-spored, 28–33 × 5–6 μ; cheilocystidia and pleurocystidia similar, scattered to numerous, (22) 30–38 × 9–12 (14) μ, broadly clavate, enlarged portion covered with short obtuse projections, hyaline; gill trama pale vinaceous brown in iodine, subhymenium not gelatinous and gill edge not gelatinizing; pileus trama with a thin adnate nongelatinous pellicle, hypoderm sharply differentiated, its cells 20–40 × 15–25 μ and occupying nearly half the thickness of the pileus, the remaining tissue floccose and filamentous, all except the pellicle faintly vinaceous brown in iodine.

Habit, habitat, and distribution.—Gregarious in large numbers under conifers, particularly *Pinus strobus*, in the eastern United States. It is very common during late summer or early fall, and is apparently not rare during the spring months along the Pacific coast. I have collected it in Nova Scotia and Ontario in Canada and have examined material from Alabama, North Carolina, Tennessee, Pennsylvania, New York, Michigan, Washington, Oregon, and California in the United States.

Material studied.—Smith, 32–91, 32–270, 32–652, 298, 655, 1614, 2513, 3155, 3327, 3464, 3480, 4288, 6994, 7508, 7937, 7974, 8192, 8246, 8284, 9086, 10907, 14432, 14463, 14728, 14804, 14805, 16311, 16380, 16381. Burke, Alabama, 1942. Hesler, 8002, 10945. Mains, 32–124, 32–413, 32–577, 32–624, 5022. Overholts, 19697. Wehmeyer, 656, 656a.

Observations.—To judge from my own field experience, this is one of the most common Mycenae in North America. A two-spored form also occurs but is apparently very rare. Its spores measure 10–12 × 4–5 (6) μ. For several years I have been interested in the sequence with which certain Mycenae fruited in a local conifer plantation. Both *M. vulgaris* and *M. clavicularis* are very abundant, but their fruiting periods rarely overlap. The latter regularly fruits under white pine late in August or early in September if the weather is moist. No matter how wet the weather is at this time, *M. vulgaris* does not fruit abundantly. During the 1940 season, in spite of what appeared to be almost perfect weather for the growth of the fungus, only a few depauperate carpophores of *M. vulgaris* appeared. When we have a very wet cool late September or early October *M. vulgaris* can be collected in almost any quantity, and *M. clavicularis* cannot be found or only occasional fruiting bodies appear. The same sequence was noted when collecting in the second-growth stands of conifers on the Olympic Peninsula of Washington during the season of 1935. *M. vulgaris* did not fruit until the cool weather just before the snow came.

Kauffman's Mt. Hood collection is apparently a fungus of the *M. filopes* group. The short, stubby two-spored basidia (16–18 × 7–8 μ) distinguish it sharply from *M. clavicularis*. The specimens reported as *M. clavicularis* var. *cinerea* Peck by Kauffman and Smith (1933) are referable to *M. vitilis*, a species neither Kauffman nor I understood at the time. *M. vitilis* has an elastic stipe which sometimes feels subviscid to the touch, but does not have a differentiated gelatinous pellicle. The type of *M. clavicularis* var. *luteipes* Kauffman has not been located, but from the description it appears to be a form of *M. epipterygia*, in spite of Kauffman's comments. Bisby reported this variety from Kenora, Ontario. The sulphur-yellow cap with olivaceous or green shades, yellowish gills, particularly the yellow stipe, and large spores are all too suggestive of *M. epipterygia*. In dry weather the pilei of the latter may be pruinose and dry to the touch, and the pellicle may be separable only with difficulty. In such cases the only reliable approach to the problem of its identity is to section the cap and make a microscopic examination. The gelatinous pellicle in water mounts assumes its typical form and consistency immediately.

From among Peck's collections I have notes on a specimen bearing the name "*M. clavicularis* var. *tenuior*," which is merely a slender form of the species. The other varieties mentioned by Peck were not located.

Mycena clavicularis is one of the easiest Mycenae to recognize, although it is a very ordinary-appearing gray fungus. It is known by its dry or moist *but not viscid* cap, and its decidedly viscid stipe. The first evidence one has of the latter character is the slipperiness of the stipe when one tries to pull the carpophores from their attachment. Microscopically the species is distinct because of the roughened cystidia, nongelatinous gill edge, and medium-sized spores. The color of the fruiting bodies is dominantly gray, with no yellow in the gills, and just enough in the pileus and stipe to give these parts a sordid appearance.

221. Mycena militaris Karsten

Symb. Myc. Fenn., 29: 91. 1889

Pileus 3–4 mm. broad, conic-campanulate, in age with a sharply conic umbo, small forms often papillate, glabrous, viscid, "fuscous" to blackish on the disc, "hair brown," "wood brown," "avellaneous," or pale gray on the margin, pellucid-striate when moist, sulcate at maturity, in age pallid cinereous with a blackish umbo; flesh thin and pliant, grayish, odor and taste slightly farinaceous; lamellae gray, subdistant, bluntly adnate or with a small decurrent tooth, moderately broad, edge whitish; stipe 2.5–3 cm. long, 1 mm. thick, yellowish gray below, paler above, pallid over all and somewhat translucent in age, very viscid, base bulbous.

Spores 8–10 × 3–4 μ, narrowly ellipsoid, amyloid; basidia four-spored; gill trama with a subgelatinous subhymenium, the floccose central portion vinaceous brown in iodine; pileus trama with a thick gelatinous pellicle nearly half the thickness of the trama, beneath it a zone of cells having dark contents, the remaining portion filamentous; stipe with a thick outer gelatinous layer, the inner tissue purplish red in iodine.

Habit, habitat, and distribution.—Gregarious under conifers, early in September; Michigan and New York. Rare.

Material studied.—Smith, 33–896, 903.

Observations.—The collections cited contain smaller carpophores than those originally described by Karsten, but a difference in size in a case like this is not necessarily significant. My specimens were few and gave the impression of being an off-season form. The conic to campanulate pileus, the spores, and the bulbous base of the stipe, as Karsten pointed out, separate the species from *M. vulgaris*. No reddish spots were noticed in my material, but, since both collections were made during comparatively dry weather, it does not seem advisable to place much emphasis upon the absence of these spots. The spore size as obtained from dried specimens is apt to intergrade more with that of *M. vulgaris* than is that of material from deposits. Because of the possibility of intergradations of *M. militaris* with *M. vulgaris* Karsten's species should be critically studied some season when material is abundant in order to evaluate its distinctive characters more fully.

222. Mycena vulgaris (Fr.) Quélet

Champ. Jura et Vosges, p. 108. 1872

Agaricus vulgaris Fries, Syst. Myc., 1: 156. 1821.
Prunulus vulgaris Murrill, North Am. Flora, 9: 326. 1916.
Prunulus melleidiscus Murrill, *ibid.*, p. 325.
Mycena melleidisca Murrill, Mycologia, 8: 221. 1916.

Illustrations:

Plate 93 A; Text fig. 52, nos. 3, 5 (p. 419).
Bresadola, Icon. Mycol., 6, pl. 254, fig. 1.
Konrad et Maublanc, Icon. Sel. Fung., 3, pl. 228 I.
Lange, Flora Agar. Dan., 2, pl. 58, B.
Ricken, Die Blätterpilze, 2, pl. 109, fig. 8.

Pileus 8–15 mm. broad, usually obtuse to convex when young, sometimes with a sharp conic umbo, becoming broadly convex and with or without a subacute or obtuse umbo, finally plane, the disc nearly always depressed in age, margin straight or only slightly incurved at first, soon spreading and often wavy in age, surface glabrous and very viscid when moist, pellicle completely separable, conspicuously translucent-striate to the disc, striations dark, sulcate when faded, color "fuscous" more or less over the disc and "smoke gray" near the margin, gradually fading to "drab" over the center and finally sordid smoke gray or yellowish gray over all, not hygrophanous; flesh very thin but pliant, pallid grayish, taste slightly disagreeable, odor slight when flesh is crushed, subnauseous to subraphanoid; lamellae bluntly adnate, soon arcuate, and in age distinctly decurrent, close, 13–17 reach the stipe, one or two tiers of lamellulae, sometimes forking near the base, moderately broad, up to 2 mm., whitish to pale grayish, edges even; stipe 2–3 cm. long, 1 mm. ± thick, cartilaginous, equal, slightly strigose at the base, tubular, cartilaginous, glabrous, glutinous or merely viscid, concolorous with the pileus or paler above.

Spores drop-shaped to subellipsoid, 6–8 (9) × 3.5–4 μ, amyloid; basidia four-spored, 28–32 × 6–7 μ; gill edges gelatinous, with clavate-roughened cheilocystidia embedded in the matrix (these are often very indistinct); no pleurocystidia; gill trama evenly floccose, subhymenium not gelatinous, purplish brown in iodine; pileus trama with a thick separable gelatinous pellicle, the remainder homogeneous, floccose and becoming purplish brown in iodine; stipe with a conspicuous outer layer of gelatinous hyphae, the inner portion of large hyphae which become purplish red in iodine.

Habit, habitat, and distribution.—Common during the late fall on needle beds under most species of conifers, often in great quantity. Material from Tennessee, Maryland, New York, Michigan, Colorado, Idaho, Washington, Oregon, and California in the United States and Manitoba, Nova Scotia, and Ontario in Canada has been studied.

Material studied.—Smith, 32–271, 32–469, 32–470, 32–799, 3275, 3452, 4854, 4936, 5078, 5346, 7935, 8063, 8451, 8616, 8830, 15803, 17067, Rock River, Michigan, 1929. Atkinson, New York (but not 9853). Hesler, 14235. Kanouse, August 16, 1925. Kauffman, Maryland, Michigan, Colorado, Montana, Idaho, Oregon, Washington. Mains, 32–797, 32–799, 34–98, Michigan (two collections), 1935. Overholts, 2051. Wehmeyer, 534a.

Observations.—According to my experience, this species fruits in cool wet weather and is often most abundant in the warmer wet periods between heavy frosts late in November and December. In cold years, far to the north or high in the mountains, it could reasonably be expected to fruit earlier. I have not seen it during August in the United States. Overholts (2051) has one collection from Tolland, Colorado, dated July 27, 1914.

Murrill's description of *M. melleidisca* does not furnish one with any reasonably good characters by which to distinguish the species from *M. vulgaris*. During unfavorable seasons or at the end of a very good season I have frequently found large numbers of small fruiting bodies in which the yellowish-gray color was rather pronounced. No correlation between small size and the yellowish-gray color, however, has been established, and these differences are here regarded as seasonal variations. A microscopic examination of Murrill's type failed to furnish a distinctive character. The pileus possesses a thick gelatinous pellicle, the gill edges are very gelatinous, the cheilocystidia are indistinct, as in *M. vulgaris*, and the spores measure 7–8.5 × 4 μ. In view of this information *M. melleidisca* is reduced to synonymy with *M. vulgaris*.

223. Mycena pelliculosa (Fr.) Quélet

Champ. Jura et Vosges, p. 343. 1873

Agaricus pelliculosus Fries, Epicr. Syst. Myc., p. 116. 1838.

Illustrations: Text fig. 53, nos. 2–3, 5 (p. 438).

Pileus 5–10 mm. broad (probably larger when growing luxuriantly), convex with an abrupt and rather broad depression on the disc, glabrous, viscid, fuliginous, fading to pale gray and at times assuming brownish tints, striate, striae dark and conspicuous; flesh thin, pliant and cartilaginous, grayish, odor and taste not distinctive; lamellae arcuate-decurrent, narrow to only moderately broad, distant, white, edges even and concolorous with the faces; stipe 1–2 cm. long, 1 mm. thick, equal, tubular, very cartilaginous-pruinose over all but soon naked and viscid, concolorous with the pileus or paler, base slightly strigose.

Spores ellipsoid to ovoid, (7) 8–10 × 4–5 μ, amyloid; basidia four-spored, 22–26 × 7–8 μ; cheilocystidia 20–40 × 4–10 μ, numerous, not gelatinizing completely (although the gill edge itself becomes distinctly gelatinized), clavate to nearly cylindric and often contorted, the apices with few to many short rodlike projections, sometimes the head or enlarged portion is branched and the branches bear rodlike processes; pleurocystidia numerous and of two types that intergrade with each other, the first type similar to the cheilocystidia in shape, size, and markings, the second composed of subfusoid individuals with smooth or undulating walls or even with obtuse humplike projections over the apex in some, 28–34 × 8–10 μ, hyaline, hardly projecting above the basidia; gill trama with a floccose central strand flanked by the broad gelatinous subhymenium on either side, floccose strand pale vinaceous brown in iodine; pileus trama divided into three zones, surface zone a gelatinous pellicle one half to two thirds the thickness of the entire trama, below this a narrow central layer of floccose tissue, the third zone formed by the gelatinous subhymenium, the floccose zone faintly brownish in iodine; stipe tissue with a thick gelatinous outer layer, the inner tissue dark vinaceous brown in iodine.

Habit, habitat, and distribution.—Gregarious on humus under conifers during August; Michigan. Very rare.

Material studied.—Smith, 33–706, 758. Mains, 34–53.

Observations.—As in *M. odorifera*, the thickness of the gelatinous layers varies greatly, depending on moisture conditions. At the time I studied fresh material the pleurocystidia escaped my observation. In sections of the dried material revived in KOH, stained in phloxine, and crushed slightly by pressure on the cover glass, both types were readily visible. The gill edges of this species became gelatinized and somewhat inflated, but not to so great an extent as in many species of this subgenus.

Mycena pelliculosa resembles *M. vulgaris* in many respects but is distinguished by its pleurocystidia, its time of fruiting, and the slightly larger spores. The difference in spore size was not appreciable in the dried material from which text figures 52 and 53 were drawn. The material examined indicates that the pileus of *M. pelliculosa* is more regularly depressed in young or freshly matured fruiting bodies, whereas in that of *M. vulgaris* the disc usually becomes depressed in age.

224. Mycena odorifera (Pk.) Saccardo

Syll. Fung., 5: 295. 1887

Agaricus odorifer Peck, Ann. Rep. New York State Mus., 30: 39. 1878.
Prunulus odorifer Murrill, North Am. Flora, 9: 324. 1916.

Illustrations:

Plate 95 B; Text fig. 53, nos. 6–7 (p. 438).
Atkinson, Stud. Am. Fungi (1900 edition), fig. 97 (as *M. vulgaris*).

Pileus 4–10 (15) mm. broad, obtusely conic to convex, becoming broadly convex and sometimes with a small conic umbo, usually nearly plane with a somewhat depressed disc in age, the margin incurved slightly at first or nearly straight, spreading in age, surface pruinose and hardly translucent-striate at first, often apparently not viscid to the touch at first but soon becoming so, bluish to fuscous gray over the disc, the margin whitish, in age becoming sordid gray or brownish over all; flesh thin but relatively tough and cartilaginous, pallid grayish to whitish, taste slight and hardly distinctive, odor very pronounced (developing within a short time after the fruiting bodies have been collected and persisting in the dried material), fragrant (somewhat resembling that of *Armillaria caligata*); lamellae adnate but soon arcuate and in age distinctly decurrent, close to sub distant, narrow to moderately broad, tapering toward the margin of the pileus, white to pale gray, edges even and pallid; stipe 1–2 (3) cm. long, 1–1.5 mm. thick, hollow, elastic and cartilaginous, equal, base slightly white-strigose, remainder at first covered with a dense coating of coarse white pruina, soon glabrous and glutinous to the touch, dark gray to bluish gray, becoming pallid.

Spores broadly ellipsoid, 7–9 × (3) 4–4.5 μ, amyloid, reaction strong; basidia four-spored, 28–30 (35) × 5–6 μ; cheilocystidia filamentous, smooth, 38–46 × 6–7 μ; pleurocystidia not differentiated; gill trama vinaceous brown in iodine, subhymenium hyaline in iodine but not truly gelatinous; pileus trama with a thick gelatinous pellicle, floccose beneath it, floccose portion vinaceous brown in iodine; stipe with an outer gelatinous sheath, interior reddish brown in iodine.

Habit, habitat, and distribution.—Gregarious to scattered on fallen leaves of conifer needles during late spring or summer; Alabama, Tennessee, New York, Michigan, and Washington in the United States and Ontario in Canada. Rare.

Material studied.—Smith, 33-592, 33-695, 709, 709a, 735, 1758, 6509, 7358, 7476, 9780, 10080, 10119, August 31, 1936, Lake Timagami, Ontario. Atkinson, 9853. Burke, Alabama. Hesler, 11483.

Observations.—The nongelatinous gill edges, filamentous cheilocystidia, and, usually, the persistent fruity odor make this a very distinctive species. When young neither the cap nor the stipe feels viscid to the touch because of the presence of a thin coating of nongelatinous hyphae. These either wear away or gelatinize, because in age the stipe is often covered with gluten much as in *M. rorida.* The gluten usually becomes aggregated toward the base of the stipe or in wet weather is washed off almost entirely. In dry weather it may not develop to the point of being readily visible. In all the variations noted, however, a gelatinous layer was readily observed under the microscope on cross sections of fresh stems. The thickness of the pellicle on the pileus varies in much the same manner as that of the stipe.

225. Mycena quinaultensis Kauffman in Smith

Mycologia, 27: 589. 1935

Illustrations: Plates 94 B, 96; Text fig. 53, nos. 1, 4 (p. 438).

Pileus 1–4 cm. broad, obtusely conic to obtusely campanulate when young, occasionally almost convex, becoming campanulate to expanded, nearly always obtusely umbonate, frequently with a slight

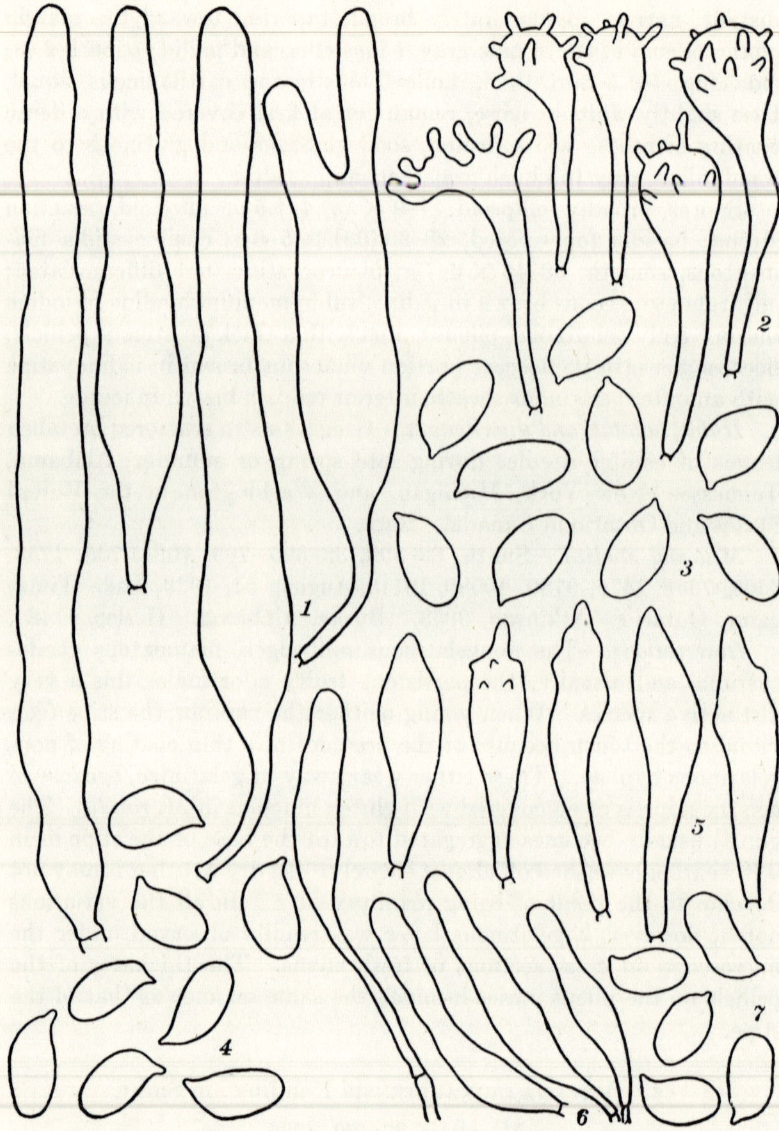

FIG. 53. *M. quinaultensis:* 1, pleurocystidia; 4, spores. *M. pelliculosa:* 2, cheilocystidia; 3, spores; 5, pleurocystidia. *M. odorifera:* 6, cheilocystidia; 7, spores

papilla, margin appressed against the stipe when young, often wavy and recurved in age, surface viscid to glutinous when young, glabrous, translucent-striate to the disc, sulcate to wrinkled in age, color at first entirely fuscous black, soon becoming paler from the margin toward the disc and "buffy brown," finally "tilleul buff," the umbo fading but remaining darker than the margin; flesh thin, pliant, whitish, odor and taste hardly distinctive; lamellae ascending and adnate, developing a slight tooth, subdistant to distant, 15–18 reach the stipe, narrow but slightly ventricose at times (3 mm.), whitish when young, soon pallid to grayish, edges even; stipe 4–7 cm. long, 1.5–2.5 (3) mm. thick, equal, rigid, hollow, glabrous, viscid to glutinous when moist, naked at the apex, concolorous with the pileus, the apex often paler, base sparsely white-mycelioid.

Spores narrowly ellipsoid, 6.5–8 × 3–3.5 μ, amyloid; basidia four-spored, 30–32 × 4–5 μ; cheilocystidia and pleurocystidia abundant, subcylindric or slightly inflated toward the base, hyaline, thin-walled except for slight irregular thickenings toward the base, 80–110 × 5–10 μ; gill trama yellowish or slightly brownish in iodine; pileus trama with a very gelatinous thin pellicle, beneath this a layer of more or less vesiculose cells, the remainder filamentous and of narrower hyphae, not becoming vinaceous brown in iodine; stipe tissue yellowish in iodine or only faintly brownish.

Habit, habitat, and distribution.—Densely gregarious on needle beds under western red cedar, Douglas fir, and redwood; Washington, Oregon, and California. Sporadic, occurring during both spring and fall but most abundant in the fall.

Material studied.—Smith, 2542, 2573, 3751, 7773, 7808, 8048, 8136, 8175, 8184, 8264, 9073, 9457, 13519, 13916, 14511, 14538, 16207, 16943, 17262. 1925, Brown, Washington. Kauffman, 1925, Lake Quinault, Washington.

Observations.—The viscid layer on the pileus and stipe is both very thin and very gelatinous. In wet weather it may be washed away completely, with the result that old specimens are sometimes difficult to place in the proper section. The layer is very easily removed when the stipe is sectioned and difficult to demonstrate for that reason. Dr. Kühner, to whom specimens were sent, reported that he could not find the gelatinous layer on the stipe. In my description I had relied on Kauffman's notes in placing the species in this subgenus. When the type was sectioned, however, I obtained the same results as Kühner, and was inclined to doubt Kauffman's observations. On western collecting trips during both 1937 and 1939 I found the species rather frequently, and in every collection the stipes of young fruiting bodies were decidedly viscid. The gelatinous layer in question can be demonstrated under the microscope from fresh specimens, but not at all or only with great care from dried material.

During the course of my study of the type I found that the collection was a mixture of two species, *M. tenax* and *M. quinaultensis*, but it was evident that Kauffman's notes were taken from carpophores of only the one species, *M. quinaultensis*. Since there is no confusion of concept, Article 64 of the International Rules of Nomenclature does not apply, and there is no reason for changing the name.

The collection reported from Nova Scotia by Smith and Wehmeyer as *M. quinaultensis* has been found to be *M. tenax*. The former is readily distinguished from the latter by the different gill edge, the very thin gelatinous pellicle on the pileus, and the lack of a disagreeable odor.

226. Mycena tenax A. H. Smith

Mycologia, 28: 414. 1936

Illustrations:

Plate 97; Text fig. 52, nos. 4, 6 (p. 419).

Smith, Mycologia, 28, fig. 2, no. 1; fig. 3, nos. 4–5.

Pileus 1–3 cm. broad, oval to convex at first, in age broadly conic or obtusely umbonate, disc flattened or slightly depressed at maturity, margin appressed against the stipe when young and somewhat flared in age, with a thick tenacious separable pellicle, surface glabrous, lubricous to subviscid, in age striate to the abruptly translucent-watery disc, opaque when faded and somewhat sulcate, color "fuscous" to "hair brown" and with a pale-grayish margin when young, becoming pale watery gray in age; flesh pliant, tough, pallid, odor and taste strongly rancid-farinaceous or somewhat resembling those of a green watermelon or a watermelon rind; lamellae adnate or slightly toothed, close, 24–27 reach the stipe, narrow (2–3 mm.), pallid to grayish, edges even; stipe 5–7.5 cm. long, 2–3 mm. thick, equal, hollow, tenacious, glabrous, subviscid to viscid, pruinose toward the apex, base white-strigose, concolorous with the pileus or paler.

Spores narrowly ellipsoid, pointed at one end, 6.5–8 $\times$ 3.5–4 μ, amyloid; basidia four-spored, 30–34 $\times$ 5–6 μ; cheilocystidia clavate and with fingerlike prolongations over the apex, gelatinizing as in *M. vulgaris*, hyaline; pleurocystidia scattered, 60–70 × 8–12 μ, narrowly fusiform to subcylindric with abruptly pointed apices, some with slightly thickened walls; gill trama vinaceous brown in iodine; pileus trama with a typical thin subgelatinous pellicle over the surface, then a region of floccose tissue with enlarged hyphal cells (appearing parenchymatous in tangential section), below this a region 75–120 μ thick of gelatinous hyphae, the remainder floccose, all floccose tissue vinaceous brown in iodine; stipe tissue vinaceous brown in iodine except for the outer gelatinous layer.

Habit, habitat, and distribution.—Densely gregarious under Douglas fir, Sitka spruce, and redwood; Washington, Oregon, and California. In Ontario and Nova Scotia in Canada and in New York it usually occurs under red spruce or balsam fir. It is very common in the fall but apparently rare in the spring in the western United States.

Material studied.—Smith, 3409, 3433, 3456, 3503, 3936, 4761, 7968, 7978, 8016, 8191, 8250, 8789, 9483, 14431, 17649. Kauffman, October 1, 1925, Washington (as *M. rugosoides*). Zeller, 9037.

Observations.—This is a very distinctive Mycena. It can be quickly recognized by its taste, very tenacious consistency, gray color (no yellow or green present), and viscid stipe. With the possible exception of *M. viscosa*, I know of no other gray Mycena having a viscid stipe which has such a bad taste. While testing the constancy of the taste of numerous carpophores one morning I was taken with a mild attack of dizziness and headache. As a result of this experience, I suspect the species of being poisonous. Microscopically the combination of the gelatinous gill edges, the large pointed pleurocystidia, and the four layers of tissue in the pileus make it equally distinctive.

SUBGENUS MYCENELLA

IN THIS subgenus are placed all species with roughened or ornamented spores. All are grayish and Mycena-like in their appearance. The subgenus is made up of two unrelated sections. The section Nodulosae is composed of a small number of very closely related species characterized by their finely pubescent stipes, large cystidia, and nodulose spores. The section Pseudoechinulatae contains *M. bisphaerigera* and *M. cineraria*. The Nodulosae may very well have been derived from the section Typicae of Eumycena and the Pseudoechinulatae from the section Omphaliariae.

The species of both sections are very rare, and apparently are more abundant during wet than during dry seasons. *M. bisphaerigera* has been found most frequently just after an extended dry spell has been broken.

KEY TO SPECIES

1. Spores nodulose 3
1. Spores echinulate or appearing so 2

2. Spores with smooth outer wall, secondary wall thickened and with numerous fine pores, which give it a falsely echinulate appearance, outer wall amyloid .... 231. *M. bisphaerigera*
2. As above, but spores not amyloid ("outer wall" absent); cystidia present .... 232. *M. cineraria*
2. Spores truly echinulate, with very fine projections, nonamyloid (see excluded species for *M. farinacea* Murrill)

3. Pleurocystidia simple or, at most, forked once at apex, often incrusted. 4
3. Pleurocystidia branched at apex; pileus 4–10 mm. broad; stipe not radicating .... 227. *M. margaritispora*

4. Stipe lacking well-differentiated pseudorhiza ...... 228. *M. trachyspora*
4. Stipe with pseudorhiza 5

5. Pileus trama corticated by palisade of clavate to basidium-like somewhat irregular cells .... 229. *M. nodulosa*
5. Pileus trama with thin pellicle, cells of which give off numerous short rodlike projections .... 230. *M. bryophila*

SECTION NODULOSAE

227. Mycena margaritispora Lange

Dansk Bot. Arkiv, (5): 37. 1914

Illustrations:

Text fig. 54, nos. 1–2 (p. 444).
Lange, Flora Agar. Dan., 2, pl. 58 D, D′.
Smith, Am. Journ. Bot., 22, pl. 3, fig. 2.

Pileus 4–10 (15) mm. broad, obtusely conic, becoming campanulate or umbonate and the margin recurved, surface appearing dry and densely pruinose at first, becoming glabrous and somewhat polished while still moist, fading and becoming somewhat glaucous, deep "fuscous" on the disc or a deep glaucous gray at first, margin pallid, fading through "army brown" to yellowish gray along the margin, translucent-striate when moist, slightly sulcate in age, margin regular; flesh thin, pallid, tough, odor and taste not distinctive; lamellae adnate, usually seceding in age, rather narrow, subdistant to moderately close, white or stained yellowish in age, pruinose under a lens, edges even but pruinose; stipe 2–5 cm. long, 0.5–1 mm. thick, straight or flexuous, often decumbent, cartilaginous and pliant, equal, base strigose, covered with a dense coating of long cystidium-like hairs, giving it a velvety appearance, deep fuscous over all or the apex paler, often stained yellowish in age.

Spores (5) 6–8 μ, globose to subglobose, verrucose, nonamyloid, basidia two-spored; pleurocystidia and cheilocystidia 40–70 $\times$ 8–15 μ, abundant, ventricose with long necks, simple or the apices more or less branched into three or many simple or compound branchlets, when fresh more or less covered with a granulose substance; gill trama homogeneous, pale yellow in iodine; pileus trama homogeneous, pellicle well formed, its cells covered with numerous short rodlike projections, pilocystidia numerous, usually smooth, 40–85 $\times$ 9–12 μ, hyaline, fusoid-ventricose, tramal body filamentous and yellowish in iodine; caulocystidia 36–90 $\times$ 7–10 μ, straight or flexuous, the apices usually contorted, branched, or incrusted.

Habit, habitat, and distribution.—Single or in groups of two or three, on old wood, chip dirt, and debris, in the summer and fall. Locally it is most often encountered on the black humus near stumps in elm-maple swamps after the humus has been dried out and then remoistened by heavy rain. The species is known from Tennessee,

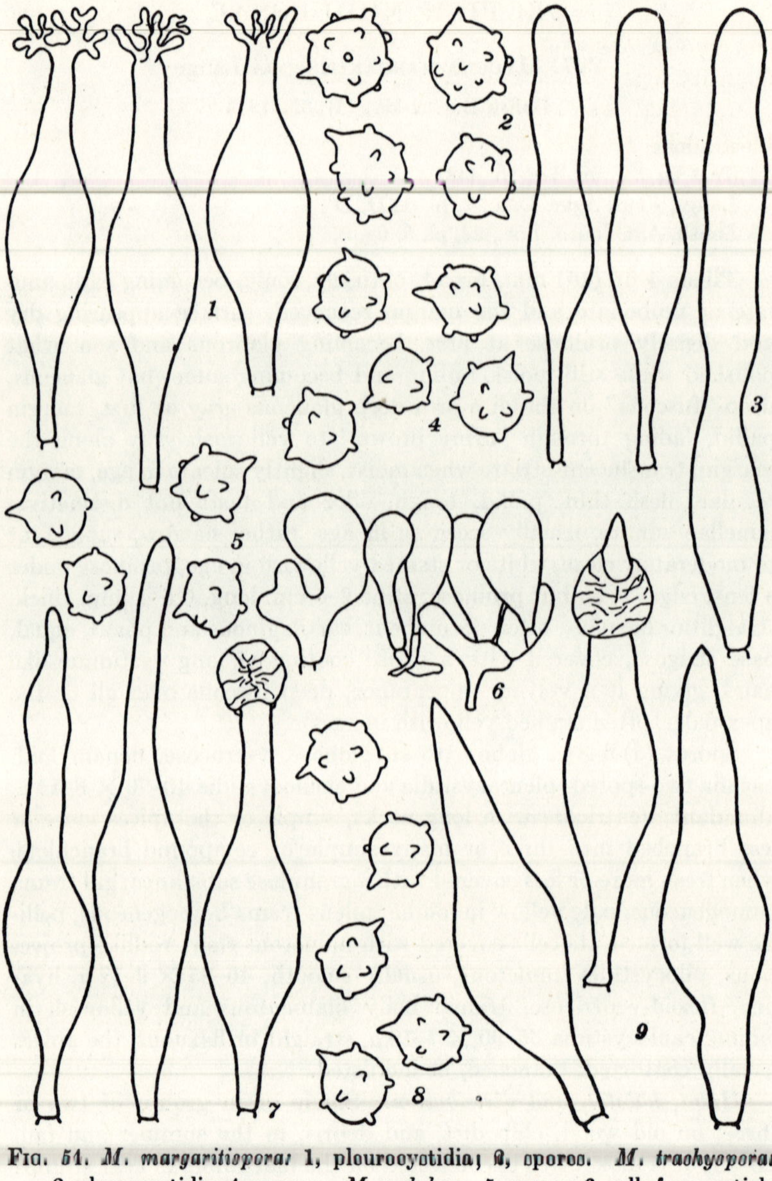

FIG. 51. *M. margaritispora:* 1, pleurocystidia; 2, spores. *M. trachyspora:* 3, pleurocystidia; 4, spores. *M. nodulosa:* 5, spores; 6, cells from cuticle of pileus; 7, pleurocystidia. *M. bryophila:* 8, spores; 9, pleurocystidia

New York, Michigan, and Washington in the United States and Ontario in Canada. Atkinson (24158) found it in considerable quantity near Ithaca, New York. Rare.

Material studied.—Smith, 32–456, 32–529, 33–705, 60, 446, 452, 1442, 1478, 1628, 4333, 4934, 6346, 6397, 7258, 7693, 10562, 10604, 10706, 10765, 14295. Atkinson, 24158, 24334 (as an undescribed species). Hesler, 10173.

Observations.—The diagnostic characters of this fungus are the branched cystidia, the lack of a pseudorhiza, and the tendency to assume ochraceous colors in age. Lange's name has been used for the American collections because of their perfect agreement with his description and illustration. Kühner believes that *M. lasiosperma* Bresadola is the same, and uses that name because of its priority. He apparently was strongly influenced by the similarity of the cystidia in the two. Bresadola, however, described *M. lasiosperma* as larger, with a pseudorhiza, and as growing more or less cespitose on wood. Size alone is a poor character, but the three mentioned above, taken together, should not be disregarded. I have examined about two dozen American collections, and they consistently differ from Bresadola's description as noted.

228. Mycena trachyspora Rea

Trans. Brit. Myc. Soc., 12: 216. 1926

Mycena Cooliana Oort, Med. Neder. Mycol. Vereen, 16–17: 248. 1928.

Illustrations:

Plate 98 A; Text fig. 54, nos. 3–4.
Smith, Mycologia, 28, 412, fig. 1, no. 6.

Pileus (6) 10–20 (30) mm. broad, conic with an appressed margin at first, soon campanulate, becoming expanded and umbonate, surface appearing dry and densely hoary at first, polished at maturity, disc "fuscous" to "bone brown" (blackish brown) at first, the margin pallid to grayish, becoming paler over all in age, and often with a tinge of ocher, faintly striate at first; flesh firm and cartilaginous, thin, pallid or grayish, odor and taste not distinctive; lamellae moderately close, narrow to moderately broad, oval in outline, narrowly adnate, pale grayish to whitish, the margin pallid, densely pruinose under a lens; stipe 4–12 cm. long, 1–2.5 mm. thick, equal, tough and cartilaginous, pallid gray above, darker and more brownish below, evenly covered with a dense layer of cystidia (appearing pubescent under a lens).

Spores 4–6.5 μ, globose to subglobose, verrucose, yellowish in iodine; basidia four-spored; pleurocystidia and cheilocystidia similar and abundant, 40–90 × 8–12 μ, fusoid-ventricose to nearly cylindric, with a long neck and obtuse apex, smooth or somewhat incrusted; gill trama homogeneous, yellowish in iodine, pileus trama homogeneous below a well-differentiated pellicle, yellowish in iodine, numerous short hairlike branches arising from the cells of the hyphae of the pellicle, producing a turflike surface covering, occasional cystidia also projecting from the pellicle.

Habit, habitat, and distribution.—Single to scattered on humus under redwood, fall; California. Fairly abundant during one season.

Material studied.—Smith, 33–962, 218, 954, 3718, 3731, 3739, 3756, 3784, 3884, 9042, 10672, 10775.

Observations.—The details of the pellicle are very difficult to ascertain even with the aid of an oil-immersion lens. However, the turflike covering produced by the branches of the cells forming the pellicle is readily visible under the usual magnifications. The cells of the pellicle are not arranged in a palisade as in *M. nodulosa* and are even less conspicuous. In age the slender projections forming the turflike zone may collapse, and under such conditions the pellicle does not appear at all distinctive. As Kühner has suggested, the presence of an incrustation on the cystidia and its location are not valuable characters in this particular group of species. *M. trachyspora* is readily distinguished from *M. bryophila* by the absence of a pseudorhiza. Kühner has placed my specimens under *M. bryophila.* After restudying them in the light of his conclusions, I find it impossible to agree that they represent only one species.

229. Mycena nodulosa A. H. Smith

Mycologia, 28: 411. 1936

Illustrations:

Text fig. 54, nos. 5–7 (p. 444).
Smith, Mycologia, 28, fig. 3, 1–3.

Pileus 1–3.5 cm. broad, obtusely conic, becoming campanulate or expanded and with a conic umbo, margin appressed against the stipe when young, striate to the disc when moist, surface at first appearing hoary because of numerous projecting cystidia (very finely pubescent under a lens), soon glabrous (the cystidia collapsing), "fuscous" to black on the umbo, margin dark or light watery gray, fading to a pallid sordid gray; flesh soft and pliant, pale fuscous, very thin, in extreme age sometimes staining reddish brown, odor and taste not distinctive; lamellae deeply adnexed to broadly rounded and depressed-adnate, moderately close, 27–33 reach the stipe, broad (5 mm.), whitish to glaucous gray and densely pruinose under a lens from cystidia, spotted reddish brown in age; stipe 6–8 cm. long, (1) 2–3 mm. thick, with a pseudorhiza 4–8 cm. long, evenly covered with a white-pruinose pubescence of hyaline cystidia, pliant, hollow, with a sharply differentiated cartilaginous rind, fuscous at first, becoming pale gray to whitish above.

Spores globose, 6–7 μ, rough, with aculeae 2–3 $\times$ 0.5–2 μ scattered over the surface, nonamyloid; basidia four-spored, 28–34 $\times$ 7–8 μ; cheilocystidia and pleurocystidia similar and abundant, unbranched, fusoid-ventricose with a greatly elongated neck, apices usually incrusted, 60–80 $\times$ 10–14 μ; gill trama yellowish in iodine; pileus trama corticated by a palisade layer of clavate basidium-like or contorted cells measuring 18–22 $\times$ 5–9 μ, elongated cystidia 100–150 $\times$ 7–9 (10) μ projecting from this layer at intervals, the tissue beneath homogeneous, and composed of compactly arranged filamentous hyphae, yellowish in iodine; cystidia on the stipe similar to those on the pileus, but more numerous.

Habit, habitat, and distribution.—Scattered on humus beside a trail in a mixed Douglas fir and hemlock forest near Crescent Lake, Washington, during October. It is known only from the Olympic Peninsula.

Observations.—Because of the collapsible nature of the cells forming the surface of the pileus they are difficult to demonstrate in mounts of dried material. The layer is most easily seen on young specimens. The characters which readily distinguish the species are the pseudorhiza, the reddish stains which develop on the gills, and the palisade layer forming the cuticle of the pileus.

230. Mycena bryophila Voglino

Atti R. Ist. Veneto Sci., Lett. ed Arti, Venezia, 6 ser., IV: 617. 1886

Illustrations:

Text fig. 54, nos. 8–9 (p. 444).
Lange, Flora Agar. Dan., 2, pl. 58 F.
Smith, Am. Journ. Bot., 22, pl. 3, fig. 6 (as *M. lasiosperma* Bresadola).

Pileus (5) 10–20 mm. broad, conic to campanulate, becoming expanded-umbonate, the margin usually remaining somewhat decurved, blackish to fuscous around the disc or on the umbo, nearly avellaneous on the margin, fading to grayish in age with the disc darker, sometimes slightly translucent-striate when mature, pruinose from projecting cystidia when young, glabrescent and subviscid at maturity; flesh very cartilaginous and tough, pallid grayish, no odor, taste slightly disagreeable; lamellae narrowly adnate or attached only by a tooth, subdistant, moderately broad, pure white or pale cinereous, intervenose, pruinose from projecting cystidia; stipe 2–7 cm. long, 1–2 mm. thick, with a well-developed pseudorhiza, tough and cartilaginous, flexuous, hollow, densely pruinose-pubescent, pallid to brownish above, blackish or sordid gray below, sometimes sordid ochraceous toward the base in age.

Spores subglobose, 7–8.5 μ, covered with scattered blunt, short rodlike protuberances, yellowish in iodine; basidia two-spored; pleurocystidia very abundant, 60–85 $\times$ 10–16 μ, fusoid-ventricose with a long neck, smooth or incrusted; cheilocystidia similar but shorter and more obese; gill trama homogeneous, pale yellow in iodine; pileus trama homogeneous beneath a thin pellicle, the hyphae of which are covered with numerous short rodlike projections, pilocystidia similar to the pleurocystidia, originating in the pellicle; stipe covered with long cystidium-like hairs.

Habit, habitat, and distribution.—Single or in clusters of two or three carpophores on humus and debris, usually around stumps and logs; New York and Michigan. Rare.

Material studied.—Smith, 32–473, 32–571, 33–893, 6482, 7257, 7627, 15259, 17253, 17956, Michigan, October 5, 1932. Kauffman, August 14, 1912, Michigan. Peck, New York (as a variety of *M. galericulata*).

Observations.—The distinguishing characters of this fungus are: first, its pellicle of slender hyphae covered with short projections; second, the smooth cystidia; and, third, the presence of a well-developed pseudorhiza. The pruinose-pubescent stipe and pileus, tough consistency, and rough spores are not specific in this group. Kühner lists *M. lasiosperma* Bresadola sensu Kauffman, *M. trachyspora* Rea sensu Smith, and *M. nodulosa* Smith, the last questionable, as synonyms of *M. bryophila*. Since Bresadola illustrated *M. lasiosperma* as having curiously branched cystidia, a species in which they are smooth should not be referred to it, as has been done by Kauffman and others. The choice of Voglino's name appears to be satisfactory enough for the fungus described above, but not for the material I have placed in Rea's species or for *M. nodulosa.*

SECTION PSEUDOECHINULATAE

231. **Mycena bisphaerigera** (Lange), comb. nov.

Omphalia bisphaerigera Lange, Dansk Bot. Arkiv, 6 (5): 9. 1930.
Fayodia bisphaerigera Kühner, Bull. Soc. Linn. Lyon, 9: 68. 1930.

Illustrations:
Plate 98 B, Text fig. 55, no. 7 (p. 451).
Lange, Flora Agar. Dan., 2, pl. 59 H.

Pileus 5–20 mm. broad, obtusely conic to convex, sometimes with an umbo, the margin usually slightly incurved at first, fuliginous to watery gray (near "olive brown" or "buffy brown"), disc usually darker, hygrophanous, fading to sordid pallid cinereous or nearly white at times, when moist striate to the disc, opaque when faded; flesh pale grayish, cartilaginous and flaccid to rather soft and fragile, watery when moist, odor and taste mild (in the four-spored form), farinaceous (in the two-spored form); lamellae bluntly adnate or with a decurrent tooth in age, sometimes arcuate, close to subdistant, 15–25 reach the stipe, one or two tiers of lamellulae, narrow to rather broad (2.5–4 mm.), pallid grayish, stipe (2) 3–7 cm. long, 1–2.5 mm. thick, equal, hollow, cartilaginous but fragile, rigid, flexuous or undulating, pale gray to nearly white, glabrous, somewhat translucent, base with only a few scattered mycelial hairs.

Spores 7–9 (10) μ on two-spored forms, 5.5–7 μ on four-spored carpophores, globose, appearing echinulate under a 3-mm. objective, actually with a smooth outer wall which is strongly amyloid, a thickened secondary wall with very minute pores through it giving the spore a falsely echinulate appearance, and finally a smooth inner membrane, the two inner walls nonamyloid; pleurocystidia and cheilocystidia not differentiated or the latter subfusoid, smooth, and 25–30 × 7–9 μ; gill trama homogeneous, interwoven sordid pale brown in iodine; pileus trama homogeneous beneath a thin pellicle, sordid brown in iodine.

Habit, habitat, and distribution.—Single to gregarious under conifers or in bogs; New York, Michigan, Washington, Oregon, and California. Rather rare and occurring during late summer and fall.

Material studied.—Smith, 379, 771, 911, 963, 1104, 1106, 1133, 1164, 2489, 3214, 3334, 3544, 3791, 4045, August 30, 1934, New York; October 10, 1934, Michigan. Bailey, 141. Isle Royale National Park.

Observations.—The shape of the pileus varies considerably but is generally Mycena-like. The gray stipe, pileus, and gills and the flexuous stipe all remind one of *M. cinerella*, which is, of course, readily separated by its smooth unornamented spores. It seems more logical to place this species in Mycena than in any other Friesian genus. If one wishes to give it generic rank, Fayodia, in which Kühner originally placed it, should be used, but that genus should then be limited to species with the falsely echinulate type of spore.

232. **Mycena cineraria,** sp. nov.

Illustration: Text fig. 55, nos. 1, 4.

Pileus 1–2.5 cm. latus, convexus demum late convexus, glaber, valde striatus, subfuscus, hygrophanus, demum cinereus; lamellae subdistantes, latae, late adnatae vel decurrentes, pallidae; stipes 3–4 cm. longus, 1–1.5 mm. crassus, aequalis, glaber, pallidus; sporae 6–7 × 4–4.5 μ, pseudo-echinulatae; cystidia 44–66 × 9–12 μ, subcylindrica. Specimen typicum in Herb. Univ. Mich. conservatum. Legit A. H. Smith, n. 17099, prope Mt. Angeles, Olympic Mts., Wash., Sept. 21, 1941.

Pileus 1–2.5 cm. broad, convex with a straight or slightly bent-in margin, becoming broadly convex to plane or very slightly depressed over the disc, surface moist and glabrous, conspicuously striate to the disc before fading, "Saccardo's umber" or darker and more grayish brown when moist, hygrophanous, becoming opaque and more or less cinereous when faded; flesh watery-fragile, concolorous with cap, no odor, taste not recorded; lamellae close to subdistant, broad, 16–20 reach the stipe, one or two tiers of lamellulae, broadly and bluntly adnate but soon short-decurrent, "tilleul buff" (pallid), edges even but faintly fimbriate under a lens; stipe 3–4 cm. long, 1–1.5 mm. thick, tubular, strict, whitish to pallid, polished, the base with a thin covering of white mycelium.

Spores 6–7 × 4–4.5 μ, broadly ellipsoid, nonamyloid, smooth but appearing minutely echinulate under an oil-immersion lens because of fine pores in the spore wall; basidia four-spored; pleurocystidia and cheilocystidia scattered to fairly abundant; obtuse, subcylindric,

FIG. 55. *M. cineraria:* 1, pleurocystidia and cheilocystidia; 4, spores. *M. bisphaerigera:* 7, spores. *M. avellaneigrisea:* 2, cheilocystidia; 3, spores. *M. fumosiavellanea:* 5, spores; 6, cheilocystidia. *M. gracilipes:* 8, cheilocystidia; 9, spores. *M. pubescens:* 10, spores

the outlines usually a little irregular, 44–66 × 9–12 μ; gill trama homogeneous and nonamyloid; pileus trama homogeneous, pellicle not differentiated, nonamyloid.

Habit, habitat, and distribution.—Scattered on wet soil among liverworts and mosses, September 21, 1941; Mt. Angeles, Olympic Mountains, Washington. Known only from the type collection.

Observations.—Under ordinary magnifications it is difficult to see clearly the markings of the spore wall. Under higher powers, however, the details are readily visible. The structure of the wall is similar to that of *M. bisphaerigera* except that the outer amyloid epispore is lacking and the secondary wall is not as thick. The close relationship of the two species is obvious, however. *M. cineraria* is readily distinguished by its large cystidia. In stature it is almost identical with Lange's illustration of *Omphalia picta* var. *concolor.* Its colors are dull and intermediate between the true browns and the grays and hence cannot be considered of more than secondary importance.

MURRILL'S TROPICAL AND SUB-TROPICAL SPECIES

IN THE *North American Flora* Murrill described a number of species from tropical North America, but did not give much information on their microscopic characters. Without such information it is impossible to identify these species accurately. I have studied his type specimens and in the following account present the data obtained for each. A key based primarily on microscopic characters to supplement that of Murrill for their macroscopic features is given below. Since I have had no firsthand experience with tropical agarics, it did not seem wise to try to include these species in the body of this work. Certain of them may very easily belong in other genera, but to transfer them now, without more detailed knowledge of both Mycena and Marasmius as these genera are represented in the tropics, might not do more than add to the existing confusion.

In the following text the species are arranged in alphabetical order.

KEY TO SPECIES

1. Stipe with gelatinous sheath or outer layer 2
1. Stipe not as above 3

2. Spores 4–5 × 2.5 μ .... *M. trojana*
2. Spores 8–9 × 5–6 μ .... *M. fumosiavellanea*

3. Pileus with thick gelatinous pellicle 4
3. Pileus not as above 5

4. Cheilocystidia obtuse, apices covered with short rodlike projections *M. argillacea*
4. Cheilocystidia smooth, 12–20 μ broad .... *M. margarita*

5. Cuticle of pileus hymeniform; spores 4–5 × 3–3.5 μ ..... *M. pubescens*
5. Pileus not as above.. 6

6. Spores small, ellipsoid to globose, 3–5 μ 7
6. Spores larger, ellipsoid 8

7. Fruiting body greenish gray; spores *globose* and amyloid *M. viridigrisea*
7. Fruiting body some other color (pallid gray or avellaneous), spores amyloid, *ellipsoid* (see discussion of *O. myceniformis*) *Omphalia myceniformis*, *O. convexa*, *O. roriduliformis*, *O. subavellanea*
7. Fruiting body lateritious; spores not amyloid .... *M. latericia*

8. Spores 12–16 × 4–5 μ, nonamyloid (4-spored) ...... *M. avellaneigrisea*
8. Spores 11–14 × 5.5–7 μ, nonamyloid (2-spored) .... *O. petasiformis*
8. Spores smaller, 7–11 μ long 9

9. Cheilocystidia long and filamentous, 51–73 × 3–5 μ .... *M. testacea*
9. Cheilocystidia not as above 10

10. Spores 8–11 × 6–7 μ, amyloid; cheilocystidia with brownish content when revived in KOH .... *M. gracilipes*
10. Spores narrower, 4–5 μ 11

11. Lamellae adnate to adnexed 12
11. Lamellae decurrent 13

12. Cap 2 cm. broad; spores 7–9 × 5–5.5 μ; cheilocystidia smooth *M. carbonicola*
12. Cap 6 mm. broad, spores 7–8 × 4 μ; cheilocystidia not differentiated *M. cinereiavellanea*

13. Gills violaceous brown when dried .... *O. subscyphoides*
13. Gills not distinctive in dried condition 14

14. Spores amyloid.. *O. cuticolor*
14. Spores nonamyloid 15

15. Stipe long, 4 cm. or more .... *O. tepeitensis*
15. Stipe short, about 1.5 cm. .... *O. cuspidatella*

MYCENA ARGILLACEA Murrill

Mycologia, 8:220. 1916

Prunulus argillaceus Murrill, North Am. Flora, 9: 341. 1916.

"Pileus small, delicate, conic to campanulate, reaching 1 cm. broad; surface striate, glabrous, avellaneous, margin entire, concolorous, appressed when young: lamellae adnate, broad, subcrowded, white: spores globose or subglobose, smooth, hyaline, 4–6 μ: stipe slender, equal, smooth, glabrous, hollow, white, 1.5 cm. long, 1 mm. thick.

"Type collected on a wet, much-decayed log in a moist virgin forest on a mountain side at Motzorongo, near Cordoba, Mexico, about 300 m. elevation, January 15, 1910, *W. A. & Edna L. Murrill 1053* (herb. N. Y. Bot. Gard.).

"Distribution: Known only from the type locality."

The type consists of one very small fruiting body. The spores measure 7–8 × 4 μ, and are borne on four-spored basidia. They are ellipsoid, smooth, and hyaline. The iodine reaction was not taken

No pleurocystidia were seen. Cheilocystidia are 22–26 × 8–12 μ, clavate, hyaline, and with the apices covered with short rodlike projections. The trama of the pileus is characterized by a gelatinous layer 125–150 μ thick. Beneath this is a well-differentiated hypoderm, the cells of which contain a homogeneous dark-brown substance. The subhymenium did not appear gelatinous. The gill trama was not distinctive in any way, and the gill edges did not gelatinize, as they do in *M. vulgaris*. The fungus is clearly a Mycena, and should be recognizable because of the microscopic characters.

Mycena avellaneigrisea Murrill

Mycologia, 8: 220. 1916

Prunulus avellaneigriseus Murrill, North Am. Flora, 9: 340. 1916.

Illustrations: Text fig. 55, nos. 2–3 (p. 451).

"Pileus hemispheric, not expanding, cespitose, 7 mm. broad and 5 mm. high; surface smooth, grayish-pulverulent, avellaneous, margin entire, concolorous, appressed when young: lamellae adnate, very distant, rather narrow, arcuate, white: spores globose, smooth, hyaline, 4–6 μ: stipe cylindric, equal, smooth, glabrous, milk-white, 2 cm. long, 2 mm. thick.

"Type collected on dead wood on a shaded bank at Chester Vale, Jamaica, 900–1200 m. elevation, December 21–24, 1908, *W. A. & Edna L. Murrill 291* (herb. N. Y. Bot. Gard.).

"Distribution: Known only from the type locality."

The type consists of two small carpophores. The spores were seen attached to sterigmata. They measure 12–16 × 4–5 μ and are smooth, hyaline, nonamyloid, narrowly ellipsoid in one view and slightly curved in the other. The basidia are four-spored and measure 37–42 × 7–9 μ. No pleurocystidia projected from the hymenium, but embedded individuals similar to the cheilocystidia were found. Cheilocystidia 37–48 × 7–11 μ, smooth, more or less basidium-like. These differ from the young basidia in the shape of the enlarged upper portion. The gill and pileus tramae are homogeneous, interwoven, of slender hyphae, and yellowish in iodine. The trama of the pileus is subgelatinous throughout and, when mounted in KOH, is characterized by a somewhat glassy appearance. The revived portions were leathery and regained their shape very well. In fact, the species could just as well be referred to Marasmius. I could not find globose spores in my mounts.

Mycena carbonicola Murrill

Mycologia, 8: 220. 1916

Prunulus carbonicola Murrill, North Am. Flora, 9: 341. 1916.

"Pileus broadly convex, scattered, reaching 2 cm. broad; surface glabrous, hygrophanous, faintly striate to the disk, ochraceous-brown, darker and slightly floccose when young, margin entire, concolorous: lamellae adnate, subventricose, subconcolorous, crowded: spores ellipsoid, smooth, hyaline, 7–8 × 5–6 μ: stipe cylindric, subequal, glabrous, shining, white, hollow, 3–4 cm. long, 2 mm. thick.

"Type collected on burned sticks in Cooper's Ranch, at the base of El Yunque Mountain, Baracoa, Cuba, March, 1903, *L. M. Underwood & F. S. Earle 423* (herb. N. Y. Bot. Gard.).

"Distribution: Known only from the type locality."

The spores of the type measure 7–9 × 5–5.5 μ and are smooth, ellipsoid, hyaline, and nonamyloid. The basidia are four-spored. No pleurocystidia were seen, and the cheilocystidia were very rare. The latter are saccate, measure 26–30 × 10–18 μ, and are smooth and hyaline. They collapse readily and are difficult to demonstrate. The tramae of the gills and pileus revived very poorly, and no reliable information could be obtained except that they were yellowish in iodine.

This appears to be a very distinct species and one which might readily be mistaken for a Galerina or a Conocybe because of its pale cinnamon-brown color. The consistency of the dried fruiting bodies is very soft and fragile—much like that of the fragile gray Mycenae.

Mycena cinereiavellanea Murrill

Mycologia, 8: 220. 1916

Prunulus cinereiavellaneus Murrill, North Am. Flora, 9: 340. 1916.

"Pileus small, delicate, conic, solitary, 6 mm. broad, 5 mm. high; surface smooth, glabrous, striate, pale-avellaneous, ashy-white on the umbo: lamellae adnexed, subcrowded, rather narrow, dull-white: spores globose, smooth, hyaline, granular, 5–6 μ: stipe cylindric, equal, whitish-pulverulent at the apex, very pale avellaneous or dull-whitish below, 2 cm. long, 1 mm. thick.

"Type collected in soil rich in humus on a bank at Monkey Hill, near Cinchona, Jamaica, 1700 m. elevation, January 4, 1909, *W. A. & Edna L. Murrill 775* (herb. N. Y. Bot. Gard.).

"Distribution: Known only from the type locality."

The type consists of a single very small fruiting body. A small portion was sectioned, and the spores were found to measure 7–8 × 4 μ. They were ellipsoid instead of globose. Spherical spores were found in the mount along with one-celled brown spores and others which were muriform and had blackish-brown walls. The hyaline ellipsoid spores occurred along the hymenium in groups of four, and from this I assume that they were produced on the basidia. No differentiated sterile organs were seen, and the pileus trama did not revive well enough to be studied. Iodine reactions were not taken.

MYCENA FUMOSIAVELLANEA Murrill

Mycologia, 8: 220. 1916

Prunulus fumosiavellaneus Murrill, North Am. Flora, 9: 340. 1916.

Illustrations: Text fig. 55, nos. 5–6 (p. 451).

"Pileus minute, delicate, conic to campanulate, gregarious, 4 mm. broad, 5 mm. high; surface glabrous, striate, avellaneous with a fumosous tint, margin entire, pallid, appressed when young: lamellae adnate, distant, broad, white: spores globose, smooth, hyaline, 5–7.5 μ: stipe very slender, equal, smooth, whitish-pulverulent, about 2 cm. long.

"Type collected on dead wood at Chester Vale, Jamaica, 900–1200 m. elevation, December 21–24, 1908, *W. A. & Edna L. Murrill 310* (herb. N. Y. Bot. Gard.).

"Distribution: Known only from the type locality."

The spores of the type measure 8–9 (10) × 5–6 μ and are broadly ellipsoid, strongly amyloid, and smooth. The basidia are mostly two-spored and measure 18–20 × 7 μ. The cheilocystidia are clavate to filamentous, roughened with short rodlike projections or the apices covered with contorted fingerlike processes, hyaline and gelatinous in age. The pileus is covered with a gelatinous pellicle 50–75 μ thick, the tissue beneath it is vinaceous brown in iodine, the subhymenium is not gelatinous but the gill edges gelatinize appreciably. The stipe is surrounded by a thick outer layer of gelatinous hyphae.

The examination of the microscopic characters of the type has revealed certain discrepancies between the specimens and Murrill's description. The spores are not globose. The error was very likely caused by the large number of immature spores found in mounts. These appear globose unless one is careful to orient them properly. The drawings (text fig. 55, no. 5 [p. 451]) were made from spores which were stained in phloxine–KOH mounts. Immature spores take this stain very poorly. Because of the gelatinous layer over the exterior of the stipe and cap and the gelatinizing gill edges, this species must be referred to the Glutinipes. The original description, however, does not offer any clue to its being placed there. The species appears to be very close to *M. vulgaris* and *M. militaris*, but its southern range and its habitat on wood are not typical for either.

Mycena gracilipes Murrill

Mycologia, 8: 220. 1916

Prunulus gracilipes Murrill, North Am. Flora, 9: 340. 1916.

Illustrations: Text fig. 55, nos. 8–9 (p. 451).

"Pileus hemispheric, rather thin, gregarious, 8 mm. broad, 5 mm. high: surface glabrous, striate, fulvous when wet, nearly avellaneous when dry, margin entire, concolorous: lamellae adnate with a slight decurrent tooth, very distant, narrow, white: spores ellipsoid, smooth, hyaline, 7–8 × 3–4 μ: stipe very long and slender, cylindric, equal, smooth, glabrous, latericeous, becoming paler on drying, more or less mycelioid at the base, 5–7 cm. long, 0.5 mm. thick.

"Type collected on a dead log at Cinchona, Jamaica, about 1500 m. elevation, December 25–January 8, 1908–9. *W. A. & Edna L. Murrill 606* (herb. N. Y. Bot. Gard.).

"Habitat: on dead logs and in leaf-mold.

"Distribution: Vicinity of Cinchona, Jamaica."

The spores of the type are rather variable in size. From a single carpophore the range was found to be 8–11 (13) × (5) 6–7 (8) μ. They are smooth, ellipsoid to subovoid, hyaline, and amyloid. Although only four-spored basidia were seen, it is likely that some are two-spored. Large spores adhering in groups of two were noted. Pleurocystidia were rare to absent; when present they were similar to the cheilocystidia. The cheilocystidia were abundant, measure 26–38 × 8–14 μ, and are either ventricose with an elongated narrow neck or typically fusoid-ventricose. When revived in KOH they were seen to have dull-brown homogeneous contents. This almost certainly indicates that the gills had colored margins—though Murrill made no mention of such a character. It is, of course, impossible to state whether the color was red, reddish brown, or brown. The gill trama was homogeneous and sordid brownish or yellow in iodine. The pileus is characterized by a distinct pellicle, the threads of which have numerous short rodlike projections scattered over their walls.

A hypoderm is apparently present (the trama did not revive well). The consistency of the remoistened fruiting body was very soft and fragile like that of specimens of *M. capillaripes*. In fact, the two are very likely closely related. To judge from the type, *M. gracilipes* has the same stature as a small specimen of *M. filopes*.

MYCENA LATERICIA Murrill

Mycologia, 8: 220. 1916 (not *M. latericia Velenovský*, České Houby, p. 303. 1920–22)

Prunulus latericius Murrill, North Am. Flora, 9: 342. 1916.

"Pileus campanulate, solitary, 1 cm. broad, 8 mm. high; surface glabrous, latericeous, margin concolorous, not striate: lamellae broad, subdistant, slightly sinuate, testaceous: spores ovoid, smooth, hyaline, $6 \times 3.5\ \mu$: stipe tapering downward, smooth, glabrous, latericeous, 2 cm. long, 2.5 mm. thick.

"Type collected in moss on the ground under tree ferns at Morce's Gap, Jamaica, 1500 m. elevation, December 29, 30, January 2, 1908–9, *W. A. & Edna L. Murrill 725* (herb. N. Y. Bot. Gard.).

"Distribution: Known only from the type locality."

The type specimens appear to be fairly well preserved but revive poorly. The spores measure 3–4 μ and are globose, or broadly ellipsoid and $4–5 \times 3.5\ \mu$. They are nonamyloid and at least some collapse very readily. Some are apparently smooth under oil immersion, but others seem to have 2–3 very short aculeae scattered over their surfaces, and some appear to be subangular. These characters were studied carefully and apparently are not reliable. I suspect that the obscure scattered aculeae and subangular appearance were caused by failure of the spores to revive completely. The only basidia seen with sterigmata were two-spored. No pleurocystidia were observed, and occasional cheilocystidia were found which were shaped like the basidia but had their apices covered with short rodlike projections. The tramae of the pileus and gills turned yellowish in iodine, but no other details could be accurately ascertained.

MYCENA MARGARITA Murrill

Mycologia, 8: 220. 1916

Prunulus margarita Murrill, North Am. Flora, 9: 340. 1916.

"Pileus broadly convex with a small umbo, solitary, 1 cm. broad; surface glabrous, distantly and regularly striate, ashy-white, almost pearly, slightly depressed about the pale-avellaneous umbo, margin very thin, entire, concolorous: lamellae adnate to a collar, subdistant, rather broad, white with a pearly tint: stipe cylindric, equal, curved, smooth, glabrous, pearly, avellaneous at the base, 1.5 cm. long, 1 mm. thick.

"Type collected on decayed wood under tree ferns at Morce's Gap, Jamaica, 1500 m. elevation, December 29, 30, January 2, 1908–9, *W. A. & Edna L. Murrill 731* (herb. N. Y. Bot. Gard.).

"Distribution: Known only from the type locality."

The type was examined, and its spores were found to be smooth, hyaline, ellipsoid, and 7–8 × 5–6 μ. The iodine reaction was not obtained. The pileus trama was characterized by a gelatinous layer 150–200 μ thick, but the gill edges did not gelatinize. The cheilocystidia were not completely visible. The projecting portion was subsaccate to broadly fusoid-ventricose and smooth. They were 12–20 μ broad and projected 26–32 μ. In the bit of stipe examined no gelatinous sheath was evident. The spores, broad cheilocystidia, and thick pellicle of the pileus should enable one to recognize this fungus when it is again collected.

MYCENA PUBESCENS Murrill

Mycologia, 8: 221. 1916

Prunulus pubescens Murrill, North Am. Flora, 9: 341. 1916.

Illustrations: Text figs. 55, no. 10 (p. 451); 56, nos. 1–6.

"Pileus convex, gregarious, 8–12 mm. broad; surface smooth, dry, minutely downy, dark-brown, margin entire, not striate, concolorous: lamellae adnate, inserted, narrow, pure-white: stipe cylindric, equal, tough, white and distinctly pubescent, especially above, brownish at the base, 3 cm. long, 1 mm. thick.

"Type collected on decayed wood at Moore Town, Jamaica, November 22, 1902, *F. S. Earle 563* (herb. N. Y. Bot. Gard.).

"Distribution: Known only from the type locality."

The type has spores 4.5–5 × 3–3.5 μ which are smooth, hyaline, ellipsoid, and nonamyloid. The basidia measure 18–20 × 3–4 μ and are four-spored. Pleurocystidia and cheilocystidia are very abundant and similar in size and shape. They measure 54–75 (98) × 8–15 (20) μ, and have slightly incrusted apices and often somewhat thick-

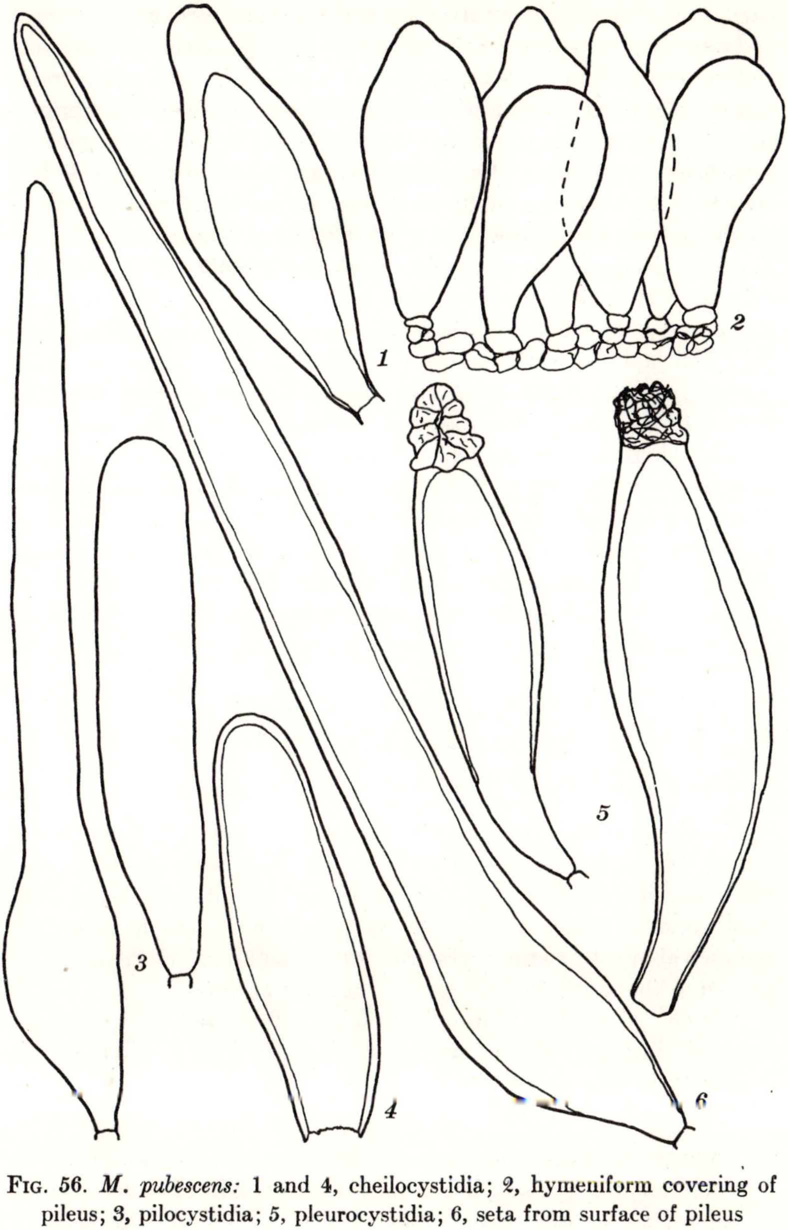

FIG. 56. *M. pubescens:* 1 and 4, cheilocystidia; 2, hymeniform covering of pileus; 3, pilocystidia; 5, pleurocystidia; 6, seta from surface of pileus

ened walls. They are ventricose, with only an inconspicuous neck and obtuse apex, or are nearly cylindric. The gill trama is gelatinous and made up of very slender hyphae (4–5 μ thick). The pileus trama is also gelatinous beneath a surface palisade of more or less upright-clavate to pear-shaped nongelatinous cells which have homogeneous dark-brown contents. Numerous hyaline setae 100–250 × 9–18 μ arise from this layer, in addition to pilocystidia which resemble those on the gills in size, shape, and markings, or are merely thin-walled. The setae have thin or only slightly thickened walls. Most of them collapse very readily and are difficult to revive. The tramae of the pileus and gills are yellowish in iodine. The fruiting bodies revive very well and are rather leathery in consistency. Numerous caulocystidia similar to the setae and pilocystidia are present over the surface of the stipe. The tissue of the stipe is yellow in iodine.

The microscopic characters of this species are very distinctive and indicate a relationship to Heliomyces. The gelatinous tramae of the gills and pileus, of course, exclude it from Mycena. There seems to me to be no point in transferring it to another genus until generic concepts in tropical agarics have been clarified on the basis of extensive anatomical studies correlated with critical observations on fresh material.

Mycena syringea Murrill

Mycologia, 8: 221. 1916

Prunulus syringeus Murrill, North Am. Flora, 9: 341. 1916.

"Pileus minute, hemispheric, solitary, 5 mm. broad, 3 mm. high; surface lilac-colored, fulvous on the disk, subgranulose in appearance, margin entire, concolorous: lamellae distant, inserted, rather broad, adnate, violet-colored: spores globose or subglobose, smooth, hyaline, about 4 μ: stipe cylindric, equal, glabrous, melleous, lilac-colored at the apex, 2 cm. long, less than 1 mm. thick.

"Type collected in humus in woods in Troy and Tyre, Cockpit Country, Jamaica, 600 m. elevation, January 12–14, 1909, *W. A. Murrill & W. Harris 1097* (herb. N. Y. Bot. Gard.).

"Distribution: Known only from the type locality."

The type has been examined and the spores have been found to be globose, smooth, hyaline, 3–4 μ, and nonamyloid. The basidia are four-spored. No differentiated cystidia were seen on the sides or edges of the gills. The gill trama is homogeneous and made up of slender interwoven hyphae that are yellowish in iodine. The pileus trama is likewise homogeneous and yellowish in iodine. Its surface is characterized by appressed hyphae, the free ends of which are saccate in shape. It is this mat of appressed hyphae with free tips which causes the pileus to appear silky or subgranulose.

Coker (1929) has reported the species from North Carolina and gives the following description:

"Cap 5–20 mm. broad, hemispheric then convex or nearly plane and with or without a small umbo, smooth to minutely silky-felted, dry or slightly viscid, pinkish violet in youth, becoming creamy violet or a pretty lilac purple all over or with the center yellowish at maturity, not striate. Flesh thin, about 0.5 mm., delicate, brittle, color of cap or paler, tasteless and odorless or with a slight odor of grapes.

"Gills color of cap or lighter, thick, moderately distant, ventricose or broadest near the stem, where they may be up to 2.5 mm., deeply and abruptly sinuate and narrowly adnexed.

"Stem 2–3 cm. long, 1–2.5 mm. thick, equal, crooked, granular scurfy above or all over, color of cap above, fading to whitish or creamy tan below, the base for about 5 mm. decorated with thick white or faintly purplish strands which run for some distance in the leaves; flesh concolorous with the surface, brittle, with a small central cylinder which is lightly stuffed or hollow.

"Spores (of No. 3474) white, smooth, subspherical, $2.6–3.2 \times 3.7–4\ \mu$.

"A pretty little species easily recognized by its violet color, coarse mycelial strands, and minute spores. We are referring it to *P. syringeus* Murrill (N. Amer. Flora 9: 341) from Jamaica, a plant known heretofore only from the type collection, because the description agrees well in every detail except that the gills are described as adnate and no mention is made of the mycelial strands at the base of the stem. It is distinguished from *Mycena pura* by smaller size, minute spores, and absence of radish taste.

"We have seven collections made around Chapel Hill during July and August. Its favorite habitat is damp woods among rotten leaves and moss, and it occurs in both coniferous and deciduous woods."

In Michigan I have found a species which fits Coker's description very well in practically all respects. I referred my collections to *Tricholoma microsporum* Ellis after a critical study of Ellis & Everhardt, North American Fungi, 2003. Since reporting the species for

Michigan I have found it in New York, Ontario, and California. It varies considerably in color when both fresh and dried. If highly colored fruiting bodies are collected they are likely to retain the purplish cast if properly dried; however, pale fresh specimens when dried nearly always lose what little color they possessed. Hence the color of herbarium specimens is not too reliable a character. Since the name Ellis gave this fungus antedates Murrill's name, it appears best to use the former. I am retaining the species in Tricholoma. Its relationships appear to me to be closer to *T. persicolor* and *T. ionides* than to *Collybia myriadophylla* or any Mycena. I hesitate to place *M. syringea* in synonymy with *T. microsporum* because the stipe of the former was described as glabrous and because the specimens were collected in Jamaica. Singer (1942) has recently studied a portion of the Ellis material and referred it to Lepiota. Further studies on a number of specimens distributed in Ellis & Everhardt, North American Fungi, 2003, should be made to determine whether or not the issue was a mixture of species.

Mycena testacea Murrill

Mycologia, 8: 221. 1916

Prunulus testaceus Murrill, North Am. Flora, 9: 341. 1916.

"Pileus thin, broadly convex, gregarious, 8 mm. broad; surface dry, tomentose, dark-testaceous, darker on the disk: lamellae adnate, pale-testaceous, crowded, narrow, sharply dentate on the edges: spores ellipsoid, smooth, hyaline, $7 \times 3\text{–}4\ \mu$: stipe cylindric, equal, subglabrous, pale-testaceous, 1.3 cm. long, 1 mm. thick.

"Type collected in humus on the ground under tree ferns at Morce's Gap, Jamaica, 1500 m. elevation, December 29, 30, January 2, 1908–9, *W. A. & Edna L. Murrill 709* (herb. N. Y. Bot. Gard.).

"Distribution: Known only from the type locality."

The type material is scanty and poorly preserved. It was possible, however, to verify the spore size given by Murrill. The spores are nonamyloid. The basidia appear to be four-spored and to measure $24\text{–}26 \times 3\text{–}4\ \mu$. The sterigmata were very fine and difficult to demonstrate. No pleurocystidia were seen. The cheilocystidia are very long and filamentous, $51\text{–}73 \times 3\text{–}4\ (5)\ \mu$, hyaline, and somewhat irregular in outline. The pileus and gill tramae turned yellow in iodine. Neither revived well enough in either KOH or chloral hydrate to allow their structure to be accurately ascertained. Among

Murrill's tropical Mycenae *M. testacea* is readily characterized microscopically by its cheilocystidia.

MYCENA TROJANA Murrill

Mycologia, 8: 221. 1916

Prunulus trojanus Murrill, North Am. Flora, 9: 339. 1916.

"Pileus broadly convex, slightly umbonate, becoming depressed at the center, cespitose, 1.5–2 cm. broad; surface smooth, dry, glabrous, white or whitish, becoming striate over the lamellae on drying, margin entire, concolorous, appressed when young: lamellae adnate, slightly decurrent at times, inserted, rather distant, broad, white: spores globose, smooth, hyaline, 5–7 μ: stipe equal, smooth, glabrous, white, 3 cm. long, 2 mm. thick.

"Type collected on decayed wood in Troy and Tyre, Cockpit Country, Jamaica, 600 m. elevation, January 12–14, 1909, *W. A. Murrill & W. Harris 866* (herb. N. Y. Bot. Gard.).

"Distribution: Known only from the type locality."

The type is well preserved and when moistened revives and is very leathery. Sections of the pileus trama show a gelatinous pellicle of loosely interwoven cells, a tramal body of loosely interwoven narrow hyphae, and a gelatinous subhymenium. All are hyaline to pale yellow in iodine. The gill trama is made up of interwoven filaments of slender hyphae, the subhymenium is somewhat gelatinous, and both are yellow in iodine. Pleurocystidia are fairly abundant. They measure 23–37 $\times$ 6–10 μ, have a long tail-like base, and a slender elongated apex, which may become irregular in outline. The mid-portion, which is buried in the hymenium, is ventricose and stains very readily in KOH mounts with phloxine added. The basidia are four-spored and measure 12–14 $\times$ 3.5 μ. The spores measure 4–5 $\times$ 2.5 μ, are blue in iodine, hyaline in KOH, smooth, and ellipsoid. The stipe tissue when carefully sectioned is found to be surrounded by a gelatinous layer.

The characters of the type specimens differ from the characters given in Murrill's description to such an extent that no one could possibly identify the species correctly without consulting the type. One might question whether or not Murrill's collection was a mixture, but an examination of the type showed that this was not the case. The gelatinous layers of the pileus and stipe indicate that both surfaces were viscid, at least when fresh and moist. There is no questioning whether or not the small spores belong to the species; they were seen attached to sterigmata and were present in great numbers. The type specimen establishes the species. Hence the discrepancies in Murrill's description should be regarded as errors of observation.

This species is obviously very closely related to *M. hondurensis* and *M. euspeirea*. From the former it differs markedly in color (being whitish as Murrill described it) and in having decurrent gills. Specimens of both species have been compared and in the dried condition are easily distinguished at sight. It is more difficult to distinguish *M. trojana* from *M. euspeirea*. In fact, the two impress me as being identical. Before reducing any species in this group to synonymy, however, one should study the type of *M. euspeirea* to determine whether it actually does belong in Glutinipes.

Mycena viridigrisea Murrill

Mycologia, 8: 221. 1916

Prunulus viridigriseus Murrill, North Am. Flora, 9: 340. 1916.

"Pileus broadly convex, minutely papillate, often becoming slightly umbilicate, at least on drying, gregarious, 1.5–2 cm. broad; surface dry, innately pubescent-fibrillose, greenish-gray, almost glaucous, margin fimbriate, concolorous, slightly striate: lamellae adnexed, very broad, subcrowded, greenish-gray: spores globose, smooth, hyaline, 5–6 μ: stipe cylindric, equal, short, smooth, glabrous, stramineous, attached by a broad circular mat of mycelium, 2 cm. long, 1–2 mm. thick.

"Type collected on a dead log at Troy and Tyre, Cockpit Country, Jamaica, January 12–14, 1909, *W. A. Murrill & W. Harris 948* (herb. N. Y. Bot. Gard.)

"Distribution: Known only from the type locality."

The type consists of a good collection of well-preserved carpophores. The spores are globose, smooth, hyaline, and amyloid, and measure 5–6 μ. The basidia are four-spored. Pleurocystidia were not seen, and the cheilocystidia were rare. They measure 20–38 × 10–20 μ, are saccate, hyaline, and smooth. It was necessary to make several mounts before they were clearly demonstrated. The gill trama and pileus trama are each homogeneous. The fruiting bodies revive well, so well, in fact, that one cannot help but wonder if the species should not be placed in Marasmius in spite of its amyloid spores.

OMPHALIA CUSPIDATELLA Murrill

Mycologia, 8: 219. 1916

Omphalina cuspidatella Murrill, North Am. Flora, 9: 349. 1916.

"Pileus convex to plane, with an abrupt small, conic, pointed umbo, cespitose, reaching 1 cm. broad; surface white, with a faint rosy tint, slightly silky, smooth, margin thin, entire, white: lamellae decurrent, arcuate, distant, white with a slightly yellowish tint: stipe slightly dilated above and tapering below, smooth, white, glabrous, 1.5 cm. long, 2 mm. thick.

"Type collected on the ground in rich moist soil under tree ferns at Morce's Gap, Jamaica, December 29–January 2, 1908–9, *W. A. & Edna L. Murrill 727* (herb. N. Y. Bot. Gard.).

"Distribution: Known only from the type locality."

I have examined the type specimen and obtained the following data: The spores measure 7–8 × 4.5–5 μ, are broadly ellipsoid, smooth, apparently hyaline in KOH, and yellowish in iodine. The basidia are four-spored and no pleurocystidia or cheilocystidia are differentiated. The tramae of the gills and pileus are each homogeneous and not otherwise distinctive. In stature the fungus resembles a robust *M. ignobilis*.

OMPHALIA CUTICOLOR Murrill

Mycologia, 8: 220. 1916

Omphalopsis cuticolor Murrill, North Am. Flora, 9: 317. 1916.

"Pileus convex to plane, slightly depressed at the center, solitary, 2 cm. broad; surface striate, dull-rosy-isabelline, hygrophanous at the center: lamellae decurrent, broad, distant, interveined, whitish: spores ovoid, much pointed at one end, smooth, hyaline, 5–7 × 3–4 μ: stipe cylindric, equal, slightly enlarged above, smooth, glabrous, concolorous, about 3 cm. long and 1 mm. thick."

Type collected on buried wood near Motzorongo, Mexico, 1,500 meters' elevation. It consists of a single carpophore of a distinctly fleshy consistency. The spores were found to be ellipsoid and no more pointed than usual in spores of this shape. They measure 7–8 × 4–5 μ, are hyaline in KOH, smooth, and amyloid. The basidia are four-spored. Pleurocystidia were not differentiated but occasional clavate-roughened cheilocystidia slightly larger than the basidia were observed. The projections are scattered and hard to locate. Upon a casual observation one would pronounce most of the cystidia smooth. The tramae of the pileus and gills were slightly reddish in iodine.

I suspect that this fungus, when restudied from fresh material, will be found to resemble *O. campanella* in many of its characters and that it should be placed in Xeromphalina if that genus is considered distinct from Omphalia.

Omphalia myceniformis Murrill

Mycologia, 8: 220. 1916

Omphalopsis myceniformis Murrill, North Am. Flora, 9: 316. 1916.

"Pileus thin, convex, not expanding, solitary, 12 mm. broad; surface glabrous, dry, stramineous with a grayish tint, margin striate, satiny, deflexed, entire: lamellae short-decurrent, inserted, ventricose, broad, subdistant, white: spores globose, smooth, hyaline, 5–7 μ: stipe long, tough, curved, smooth, glabrous, pale-reddish-brown, 3 cm. long, 1 mm. thick."

Type collected on humus in Tepeite Valley near Cuernavaca, Mexico. It consists of one well-preserved carpophore. The spores measure 3–4 × 2–2.5 μ, are hyaline in KOH, smooth, and bluish in iodine (amyloid). The basidia are four-spored. Pleurocystidia and cheilocystidia are present, measure 30–40 × 7–12 μ, and are of the smooth fusoid-ventricose type. The tramae of the pileus and gills are each subgelatinous and appear somewhat "glassy" in KOH. This species is apparently related to *O. convexa*, *O. roriduliformis*, and *O. subavellanea* by virtue of its tramal character and small amyloid spores.

In *O. convexa* the pileus and gill tramae are each distinctly subgelatinous or "glassy" in KOH, the spores 4–4.5 × 2–2.5 μ and amyloid, the basidia are four-spored, and pleurocystidia and cheilocystidia are not differentiated. The fruiting bodies revive well and remind one of those of Marasmius. *O. roriduliformis* has the same appearance and tramal characters as *O. convexa*. Its spores are 3–4 × 2.5 μ and amyloid. Cheilocystidia varying from the fusoid-ventricose type to the more or less clavate-contorted type were present. *O. subavellanea* has a pileus with a gelatinous pellicle, but the remainder of the trama of the pileus and also that of the gills are "glassy" in KOH and hence at least subgelatinous. Its spores are 3.5–4 × 2.5 μ and amyloid. Its cheilocystidia are irregular to clavate and have more or less roughened apices. These species appear to form a fairly distinct natural group which has certain characters in common with the *M. euspeirea* complex. I did not find gelatinous layers over the stipes, however, and the subgelatinous trama indicates a progression away from Mycena toward Heliomyces. *O. convexa*, *O. myceniformis*, *O. roriduliformis*, and *O. subavellanea* may all belong in one species. The differences Murrill used to separate them do not appear significant. The spore size he has given for *myceniformis* is incorrect. In *roriduliformis* he probably observed spores suspended in the mount and consequently, because of their small size, failed to orient them properly and hence erroneously described them as subglobose.

Omphalia petasiformis Murrill

Mycologia, 8: 220. 1916

Omphalopsis petasiformis Murrill, North Am. Flora, 9: 316. 1916.

"Pileus hat-shaped, with a large and prominent umbo, solitary, 2 cm. broad, 1 cm. high; surface pure-white, glabrous, faintly striate, the umbo faintly straw-colored: lamellae decurrent, distant, broad, thin, ventricose, pure-white: stipe cylindric, equal, smooth, glabrous, pure-white, 3 cm. long, 2–3 mm. thick."

Type collected on ground under tree ferns, Morce's Gap, Jamaica. The single carpophore appears to be typical of the white Mycenae. The tramae of the pileus and gills are each homogeneous and not otherwise distinctive. Pleurocystidia and cheilocystidia are not differentiated. The basidia are two-spored and have prominent sterigmata. The spores measure $11–14 \times 5.5–7\ \mu$, are hyaline in KOH, smooth, and yellowish in iodine. In the size of the spores this species appears to be quite similar to *M. McMurphyi*, but it is apparently distinct by virtue of its larger size and slightly smaller spores on two-spored basidia. When the four-spored form of *O. petasiformis* is found, its spores should be smaller, and the difference in spore size between these two should thus be sharper than is indicated by the published measurements.

Omphalia subscyphoides Murrill

Mycologia, 8: 220. 1916

Omphalina subscyphoides Murrill, North Am. Flora, 9: 348. 1916.

"Pileus convex, depressed at the center, solitary, reaching 13 mm. broad; surface dull-white, smooth, glabrous, margin irregular, concolorous: lamellae decurrent, white, distant, rather broad: stipe dilated above, smooth, glabrous, white, 2 cm. long, 2 mm. thick."

Type collected along the Tepeite River, 2,100 meters' elevation, near Cuernavaca, Mexico. The type consists of a single carpophore of rather firm consistency. When revived it is distinctly fleshy. In the dried condition the gills are a very pale violaceous brown, and the color of the cap tends toward violaceous umber. The stipe is tinged yellowish. The trama of the pileus is homogeneous, and no distinctive features were noted. The basidia are four-spored. The spores are yellowish in iodine, 7–8 (9) × 4–5 μ, smooth, and ellipsoid. Pleurocystidia and cheilocystidia are not differentiated. To judge from the color assumed by the carpophore when dried, this species should be quite distinctive among the small white agarics with more or less decurrent gills.

Omphalia tepeitensis Murrill

Mycologia, 8: 220. 1916

Omphalina tepeitensis Murrill, North Am. Flora, 9: 348. 1916.

"Pileus hemispheric, solitary, 8 mm. broad; surface smooth, glabrous, white, hygrophanous, not viscid, margin incurved, concolorous: lamellae decurrent, white, distant, rather broad: stipe unusually long, cylindric, equal, smooth, glabrous, snow-white, 4 cm. long, nearly 2 mm. thick."

Type collected on the ground in humus in a moist virgin forest along the Tepeite River, 2,100 meters' elevation, near Cuernavaca, Mexico.

The spores of the type measure 8–10 × 4–5 μ, are hyaline in KOH, smooth, and yellowish in iodine. The basidia are two-spored, and no pleurocystidia or cheilocystidia were seen. The tramae of the pileus and gills each appear homogeneous and not otherwise distinctive. The dried fruiting body resembles a carpophore of *M. pseudolactea* Kühner in stature, but when revived was more cartilaginous. It is, of course, distinguished from Kühner's species by the lack of cystidia. However, it may be merely the two-spored form of *O. cuspidatella*.

EXCLUDED AND DOUBTFUL SPECIES

Agaricus (*Mycena*) *alliaceus*. See Marasmius.
Agaricus (*Mycena*) *aureo-squamosus* Frost. Nomen nudum ?
Agaricus (*Mycena*) *esculentus*. Reported by Schweinitz (1822).
Agaricus (*Mycena*) *flavipes*. Reported by Schweinitz (1822).
Agaricus (*Mycena*) *hypnorum*. Reported by Schweinitz (1822).
Agaricus (*Mycena*) *integrellus*. Reported by Schweinitz (1822).
Agaricus (*Mycena*) *iocephalus* Berk. & Curt. Ann. and Mag. Nat. Hist., II, 12: 420. 1853. See *Marasmius iocephalus*.
Agaricus (*Mycena*) *nigripes*. See *Marasmius nigripes* (Schw.) Fries.
Agaricus (*Mycena*) *olivarius* Peck. *Omphalia olivaria* Peck. Close to *O. umbellifera* but spores of type 5–6 × 4–5 μ.
Agaricus (*Mycena*) *rubiginosus*. Reported by Schweinitz (1822).
Agaricus (*Mycena*) *setisedus*. Reported by Schweinitz (1834).
Agaricus (*Mycena*) *tenacellus*. Reported by Schweinitz (1822).
Agaricus (*Mycena*) *tener*. Reported by Schweinitz (1822).

Eomycenella echinicephala Atkinson, Bot. Gaz., 34: 37. 1902.

"Plants white, 3–8 mm. high. Pileus 0.5–0.75 mm. broad. Stipe 60–80 μ in diameter. Plants campanulate to expanded and upturned in age, trabeculae of upper surface echinulate, bearing here and there globose free branches, 10–15 μ in diameter and also echinulate, margin of pileus with free clavate branches in the form of a fringe. Cells of the trabeculae 25–30 × 6–10 μ. Hymenium plane or with few short, narrow gills. Gills when present narrowed at each end, not reaching the stipe. In the plant observed 8–10 long lamellae, 4–6 intermediate ones much shorter. Subhymenium loosely branched, obovate cells arising from the trabeculae and terminating in the basidia, or in the forms with lamellae arising from a rudimentary trama in the lamella. Basidia short clavate, abruptly narrowed into a pedicel, 9–12 × 6–9 μ, 4-spored. Spores obovate-oblong, elliptical, proximal end pointed, 6–8 × 3–4 μ, hyaline, smooth, granular. Stipe thread-like, with scattered hairs bearing a short echinulate cell on the end. Base of stipe slightly broadened."

The diagnosis is quoted from the original account. Pennington, in the *North American Flora*, Vol. 9, questionably referred this species to *Marasmius minutissimus*. I have not found a species such as this, but suspect it to belong in the vicinity of *M. cylindrospora*, from which its smaller spores readily distinguish it. Its generic position should be redetermined from a study of fresh material.

Mycena acuto-conica Clements, *in* A. F. Woods, Bot. Surv. Neb., 2: 38. 1893.

Prunulus acutoconicus Murrill, North Am. Flora, 9: 330. 1916.

"Pileus slightly membranaceous, persistently conical, acute, viscous, smooth, bright yellow; stipe long, equal, smooth, yellowish-brown; lamellae free, linear, deep ochraceous when dry; spores oblong-elliptical, 12 × 7–8 μ. Pileus 2½ cm. wide by 3 cm. high. Stipe 6–7 cm. long. Distinguished by its persistently conical, bright yellow pileus, and by the free ochraceous gills."

The original description has been quoted above. The type is deposited in the University of Nebraska Herbarium. Through the kindness of Dr. Leva B. Walker I was able to examine one of the carpophores from the type collection. Its spores are 10–12 (14) × 6–7 μ, ellipsoid, smooth, and nonamyloid. The basidia measure 40–46 × 7–9 μ and are mostly four-spored. A few bispored basidia were also found. The trama of the gills is made up of long, regularly arranged hyphal cells and is nonamyloid. The pileus trama is characterized by a gelatinous pellicle, but this is difficult to demonstrate because a great deal of sand is lodged in the pellicle and good sections are hard to obtain. The yellow viscid conic pileus, size of the carpophore, and microscopic characters all clearly indicate that the fungus is an Hygrophorus and without doubt *H. Langii* Kühner. Since Clement's name is the earliest, the combination **Hygrophorus acutoconicus** (Clements, *in* Woods), comb. nov. must be made and the following names become synonyms: *Hygrocybe Langii* Kühner, Le Botaniste, 18: 175. 1927; *Hygrophorus Rickenii* Maire, Bull. Soc. Myc. Fr., 46: 220. 1930; and *H. Langii* Dodge, Rhodora, 29: 239. 1927.

Mycena alphitophora (Berk.) Saccardo, Syll. Fung., 5: 290. 1887.

Agaricus (*Mycena*) *alphitophorus* Berkeley, Journ. Linn. Soc., 15: 48. 1877.

Prunulus alphitophorus Murrill, North Am. Flora, 9: 339. 1916.

"Minutus, totus aleuriatus niveus; pileo conico-campanulato; stipite filiformi; lamellis angustis adscendentibus.

"On small twigs, Devonshire Marsh.

"Scarcely a line high; stem ½–1 inch high, covered, as well as the pileus, with white mealy particles.

"Very delicate."

The fungus was found in Bermuda. Until the microscopic characters of the type have been studied, there is no use in trying to recognize the species. The original description is quoted.

Mycena atroalba (Bolt. ex Fr.) Gillet.

This is imperfectly known. The best European account is Ricken's, and the concept of Beardslee and Coker (1924) is not that of Ricken; the difference in spore size allows that much to be ascertained. Since information on the cystidia of the Asheville material was not given, it is impossible to be certain of the identity of the specimens.

Mycena brunneidisca Murrill, North Am. Flora, 10: 192. 1917.

Bolbitius brunneidiscus Murrill, Mycologia, 9: 320. 1917.

Mycena caesia Peck, Bull. Torr. Bot. Club, 22: 486. 1895.

Prunulus caesius Murrill, North Am. Flora, 9: 327. 1916.

"Pileus submembranous, campanulate, striate, glabrous, grayish-brown, blackish or blackish-brown in the center; lamellae thin, subventricose, grayish or bluish-gray; stem slender, hollow, glabrous, grayish-brown; spores subelliptical, slightly apiculate at one end, .0003 to .0004 in. long, .00016 to .0002 broad.

"Pileus 4–6 lines broad; stem 1–2 in. long, 0.5 to 1 line thick.

"Among sphagnum. Newfoundland. September. Rev. A. C. Waghorne.

"I have seen dried specimens only and am not certain that the bluish-gray of the lamellae is so conspicuously present in the fresh plant. The margin of the pileus is sometimes tinged with yellowish-brown."

I have examined the type and found that its microscopic characters are identical with those of *M. atroalboides* Peck. The specimens are slightly more umbonate when dried, a difference which has little taxonomic value. Apparently *M. caesia* is synonymous with *M. atroalboides*, but it is placed here in the excluded list because of the slight differences in appearance between the two types as they are preserved at Albany, New York.

Mycena caesiialba Murrill, Mycologia, 8: 220. 1920.

Prunulus caesiialbus Murrill, North Am. Flora, 9: 337. 1916.

"Pileus thin, conic to campanulate, gregarious, 1.5 cm. broad; surface dry, smooth, glabrous, caesious with an aeruginous tint, faded and much wrinkled on drying: lamellae adnate to the enlarged apex of the stipe, narrow, subcrowded, arcuate, white: spores subglobose, slightly roughened, densely granular, 7–8 μ long: stipe long and slender, pruinose to glabrous, white or avellaneous at the apex, tomentose and aeruginous at the base, 6 cm. long, 1–2 mm. thick.

"Type collected among dead leaves in Preston's Ravine near Palo Alto, California, November 25, 1911, *W. A. Murrill & L. R. Abrams 1208* (herb. N. Y. Bot. Gard.).

"Distribution: Known only from the type locality."

The type consists of one complete carpophore and some fragments of a second. I have examined it twice. The first time the fragments were studied, but the second time the carpophore was sectioned. In the fragments the spores were found to measure 7–8 × 6–7 μ, to be subglobose, smooth and borne on four-spored basidia. Pleurocystidia were scattered and similar to the cheilocystidia. The cystidia measured 32–41 × 4–8 μ, were cylindric to very narrowly fusoid in shape and thin walled. No pellicle was noted on the pileus.

In the second examination, I found the spores to be 7–9 × 5–6 (7.5) μ, broadly ellipsoid, smooth, and amyloid. The basidia were four-spored. Pleurocystidia were not located, although a careful search was made for them. The cheilocystidia were 28–39 × 4–7 μ and narrowly fusoid to subcylindric in outline. The pileus trama was characterized by a thick gelatinous pellicle.

As a result of this second study it is perfectly clear to me that the large specimen, which must be regarded as the type, is nothing more than a carpophore of *M. amicta*. Just what the fragments represented, I am not prepared to say, but apparently they are not the same as the one large carpophore. Murrill described the spores as roughened and densely granular. Obviously he mistook as roughness the appearance which the accumulation of numerous small oil droplets around the periphery of a spore gives to it—a rather easy mistake to make if the spores are not stained. Because of the likelihood that Murrill's collection represented a mixture of two species and that characters of both were included in the original description, it appears better to exclude the name than to classify it under *M. amicta* as a synonym.

Mycena californiensis (Berk. & Curt.) Saccardo, Syll. Fung., 5: 255. 1887

Agaricus (*Mycena*) *californiensis* Berkeley & Curtis, Proc. Am. Acad., 4: 112. 1858.

"Pileo ex conico breviter campanulato stipiteque gracili rufis; lamellis liberis albis rubro-marginatis.

"On oak leaves, Mare Island, California.—Differs from *A. aurantio-marginatus* in the nature of the gills, and is a more graceful species."

It cannot be recognized until the microscopic characters of the type are known.

Mycena cervinialba Murrill, Mycologia, 8: 220. 1916.

Prunulus cervinialbus Murrill, North Am. Flora, 9: 326. 1916.

"Pileus hemispheric, not fully expanding, gregarious to subcespitose, 1 cm. broad; surface dry, glabrous, conspicuously rugose, fawn-colored, margin entire, not striate: lamellae adnate with a decurrent tooth, crowded, narrow, pure white: cystidia spindle-shaped, 40–50 × 10–15 μ; stipe slender, cylindric, equal, smooth, glabrous, subconcolorous, hollow, 4–7 cm. long, 1 mm. thick.

"Type collected by the roadside in the New York Botanical Garden, June 14, 1902, *F. S. Earle 71* (herb. N. Y. Bot. Gard.).

"Distribution: Known only from the type locality."

The type is sterile. Atkinson (1918) placed the species in Galerula.

Mycena chlorantha Fries.

This name should be discarded because of confused concepts. See discussions under *M. aurantiomarginata* and *M. citrinomarginata*.

Mycena cinchonensis Murrill, Mycologia, 8: 220. 1916.

Prunulus cinchonensis Murrill, North Am. Flora, 9: 341. 1916.

"Pileus conic to campanulate, gregarious, reaching 1.5 cm. broad and 1 cm. high; surface dry, glabrous, striate, pale-isabelline or avellaneous, usually avellaneous on the disk, margin entire, pale isabelline, appressed when young: lamellae adnate, slightly decurrent at times, crowded, rather narrow, white: spores globose, smooth, hyaline, 4–5 μ; stipe cylindric, equal, smooth, glabrous, slightly pulverulent at the apex, whitish, avellaneous at the base when the pileus is avellaneous, 3–4 cm. long, about 1 mm. thick.

"Type collected on dead sticks at Cinchona, Jamaica, about 1500 m. elevation, December 25–January 8, 1908–9, *W. A. & Edna L. Murrill 550* (herb. N. Y. Bot. Gard.).

"Distribution: Known only from the type locality."

Originally the type consisted of a good collection of carpophores, but unfortunately the hymenium has been eaten off every cap by beetle larvae. Only one specimen with traces of gills was found. It was sectioned, and the only character which could be ascertained was that the cheilocystidia were of the smooth fusoid-ventricose type. In view of Murrill's errors on the spore size of other tropical species no reliance can be placed on his measurements here. Since his description contains no outstanding character by which the species can be recognized, the name should be discarded.

Mycena citricolor (Berk. & Curt.) Saccardo, Syll. Fung., 5: 263. 1887.

Agaricus (*Mycena*) *citricolor* Berkeley & Curtis, Journ. Linn. Soc., 10: 285. 1868.

Omphalopsis citricolor Murrill, North Am. Flora, 9: 316. 1916.

The following description is quoted from the *North American Flora*, 9: 316: "Pileus small, thin, pellucid, convex, gregarious, 5 mm. broad; surface lemon-yellow: lamellae decurrent, few, concolorous: stipe glabrous, filiform, concolorous, 6 mm. long."

The species cannot be recognized until its microscopic characters have been ascertained.

Mycena citrinella (Fr. ex Pers.) Quélet, Champ. Jura et Vosges, p. 108. 1872.

Agaricus citrinellus Fries, Syst. Myc., p. 155. 1821.

This species has been variously interpreted by European authors. Murrill excluded it from the *North American Flora*. There are specimens under this name in the Atkinson Herbarium (14265) and from an Ellis & Everhardt collection at the New York Botanical Garden (October 3, 1887). They resemble a small *M. epipterygia* but have spores 7–9 × 3–4 μ. The cheilocystidia are the roughened type, and about half the pileus trama (in revived mounts) is gelatinous. This form deserves further study. It may be only a depauperate form of *M. epipterygia*. Mains found the same form in northern Michigan growing along with *M. epipterygia* var. *lignicola*.

Mycena cohaerens (Fr.) Gillet. See *Marasmius cohaerens*.

Mycena collariatus (Fr.) Quélet, Champ. Jura et Vosges, p. 244. 1872.

Reported from North America, but insufficiently known.

Mycena conferruminata (Berk. & Mont.), Saccardo, Syll. Fung., 5: 258. 1887.

Agaricus (*Mycena*) *conferruminatus* Berkeley & Montagne, Syll. Crypt., 110. 1856.

"Pileo nutante campanulato purpureo-flavescente striato, margine crenulato; stipi[ti]bus fistulosis striolatis concoloribus mycelio concretis; lamellis linearibus affixis albis.

"Hab. Ad. terram fragmentis ligni corrupti conspurcatam, julio. Columbus: Sullivant, *Icon.*, no. 112.

"Desc. Pileus initio digitaliformis, mox campanulatus, membranaceus, radiatim striatus, vivus et, ex icone, praesertim centrum versus purpurascens, luteo flavoque tinctus, siccus fuscescens, centro depresso umbonulatus, 10–12 millim. altus et latus, margine demisso crenulatus. Stipites fistulosi, 5–7 centim. longi, millimetrum et ultra crassi, a basi ultra medium per mycelium copiosum album coaliti, pileo subconcolores, apice flexi, ita ut nutare videatur pileus. Lamellae postice affixae nec decurrentes, medio vix millim. latae, antice attenuatae, albae, nec confertae, nec distantes, post exsiccationem madore flexuosae. Hymenium 2 centimillim. crassum, e sporophoris clavulatis conflatum. Sporae albae, hyalinae, globosae, 0 mm., 005 diametro metientes.

Obs. Haecce species prope *A. adonidem* in systemate locanda, quorum discrimina exponere pro supervacaneo habeo."

This species has been excluded by other American authors. I have been unable to obtain any additional information on it. The portion of the type discovered among Atkinson's material was in such poor condition that no details could be obtained.

Mycena connatipes (Berk. & Curt.) Saccardo, Syll. Fung., 14: 84. 1899.

Agaricus (Mycena) connatipes Berkeley & Curtis, Ann. and Mag. Nat. Hist., III, 4: 286. 1859.

Prunulus connatipes Murrill, North Am. Flora, 9: 326. 1926.

"Dense caespitosus; pileo hemisphaerico alutaceo; stipitibus albis basi connatis; lamellis albido-flavis denti adnatis. Curt. no. 512. On the ground in dense clusters, Upper Carolina, M. A. Curtis.

"Pileus 2–8 lines across, hemispherical, buff; stems 2 inches long, $1\frac{1}{2}$ lines thick, white, confluent at the base and rooting; gills whitish yellow, adnate, subdecurrent.

"Allied to *A. proliferus* from which it differs in its hemispherical pileus, adnato-decurrent gills, etc."

The original description is quoted above. Beardslee and Coker (1924) are inclined to believe that this species is the same as *A. intertextus*. At all events, since the fungus was not even recognized later by Curtis in his catalogue of North Carolina plants, the name should be dropped.

Prunulus corrugatus (Pat.) Murrill, North Am. Flora, 9: 342. 1916.

Androsaceus corrugatus Patouillard, Bull. Soc. Myc. Fr., 16: 175. 1900.

"*A. corrugatus* n. sp.—Sur toutes sortes de bois pourris. Chapeau charnu-membraneux, orbiculaire, convexe, glabre, roux châtain, ridé vers le centre, lisse à la marge qui n'est ni striée ni plissée. Lames fauves, très serrées, adnées, inégales, minces, charnues, ondulées ou crispées. Stipe coriace, ténace, cylindracé, droit ou flexueux, glabre, luisant, roux noir, plus pâle vers le haut, étalé à la base en une membrane orbiculaire petite et concolore. Les cellules de la pellicule, sont hautes de 10 μ, dressées, serrés, et portent des protubérances allongées et rousses.

"Plante de 2–3 centim. de haut; stipe épais de 1 millim.; chapeau large de 10–20 millim.

"Analogue a *A. calopus* (Fr.) mais plus grêle et dépourvue de cystides colorées."

I have quoted the original description here because Murrill in his translation of it in the *North American Flora* left out the part which clearly characterized the species as a Marasmius of the Androsaceus group. Patoulliard's comparison of his species to *A. calopus* was a bit unfortunate, but certainly does not justify transferring the fungus to Mycena (Prunulus), as Murrill did. At present *A. calopus* Fries is regarded as a robust form of *M. inclinata*. Patouillard possibly confused *Marasmius cohaerens* with *Mycena calopus*.

Mycena cruenta (Fr.) Quélet, Champ. Jura et Vosges, p. 107. 1872.
This is a poorly known species, possibly identical with *M. haematopus*.

Mycena cymbalifera (Mont.) Saccardo, Syll. Fung., 5: 273. 1887.

Agaricus (*Mycena*) *cymbaliferus* Montagne, Syll. Crypt., p. 110. 1856.
Prunulus cymbaliferus Murrill, North Am. Flora, 9: 331. 1916.

"Pileo membranaceo convexo explanato ample umbonato margine undoso patente striatulo ex albo griseo, sicco fuscescente; stipite longissimo fistuloso fragili basi decumbente fusiformi-incrassato longe radicato fibrilloso; lamellis didymis linearibus angustissimis utrinque attenuatis haud affiixis albis.

"Hab. Ad terram inter folia, octobri. Columbus: Sullivant, *Icon*. no. 291.

"Obs. Cum cymbalis musicae militaris quoad formam pilei comparandus. Pileus $1\frac{1}{2}$–$2\frac{1}{2}$ centim. latus, umbone 6–8 millim.—Stipes cum parte ejus radiciformi 15 centim. longus, 2 millim. crassus."

This is the original description. Nothing more is known of the species. I tried to locate the type but was not successful. The species might possibly be the same as *M. fagetorum* or *M. megaspora*, but the latter is not fragile—a character specifically mentioned in the description. Consequently, since the Ohio fungus cannot be identified with any other species and has never been recognized for certain after it was described and since the description is inadequate, the name should be dropped.

Mycena dilatatus (Fr.) Gillet. Reported by Harkness and Moore (1880).

Mycena discopus (Lev.) Quélet.
Reported by Peck in 1885. Probably *M. stylobates*.

Mycena discreta (Fr.) Saccardo, Syll. Fung., 5: 260. 1887.

Agaricus discretus Fries, Elench. Fung., p. 20. 1828.

"Pileo conico umbonato densissime striato, lamellis liberis callo a stipite discretis confertis, stipite cavo ventricoso pallido.

"Nulli descripto affinis. Gregarius, in spiritu vini conservatus totus albidus l. pallescens. Stipes valde crassus sursum attenuatus, subinde conicus, basi solida, ceterum cavus, laevis, glaberrimus, 1½ unc. longus, basi ⅓ unc. usque, apice 2 lin. crassus. Pileus quam maxime membranaceus, tenerrimus, sed umbone firmo laevi aequali maxime prominente semper instructus, primo conicus, margine infracto, dein magis expansus, Coprino plicatili fere tenuior, striis delicatis longe confertissimis. Lamellae non tantum liberae, sed etiam callo, omnino ut in *A. procero* stipitem cingente remotae, confertissimae, sublineares, adscendentes. Inter specimina numerosissima fungum vidi affinem stipite 3–4 pollicari, fere aequali, pileo multo latiore obtuso, lamellis latioribus; forsan tamen varietas est. *In insula St. Crucis.* Benzon. (v. in spir. vini.)"

Mycena dissiliens (Fr.) Quélet.

This name should be discarded because of confused concepts.

Mycena echinipes (Fr.) Quélet.

American reports of this are very likely based on *M. delicatella.*

Mycena farinacea Murrill, Mycologia, 8: 220. 1916.

Prunulus farinaceus Murrill, North Am. Flora, 9: 326. 1916.

"Pileus conic to convex and at length plane, gibbous, gregarious, reaching 1.5 cm. broad; surface smooth, dry, glabrous, more or less striate, avellaneous-isabelline, margin entire, blackish (probably frozen): context pallid, the odor and taste decidedly mealy; lamellae adnexed, broad, crowded, ventricose, grayish-white: spores ellipsoid, smooth, hyaline, $7–8 \times 4–5\ \mu$: stipe cartilaginous, subequal, quite short for the genus, smooth, polished, brownish, paler at the apex, 3 cm. long, 2 mm. thick.

"Type collected in an open mossy field near the New York Botanical Garden, October 8, 1911, *W. A. Murrill & E. C. Volkert* (herb. N. Y. Bot. Gard.).

"Distribution: Known only from the type locality.

"Illustration: Mycologia 6: *pl. 137. f. 3.*"

I have examined the type and found the spores to measure $5–6 \times 4–5\ \mu$ in some caps and $7–9 \times 5–6\ \mu$ in others. In all they are finely echinulate and nonamyloid. The difference in spore size noted above is more likely caused by a difference in maturity than by a mixture of species. Only four-spored basidia were seen, and there were no differentiated pleurocystidia or cheilocystidia on the sections studied. The gill trama is homogeneous, and the pileus trama is likewise homogeneous beneath a well-developed pellicle. All parts turned yellowish in iodine. I have found this species among Atkinson's and Peck's unidentified material, but have not collected it myself. From Murrill's illustration and an examination of the specimens it is very evident that the species is a Collybia, near *C. erosa*, as Kühner has suggested. Certain of Peck's collections identified as *M. latifolia* also belong here.

Mycena flava Murrill, Mycologia, 9: 320. 1917.
This is *Bolbitius flava* Murrill, Mycologia, 9: 320. 1917.

Mycena Glatfelteri (Pk.) Murrill, North Am. Flora, 10: 190. 1917.
This is *Bolbitius Glatfelteri* Peck, Bull. Torr. Bot. Club, 30: 97. 1903.

Mycena gloiocyanea (Atk.) Murrill, North Am. Flora, 10: 191. 1917.
This is *Bolbitius gloiocyaneus* Atkinson, Ann. Myc., 6: 54. 1908.

Mycena Grantii Murrill, Mycologia, 8: 220. 1916.

Prunulus Grantii Murrill, North Am. Flora, 9: 338. 1916.

"Pileus broad, convex to plane and at length depressed, not papillate, cespitose, 3–3.5 cm. broad; surface smooth, dry, glabrous, pale-avellaneous: context thin, white, the taste mild; lamellae adnate, rather crowded, of medium breadth, thin and delicate, pure-white: spores subglobose to ellipsoid, smooth, hyaline, 6–7 × 5–6 μ: stipe equal, dry, smooth, glabrous, concolorous, hollow, with a thin rind, 6–8 cm. long, about 5 mm. thick.

"Type collected on the ground among leaves and bits of dead wood at Sequim, Washington, *J. M. Grant 332* (herb. N. Y. State Mus.).

"Distribution: Known only from the type locality."

I have never been able to recognize this species in the field. When I first examined the type the iodine reactions were not taken, and I considered the fungus to be near *M. Berkeleyi*. A second examination of the type and the application of the iodine test to the spores showed that the reactions were negative. The spores, gill trama, and pileus tissue are all nonamyloid. The pileus trama is perfectly homogeneous, with scarcely any pellicle at all. No differentiated sterile organs were found in the hymenium or on the edges of the gills at either time. The spores measure 5–5.5 × 4–5 μ and are smooth and hyaline.

More than likely the fungus can be identified with a previously described species of Collybia. There is no assurance that Grant's collection did not consist of a mixture of several species. Whatever he distributed under this name should not be considered authentic. An examination of many of his agaric collections made it evident that he had no concept of species as far as agarics were concerned. *Laccaria laccata*, for instance, turns up in his exsiccatae under more than a half-dozen different genera.

Mycena inconspicua Krieger, Mycologia, 19: 311. 1927.

"*Mycena inconspicua* n. sp.; in the nomenclature of the *North American Flora*, *Prunulus inconspicuus* (pl. 36, fig. 1).

"Pileus up to 7 mm. broad when fully expanded, broadly campanulate, with a rounded disk, brownish-ochraceous, becoming paler with age toward the margin (there almost white); latter straight from the first, striate, the striations conspicuous and reaching almost to the disk; flesh so thin as to appear membranous. Gills subdistant, unequal, white, broad for their size, rounded and ascending toward the stem where they are adnexed; edges entire, concolorous. Stem delicate, about 1.5 to 1.7 cm. long, 1.5 to 2 mm. thick, usually curved, equal or slightly thicker below, white but not shining, finely pubescent under a hand-lens, cartilaginous, tubular. Spores smooth, elliptical, one end somewhat truncate, uniguttulate, slightly pinkish within, 7.7 × 4.8 μ.

"Baltimore, Md., July 20, 1919. Grew caespitosely on naked soil at a street corner. No. *375*.

"Near *M. minutissima* Murrill and *M. paupercula* Berk., but neither."

The type has been badly damaged by insects, and few reliable microscopic data could be obtained. The only spores seen which were of the size given by Krieger were slightly angular, a fact which leads me to believe the species is a true Rhodophyllus.

Mycena integrella (Fr.) S. F. Gray.
This is *Delicatula integrella* (Fr.) Fayod.

Mycena jalapensis Murrill, Mycologia, 4: 73. 1912.
This is *Pluteolus tropicalis* Murrill, Mycologia, 4: 74. 1912 and *Bolbitius jalapensis* Murrill, Mycologia, 4: 332. 1912.

Mycena lactea (Fr.) Quélet. See discussion of *M. delicatella*.

Mycena lepiotiformis Murrill, Mycologia, 8: 220. 1916.

Prunulus lepiotiformis Murrill, North Am. Flora, 9: 322. 1916.

"Pileus minute, broadly convex, not fully expanding, solitary, 3 mm. broad; surface dry smooth, white, minutely and densely pubescent-scaly, margin entire, concolorous: lamellae subdistant, rather broad, inserted, adnexed to nearly free, white to pale-yellowish: stipe short, filiform, smooth, white above, grayish-brown and whitish pubescent below, 7 mm. long.

"Type collected on much decayed hemlock wood in woods at Unaka Springs, Tennessee, August 18–24, 1904. *W. A. Murrill 716*. Known only from the type locality."

I have examined Murrill's collection. The basidia are four-spored and measure 14–15 × 5–6 μ. The spores (only a few were found) measure about 3 × 1.5 μ. They are very minute. Iodine reactions were not obtained. No pleurocystidia or cheilocystidia were present. The structure of the gill and pileus tramae was not clearly evident in the sections, but neither appeared to possess any distinctive features. The dried specimen has the consistency of a small Lepiota, and the species very likely belongs in that genus. No vesiculose cells were found on the pileus.

Mycena lilacina (Mont.) Saccardo, Syll. Fung., 5: 257. 1887.

Agaricus (*Mycena*) *lilacinus* Montagne, Syll. Crypt., p. 110. 1856.

". . . unicolor; pileo carnoso convexo-explanato centro umbilicato lilacino laevi margine undulato fisso; stipite concolori spongioso-farcto laevi basi nudo; lamellis latis postice (in sicco) remotis antice attenuatis e roseo ochraceis pulverulentis.

"Hab. In ligno dejecto, julio. Columbus; Sullivant, *Icon.*, no. 90.

"Desc. Hymenii haec est structura: e cellulis constat breviter clavatis inter quas prominent sporophora antheridiaque. Illa autem ad medium prosiliunt et apice sterigmatibus quaternis coronantur. Haec vero apice attenuata, ea de causa fusiformia, eamdem cum sporophoris longitudinem metiuntur. Sporae oblongae, hyalinae, uni-binucleolatae, at nullo septo transverso divisae.

"Obs. Species *Agarico puro* Bull. affinis, sed lignicola, haud terricola. Ab eodem insuper pileo laevi e lilacino colore post exsiccationem alutaceo-helvolo, centro depresso, haud umbonulato, lamellis tandem a stipite ut in *A. procero* distantibus differre mihi summopere videtur."

The species has not been recognized by American investigators. A portion of Sullivant's no. 90 was found among Atkinson's material. The spores were 5–6 × 3–4 μ, and the abundant pleurocystidia were similar to those of *Pluteus cervinus*.

Mycena luteoalba (Fr.) Quélet. Not sufficiently known.

Mycena ludea (Fr.) Ricken. Reported by Kauffman from Mt. Hood. His specimens are apparently *M. delicatella*.

Mycena macrorrhiza (Berk. & Mont.) Murrill, North Am. Flora, 10: 191. 1917.

This is *Bolbitius macrorrhiza* Berkeley & Montagne, Mont. Syll. Crypt., p. 133. 1856.

Mycena mexicana Murrill, Mycologia, 4: 73. 1912.

This is *Bolbitius mexicanus* Murrill, Mycologia, 4: 332. 1912.

Mycena myceliosa Murrill, Mycologia, 8: 221. 1916.

Prunulus myceliosus Murrill, North Am. Flora, 9: 338. 1916.

"Pileus rather large and firm, convex to plane, gregarious to subcespitose, reaching 2.5 cm. broad; surface dull-brownish varying to whitish, moist, glabrous, striate, margin white, entire, appressed when young: lamellae adnate or adnexed, narrow, crowded, white: spores ellipsoid, pointed, smooth, hyaline, 5 × 3 μ: stipe very long and slender, cylindric, equal, smooth, glabrous, light-brown, mycelioid at the base and arising from copious white mycelium, about 10 cm. long, 1–2 mm. thick.

"Type collected on humus on the ground in a virgin forest near Seattle, Washington, October 20–November 1, 1911. *W. A. Murrill 223* (herb. N. Y. Bot. Gard.).

"Distribution: Known only from the type locality."

The type is well preserved, and its microscopic characters were easily obtained. The spores are 4–5 × 2.5–3 μ, nonamyloid, and borne on four-spored basidia. Pleurocystidia and cheilocystidia are abundant. Those on the sides are 62–85 × 10–16 μ and somewhat fusoid-ventricose, with long necks and obtuse apices. The cheilocystidia are shorter and fatter but also of the fusoid-ventricose type. Both the gill trama and the pileus trama are yellowish in iodine, and the latter is homogeneous beneath a palisade of pyriform cells. Numerous long pilocystidia (150 μ or more) project from this layer. These are thin-walled, hyaline, and ventricose at the bases. Caulocystidia are numerous over the stipe. They are less ventricose than the pilocystidia and appear more like hyaline setae with slightly thickened walls.

Without a doubt *M. myceliosa* is synonymous with *Collybia albipilata*. For an account of the latter and illustrations of the microscopic characters see Smith (1938). There is only one discrepancy in the microscopic details, and that is not important. In Peck's type the palisade of cells over the cap turned brown in iodine. In Murrill's type they remained pale yellowish or nearly hyaline. I have not found such differences as this to be reliable in other Collybiae. Murrill's description also bears out my conclusion in regard to the identity of his species. The narrow crowded gills, convex gray to whitish pileus, and long, slender mycelioid stipe are very convincing. He described the stipe as glabrous, it is true, but the large caulocystidia of the type clearly demonstrate that his description is in error. The mycelium around the base of the stipe is white at first but soon becomes tawny. The fungus frequently appears to be growing on humus but, if one digs far enough, he can usually find the old cone to which the stipe is attached. Because the fruiting bodies are tough and revive well, one might at first think they belonged in Marasmius.

Mycena Marasmius **Murrill, Bull. Torr. Bot. Club, 67: 154. 1940.**

Prunulus Marasmius **Murrill, *ibid.*, 148. 1940.**

"Pileus convex to expanded, slightly umbilicate at times, densely cespitose, about 1–1.5 cm. broad; surface dry, smooth, glabrous, grayish-white, margin even, entire to rimose; context very thin, white, odorless, slightly astringent; lamellae adnate, narrow, distant, inserted, white to discolored, entire; spores ellipsoid, smooth, hyaline, 1-guttulate, about 4.5–3 μ; stipe equal, smooth, glabrous, white above, ferruginous below, about 6 × 0.1–0.2 cm.

"Type collected by West, Arnold and Murrill on a much-decayed hardwood log in Kelley's Hammock, ten miles northwest of Gainesville, Fla., July 21, 1938 (*F 18277*). Also collected by the same persons at Grove Park, Fla., July 15, 1938 (*F 18270*). Very thin and partially reviving but having the general appearance of *Mycena*."

I have studied a portion of the type which Dr. Murrill sent to me for examination and have recorded the following data: The tramae of the pileus and gills are nonamyloid. The pileus is corticated by a layer of clavate-hyaline cells arranged into an upright hymeniform layer. The individual cells measure 18–28 × 7–12 μ, and occasionally bear one or two very inconspicuous projections. Pleurocystidia are not present. Cheilocystidia are abundant and more or less capitate but hyaline and smooth. They measure 9–12 μ across the apex and frequently have pedicels up to 20 μ long. The spores are nonamyloid and measure 4–5 × 3 μ. In my opinion the species should be placed in Marasmius.

Prunulus Myrciae (Pat.) Murrill, North Am. Flora, 9: 339. 1916.

Androsaceus Myrciae Patouillard, Bull. Soc. Myc. Fr., 15: 192. 1899.

"Pileus very minute, orbicular, convex-campanulate, obtuse, cespitose, 1–2 mm. broad; surface white, changing to lemon-yellow on drying and becoming tinged with russet on the disk, smooth or slightly striate, pellucid, fleshy, covered with a pellicle formed of rounded, hyaline, verrucose cells reaching 8–12 μ in diameter: lamellae white, few, quite thick, adnate to slightly decurrent, inserted, not connected by veins: spores ovoid, smooth, hyaline, 5 × 3 μ: stipe white, cylindric, delicate, almost filiform, equal or scarcely enlarged below, 4–6 mm. long, pubescent over its entire surface with cylindric, unicellular hairs which are scattered or in tufts. . . ."

This is Murrill's translation of Patouillard's description. The species was found in Guadeloupe on dying leaves of Myrcia. I have not seen authentic specimens, and believe that, in the absence of information in addition to that given by Patouillard, the fungus should be left in the genus in which it was originally described. The clavate-roughened cells forming the cuticle of the pileus are very decidedly a character of one group of species in Marasmius—those designated as Androsaceus by Patouillard.

Mycena nobilis (Pk.) Murrill, North Am. Flora, 10: 191. 1917.

Bolbitius nobilis Peck, Ann. Rep. New York State Mus., 24: 71. 1872.

Mycena palustris (Pk.) Saccardo, Syll. Fung., 5: 282. 1887.

Agaricus paluster Peck, Ann. Rep. New York State Cab., 23: 82. 1872.
Agaricus leucomyosotis Cooke & Smith in Cooke, Grevillea, 13: 57. 1885.
Collybia leucomyosotis Saccardo, Syll. Fung., 5: 220. 1887.
Prunulus paluster Murrill, North Am. Flora, 9: 331. 1916.
Collybia palustris Smith, Pap. Mich. Acad. Sci., Arts, and Letters, 21: 148. 1936.
Lyophyllum palustre Singer, Rev. de Mycol., 4: 65. 1939.
Tephrophana palustris Favre, Bull. Soc. Myc. Fr., 55: 204. 1939.

Illustrations:
Peck, Ann. Rep. New York State Cab., 23, pl. 5, figs. 6–11.
Smith, Pap. Mich. Acad. Sci., Arts, and Letters, 21, pl. 18.

If the genus Lyophyllum is to be recognized the combination given by Singer is the correct one to use.

Mycena plumbeibrunnea Murrill, Mycologia, 8: 221. 1916.

Prunulus plumbeibrunneus Murrill, North Am. Flora, 9: 338. 1916.

"Pileus convex, gregarious, 1.5–2 cm. broad; surface smooth, glabrous, opaque, pale-lead-brown, lighter near the margin, which is somewhat crenate and appressed when young: context cream-colored, without characteristic taste or odor; lamellae sinuate, rather distant, inserted, broad, ventricose, cream-colored: spores oblong-ellipsoid, smooth, hyaline, 7–9 × 4–5 μ: stipe cylindric, smooth, glabrous, concolorous with the margin of the pileus, hollow, 4–6 cm. long, 1–2 mm. thick.

"Type collected among decaying pine needles at Stanford University, California, December 6, 1901, *C. F. Baker 168* (Herb. N. Y. Bot. Gard.).

"Distribution: Known only from the type locality."

The type specimens are very poorly preserved. The spores measure 7–9 × 5–5.5 μ and are weakly amyloid. The cystidia are abundant in some caps and rare in others, but they revive very poorly because of the poor preservation of the fruiting bodies. Only a few that were reliable as to shape and size could be found, but it was apparent that the cheilocystidia and pleurocystidia were alike. The gills had been badly parasitized, and many cystidium-like hyphae apparently belonging to a phycomycete were observed in most mounts. It is impossible to determine from the type whether the species should be referred to *M. scabripes*, *M. stannea* sensu Smith, or *M. aetites* sensu Kühner.

Mycena prolifera (Fr.) Gillet.

Reported by several investigators but no clear species-concept exists for the fungus in Europe.

Mycena pulcherrima (Pk.) Saccardo, Syll. Fung., 5: 288. 1887.

Agaricus pulcherrimus Peck, Ann. Rep. New York State Cab., 23: 83. 1872.

"Pileus subcampanulate or convex, rarely subumbonate, striatulate, dull yellow, the disk a little darker; lamellae broad, subventricose, attached to the stem, not uncinate, yellowish; stipe slender, hollow, concolorous, white at the base; plant gregarious.

"Height 1′–1.5′, breadth of pileus 3″–5″.

"Under pine trees. Center. October.

"Related to *A. acicula*, but that has a smaller scarlet colored pileus."

I have not seen fresh material of the species, and have been unable to locate the type. The original description is quoted here. As previously stated (Smith, 1935), the specimens Kauffman referred to *M. pulcherrima* are identical with those of *M. roseipallens*, and there is no good reason for attributing reddish colors to the former. To judge *M. pulcherrima* by its habitat, it could scarcely be considered a faded *M. roseipallens*. Since it was described as larger than *M. acicula*, Peck's species can hardly be the same as *M. oregonensis*, which, in addition, is bright yellow and has colored gill margins. To me the most likely possibility is that *M. pulcherrima* represents a small *M. flavoalba*. However, this cannot be proved. Consequently, the best course is to regard it as doubtful, at least until the type is found or fresh material is collected and critically studied.

Mycena pulchrifolia (Pk.), Murrill, North Am. Flora, 10: 191. 1917.

This is *Coprinus pulchrifolius* Peck, Ann. Rep. New York State Mus., 29: 41. 1878.

Mycena pumila (Bull.) Rostrup. Reported by Dearness (1923).

Mycena radius (Johnston) Saccardo, Syll. Fung., 9: 37. 1891.

Agaricus (*Mycena*) *radius* Johnston, Bull. Minn. Acad., 1: 218. 1877.

"Pale yellow Mycena.

"Pileus translucent, finely striated, from margin to center, disc pale yellow, or lead color, stem thread-like, hollow, white; gills extremely narrow, regular. Pileus expanded, one to two lines broad; stem 1 to 1½ inches high.

"Found on the ground under an inverted tumbler. May."

This species has never been recognized and very likely never can be unless the type is located.

Mycena roridula (Berk. & Curt.) Saccardo, Syll. Fung., 5: 297. 1887.

Agaricus (*Mycena*) *roridulus* Berkeley & Curtis, Journ. Linn. Soc., 10: 285. 1868.

Prunulus roridulus Murrill, North Am. Flora, 9: 339. 1916.

"Pileo e convexo plano tenero albido viscido, umbone fusco; stipite aequali fistuloso glabro subfusco; lamellis adnatis puberulis latiusculis albis, interstitiis subtrabeculatis.

'On rotten wood. October. Pileus ¼ inch across; stem 1 inch high, ⅓ line thick."

I have not seen authentic material. The species cannot be recognized until data on the microscopic characters of the type become available.

Mycena setosa (Fr.) Gillet.

The name has been used in this country by Beardslee and Coker and by Kauffman; see comments under *M. osmundicola*.

Mycena sordida (C. G. Lloyd) Murrill, North Am. Flora, 10: 190. 1917.

This is *Bolbitius sordidus* C. G. Lloyd, Myc. Notes, 18. 1899.

Mycena stipularis (Fr.) Quélet.

Kühner has placed this among the incompletely known species, and I have found no fresh material answering to the Friesian description.

Mycena strobilina (Fr.) Gillet.

Reported for North America, but data concerning it are insufficient.

Mycena subpulverulenta Murrill, Mycologia, 8: 221. 1916.

Prunulus subpulverulentus Murrill, North Am. Flora, 9: 339. 1916.

"Pileus minute, delicate, conic to campanulate, gregarious to subcespitose, 4 mm. broad; surface striate, pulverulent, snow-white with a creamy tint, margin entire, appressed when young: lamellae adnate, subcrowded, white: stipe filiform, equal, smooth, subglabrous, white, mycelioid at the base, 1–1.5 cm. long, less than 1 mm. thick.

"Type collected on decayed sticks in a moist virgin forest along the Tepeite River, near Cuernavaca, Mexico, 2100 m. elevation, December 28, 1909, *W. A. & Edna L. Murrill, 473* (herb. N. Y. Bot. Gard.).

"Distribution: Known only from the type locality."

The type has been examined but no information could be obtained from it. None of the pilei examined yielded spores, and the trama and hymenium revived so poorly that no accurate observations were possible. In view of this situation the only satisfactory disposition that can be made of the species is to discard it.

Mycena tenuis (Fr.) Gillet. Reported by Moffat (1909).

Mycena tinctura Kauffman, Pap. Mich. Acad. Sci., Arts, and Letters, 5: 138. 1926.

"Pileus 1–2.5 cm. broad, subfleshy, conic-campanulate, obtusely umbonate, "pale pinkish buff" to "drab gray" (R.), umbo "fuscous," at length tinged avellaneous, striate-subplicate, striae subdistant, subhygrophanous, scarcely fading, glabrous, margin at first straight; flesh submembranous, equal, concolor. Gills ascending, adnate-seceding, rather narrow, without decurrent tooth, *subdistant*, distinct, white, at length gray-tinted, edge entire. Stem 4–7 cm. long, 1–2 mm. thick, equal, even, glabrous, naked at the apex, pale "ecru-drab" to "wood brown" (R.), terete or twisted, sometimes compressed-sulcate, cartilaginous, *tough*, strict and elastic, *with a watery juice, becoming dark vinaceous when crushed.* Odor and taste none. Spores 9–11 (12) × 5–6 (7) μ, ellipsoid, acute at apiculate end, hyaline, smooth; cystidia none; sterile cells on edge of gills saccate but indistinct; surface layer of pileus corticate, composed of 3 to 4 rows of subglobose, vesiculose, rather large cells.

"On decayed wood and needle beds of hemlock. Cespitose to solitary. Mt. Hood, Oregon. October 7 and 16. Collected by C. H. Kauffman.

"It belongs to the section Lactipedes by virtue of the rather copious watery juice of the stem, which stains the crushed flesh wine color. In other respects, it could be referred to the Filipedes. In age the drab and fuscous colors of the cap may fade. In addition to the juicy stem it is distinguishable from its relatives by the lack of odor, the narrow, subdistant gills, the spore characters, and the absence of cystidia. When growing on wood the stems may be dwarfed."

Among Kauffman's collections there are three, all from Mt. Hood, Oregon, which he identified as this species and upon which he based his description and comments. The first is dated October 7; the second, October 12, and the last, October 16. I have sectioned the carpophores in all three.

The cheilocystidia in the first collection are fusoid-ventricose and smooth, (25) 30–40 × 9–12 μ, embedded in the gill edge and with only the tapered neck projecting. The spores measure 6–8 × 4 μ and are strongly amyloid. The basidia are 28–30 × 5–6 μ. There are about five dried carpophores in the collection, and they impress one as belonging to the old section Fragilipedes of Fries.

The second collection is the short-stiped form Kauffman mentioned in his discussion. Its cheilocystidia measure 28–37 (42) × 7–12 μ, are hyaline, clavate, and with the upper portions covered with short straight or flexuous rodlike projections. The basidia are 30–32 × 5–6 μ and are four-spored. The spores are 7–9 (10) × 4–5 μ, ellipsoid, and strongly amyloid. The species is clearly in the Friesian section Rigidipedes.

The last collection (October 16) is the one photographed and illustrated on his plate 9. Its cheilocystidia are clavate and hyaline, and their upper portions are covered with rodlike projections. They are very abundant and form a broad sterile band along the gill edge. They are 20–24 × 8–12 μ. The basidia are two-spored and measure

18 × 7–8 μ. The spores are 9–11 × 5–6 μ and weakly amyloid. These specimens clearly belong in the Friesian section Filipedes, and are very likely two-spored specimens of *M. filopes* or *M. iodiolens*.

Kauffman's description was obviously compiled from his notes on all three of these collections, and since each collection represents a different species, it is necessary, according to Article 64 of the *International Rules of Botanical Nomenclature*, to discard the name *M. tinctura*. The October 12 collection is probably *M. maculata*. It was this collection in which Kauffman noticed the juice that stained the stipe wine-red. The October 7 collection is in the *M. aetites* group, and with the information at hand it cannot be stated for certain whether it should be placed in *M. stannea*, *M. aetites*, or *M. subvitrea*.

Mycena tintinnabulum (Fr.) Quélet. See *M. semivestipes* for comments.

Mycena viscidipes Murrill, Mycologia, 10: 177. 1916.

Prunulus viscidipes Murrill, *ibid.*

"Pileus hemispheric to expanded, becoming somewhat depressed at the center, gregarious to subcespitose, 1–2 cm. broad; surface hygrophanous but not viscid, glabrous, distinctly striate, fuliginous on the disk, pale-avellaneous to almost white toward the entire, straight margin; context exceedingly thin, pallid; lamellae arcuate-subdecurrent behind, ventricose in front, pallid, entire; spores ellipsoid, smooth, hyaline, 6–7 × 4 μ; stipe slightly tapering upward, dull-gray, glabrous above, tomentose below, very viscid, 4–5 cm. long, 1.5–2.5 mm. thick.

"Type collected by W. A. Murrill among leaves and sticks on the ground in deciduous woods in the New York Botanical Garden, July 30, 1915. Known only from the type locality. This is a dainty little plant with brownish, striate pileus and viscid stipe. It seems to have a preference for oak leaves.

"Illustrations: Murrill, Mycologia, 10; pl. 8, fig. 1. 1918."

I have been unable to locate the type. The illustration, however, strongly suggests *M. odorifera*, which, when young and fresh, may appear to be moist and hygrophanous. The viscidity becomes evident only after the caps have been moistened. Rather frequently the odor of *M. odorifera* is faint and easily overlooked. Hence, unless one paid particular attention to these characters when the fresh specimens were in hand, they could very easily be misinterpreted. It seems best to regard *M. viscidipes* as a doubtful species, at least until the type can be found and the microscopic characters determined.

Mycena vitrea (Fr.) Quélet. Reported by Stover (1912).

Omphalia clavata Peck, Ann. Rep. New York State Mus., 51: 285. 1898.

Omphalopsis clavata Murrill, North Am. Flora, 9: 313. 1916.

"Pileus thin, convex, becoming nearly plane, glabrous, pallid or subcinereous, the margin decurved; lamellae narrow, distant, very decurrent, pallid; stem long, slender, glabrous, stuffed, commonly enlarged at the top, slightly villous-tomentose at the base, pallid; spores globose, .0002 to .00024 in. broad.

"Pileus 2 to 4 lines broad; stem about 1 in. long, 5 lines thick.

"Dead prostrate trunks of arbor-vitae, *Thuja occidentalis* . . . .

"The base of the stem is clothed with a few long loose whitish filaments, and the thickened upper part is often fluted by the long decurrent lamellae. The clavate form given to the stem by this enlargement is suggestive of the specific name."

I have examined the type, but did not make tests with iodine. The spores are smooth, measure 5–6 × 4–5 μ, and are globose to subglobose. The basidia are four-spored. Pleurocystidia and cheilocystidia, which are abundant, are of the fusoid-ventricose type and smooth. They measure 34–52 × 7–12 μ. The hyphae of the surface of the pileus are quite compact, and numerous pilocystidia project from them.

Superficially this species appears to resemble the members of the section Omphaliariae, but an examination of the type indicates that it may belong in Marasmius. Further study, especially of fresh material, is necessary to determine its true relationships.

Omphalia cuspidata Quélet.

Reported from Asheville, North Carolina, by Beardslee, but microscopic details were not given. There are numerous species in North America with the stature illustrated by the published photograph.

Omphalia niveicolor Murrill, Mycologia, 8: 219. 1916.

Omphalina niveicolor, North Am. Flora, 9: 348. 1916.

Since the type of this species has been destroyed by insects and Murrill did not give any microscopic characters, it is impossible to distinguish it from a number of other small white agarics with decurrent gills. The following is Murrill's description:

"Pileus convex to nearly plane, gregarious, 3–5 mm. broad, rarely reaching 8 mm., surface snow-white, smooth, glabrous, margin incurved when young, concolorous: lamellae decurrent, distant, narrow, forked, snow-white, becoming slightly rose-tinted with age, the edges rather obtuse when young but thin at maturity: stipe snow-white, smooth, glabrous, very slender, 6 mm. long."

The specimen was collected on a wet log in Cuernavaca, Mexico.

BIBLIOGRAPHY

ARENBERG, FLORENCE F. 1941. Agaricaceae from the Medicine Bow Mountains of Wyoming. Mycologia, 33: 50–53.

ARNOLD, J. D. 1935. A Comparative Study of Certain Species of Marasmius and Collybia in Culture. Mycologia, 27: 388–416.

ATKINSON, GEORGE FRANCIS. 1900. Studies of American Fungi. Mushrooms, Edible, Poisonous, Etc., pp. 1–275. Andrus and Church, Ithaca, New York.

—— 1902. Three New Genera of the Higher Fungi. Bot. Gaz., 34: 36–43.

—— 1908. Notes on Some New Species of Fungi from the United States. Ann. Myc., 6: 54–62.

—— 1918. The Genus Galerula in North America. Proc. Am. Phil. Soc., 57: 357–374.

—— AND SCHRENK, HERMANN. 1893. Some Fungi of Blowing Rock, N. C. Journ. Elisha Mitchell Sci. Soc., 9: 95–108.

BEARDSLEE, H. C. 1917. Notes on a Few Species of Asheville Fungi. Mycologia, 9: 30–33.

—— 1934. New and Interesting Fungi. *Ibid.*, 26: 253–260.

—— AND COKER, W. C. 1924. The Mycenas of North Carolina. Journ. Elisha Mitchell Sci. Soc., 40: 49–91.

BELL, GARNET S. 1933. List of the Larger Fungi, Toronto Region. Trans. Roy. Canad. Inst., 19: 275–299.

BERKELEY, M. J. 1836. Fungi, Vol. V (or Vol. II of Dr. Hooker's British Flora), Part II, pp. 1–386, of The English Flora of Sir James Edward Smith. London.

—— 1845. Decades of Fungi, vii–x. London Journ. Bot., 4: 299–313.

—— AND BROOME, C. E. 1876. Notices of British Fungi. Ann. and Mag. Nat. Hist., Ser. 4, 17: 130.

—— 1877. Enumeration of Fungi Collected during the Expedition of H.M.S. "Challenger." Bermuda. June, 1873. Journ. Linn. Soc., 15: 48–52.

—— AND CURTIS, M. A. 1853. Centuries of North American Fungi. Ann. and Mag. Nat. Hist., Ser. 2, 12: 417–435.

BERKELEY, M. J., CURTIS, M. A. 1859. (Same title.) *Ibid.*, Ser. 3, 4: 284–296.

—— —— 1860. Characters of New Fungi, Collected in the North Pacific Exploring Expedition by Charles Wright. Proc. Am. Acad. Arts and Sci., 4: 111–130.

—— —— 1860. Fungi Cubenses (Hymenomycetes). Journ. Linn. Soc., 10: 280–341.

BISBY, G. R., BULLER, A. H. R., AND DEARNESS, JOHN. 1929. The Fungi of Manitoba, pp. 1–194. Longmans, Green and Co. London, New York, Toronto.

—— —— —— FRASER, W. P., AND RUSSELL, R. C. 1938. The Fungi of Manitoba and Saskatchewan, pp. 1–189. Ottawa.

BISSONNETTE, T. H. 1915. List of Georgian Bay Fleshy Fungi and Myxomycetes. Canada Biol. Board. Contrib. Canad. Biol., pp. 213–220.

BRESADOLA, J. 1882. Fungi Tridentini, Vol. I, Fasc. 2. Tridenti.

—— 1928. Iconographia Mycologica, Vols. V–VI. Milano.

BULLER, A. H. REGINALD. 1934. Researches on Fungi, Vol. VI. Longmans, Green and Co., London, New York, Toronto.

BUNDY, W. E. 1883. A Partial List of the Fungi of Wisconsin, with Descriptions of New Species. Geology of Wisconsin, I: 396–401.

CEJP, K. 1929a. Rivise Středoevropských Druhů Skupiny Mycena-Omphalia. Publ. Fac. Sci. Univ. Charles, XCVIII: 1–86.

—— 1929b. (Same title.) *Ibid.*, C: 1–91.

—— 1930. (Same title.) *Ibid.*, CIV: 1–157.

CHARLES, VERA K. 1931. Some Common Mushrooms and How to Know Them. U. S. Dept. Agric., Circular 143.

CLELAND, JOHN BURTON. 1934. Toadstools and Mushrooms and Other Larger Fungi of South Australia. Adelaide.

CLEMENTS, FREDERIC E. 1910. Minnesota Plant Studies IV. Minnesota Mushrooms, pp. 1–169. University of Minnesota, Minneapolis.

COKER, W. C. 1929. Notes on Fungi. Journ. Elisha Mitchell Sci. Soc., 45: 164–178.

COLE, EMMA L. TAYLOR. 1910. Guide to the Mushrooms, pp. 1–206. Charles K. Reed, Worcester, Massachusetts.

COOKE, M. C. 1885. New British Fungi. Grevillea, 13: 57–61.

—— 1889–91. Illustrations of British Fungi, Vol. VIII.

CURTIS, M. A. 1867. A Catalogue of the Indigenous and Naturalized Plants of the State. Geol. and Nat. Hist. Surv. of North Carolina, Part III. Raleigh, North Carolina.

DAVIDSON, JEAN E. 1930. Notes on the Agaricaceae of Vancouver (B.C.), District–I. Mycologia, 22: 80–93.

DAY, DAVID F. 1882–83. The Plants of Buffalo and Its Vicinity. Bull. Buffalo Soc. Nat. Sci., 4 (3–4): 1–290.

DEARNESS, JOHN. 1923. Report of the Canadian Arctic Expedition, 1913–18, Vol. IV. Botany. Part C, Fungi. Ottawa. 24 pp.

DODGE, B. O. 1914. A List of Fungi, Chiefly Saprophytes, from the Region of Kewaunee County, Wisconsin. Trans. Wis. Acad. Sci., 17: 806–845.

DOUGLAS, GERTRUDE E. 1918. The Development of Some Exogenous Species of Agarics. Am. Journ. Bot., 5: 36–54.

EARLE, F. S. 1909. The Genera of the North American Gill Fungi. Bull. New York Bot. Gard., 5: 373–451.

FAIRMAN, CHARLES E. 1893. Hymenomycetae of Orleans County, N. Y. Proc. Rochester Acad. Sci., 2: 154–167.

FAVRE, JULES. 1939. Champignons rares ou peu connus des Hauts-Marais jurassiens. Bull. Soc. Myc. Fr., 55: 196–219.

FAYOD, V. 1889. Prodrome d'une histoire naturelle des agaricinés. Ann. Sci. Nat., Série 7, 9: 181–411.

FRIES, E. 1821. Systema Mycologicum, Vol. I. Lundae.

—— 1828. Elenchus Fungorum Sistens Commentarium in Systema Mycologicum. Vol. I: 238. Gryphiswaldiae.

—— 1836–38. Epicrisis Systematis Mycologici. Upsaliae.

—— 1867. Icones Selectae Hymenomycetum Nondum Delineatorum I. Holmiae.

—— 1874. Hymenomycetes Europaei. Upsaliae.

FROST, CHAS. C. 1869. Further Enumeration of New England Fungi. Proc. Boston Soc. Nat. Hist., 12: 77–81.

GILLET, C. C. 1874. Les Hyménomycètes ou description des tous les champignons qui croissent en France. . . . Alençon.

—— 1878. Les Champignons qui croissent en France . . . , Vol. VI. Paris.

GLATFELTER, N. M. 1906. Preliminary List of Higher Fungi Collected in the Vicinity of St. Louis, Mo., from 1898 to 1905. Trans. Acad. Sci. St. Louis, 16: 33–94.

Graham, V. O. 1944. Mushrooms of the Great Lakes Region, pp. 1–390. 49 pls. Chicago Acad. Sci. and the Chicago Nat. Hist. Mus. Chicago.

Gray, S. F. 1821. A Natural Arrangement of British Plants, 1: 596–676. London.

Greene, Wesley. 1907. Plants of Iowa. Bull. State Hort. Soc., pp. 1–240. Des Moines.

Guba, E. F., and Seeler, E. V., Jr. 1942. Third List of Fungi of Nantucket. Rhodora, 44: 167–175.

Güssow, H. T., and Odell, W. S. 1927. Mushrooms and Toadstools. An Account of the More Common Edible and Poisonous Fungi of Canada, pp. 1–274. Ottawa.

Hard, M. E. 1906. *Mycena haematopa* Pers. Mycol. Bull., 69–70: 275.

—— 1908. The Mushroom, Edible and Otherwise. Its Habitat and Its Time of Growth, pp. 1–602. Columbus, Ohio.

Harkness, H. W., and Moore, A. M. 1880. Catalogue of the Pacific Coast Fungi. Calif. Acad. Sci., 45 pp.

Henry, L. K. 1939. Twenty-eight Common Gilled Mushrooms of Western Pennsylvania. Trillia, 10: 82–100.

Herbst, William. 1899. Fungal Flora of the Lehigh Valley, Pa., pp. 1–230. Berkemeyer, Keck and Co. Allentown, Pennsylvania.

Hesler, L. R. 1937a. A Preliminary List of the Fungi of the Great Smoky Mountains National Park. Castanea, 2: 45–58.

—— 1937b. Notes on Southern Appalachian Fungi, II. Journ. Tenn. Acad. Sci., 12: 239–254.

Hess, A. W., and Vandivert, H. 1900. Basidiomycetae of Central Iowa. Proc. Iowa Acad. Sci., 7: 183–186.

Höhnel, Franz von. 1913. Fragmente zur Mykologie (XV Mitteilung, nr. 793 bis 812). Sitzungsb. Kaiserl. Acad. Wissensch. Wien, Math.-naturw. Klasse, 122: 256–278.

House, Homer D. 1920. Notes on Fungi, VI. New York State Mus. Bull., 219–220: 233–237.

Imai, Sanshi. 1938. Studies on the Agaricaceae of Hokkaido. Journ. Fac. Agric. Hokkaido Imperial University, 43: 1–178.

Johnston, A. E. 1877. The Mycological Flora of Minnesota. Bull. Minn. Acad. Nat. Sci., 1: 203–302.

Josserand, Marcel. 1930. Note sur deux mycènes: *Mycena flavo-alba* (Fr.) Q. et *M. floridula* (Fr.) Q. Bull. Soc. Myc. Fr., 46: 38–42.

JOSSERAND, MARCEL. 1933. Notes critiques sur quelques champignons de la région lyonnaise. *Ibid.*, 49: 340–376.

—— 1937. Description de quelques omphales blanches. Ann. Soc. Linn. Lyon, 80: 77–96.

—— ET MAIRE, R. 1931. Sur l'*Omphalia marginella* (Pers.) Josserand et Maire (*Omphalina marginella* Quélet). Bull. Soc. Linn. Lyon, 10: 115–119.

—— AND SMITH, ALEXANDER H. 1937. Notes on the Synonymy of French and American Agarics–I. Mycologia, 29: 717–724.

KALCHBRENNER, C. 1873. Icones Selectae Hymenomycetum Hungariae. Pestini.

KARSTEN, P. A. 1879. Rysslands, Finlands och den Skandinaviska Halföns Hattsvampar. Mustalia. (Also cited as Bidr. Finl. Nat. Folk, 32: 1–571.)

—— 1883. Icones Selectae Hymenomycetum Fenniae. Helsinforsiae.

—— 1890. Symbolae ad Mycologiam Fennicam. Pars XXIX. Medd. Soc. pro Faune et Flora Fenn., 16: 84–106.

—— 1892. Fragmenta Mycologica. Hedwigia, 31: 218.

—— 1893. Symbolae ad Mycologiam Fennicam. Pars XXXII. Acta Soc. pro Fauna et Flora Fenn., 9: 1–11.

KAUFFMAN, C. H. 1906. Unreported Michigan Fungi from Petoskey, Detroit and Ann Arbor for 1905. Rep. Mich. Acad. Sci., 8: 26–37.

—— 1907. Unreported Michigan Fungi for 1906. *Ibid.*, 9: 83–87.

—— 1911. Unreported Michigan Fungi for 1910, with Outline Keys of the Common Genera of Basidiomycetes and Ascomycetes. *Ibid.*, 13: 215–249.

—— 1915a. The Fungi of North Elba. New York State Mus. Bull., 179: 79–104.

—— 1915b. Unreported Michigan Fungi for 1911, 1912, 1913 and 1914. Rep. Mich. Acad. Sci., 17: 194–216.

—— 1917. Tennessee and Kentucky Fungi. Mycologia, 9: 159–166.

—— 1918. The Agaricaceae of Michigan. Lansing, Michigan.

—— 1921. The Mycological Flora of the Higher Rockies of Colorado. Pap. Mich. Acad. Sci., Arts, and Letters, 1: 101–150.

—— 1926. The Fungus Flora of Mt. Hood, with Some New Species. *Ibid.*, 5: 115–148.

KAUFFMAN, C. H. 1930. The Fungous Flora of the Siskiyou Mountains in Southern Oregon. *Ibid.*, 11: 151–210.

—— AND SMITH, ALEXANDER H. 1933. Agarics Collected in the Vicinity of Rock River, Michigan, in 1929. *Ibid.*, 17: 153–200.

KELLERMAN, W. A., AND WERNER, WILLIAM C. 1893. Catalogue of Ohio Plants. Geology of Ohio, Vol. VII, Part II, Botany. Pp. 56–403.

KONRAD, PAUL, ET MAUBLANC, ANDRÉ. 1924–32. Icones Selectae Fungorum, Vol. II. Paris.

—— —— 1924–33. (Same title.) *Ibid.*, Vol. III.

KRIEGER, L. C. C. 1927. New or Otherwise Interesting Agaricaceae from the United States and Canada. Mycologia, 19: 308–314.

—— 1935. A Popular Guide to the Higher Fungi (Mushrooms) of New York State. New York State Mus. Handbook, 11. Albany, New York.

KÜHNER, ROBERT. 1926. Contribution à l'étude des Hyménomycètes et spécialement des Agaricacés. Le Botaniste, 17: 1–244.

—— 1930. Un nouveau groupe d'Agarics leucospores. Bull. Soc. Linn. Lyon, 9: 67–69.

—— 1931. Utilisation de la réaction iodée dans la classification des Mycena. *Ibid.*, 10: 122.

—— 1936. Sur la réaction à l'iode des parois des hyphes du carpophore des *Mycena*. Comptes rendus, Acad. Sci., Paris, 203: 1287.

—— 1937. Les Noyaux et leurs divisions dans les articles du carpophore et particulièrement dans Basides des *Mycena*. *Ibid.*, 204: 62.

—— 1938. Le Genre Mycena (Fries). Encyc. Myc., 10: 1–710.

LANGE, J. E. 1914. Studies in the Agarics of Denmark. Part I. Dansk Bot. Arkiv, 1 (5): 1–40.

—— 1930. (Same title.) Part VIII. *Ibid.*, 6 (5): 1–61.

—— 1933. (Same title.) Part IX. *Ibid.*, 8 (3): 1–46.

—— 1934. Mycofloristic Impressions of an European Mycologist in America. Mycologia, 26: 1–12.

—— 1936. Flora Agaricina Danica, Vol. II. Copenhagen.

LASCH, GOTF. 1828. Enumeratio Agaricorum Marchiae Brandenburgicae. Continuatio hujus enumerationsis et finis. Linnaea, 3: 378–430.

LEA, THOMAS G. 1849. Catalogue of Plants, Native and Naturalized, Collected in the Vicinity of Cincinnati, Ohio, during the Years 1834–1844, pp. 1–77. Published by Isaac Lea. Printed by T. K. and P. G. Collins, Philadelphia.

LLOYD, C. G. 1899. Mycological Notes. Mycological Writings, I: 18 (no. 41).

—— 1901. (Same title.) *Ibid.* (no. 129).

LONGYEAR, B. O. 1904. A Preliminary List of the Saprophytic Fleshy Fungi Known to Occur in Michigan. Rep. Mich. Acad. Sci., 4: 113–124.

LUNDELL, SETH. 1932. Bitrag til Uppsalatraktens Hymenomycet-flora, I. Vrådsätra Naturpark. Sv. Vet.-Akad. Skr., 22: 1–35.

MCCLATCHIE, A. M. 1897. Seedless Plants of Southern California. Protophytes–Pteridophytes. Proc. South. Calif. Acad. Sci., 1: 338–394.

MCDOUGAL, W. B. 1917. Some Edible and Poisonous Mushrooms. Bull. Ill. State Lab. Nat. Hist., 11 (7): 413–555.

—— 1925a. Mushrooms, a Handbook of Edible and Inedible Species, pp. 1–151. Cambridge, Massachusetts.

—— 1925b. A New Mushroom. Trans. Ill. State Acad. Sci., 17: 84.

MCILVAINE, CHARLES, AND MACADAM, ROBERT. 1912. One Thousand American Fungi, 740 pp. Bobbs-Merrill Co., Indianapolis.

MACKAY, A. H. 1907. Fungi of Nova Scotia; a Provisional List. Proc. and Trans., Nova Scotian Inst. Nat. Sci., 11 (1902–03): 122–143.

—— 1908. Fungi of Nova Scotia; First Supplementary List. *Ibid.*, 12 (1906–07): 119–138.

MAIRE, RENÉ. 1910. Notes critiques sur quelques champignons récoltes pendant la session de Dijon de la Société Mycologique de France (Octobre, 1909). Bull. Soc. Myc. Fr., 26: 159–198.

MANEVAL, WILLIS E. 1937. A List of Missouri Fungi. Univ. Mo. Stud., 12 (3): 1–150.

MARSHALL, NINA L. 1905. The Mushroom Book. A Popular Guide to the Identification and Study of Our Commoner Fungi with Special Emphasis on the Edible Varieties, pp. 1–170. Doubleday, Page and Co., New York.

MILLSPAUGH, C. F. 1892. Preliminary Catalogue of the Flora of West Virginia. Bull. West Virginia Agric. Exper. Station, 24. Morgantown, West Virginia.

—— AND NUTTALL, L. W. 1896. Flora of West Virginia. Field Columbian Museum Publication, 9. Bot. Ser. I (2), pp. 1–276.

MOFFATT, W. S. 1909. The Higher Fungi of the Chicago Region. Part I. The Hymenomycetes. Natural History Survey, Bull. VII, Part I, pp. 1–156. Chicago Acad. Sci.

MOHR, CHARLES. 1901. Plant Life of Alabama. Geol. Surv. Ala. Alabama Edition, pp. 1–921 (Oct.). Montgomery, Alabama.

MONTAGNE, J. F. 1856. Sylloge Generum Specierumque Cryptogamarum. Paris.

MORGAN, A. P. 1883. The Mycologic Flora of the Miami Valley, Ohio. Journ. Cincinnati Soc. Nat. Hist., 6: 54–81.

MORSE, ALBERT P. 1918. List of the Water-Color Drawings of Fungi by George E. Morris. Peabody Museum, Salem, Massachusetts.

MURRILL, W. A. 1912. The Agaricaceae of Tropical North America-V. Mycologia, 4: 72–83.

—— 1912. Illustrations of Fungi-XI. *Ibid.*, 4: 163–169.

—— 1913. (Same title.) XV. *Ibid.*, 5, pl. 92, fig. 8.

—— 1915a. (Same title.) XX. *Ibid.*, 7: 115–120.

—— 1915b. (Same title.) XXI. *Ibid.*, 7: 163–167.

—— 1915c. Preliminary List of Upper St. Regis Fungi. *Ibid.*, 7: 297–306.

—— 1916a. Omphalopsis, Galactopus, Prunulus, Omphalina and Gymnopus. North Am. Flora, 9: 310–376.

—— 1916b. Pleurotus, Omphalia, Mycena and Collybia Published in North Am. Flora. Mycologia, 8: 218–221.

—— 1917a. The Taxonomy of the Agaricaceae. Am. Journ. Bot., 4: 315–326.

—— 1917b. Some Fungi Collected in Virginia. Mycologia, 9: 34–36.

—— 1917c. Illustrations of Fungi—XXVII. *Ibid.*, 257–260.

—— 1917d. North Am. Flora, 10: 190–193.

—— 1917e. Two New Species of Fleshy Fungi. Mycologia, 9: 40–41.

—— 1918. Illustrations of Fungi—XXIX. *Ibid.*, 11: 177–181.

—— 1919. Cuban Polypores and Agarics. *Ibid.*, 11: 22–32.

—— 1938. New Florida Agarics. *Ibid.*, 30: 359–371.

—— 1939. Additions to Florida Fungi—III. Bull. Torr. Bot. Club, 66: 29–37.

—— 1940a. (Same title.) III. *Ibid.*, 67: 145–154.

—— 1940b. (Same title.) IV. *Ibid.*, 227–235.

ODELL, W. S. 1926. List of Mushrooms and Other Fleshy Fungi of the Ottawa District. Bull. Victoria Memorial Museum, 43, 15 pp.

OORT, A. J. P. 1928. De Nederlandsche Mycena's Mededeelingin von de Nederlandsche Mycologische Vereeniging, 16–17: 184–255.

Paige, F. W. 1928. A List of the Fleshy Fungi from Webster County, Iowa. Iowa State Coll. Journn. Sci., 2: 117–135.

Patouillard, N. 1899. Champignons de la Guadeloupe. Bull. Soc. Myc. Fr., 15: 191–209.

—— 1900. (Same title.) *Ibid.*, 16: 175–188.

Pearson, A. A. 1918. A New Mycena. Trans. Brit. Myc. Soc., 6: 135–136.

Peck, Chas. H. 1872b. Report of the Botanist. Ann. Rep. New York State Cab., 23: 27–135.

—— 1872b. (Same title.) *Ibid.*, 24: 41–108.

—— 1873a. V. Descriptions of New Species of Fungi. Bull. Buffalo Soc. Nat. Sci., 1: 41–72.

—— 1873b. Report of the Botanist. Ann. Rep. New York State Mus., 25: 57–123.

—— 1874. (Same title.) *Ibid.*, 26: 35–91.

—— 1875. (Same title.) *Ibid.*, 27: 73–116.

—— 1876. (Same title.) *Ibid.*, 28: 31–88.

—— 1878a. (Same title.) *Ibid.*, 29: 29–82.

—— 1878b. (Same title.) *Ibid.*, 30: 23–78.

—— 1879a. (Same title.) *Ibid.*, 31: 19–60.

—— 1879b. (Same title.) *Ibid.*, 32: 17–72.

—— 1885. (Same title.) *Ibid.*, 38: 77–138.

—— 1886. (Same title.) *Ibid.*, 39: 30–73.

—— 1887. Descriptions of New Species of New York Fungi. Bull. New York State Mus., 1 (2): 5–24.

—— 1888. Report of the Botanist. Ann. Rep. New York State Mus., 41: 51–122.

—— 1891. Fungi of Maryland. *Ibid.*, 44: 5–75.

—— 1893a. Report of the State Botanist. *Ibid.*, 45: 5–42.

—— 1893b. (Same title.) *Ibid.*, 46: 1–69.

—— 1895a. New Species of Fungi. Bull. Torr. Bot. Club, 22: 198–211.

—— 1895b. (Same title.) *Ibid*, 22: 485–493.

—— 1897. Ann. Rep. New York State Botanist for 1895. Ann. Rep. New York State Mus., 49: 17–83.

—— 1898. Report of the State Botanist. *Ibid.*, 51: 267–321.

PECK, CHAS. H. 1899. Plants of North Elba. Bull. New York State Mus., 6 (28): 1–266.

—— 1901. Report of the State Botanist. Ann. Rep. New York State Mus., 54: 131–195.

—— 1903. (Same title.) Bull. New York State Mus., 67: 3–194.

—— 1905. New Species of Fungi. Bull. Torr. Bot. Club, 32: 77–81.

—— 1907a. Report of the State Botanist. Bull. New York State Mus., 116: 1–98.

—— 1907b. New Species of Fungi. Bull. Torr. Bot. Club, 34: 97–104.

—— 1907c. (Same title.) *Ibid.*, 34: 345–349.

—— 1912. Report of the State Botanist. Bull. New York State Mus., 157: 5–139.

—— 1913. (Same title.) *Ibid*, 167: 6–137.

—— 1915. (Same title.) *Ibid.*, 176: 5–77.

POLLOCK, JAS. B., AND KAUFFMAN, C. H. 1905. Michigan Fungi. Rep. Mich. Acad. Sci., 7: 57–67.

POVAH, A. H. W. 1935. The Fungi of Isle Royale, Lake Superior. Pap. Mich. Acad. Sci., Arts, and Letters, 20: 113–156.

QUÉLET, LUCIEN. 1872. Les Champignons du Jura et des Vosges, pp. 1–320, pls. 1–24.

—— 1873. (Same title.) Pp. 321–424, pls. 1–5.

—— 1875. (Same title.) Pp. 1–128, pls. 1–4.

—— 1876–1901. (Same title.) Suppls. 4–22 (published as articles in Soc. Bot. Fr., Assoc. Fr. Sci., and other journals).

REA, CARLETON. 1922. British Basidiomycetae. Cambridge.

—— 1926. Appendix to *British Basidiomycetae*. Trans. Brit. Myc. Soc., 12: 205–230.

REDDICK, DONALD. 1908. A Preliminary List of the Hymenomycetes or Mushrooms of Indiana. Ann. Rep. Dept. Geol. and Nat. Resources Ind., 32: 1193–1231.

RICKEN, ADALBERT. 1915. Die Blätterpilze. Leipzig.

RIDGWAY, R. 1912. Color Standards and Color Nomenclature. Washington, D. C.

SACCARDO, P. A. 1887–1931. Sylloge Fungorum, Vols. V–XXV.

—— PECK, CHAS. H., AND TRELEASE, WILLIAM. 1904. The Fungi of Alaska, pp. 13–53. Alaska, V. Cryptogamic Botany (Harriman Alaska Expedition). Doubleday, Page and Co., New York.

Schroeter, J. 1889. Die Pilze Schlesiens. *In* Cohn, Kryptogamen Flora von Schlesien.

Schweinitz, Lewis David de. 1822. Synopsis Fungorum Carolinae Superioris. Schrift. d. Naturf. Ges. zu Leipzig, 1: 20–131.

—— 1834. Synopsis Fungorum in America Boreali Media Degentium. Trans. Am. Philos. Soc., N. S., 4: 141–315.

Seaver, Fred J. 1916. Bermuda Fungi. Mem. New York Bot. Gard., 6: 501–511.

—— and Shope, Paul F. 1936. Some Rocky Mountains Basidiomycetes. Univ. Colo. Stud., 23: 189–197.

Secretan, L. 1833. Mycographie suisse, Vol. II. Genève.

Singer, Rolf. 1931. Pilze aus dem Kaukasus II. Beih. Bot. Centralbl., 48: 513–542.

—— 1936a. Das System der Agaricales. Ann. Myc., 34: 286–378.

—— 1936b. Bemerkungen über einige Basidiomyceten. *Ibid.*, 34: 423–434.

—— 1937. Notes sur quelques Basidiomycètes, III. Rev. de Mycologie, 2: 226–242.

—— 1938. (Same title.) IV. *Ibid.*, N. S., 4: 187–199.

—— 1939. (Same title.) V. *Ibid.*, 4: 64–72.

—— 1942. Type Studies on Agarics. Lloydia, 5: 97–135.

—— and Smith, Alexander H. 1943. A Monograph on the Genus Leucopaxillus Boursier. Pap. Mich. Acad. Sci., Arts, and Letters, 28: 85–132.

Smith, Alexander H. 1934a. Investigations of Two-Spored Forms in the Genus Mycena. Mycologia, 26: 305–331.

—— 1934b. Unusual Agarics from Michigan. Pap. Mich. Acad. Sci., Arts, and Letters, 19: 205–216.

—— 1935a. Studies in the Genus Mycena, I. Am. Journ. Bot., 22: 858–877.

—— 1935b. Studies in the Genus Mycena, II. Mycologia, 27: 586–604.

1935c. Unusual Agarics from Michigan, II. Pap. Mich. Acad. Sci., Arts, and Letters, 21: 171–183.

—— 1936a. Studies in the Genus Mycena, III. Mycologia, 28: 410–430.

—— 1936b. Unusual Agarics from Michigan, III. Pap. Mich. Acad. Sci., Arts, and Letters, 21: 147–161.

—— 1937a. Studies in the Genus Mycena, IV. Mycologia, 29: 338–354.

SMITH, ALEXANDER H. 1937b. Notes on Agarics from the Western United States. Bull. Torr.Bot. Club, 64: 477–487.

—— 1938a. New and Unusual Agarics from North America, I. Mycologia, 30: 20–41.

—— 1938b. "Flora Agaricina Danica." *Ibid.*, 30: 599–600.

—— 1939a. Studies in the Genus Mycena, V. *Ibid.*, 31: 267–285.

—— 1939b. Notes on Agarics from British Honduras. Contrib. Univ. Mich. Herb., No. 1, 28 pp., 2 pls.

—— 1941a. Studies of North American Agarics, I. *Ibid.*, No. 5, 73 pp., 32 pls.

—— 1941b. New and Unusual Agarics from Michigan, II. Pap. Mich. Acad. Sci., Arts, and Letters, 26: 61–68.

—— AND HESLER, L. R. 1940. New and Unusual Agarics from the Great Smoky Mountains National Park. Journ. Elisha Mitchell Sci. Soc., 56: 302–324.

—— AND WEHMEYER, LEWIS E. 1936. Contributions to a Study of the Fungous Flora of Nova Scotia. II. Agaricaceae and Boletaceae. Pap. Mich. Acad. Sci., Arts, and Letters, 21: 163–197.

SMITH, HURON H. 1931. Mushrooms of the Milwaukee Region. Milwaukee Pub. Mus. Field Guide, No. 1. Bot. Ser.

SPEGAZZINI, CARLOS. 1883. Fungi Guaranitici. Pugillus I. Anal. Soc. Ci. Argent., 16: 242–248.

SPRAGUE, C. J. 1856. Contribution to New England Mycology. Proc. Boston Soc. Nat. Hist., 5: 323–332.

—— 1859. (Same title.) *Ibid.*, 6: 315–321.

STOVER, W. G. 1912. The Agaricaceae of Ohio. Proc. Ohio Acad. Sci., 5: 462–577.

SWARTZ, DELBERT. 1933. Studies of Arkansas Fungi. Am. Mid. Nat., 14: 714–719.

THOMAS, W. S. 1928. Field Book of Common Gilled Mushrooms, 332 pp. G. P. Putnam's Sons, New York.

TUCKERMAN, E., AND FROST, C. C. 1875. A Catalogue of Plants Growing without Cultivation within Thirty Miles of Amherst College, 98 pp. Amherst.

UNDERWOOD, L. M., AND EARLE, F. S. 1897. A Preliminary List of Alabama Fungi. Ala. Agric. Exper. Station, Bull. 80, 283 pp.

VAN HOOK, J. M. 1910. Indiana Fungi. Proc. Ind. Acad. Sci., pp. 205–212.

VAN OVEREEM, C., UND WEESE, J. 1923–26. Icones Fungorum Malayensium, Vols. I–VI.

VELENOVSKÝ, JOSEF. 1920–22. České Houby.

VOGLINO, PIETRO. 1886. Ricerche Analitiche sugli Agaricini della Venezia. Atti R. Ist. Veneto Sci., Lett. ed Arti, Venezia, 6 ser., 4: 603–656.

WHITE, EDWARD A. 1905. A Preliminary Report on the Hymeniales of Connecticut. Bull. Conn. State Geol. and Nat. Hist. Surv., Bull. 3, 80 pp.

—— 1910. Second Report of the Hymeniales of Connecticut. *Ibid.*, Bull. 15, 70 pp.

WHITE, V. S. 1902. Some Mt. Desert Fungi. Bull. Torr. Bot. Club, 29: 550–563.

PLATE 1

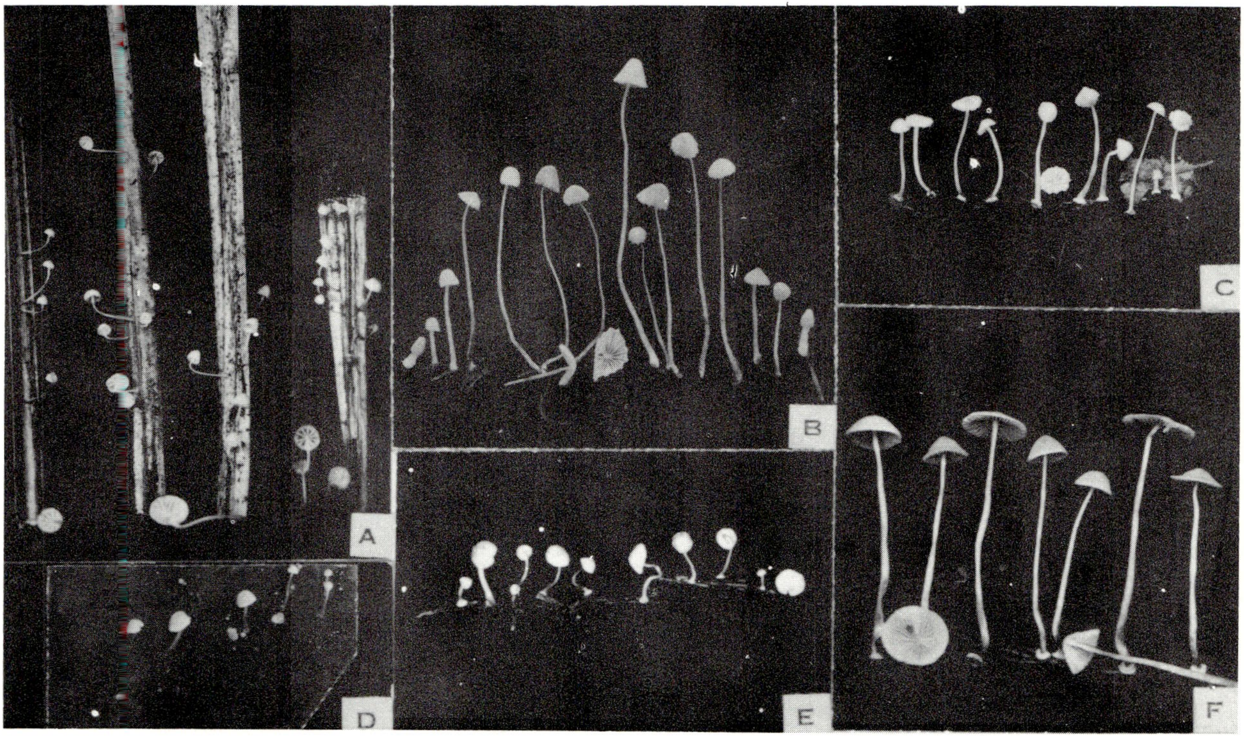

A, *M. juncicola* (Fr.) Gillet; B, *M. osmundicola* Lange; C, E, F, *M. stylobates* (Fr.) Quélet; D, *M. tenerrima* (Berk.) Quélet. All $\times$ 1

PLATE 2

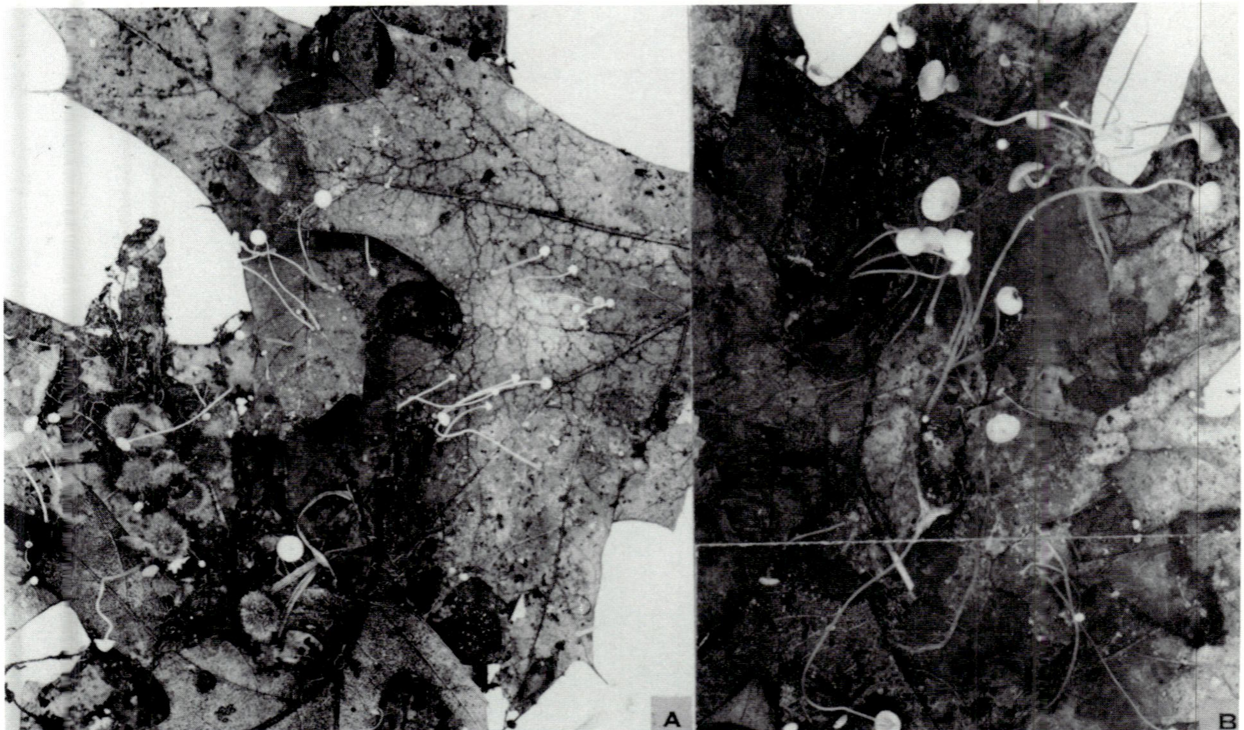

A, *M. mucor* (Fr.) Quélet; B, *M. capillaris* (Fr.) Gillet. Both × 1

PLATE 3

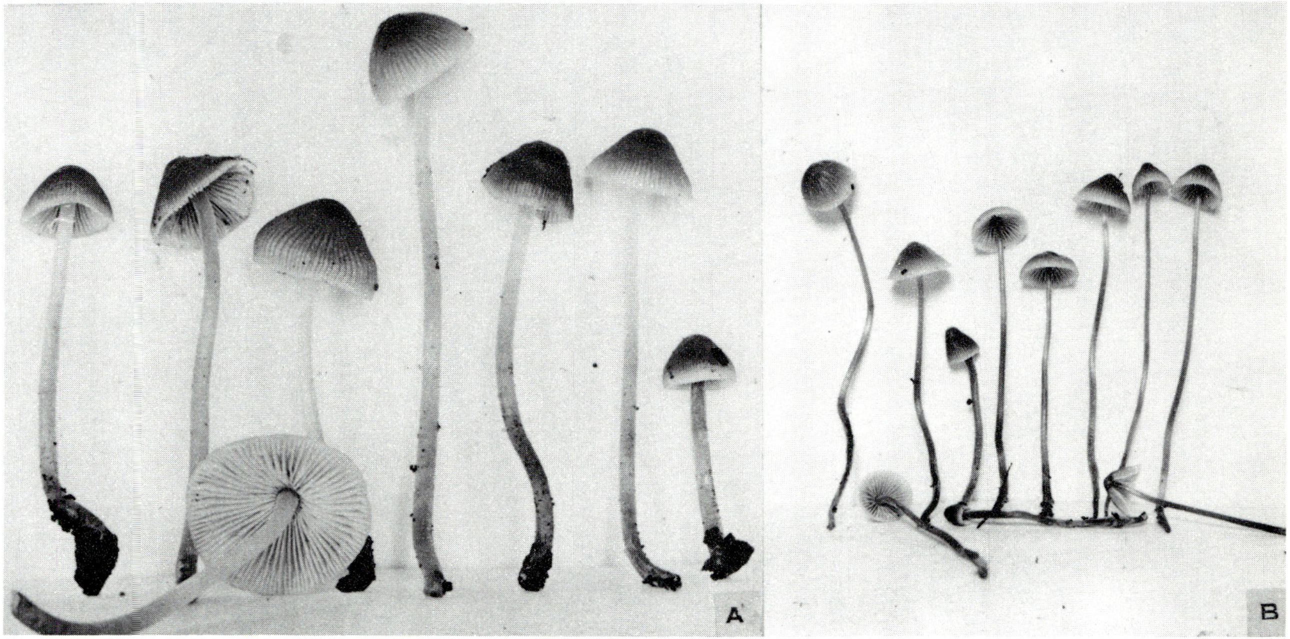

A, *M. amicta* (Fr.) Gillet (large form); B, *M. amicta* (Fr.) Gillet (small form). Both × 1

PLATE 4

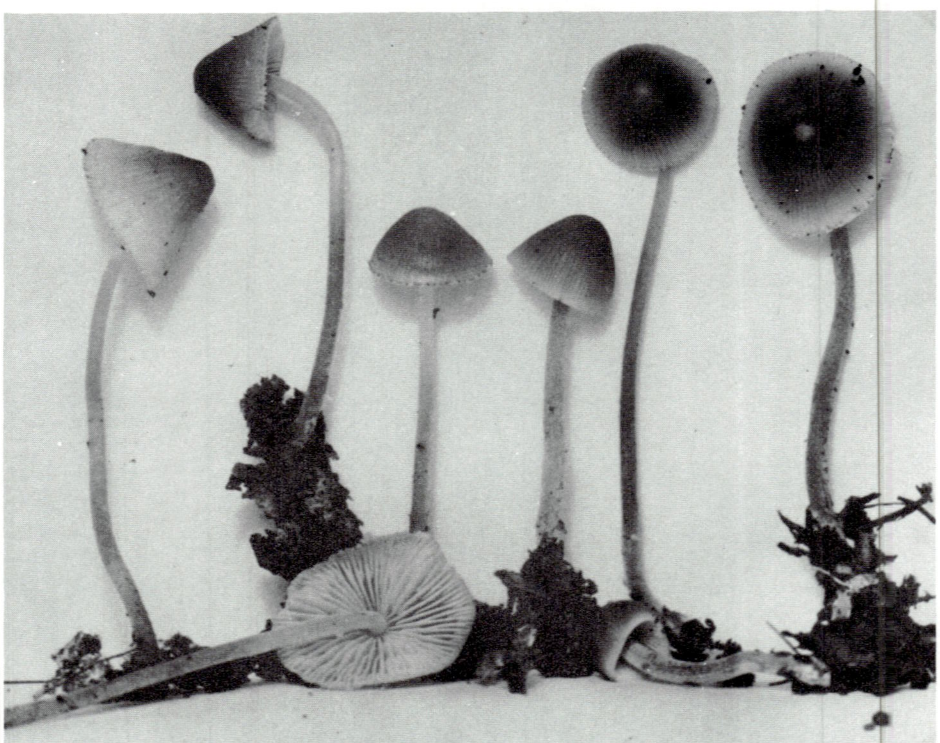

M. subcaerulea (Pk.) Saccardo. × 1

PLATE 5

M. corticola (Fr.) Quélet. × 1

PLATE 6

M. madronicola Smith. × 1

PLATE 7

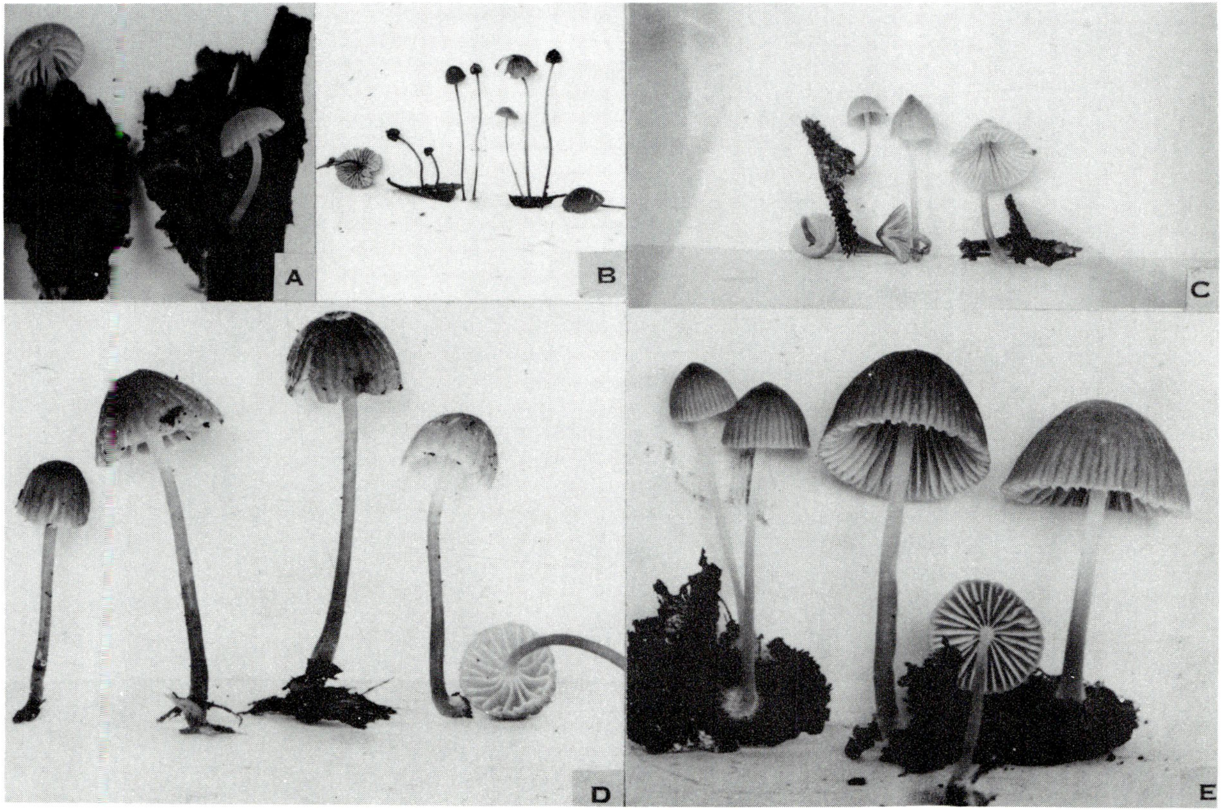

A, *M. subcana* Smith; B, *M. oregonensis* Smith; C, *M. subcana* Smith; D, *M. corticalis* Smith; E, *M. subcana* Smith (large form). All × 1

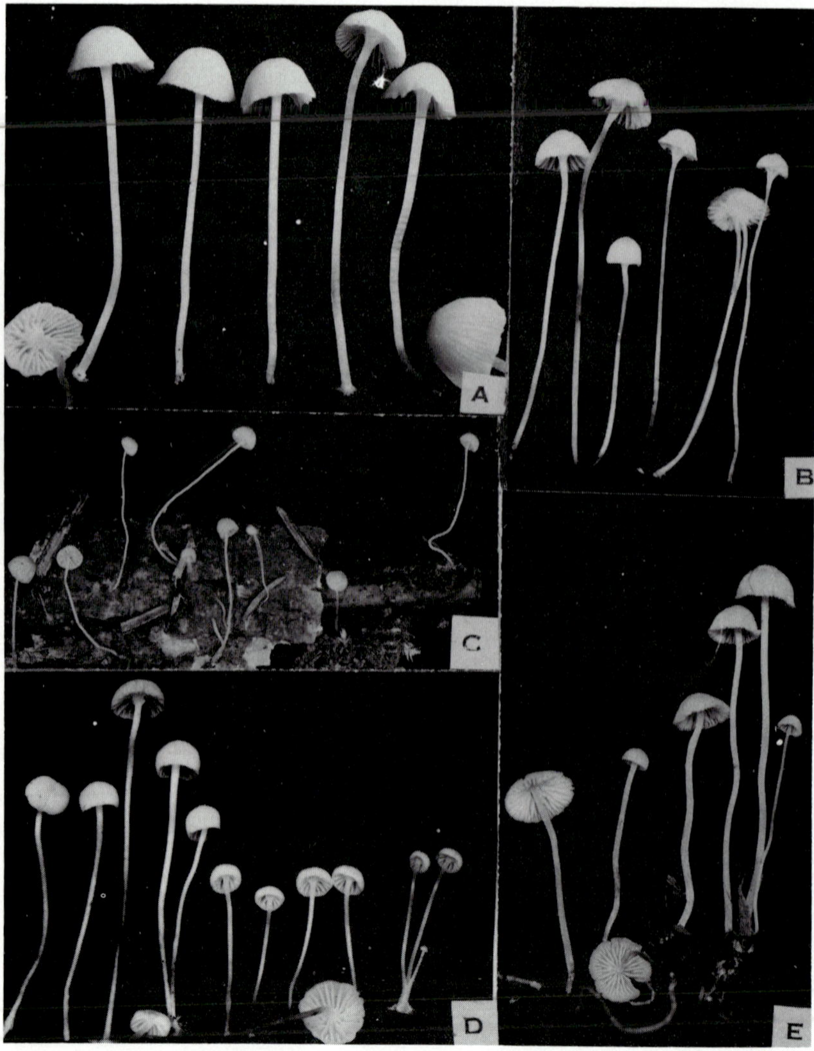

A, *M. delectabilis* (Pk.) Saccardo; B, *M. gracilis* (Quél.) Kühner; C, *M. mirata* (Pk.) Saccardo; D, *M. gracilis* (Quél.) Kühner (common form); E, *M. acicula* (Fr.) Quélet (large form). All × 1.

PLATE 9

M. tenella (Fr.) Quélet (robust form). × 1

PLATE 10

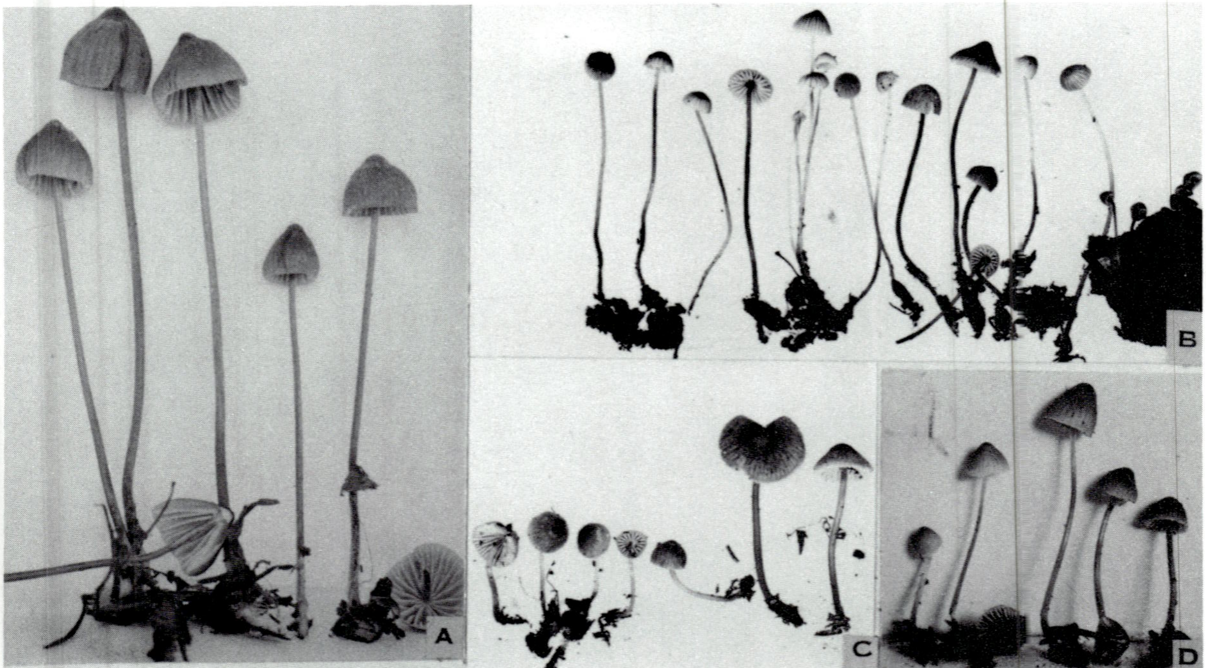

A, *M. galopus* (Fr.) Quélet; B, *M. supina* (Fr.) Gillet; C, *M. erubescens* von Höhnel (short form); D, *M. subsupina* Smith. All × 1

PLATE 11

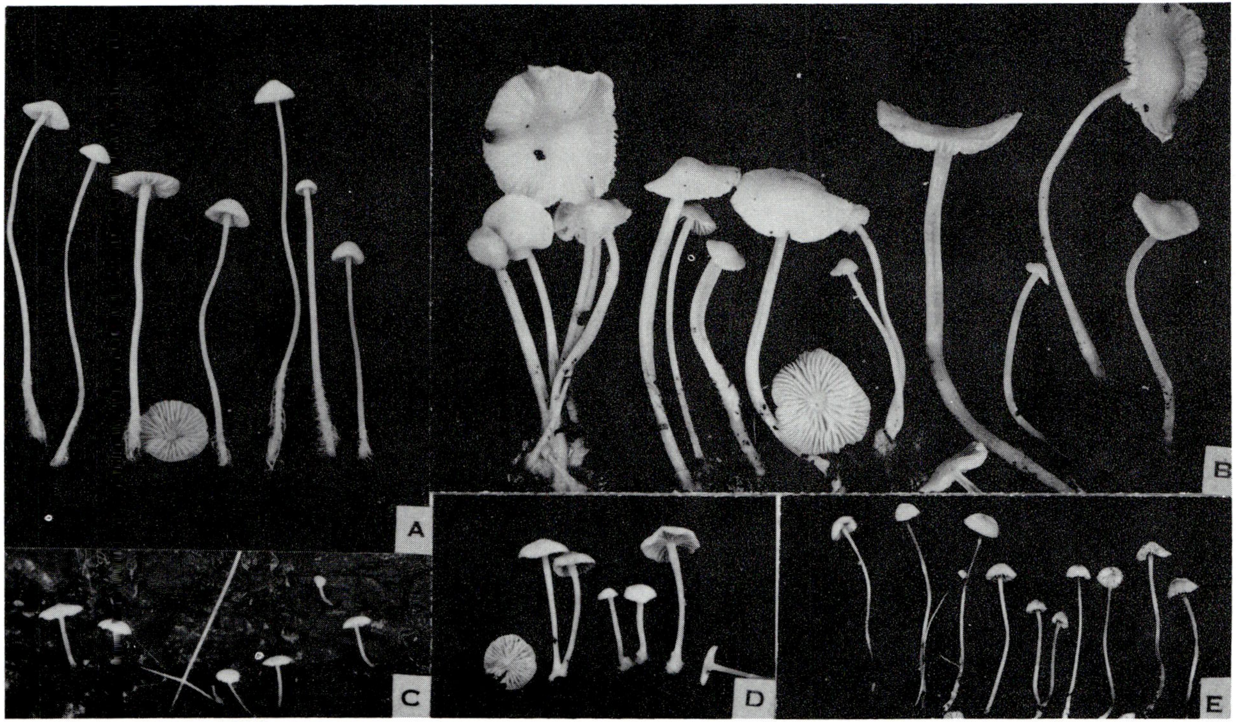

A, *M. pseudolactea* Kühner; B, *M. roseocandida* (Pk.) Saccardo; C, *M. olida* var. *americana* Smith (small form)
D, *M. delicatella* (Pk.) Smith; E, *M. acicula* (Fr.) Quélet (typical form). All × 1

PLATE 12

M. erubescens von Höhnel (typical form). × 1

PLATE 13

M. haematopus (Fr.) Quélet (typical form). × 1

PLATE 14

A, *M. haematopus* (Fr.) Quélet; B, *M. Atkinsoniana* Smith.
Both × 1

PLATE 15

M. haematopus var. *marginata* Lange. × 1

PLATE 16

A, *M. subsanguinolenta* Smith; B, *M. sanguinolenta* (Fr.) Quélet. Both × 1

PLATE 17

M. gypsea (Fr.) Quélet. × 1

PLATE 18

A, *M. monticola* Smith; B, *M. olida* var. *americana* Smith (small form).
Both × 1

PLATE 19

A, *M. flavoalba* (Fr.) Quélet; B, *M. adonis* (Fr.) Quélet. Both × 1

PLATE 20

M. roseipallens Murrill. ×1

PLATE 21

M. amabilissima (Pk.) Saccardo: A, typical form; B, large form. Both × 1

PLATE 22

M. subaquosa Smith. × 1

PLATE 23

M. pura (Fr.) Quélet. × 1

PLATE 24

A, *M. pura* (Fr.) Quélet; B, *M. Kühneriana* Smith. Both × 1

PLATE 25

M. pelianthina (Fr.) Quélet. × 1

PLATE 26

A, *M. aurantiomarginata* (Fr.) Quélet (typical form); B, *M. strobilinoides* Peck (small form); C, *M. rosella* (Fr.) Quélet (small form). All × 1.

PLATE 27

M. aurantiomarginata (Fr.) Quélet (large form). × 1

PLATE 28

M. elegans (Fr.) Quélet (large form). × 1

PLATE 29

A, *M. flavescens* Velenovský; B, *M. elegantula* Peck (solitary form). Both × 1

PLATE 30

M. rosella (Fr.) Quélet (large form). × 1

PLATE 31

M. purpureofusca (Pk.) Saccardo. × 1

PLATE 32

M. elegantula Peck (cespitose form). × 1

PLATE 33

M. rubromarginata (Fr.) Quélet (typical form). × 1

PLATE 34

M. citrinomarginata Gillet. × 1

PLATE 35

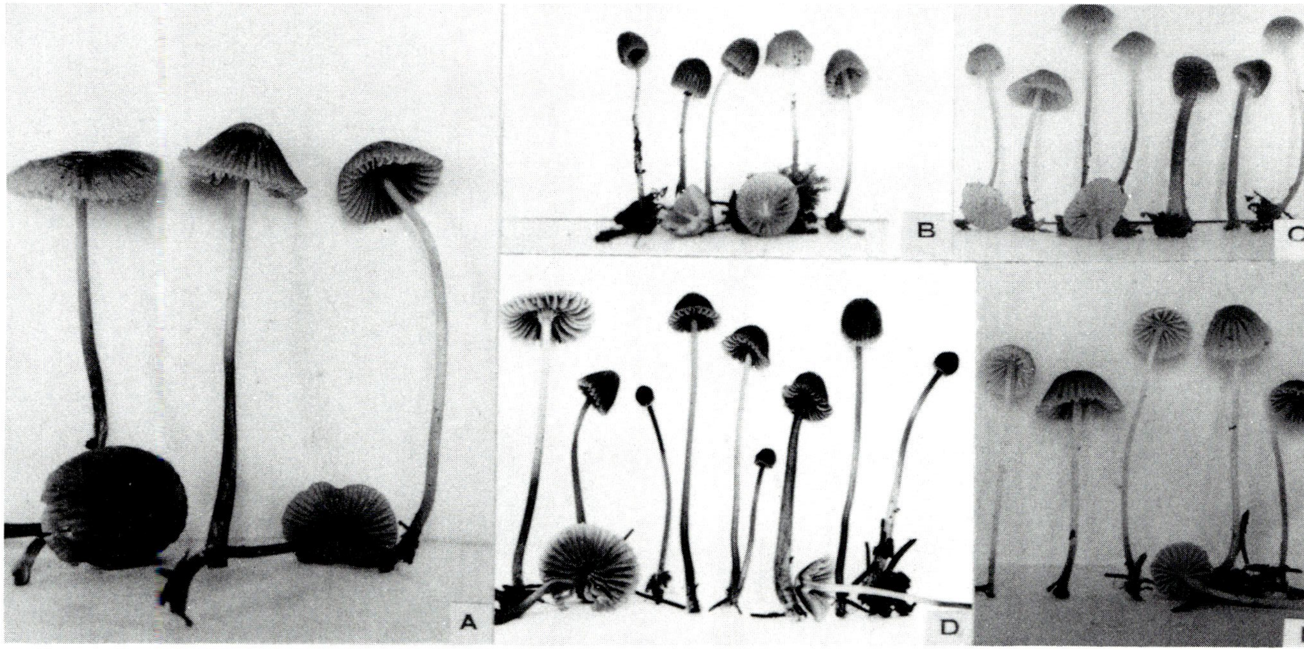

A, B, C, D, *M. capillaripes* Peck: A, large form; B, two-spored form from oak woods; C, small form; D, typical form.
E, *M. alcaliniformis* Murrill. All × 1

PLATE 36

M. scabripes Murrill (robust form). × 1

PLATE 37

M. pseudotenax Smith (typical form). × 1

PLATE 38

M. Abramsii Murrill. $\times 1$

PLATE 39

M. leptocephala (Fr.) Gillet (large form). × 1

PLATE 40

A, *M. praelonga* (Pk.) Saccardo; B, *M. pectinata* Murrill (small form); C, *M. leptocephala* (Fr.) Gillet (small form). All × 1

PLATE 41

M. subvitrea Smith. × 1

PLATE 42

M. stannea (Fr.) Quélet (typical form). × 1

PLATE 43

M. stannea (Fr.) Quélet (form with fragile stipe). × 1

PLATE 44

M. plumbea (Fr.) Saccardo (typical form). × 1

A, *M. plumbea* (Fr.) Saccardo (short-stiped form); B, *M. pectinata* Murrill (typical form). Both × 1

PLATE 46

M. plumbea var. *robusta* Smith. × 1

PLATE 47

A, *M. atroalboides* (Pk.) Saccardo; B, *M. subplicosa* Karsten; C, *M. atrocyanea* (Fr.) Gillet. All × 1

PLATE 48

M. latifolia (Pk.) Saccardo: A, pleurocystidia rare; B, pleurocystidia abundant. Both × 1

PLATE 49

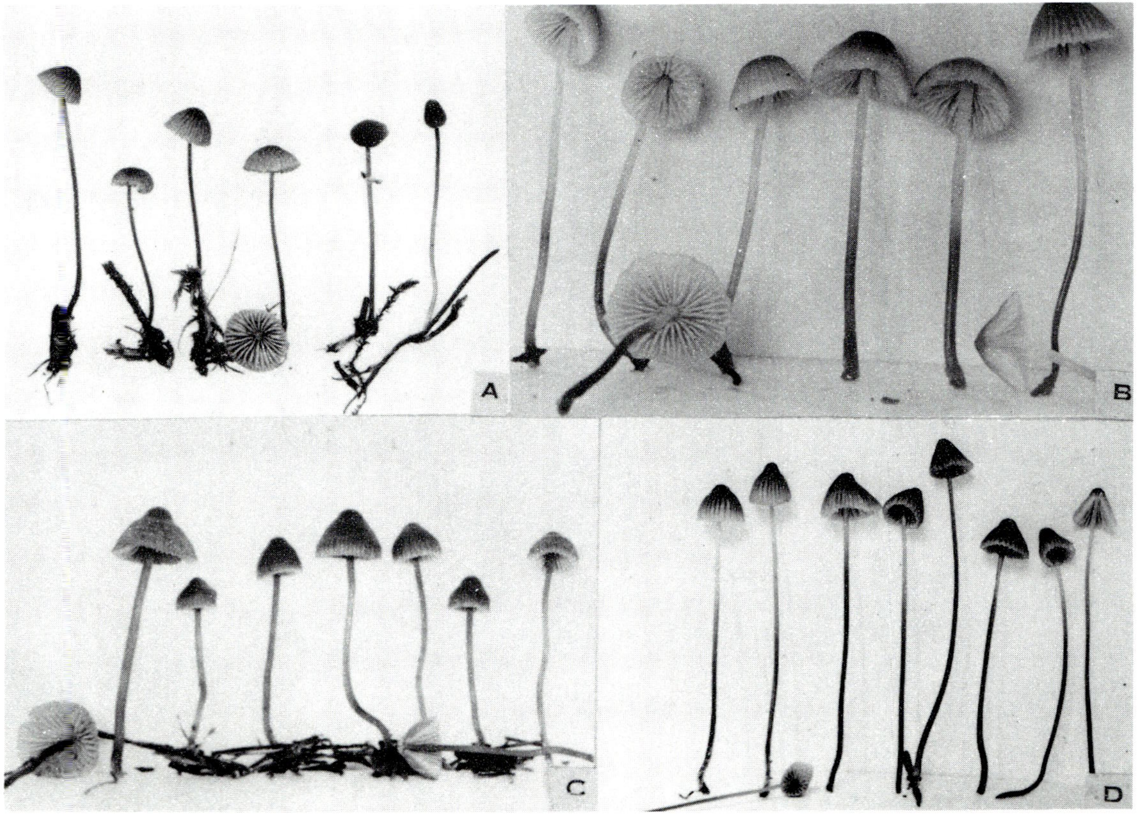

A, *M. urania* (Fr.) Gillet; B, *M. pusilla* Smith; C, *M. psammicola* (Berk & Br.) Saccardo; D, *M. olivaceobrunnea* Smith. All × 1

PLATE 50

A, *M. metata* (Fr.) Quélet; B, *M. plicosa* (Fr.) Gillet. Both × 1

PLATE 51

M. piceicola Smith (typical form). × 1

PLATE 52

M. plicosa (Fr.) Gillet (large form). × 1

PLATE 53

A, *M. maculata* Karsten; B, *M. hiemalis* (Fr.) Quélet; C, *M. atroalboides* (Pk.) Saccardo; D, *M. occidentalis* Murrill. All × 1

PLATE 54

A, *M. filopes* (Fr.) Quélet; B, *M. iodiolens* Lundell. Both × 1

PLATE 55

M. iodiolens Lundell (typical form). × 1

PLATE 56

M. vitilis (Fr.) Quélet (typical form). × 1

PLATE 57

M. pullata (Berk. & Cooke) Saccardo (large form). × 1

PLATE 58

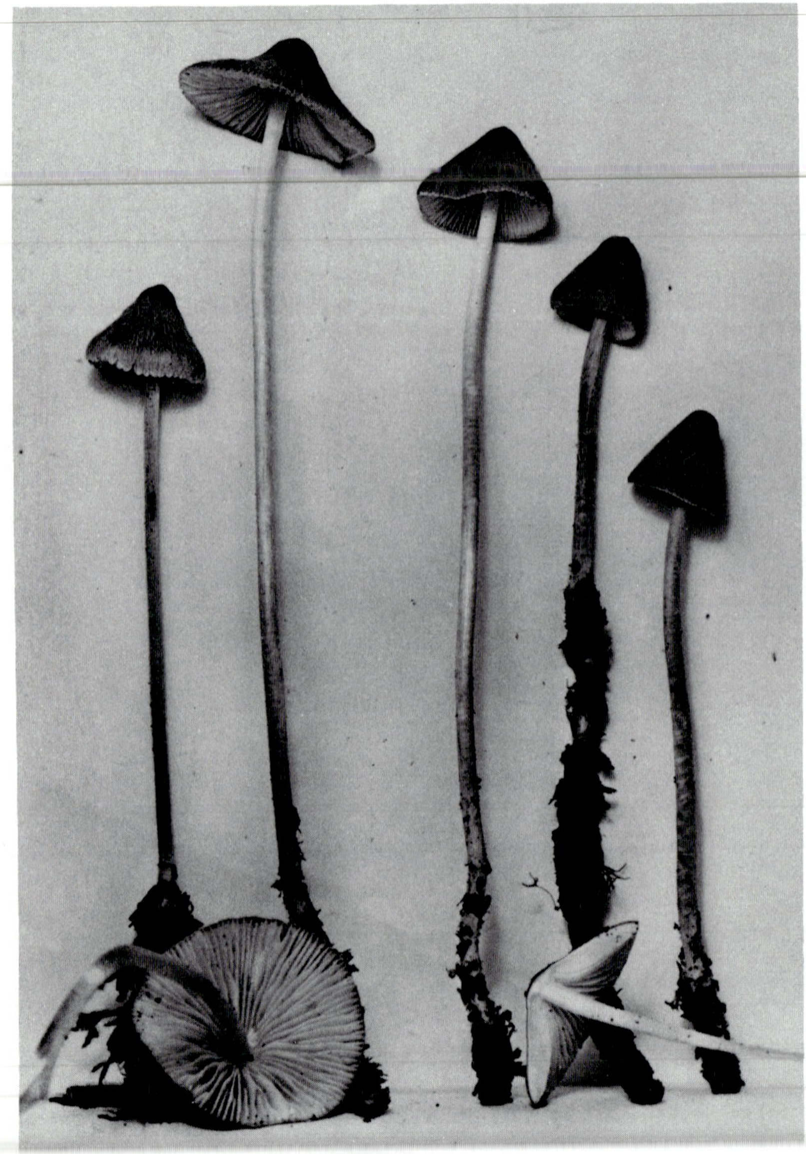

M. polygramma (Fr.) Quélet (slender form). × 1

PLATE 59

M. polygramma (Fr.) Quélet (typical form). × 1

PLATE 60

M. megaspora Kauffman (from sphagnum). × 1

PLATE 61

M. megaspora Kauffman (from muck land). × 1

PLATE 62

M. fagetorum **(Fr.) Gillet (typical form). × 1**

PLATE 63

M. borealis Smith. × 1

PLATE 64

M. algeriensis Maire (slender form). × 1

PLATE 65

M. algeriensis Maire (long-stiped form). × 1

PLATE 66

M. algeriensis Maire (typical form). × 1

PLATE 67

M. niveipes Murrill (buttons). × 1

PLATE 68

M. niveipes Murrill (nearly mature). × 1

PLATE 69

M. macrocystidiata Singer (large form). × 1

PLATE 70

M. alcalina (Fr.) Quélet. × 1

PLATE 71

M. alcalina (Fr.) Quélet (odorless form). × 1

PLATE 72

M. incarnatifolia Smith. × 1

PLATE 73

M. laevigata (Lasch) Quélet. × 1

PLATE 74

M. semivestipes (Pk.) Smith. × 1

PLATE 75

M. parabolica (Fr.) Quélet sensu Kauffman. × 1

PLATE 76

M. rubrotincta Smith. × 1

PLATE 77

M. inclinata (Fr.) Quélet (typical form). × 1

PLATE 78

M. inclinata (Fr.) Quélet (part of a large cluster). × 1

M. inclinata (Fr.) Quélet (pale slender form). × 1

PLATE 80

M. maculata Karsten (cespitose, western form). $\times 1$

PLATE 81

M. occidentalis Murrill (typical form). × 1

PLATE 82

M. hemisphaerica Peck (small form). × 1

PLATE 83

M. galericulata (Fr.) Quélet. × 1

PLATE 84

M. rugulosiceps (Kauffman) Smith. × 1

PLATE 85

M. radicatella (Pk.) Saccardo. × 1

PLATE 86

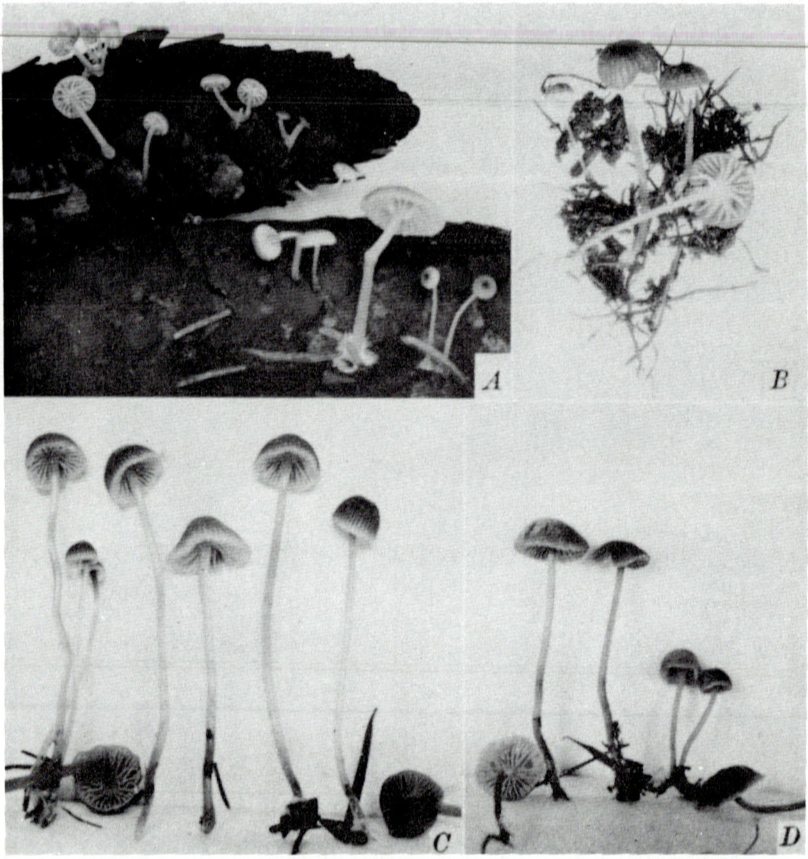

A, *M. speirea* (Fr.) Gillet; B, *M. concolor* (Lange) Kühner; C, *M. cinerella* Karsten; D, *M. thujina* Smith. All × 1

PLATE 87

A, *M. floccipes* (Fr.) Kühner; B, *M. Kauffmanii* Smith; C, *M. misera* (Fr.) Smith. All × 1

PLATE 88

M. pseudoclavicularis Smith. × 1

PLATE 89

A, *M. ulmicola* Smith, × 1; B, *M. rorida* (Fr.) Quélet, × 2

PLATE 90

M. Leaiana (Berk.) Saccardo. × 1

PLATE 91

M. griseoviridis Smith. × 1

PLATE 92

M. viscosa (Sec.) Maire. × 1

PLATE 93

A, *M. vulgaris* (Fr.) Quélet; B, *M. epipterygioides* Pearson; C, *M. clavicularis* (Fr.) Gillet. All × 1

PLATE 94

A, *M. clavicularis* (Fr.) Gillet; B, *M. quinaultensis* Kauffman.
Both × 1

PLATE 95

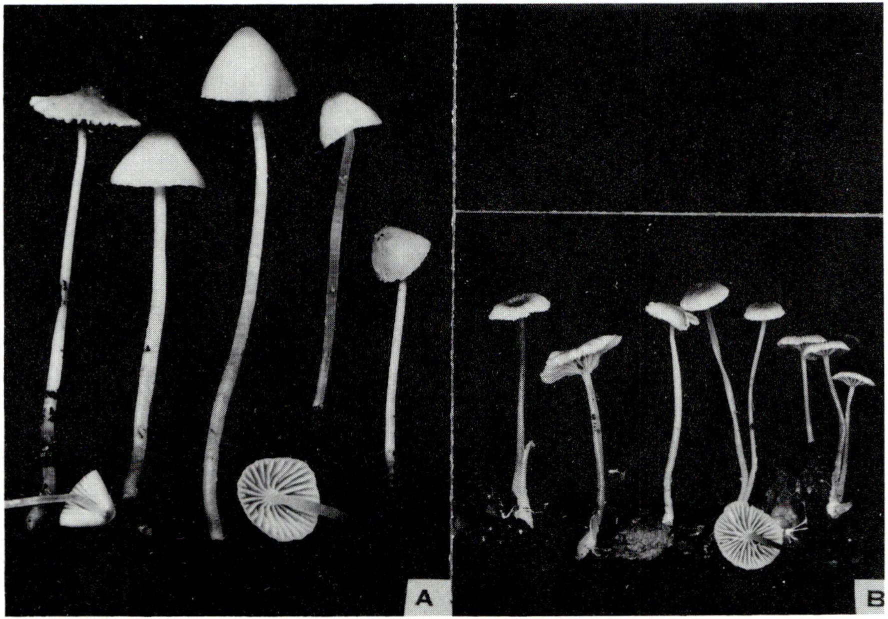

A, *M. epipterygia* (Fr.) Quélet; B, *M. odorifera* (Pk.) Saccardo. Both × 1

PLATE 96

M. quinaultensis Kauffman. × 1

PLATE 97

M. tenax Smith. × 1

PLATE 98

A, *M. trachyspora* Rea; B, *M. bisphaerigera* (Lange) Smith. Both × 1

PLATE 99

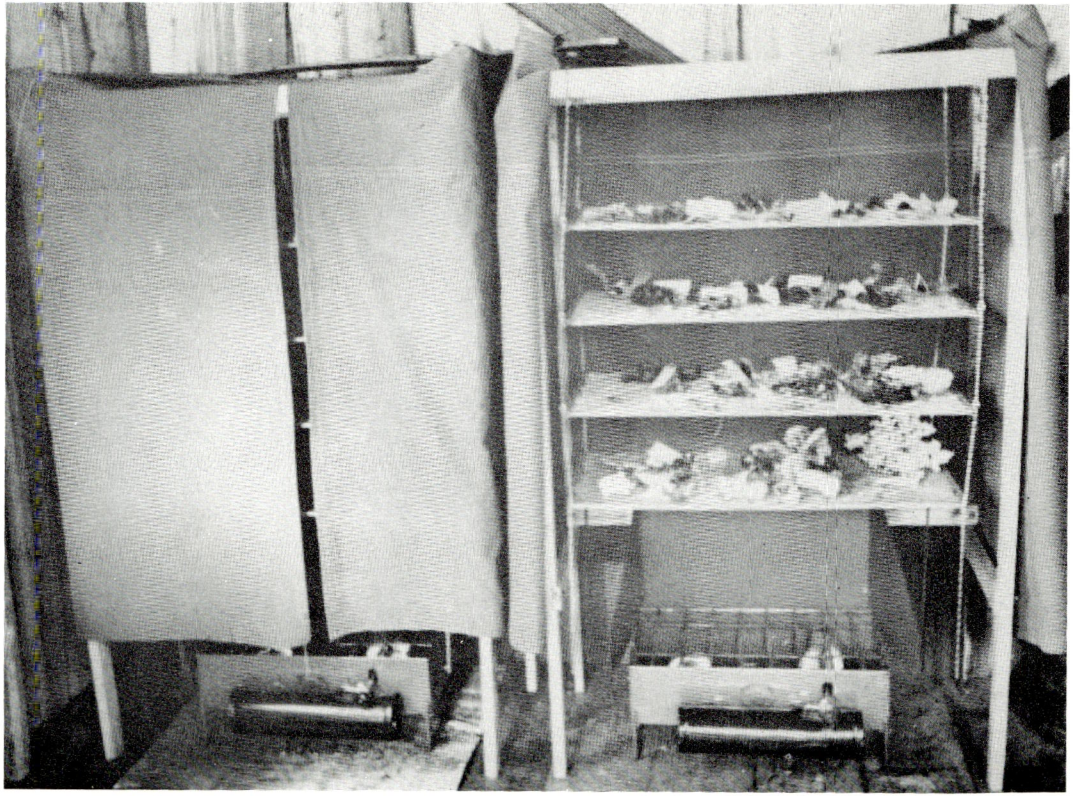

Photograph by E. B. Mains

Drying racks

INDEX